LASER PULSE PHENOMENA AND APPLICATIONS

Edited by **Dr. F. J. Duarte**

Laser Pulse Phenomena and Applications
Edited by Dr. F. J. Duarte

Published by InTech
Janeza Trdine 9, 51000 Rijeka, Croatia

Publishing Process Manager Jelena Marusic

Technical Editor Teodora Smiljanic

Cover Designer InTech Design Team

Additional hard copies can be obtained from orders@intechopen.com

Laser Pulse Phenomena and Applications, Edited by Dr. F. J. Duarte
p. cm.
ISBN 978-953-307-405-4

For EU product safety concerns:
IN TECH d.o.o., Prolaz Marije Krucifikse Kozulić 3, 51000 Rijeka, Croatia,
info@intechopen.com or visit our website at intechopen.com.

Meet the editor

F.J. Duarte is a laser physicist based in Western New York, USA. He is the author and editor of several well-known books on tunable lasers including Dye Laser Principles (Academic, New York, 1990) and Tunable Laser Optics (Elsevier Academic, New York, 2003). His most recent edited work is Tunable Laser Applications, 2nd Edition (CRC, New York, 2009).

Dr. Duarte has made key experimental and theoretical contributions to the field of narrow-linewidth tunable laser oscillators. These include original oscillator architectures and the generalized multiple-prism grating dispersion theory. He has also pioneered the use of Dirac's quantum notation in the description of generalized N-slit interference and classical optics phenomena. Currently, his research focuses on further developments of dispersive narrow-linewidth laser oscillators and very large N-slit laser interferometers. Dr. Duarte's contributions are cited in some 130 laser and optics books including several classics. He received the Engineering Excellence Award from the Optical Society of America, is a Fellow of the Australian Institute of Physics, and a Fellow of the Optical Society of America.

Contents

Preface

Pulsed lasers are available in the gas, liquid, and the solid state. These lasers are also enormously versatile in their output characteristics yielding emission from very large energy pulses to very high peak-power pulses. Pulsed lasers are equally versatile in their spectral characteristics. This volume includes an impressive array of current research on pulsed laser phenomena and applications. *Laser Pulse Phenomena and Applications* covers a wide range of topics from laser powered orbital launchers, and laser rocket engines, to laser-matter interactions, detector and sensor laser technology, laser ablation, and biological applications.

F. J. Duarte

Rochester
New York, USA

Part 1

Laser Emission and Propulsion Phenomena

1

Pulse-Laser Powered Orbital Launcher

Hiroshi Katsurayma[1], Kimiya Komurasaki[2] and Yoshihiro Arakawa[2]
[1]*Yamaguchi University*
[2]*The University of Tokyo*
Japan

1. Introduction

Some innovative plans in space development have been suspended because of high transportation costs of conventional launching systems. For example, the Japanese H2A rocket will cost $400 billion to launch a 1 GW output Solar Power Satellite (SPS) whose weight is 10^4 ton (Collins, 1993). A pulse-laser powered orbital launcher is a potential alternative to reduce those costs: a large payload ratio can be achieved because energy is provided from a ground-based laser and atmospheric air is used as a propellant.

Figure 1 shows an air-breathing pulse-laser powered vehicle ("Lightcraft") (Myrabo, 2001) with representative scales for 100 MW-class laser input. The vehicle forms plasma by focusing transmitted laser beams using a parabolic spike nozzle. The plasma absorbs the following part of the laser pulse while expanding outward. The resulting blast wave reflects on the nozzle surface and generates impulsive thrust.

Figure 2 shows a schematic view of a pulse-laser powered launching system from the ground to a Geosynchronous Earth Orbit (GEO). In the initial stage of the launch, the vehicle closes its inlet and takes air from its rear area. This flight mode is called "pulsejet mode". The inlet is opened and air is taken from the front end of the vehicle when the vehicle is accelerated sufficiently to be able to use ram-compression. This flight mode is called "ramjet mode". The inlet is again closed and on-board hydrogen is used as a propellant when the vehicle cannot obtain sufficient air at high altitudes. This flight mode is called "rocket mode." Through these three modes, the vehicle is accelerated to reach orbital velocity.

Several researchers have studied the feasibility for orbital launch using several laser propulsion systems (Toki, 1991; Kare, 1986; Humble et al., 1995; Phipps et al., 2000) , but most of them calculate flight trajectories using the thrust modeled with a fixed energy conversion efficiency. However, in an air-breathing propulsion system, the energy conversion efficiency evidently depends on its flight trajectory.

The present article introduces our realistic performance modeling in three flight modes: the performance in the pulsejet mode is modeled using measured energy conversion efficiency (Mori et al., 2004, a) and Computational Fluid Dynamics (CFD) analysis (Katsurayama et al., 2008); the performance during the ramjet mode is computed using CFD analysis (Katsurayama et al., 2003); and the performance in the rocket mode is obtained analytically with the energy conversion efficiency computed using a thermochemical equilibrium calculation (Katsurayama et al., 2003). In addition, a transfer trajectory to the GEO is proposed. The launch trajectory to its geosynchronous transfer orbit (GTO) is computed using these realistic thrust models. Finally, the feasibility of the pulse-laser

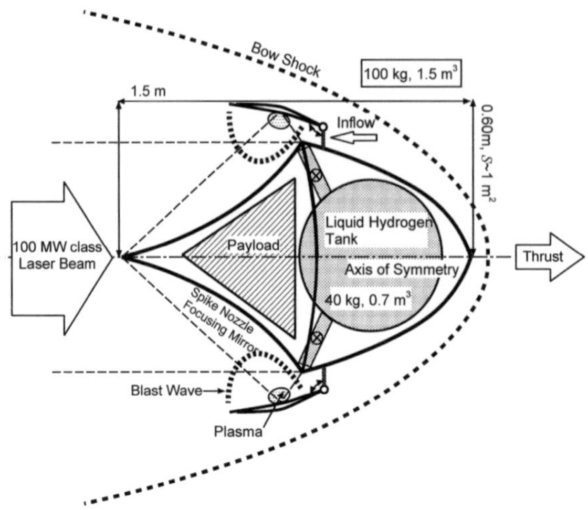

Fig. 1. Air-breathing pulse-laser powered vehicle(Myrabo, 2001).

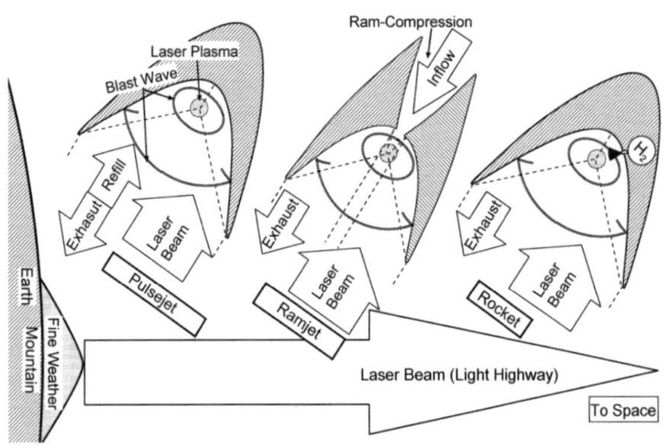

Fig. 2. A pulse-laser powered orbital launching system.

powered orbital launcher is discussed through estimation of its achievable payload mass per unit beam power and costs (Katsurayama et al., 2009).

2. Performance modeling of pulse-laser powered vehicle

2.1 Momentum coupling coefficient and blast wave energy conversion efficiency

A momentum coupling coefficient C_m, which is the ratio of cumulative impulse to laser energy per pulse E_L, is used as a performance indicator for laser propulsion. It is defined as

$$C_m = \int_{I\ pulse} F dt \Big/ E_L, \tag{1}$$

where F denotes thrust. The laser energy absorbed in a gas is converted to the blast wave energy E_{bw}, which is the source energy necessary to drive an equivalent blast wave in a calorically perfect gas Ushio et al. (2008):

$$E_{bw} = \int_{V_{bw}} \left[\rho \left(e^{t+r} + 1/2u^2 \right) - \rho_0 \left(e^{t+r} + 1/2u^2 \right)_0 \right] dV, \tag{2}$$

where V_{bw} is the volume surrounded by the blast wave, and ρ and $1/2u^2$ are the density and kinetic energy. Subscript 0 indicates the properties before laser incidence. Here, e^{t+r} is the sum of internal translational and rotational energy. On the other hand, the vibrational and electric excitation energy that are excited because of laser absorption are excluded from E_{bw} because they are newly stored in molecules and can not achieve pressure work as well as chemical potential energy. Because only E_{bw} contributes to F, C_m is proportional to the blast wave energy conversion efficiency η_{bw}

$$\eta_{bw} = E_{bw}/E_L. \tag{3}$$

Therefore, it is necessary to model the performance.

2.2 Explosion source model

According to our previous experiment (Mori et al., 2004, a;b; Mori et al., 2002) with a CO_2 TEA laser, approximately 95 % of E_L is absorbed in the form of the Laser Supported Detonation (LSD) wave (Raizer, 1977), and approximately its 45 % is converted to drive a blast wave. The remainder energy is confined in the form of chemical potential and electric excitation energy into rarefied plasma left near the focus: it is inconvertible to thrust, and is gradually lost in the form of radiation or dissipative heat flux to the surrounding. Therefore the remainder energy is excludable to reproduce this adiabatically expanding blast wave; the energy converted to the blast wave E_{bw} can be assumed to be equivalent to an instantaneous point explosion energy necessary to drive a blast wave with the same strength.

Mori et al., 2004 (a) has investigated η_{bw} by comparing measured shock speed with that calculated using a similarity solution (Kompaneets, 1960) under the assumption of ideal air. The resulting η_{bw} in the standard atmosphere was 0.43±0.04, which was insensitive to E_L within the tested range of 4.0–12.8 J (Mori et al., 2004, a).

Although Wang et al. (2002) has computed the propagation of a LSD wave to investigate this energy conversion mechanism, such a computation is too expensive for our purpose to model ther thrust performance resulting from the adiabatic expansion of the blast wave whose time scale (the order of 100 μs) are much longer than that of energy absorption process (3.5 μs). In the performance modeling using our CFD analyses, the blast wave is driven by a pressurized explosion source with η_{bw}=0.43, whose radius is 1 mm and density is equal to that in the ambient atmosphere. Such an explosion source method (Steiner et al., 1998; Jiang et al., 1998; Liang et al., 2001; 2002) is familiar to simulate blast wave propagation whose time scale is much longer than that of energy input process.

Measured (Mori et al., 2004, a) and computed (Katsurayama et al., 2008) propagation of a blast wave in free space are compared to validate the blast wave reproducibility of this explosion source. Figure 3 shows the histories of the shock front radius R_{bw} and Mach number M_{bw} of the blast wave in the case of E_L=5.4J. The CFD can reproduce the measured R_{bw} and M_{bw} after laser heating. Thereby the source model is used for computation to predict the thrust performance, and it was located at a laser focus.

2.3 Conical laser pulsejet

The thrust generation processes in a laser pulsejet with a conical nozzle were simulated to validate the performace modeling using our CFD analyses (Katsurayama et al., 2008). Figure 4 schematically shows sequential thrust generation processes in the laser pulsejet with a simple conical nozzle of its half apex angle α. In an energy absorption process (a), plasma is produced near the laser focus. The plasma absorbs laser energy in the form of a LSD wave. The large

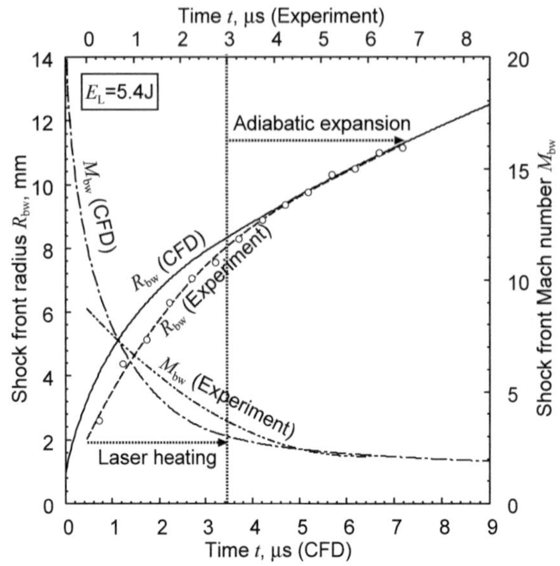

Fig. 3. Comparison of the CFD and the experiment on R_{bw} and M_{bw} in the explosion in free space. (CFD: η_{bw}=0.43; Experiment (Mori et al., 2004, a) : η_{bw}=0.43±0.04)

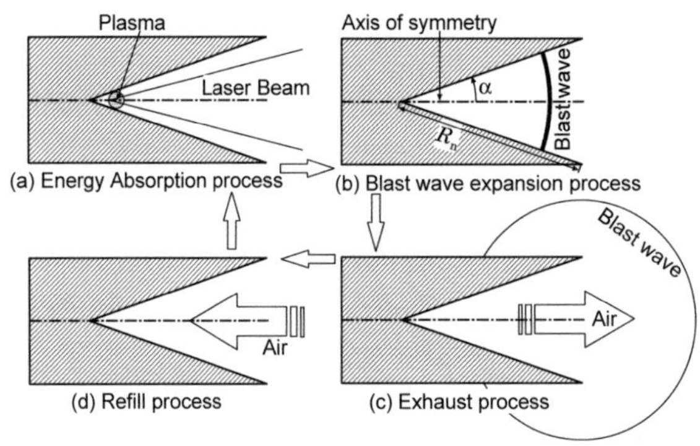

Fig. 4. Schematic of a laser pulsejet engine cycle.

part of the absorbed energy is used to drive a high-pressure blast wave in the surrounding air. In a blast wave expansion process (b), the blast wave imparts an impulsive thrust directly to a nozzle wall; thereby, main thrust is produced. In exhaust (c) and refill (d) processes, the air in the nozzle is exhausted and fresh air is taken in. Additional thrust will be produced in these processes.

Because the exhaust-refill prcoesses results from an adiabatically expanding blast wave after laser heating, an explosion source is used to drive the blast wave instead of solving a laser abosrption process. Figures 5 and 6 respectively show the thrust history and corresponding pressure contours of a conical laser pulsejet. Δp in the captions of pressure contours shows the interval of the contours. An explosion starts at t=0 (see Fig. 6(a)). A shock wave reaches the nozzle exit at t_0 μs as shown in Fig. 6(b). F_0 is the thrust at $t=t_0$. After $t=t_0$, the heated air starts to be exhausted and thrust decreases gradually. At $t=t_1$ (see Fig. 6(c)), thrust becomes zero. Thrust becomes negative because of the rarefaction wave behind the shock wave and

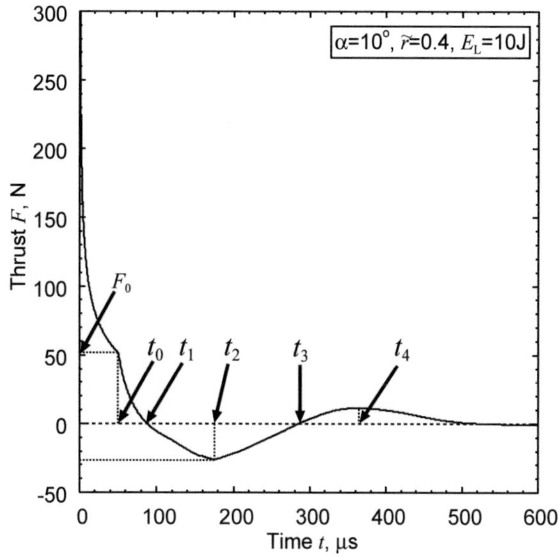

Fig. 5. Thrust history. (η_{bw}=0.43)

thrust takes a minimum value at t=t_2. The gauge pressure becomes negative for the entire region inside of the nozzle, as seen in Fig. 6(d). At t=t_3 (see Fig. 6(e)), thrust reverts to zero; subsequently, thrust has a second peak at t=t_4. After t=t_4, thrust oscillates and the oscillation attenuates gradually.

Figure 7 shows measured (Mori et al., 2004, b) and computed relationships between C_m and α. The computation reproduces the C_m decreasing tendency. The processes until the shock front of a blast wave reaches the nozzle exit are similar regardless of the nozzle apex angle. However, after the blast wave leaves the nozzle edge, the behavior of the rarefaction wave induced behind the shock wave depends greatly on the nozzle apex angle. In the case of small apex angles, succsessive refilling mechanisms with vortices are activated by the prominently evolved rarefaction wave, in contrast, in the case of large apex angles, this mechanims do not appear due to the moderate evolution of the rarefaction wave. This difference was found to result in the decreasing tendency of the momentum coupling coefficient. More detailed descriptions of these phenomena is found in Katsurayama et al. (2008).

Because the computed C_m is in good agreement with measured data in the cases of small α, our CFD analyses have the capability for predicting the thrust performace of a pulse-laser powerd launcher. Although the deviation from the measurement increases with increasing α, this unpredictability is insignificant because the cause is attributable to the not-optimized geomertical relation between the LSD propagation distance and nozzle length (Katsurayama et al., 2008).

2.4 Laser ramjet

C_m during the laser ramjet mode was computed in our previous CFD analyses (Katsurayama et al., 2003). A blast wave is again driven by an explosion source model with η_{bw} which depends on atmospheric pressure (Katsurayama et al., 2009).

Figure 8 shows a part of computational results, and these results (Katsurayama et al., 2003) have showen that C_m in the ramjet mode is insensitive to laser energy E_L and depends only on flight conditions. Therefore, the performance is obtained from the map of C_m, which is

(a) $t=0\,\mu s$
($p_{\text{max}}=3.77\times10^3$ atm, $p_{\text{min}}=1.00$ atm, $\Delta p=1.26\times10^2$ atm)

(b) $t=50\,\mu s \cong t_0$
($p_{\text{max}}=3.99$ atm, $p_{\text{min}}=1.00$ atm, $\Delta p=9.98\times10^{-2}$ atm)

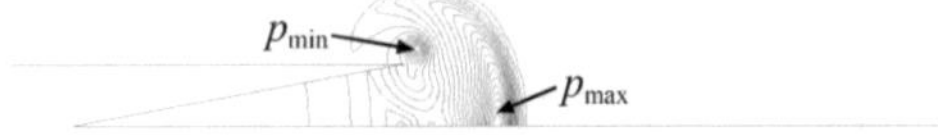

(c) $t=88\,\mu s \cong t_1$
($p_{\text{max}}=2.27$ atm, $p_{\text{min}}=4.32\times10^{-1}$ atm, $\Delta p=6.14\times10^{-2}$ atm)

(d) $t=176\,\mu s \cong t_2$
($p_{\text{max}}=1.18$ atm, $p_{\text{min}}=3.70\times10^{-1}$ atm, $\Delta p=2.70\times10^{-2}$ atm)

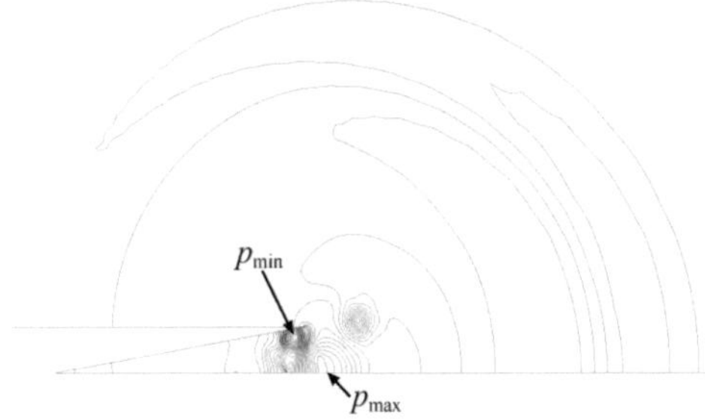

(e) $t=284\,\mu s \cong t_3$
($p_{\text{max}}=1.18$ atm, $p_{\text{min}}=4.79\times10^{-1}$ atm, $\Delta p=2.34\times10^{-2}$ atm)

Fig. 6. Typical pressure contours of the conical laser pulsejet.

constructed using CFD analyses under the conditions of 16 pairs of atmospheric density ρ_∞ and flight Mach number M_∞. The map of C_{m} is shown in Fig. 9 with the fitting function of M_∞ and ρ_∞

$$C_{\text{m}}\left(M_\infty,\rho_\infty\right) = \left[0.04M_\infty^2 - 1.81M_\infty + 19.00\right] \times \left[0.35\left(\log_{10}\rho_\infty\right)^2 + 5.01\left(\log_{10}\rho_\infty\right) + 13.06\right], \quad (4)$$

which is used to calculate the flight trajectory. The value of C_{m} decreases with decreasing ρ_∞ because of the decrease in a captured mass flow rate. It also decreases with increasing M_∞ because the blast wave quickly leaves the nozzle because of the increase in the engine flow speed.

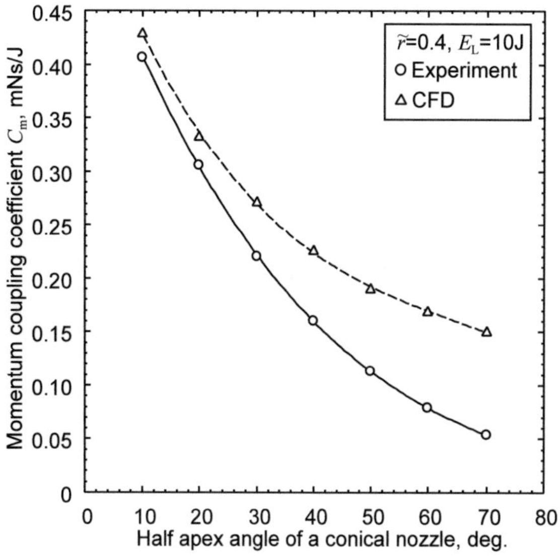

Fig. 7. Relationship between C_m and α. (CFD: η_{bw}=0.43; Experiment: η_{bw}=0.43±0.04)

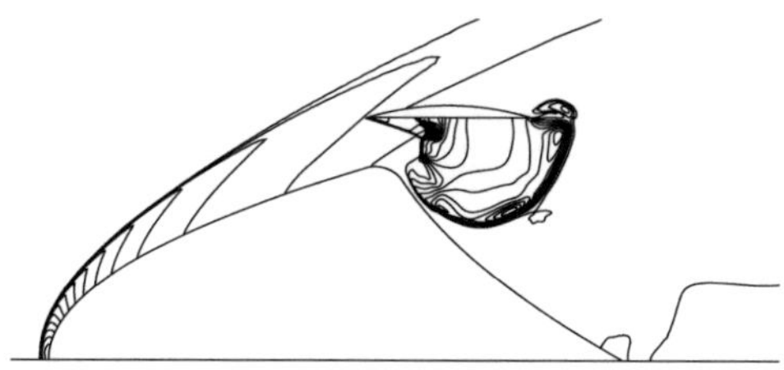

(a) At t=12 μs. (p_{max}=2.27 atm, p_{min}=2.1×10^{-2} atm, Δp=0.11 atm)

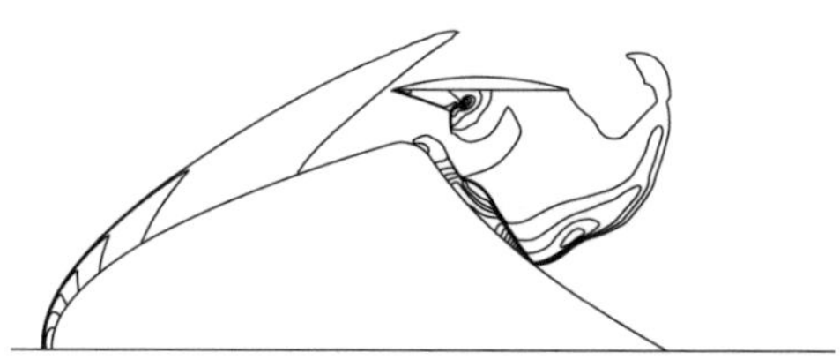

(b) At $t = 20$ μs. (p_{max}=4.63atm, p_{min}=2.1×10^{-2} atm, Δp=0.23 atm)

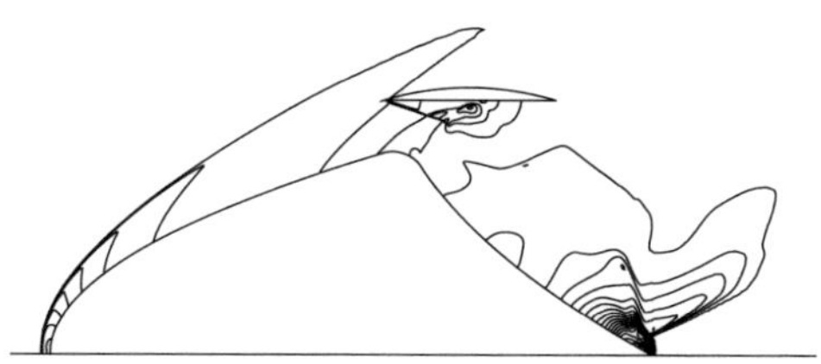

(c) At $t = 38$ μs. (p_{max}=4.27 atm, p_{min}=2.0×10^{-2} atm, Δp=0.21 atm)

Fig. 8. Typical pressure contours of the laser ramjet: E_L=400 J, H=20 km and M_∞=5.

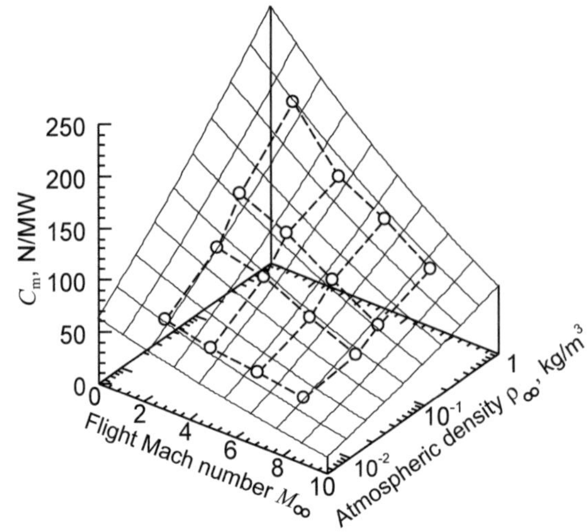

Fig. 9. C_m map of the laser ramjet.

2.5 Laser rocket

Because the vehicle flies in a near-vacuum environment in the rocket mode, its thrust depends only on the nozzle expansion ratio. The self-similar solution in a conical nozzle (Simons et al., 1977) is available to estimate C_m. Assume that the blast wave propagates in the hydrogen propellant expanding to a vacuum. In such a case, C_m is modeled as

$$C_m = \sqrt{2\dot{m}_p \eta_{bw,rocket} / P_L} \sin^2 \alpha \Big/ \left[2\pi \left(1 - \cos\alpha \right) \right], \tag{5}$$

where $\dot{m}_p$ and P_L respectively represent the mass flow rate of hydrogen propellant and time-averaged laser power. The apex angle of the nozzle cone α is set to 30°which is almost equivalent to the value of the Lightcraft used in the CFD analysis of the ramjet mode.
The value of η_{bw} in the rocket mode is calculated analytically by solving the propagation of a LSD in hydrogen propellant using our previous numerical model (Katsurayama et al., 2003) obtained by combining a plain Chapman-Jouguet detonation relation and chemical equilibrium calculation, resulting in $\eta_{bw,rocket} = 23.5\%$.

3. Feasibility of pulse-laser orbital launcher

3.1 Light highway

The pulse-laser powered orbital launcher requires a laser base with average output of 100 MW~1 GW. For development of such a high-powered laser base, the cost of a laser transmitter is expected to predominate over the costs required for other systems such as cooling and power supply (Kare, 2004). Thereby, the cost-reduction of the laser transmitter is indispensable for the launching system. An optical phased array with diode lasers (Kare, 2004; Komurasaki et al., 2005) will reduce the cost because existing laser technology is applicable and mass-production effect is expectable. Moreover, launch using only a laser base is realistic. Furthermore, several obstacles on beam transmission should be assessed briefly to discuss the launching system feasibility. If the beam is a Gaussian beam whose respective aperture and wavelength are 1 m and 1 μm, the total spreading angle is 1.2×10^{-6} rad. The energy loss caused by beam diffraction is acceptable because the beam radius spreads to only 1.19 m at the transmission distance of 500 km, which is the typical value required for the launching system. The beam spread, attenuation, and refraction caused by nonlinear effects of the atmosphere, such as the variation of atmospheric refraction index, thermal blooming, Rayleigh, Raman and Mie scattering, have been analyzed for the ORION project (Phipps et al., 1996; Cambell, 1996), which is intended to remove debris on orbits using a pulse laser. Nonlinear effects are inferred to be negligible for transmission with a beam power density under a threshold determined by a laser pulse width and wavelength.

To avoid the whipping phenomenon of the beam center caused by atmospheric turbulence, the launch site should be built on the top of a mountain where the weather is expected to always be fine and for which scintillation caused by atmosphere is small. A location where an astronomical observatory has been built is suitable if a vehicle is launched at a time without clouds or turbulence. For example, Mauna Kea in Hawaii has no cloud cover for 90% of the days in a year.

In addition, an ongoing study (Libeau et al., 2002) proposes a vehicle shape with which the vehicle can maintain its center axis parallel to the beam direction by generating thrust vector and torque automatically in the direction that allows it to retain aerodynamic stability. Both directions of the flight and beam can be maintained as vertical to the ground using this technology. Therefore, vehicle tracking and beam pointing are unnecessary for the launching system. The vehicle can be transferred to space along a "light highway" (Myrabo, 2001) constructed vertically from the ground, as shown in Fig. 2.

3.2 Proposed trajectory to GEO

Considering requirements for beam transmission, the vehicle is accelerated vertically in a short distance along the light highway. Figure 10 shows a trajectory to a GEO through the pulse-laser powered launcher and an upper-stage propulsion system for the Hohmann transfer. The vehicle is launched from the equator through the pulse-laser powered launcher; it is accelerated rapidly to Δv_{L}. The vehicle then reaches an apogee point beyond the GEO through inertial flight to use the Δv_{L} efficiently. The structure weight of laser propulsion is detached at the apogee point, and the upper-stage propulsion system is burned. The detached structure is attracted to the earth and it is incinerated during reentry into the atmosphere.

Figure 11 shows the variation of Δv_{L} and the velocity increment Δv_{H} required for the Hohmann transfer with cut-off velocity v_{c}, which is the flight velocity when laser propulsion is terminated. The required Δv_{L} and Δv_{H} are 18.85 and 1.84 km/s, respectively, if v_{c}=10.6km/s is chosen.

Electric propulsion is available for the upper-stage propulsion system by virtue of an abundant electricity supply from the cells if solar cells are transferred to construct an SPS. Using the Hall thruster, whose specific impulse $I_{\mathrm{sp,EP}}$ is 2000 s, the transfer to the GEO takes a spiral trajectory. As a result of calculating them by solving equations of motion (Spencer et al., 1995; Kluever et al., 1998) without trajectory optimization, the effective velocity increment $\Delta v_{\mathrm{spiral}}$ (Spencer et al., 1995) through the upper-stage propulsion is estimated as 3.68 km/s. The resulting payload ratio λ_{u} of the upper stage is

$$\lambda_{\mathrm{u}} = 1 - \int_{\mathrm{spiral}} \frac{2\eta_{\mathrm{EP}} \left(P_{\mathrm{EP}}/m_{\mathrm{u},0}\right)}{\left(g I_{\mathrm{sp,EP}}\right)^2} \mathrm{d}t \Bigg/ \left(1 - \varepsilon_{\mathrm{EP}}\right) = 0.83. \tag{6}$$

where the propulsion system efficiency η_{EP} and the structure weight coefficient $\varepsilon_{\mathrm{EP}}$ are assumed respectively as 65% (Kluever et al., 1998) and 0.1. The ratio of power to the initial

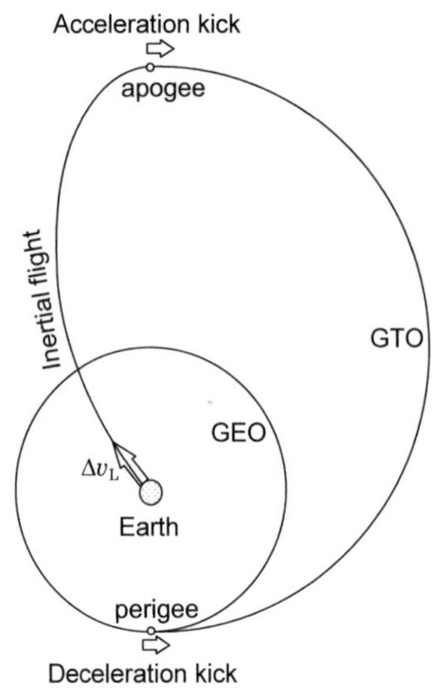

Fig. 10. Proposed trajectory to GEO.

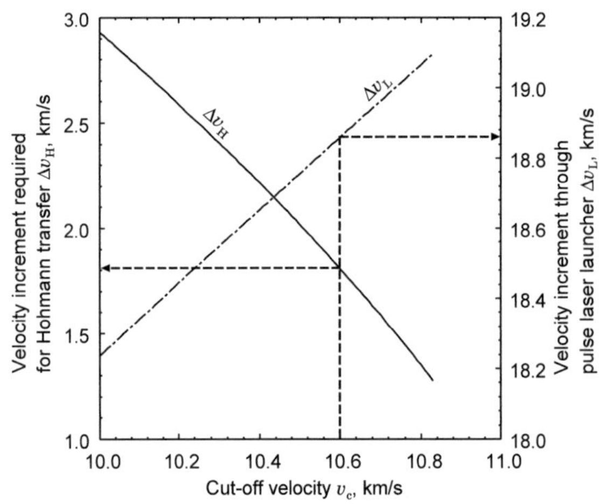

Fig. 11. Variation of the velocity increment through the pulse-laser powered launcher and the velocity increment required for the Hohmann transfer with cut-off velocity.

upper-stage weight ($P_{EP}/m_{u,0}$) is set to 100 W/kg in view of the ratio of power to weight of the 1 GW-output SPS.

v_c=10.6km/s and λ_u=0.83 are used in the calculation of launch trajectories described in § 3.4.

3.3 Mode switching criteria

The pulsejet mode should be switched to the ramjet mode when the blast wave becomes free from propagation over the inlet because of the vehicle acceleration. However, numerous CFD analyses are necessary to model the timing correctly because it depends on E_L and flight conditions. The present model therefore chooses the safest timing: the mode is switched when the flow in the vehicle is free from choking by laser heating. An engine cycle, as shown

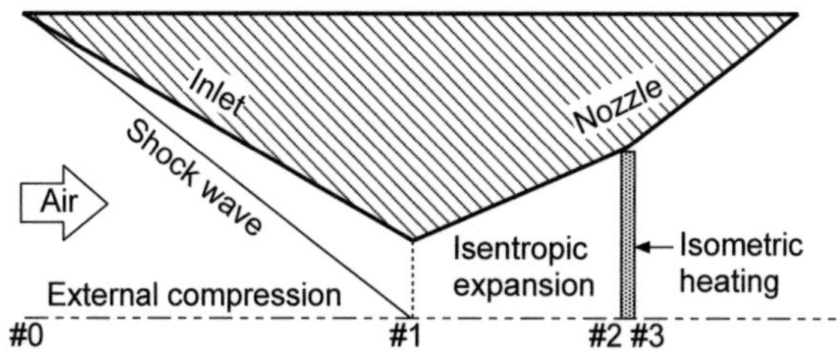

Fig. 12. Engine cycle for assessing heat-choking.

in Fig. 12, is analyzed to assess heat-choking. In this cycle, the air is taken in at location #0 through an effective inlet area

$$A_0 = S_\mathrm{v} \times \mathrm{C.A.R.}, \tag{7}$$

where S_v and C.A.R. are the maximum cross sections of the vehicle and the capture area ratio. The captured air is compressed externally from #0 to #1 as

$$u_1 = \eta_\mathrm{d} u_0 \tag{8}$$

where u and η_d are the flow velocity of the air and diffuser efficiency. η_d and C.A.R. are obtained using CFD analysis on the Lightcraft. With increasing M_∞, η_d decreases from 0.72 to 0.62 and C.A.R increases from 0.31 to 0.71 (Katsurayama et al., 2004). The air is then expanded isentropically from #1 to #2 to delay heat-choking at #3. Finally, it is laser-heated isometrically from #2 to #3. The Mach number at #3 is calculated as

$$M_3 = u_3 \Big/ \sqrt{\gamma R T_3} = u_2 \Big/ \sqrt{\gamma R \left[T_2 + \eta_\mathrm{bw} \eta_\mathrm{trans} P_\mathrm{L} / \left(C_\mathrm{p} \dot{m}_\mathrm{air}\right)\right]}$$

Herein, T and $\dot{m}_\mathrm{air}$ is temperature and the captured mass flow rate. Respectively, R, C_p and γ denote the gas constant, the constant pressure specific heat, and the specific heat ratio of ideal air. Furthermore, η_trans is the transmission efficiency of the laser beam. The pulsejet mode is switched to the ramjet mode when

$$M_3 = 1. \tag{9}$$

In the ramjet mode, the mass flow rate taken from the inlet decreases with altitude because of the decrease in the atmospheric density. Finally, the acceleration of the vehicle becomes zero because of the balance between thrust and aerodynamic drag, at the time when the flight mode is switched to the rocket mode.

3.4 Computed launch trajectory and payload ratio

A launch trajectory to the GEO is calculated by solving the following equation of motion by the 4th order Runge-Kutta scheme.

$$m_\mathrm{v} \frac{\mathrm{d}v}{\mathrm{d}t} = \overline{F} - \frac{1}{2}\rho_\infty v^2 S_\mathrm{v} C_\mathrm{d} - m_\mathrm{v} g_0 \left[R_\mathrm{E} / \left(R_\mathrm{E} + h\right)\right]^2 \tag{10}$$

Therein, S_v, v and m_v are the maximum cross sections of the vehicle, the flight velocity, and the vehicle weight. Also, g_0, R_E and h respectively indicate the gravity acceleration on the ground, the radius of the earth and the flight altitude. An aerodynamic drag coefficient C_d is obtained using the CFD analysis on the Lightcraft; it varies from 0.15 to 0.64 depending

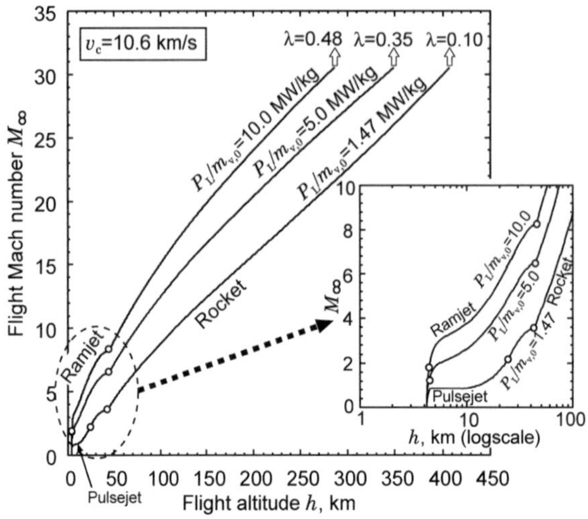

Fig. 13. Flight Mach number vs. flight altitude.

on M_∞ (Katsurayama et al., 2004). Flight conditions are determined by tracing the trajectory. Time-averaged thrust $\overline{F}$ in each mode is estimated using C_m modeled in § 2 as

$$\overline{F} = C_m \eta_{trans} P_L. \tag{11}$$

Herein, if the laser transmitter is the phased array with an effective broad aperture, the fraction of beamed energy contained in a main-lobe of beam is theoretically predictable. It is independent of the transmission distance; η_{trans} is set to its predicted value of 72 % (Komurasaki et al., 2005) on the assumption that the nonlinear effects attributable to the atmosphere negligibly affect the beam transmission.

As the result of the trajectory calculation, the payload ratio λ is estimated as

$$\lambda = \left\{ 1 - \int_{\text{Rocket mode}} \dot{m}_p dt \Big/ \left[m_{v,0} \left(1 - \varepsilon_L\right)\right] \right\} \lambda_u, \tag{12}$$

where $m_{v,0}$ is an initial vehicle mass and the upper stage payload ratio λ_u=0.83 is used (Katsurayama et al., 2009). The structure weight coefficient of the pulse-laser powered vehicle ε_L is set to 0.1 in view of the simple structure of the vehicle.

The vehicle is launched at h=4200 m equal to the altitude of Mauna Kea. The calculation is terminated when the vehicle is accelerated to the cut-off velocity (Katsurayama et al., 2009) v_c=10.6 km/s. Figure 13 shows the trajectories and λ for several specific beam power $P_L/m_{v,0}$. The ramjet period and λ decrease with decreasing $P_L/m_{v,0}$. The ramjet mode becomes unavailable at $P_L/m_{v,0}<1.47$ MW/kg. The beam transmission requests the longest distance of 400 km and the longest duration of 160 s.

Figure 14 shows the variations of the payload mass per 1 MW beam power m_{pl}/P_L and λ with $P_L/m_{v,0}$. Although λ increases with $P_L/m_{v,0}$, m_{pl}/P_L has the maximum 0.084 kg/MW at $P_L/m_{v,0}$ =2.42 MW/kg. Therefore, this

$$\left(P_L/m_{v,0}\right)_{opt} = 2.42 \tag{13}$$

is the optimum condition of minimizing the building cost of a laser transmitter.

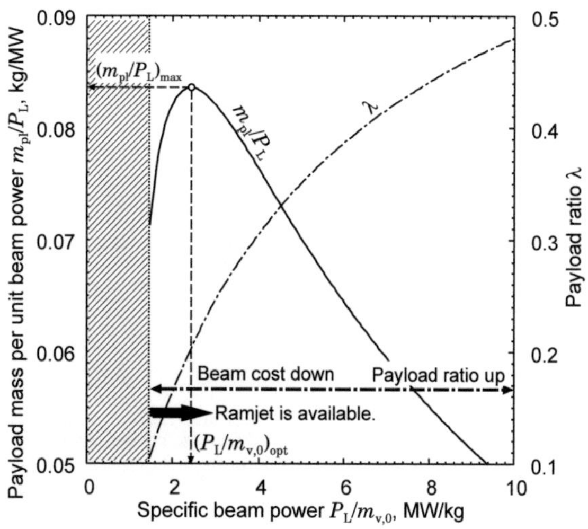

Fig. 14. Variation of payload mass per unit beam power and payload ratio with specific beam power.

C_{LT}, \$/W	C_E, \$/kWh	C_{VP}, % \$/kg	η_{DL}, %
10	0.06	1,000	40

Table 1. Costs and efficiency considered in cost estimation.

3.5 Launch cost

This section estimates the cost of the mission in which a 10^4-ton SPS is transferred to a GEO through many pulse-laser powered orbital launches. Although the cost estimation of a laser transmitter with 100 MW∼1 GW-output is difficult, the most promising technology for high power output is to coherently couple lasers such as solid, fiber, and diode lasers (Shirakawa et al., 2002; Sanders et al., 1994). The cost of the laser transmitter can be estimated as shown in Table 1 if the coherent coupling is applicable to the array of diode lasers whose production cost per unit of power is currently the cheapest. The cost of the laser transmitter C_{LT} is assumed to be a current diode laser price per unit output power (Kare, 2004). Other costs, such as that for the cooling system, power supply and maintenance, are not considered because they are negligibly small compared with C_{LT} (Kare, 2004). The general electricity cost in Japan is C_E. The vehicle production cost C_{VP} is the cost per unit vehicle weight estimated for the Lightcraft whose dry mass and diameter are 100 kg and 1.4 m (Richard et al., 1988). η_{DL} is the energy conversion efficiency of a general diode laser (Kare, 2004). To compare costs of the pulse-laser powered launcher and an existing commercial launcher, the launch cost is defined in the form of redeeming the cost of the laser transmitter.

$$\frac{\text{Launch cost}}{\text{Payload mass}} = \frac{\left[C_{LT}/n_L + (C_E/\eta_{DL})\, t_{flight}\right] P_L + C_{VP} m_{v,0}}{m_{pl}}$$

$$= \frac{C_{LT}/n_L}{\left(m_{pl}/P_L\right)_{max}} + \frac{(C_E/\eta_{DL})\, t_{flight}}{\left(m_{pl}/P_L\right)_{max}} + \frac{C_{VP}}{\lambda} \quad (14)$$

For that calculation, a payload is divided and transferred through n_L launchings. Furthermore, $\left(m_{pl}/P_L\right)_{max}$ (=0.084) is used on the basis of using $(P_L/m_{v,0})_{opt}$, on the condition of which

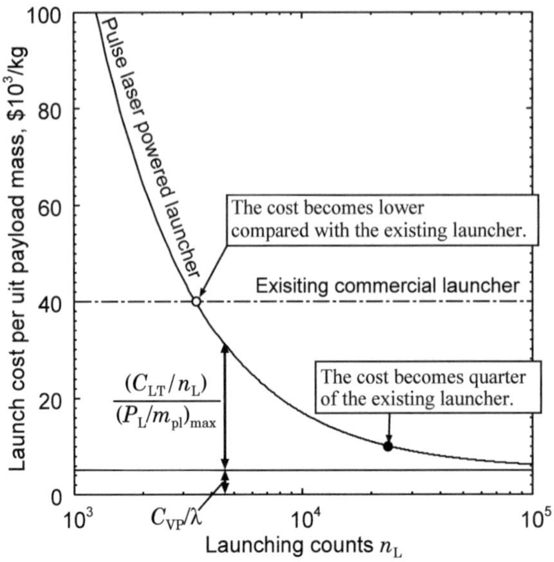

Fig. 15. Variation of launch cost per unit payload mass with $\boldsymbol{n}_{\mathrm{L}}$.

the resulting λ and flight time t_{flight} are 0.20 and 112 s. The first term in the second line of Eq. (14) is the redemption of the laser transmitter. The second and third terms correspond to the electricity and vehicle production cost. Labor costs are omitted because a generally published launch cost, e.g. $80 million per launch of the Japanese H2A rocket, includes only the rocket's production cost.

Figure 15 shows the variations of the launch cost per unit payload mass with n_L of the pulse-laser powered orbital launcher and an existing commercial launcher. The cost of the existing launcher is assumed to be $80 million per launch. The cost of the laser transmitter is predominant when n_{L} is less than 10^3. On the other hand, the electricity cost is negligible. The launch cost decreases with increasing n_{L} and the cost of the laser transmitter is recovered at 3,500 launches compared with the existing launcher. The cost becomes a quarter of the existing launcher at n_{L}=24,000. When n_{L} is greater than 10^5, the cost of the laser transmitter is completely recovered and the cost comprises only the vehicle production cost of approximately $5,000 per unit of payload mass. The electricity cost remains negligibly small over $n_{\mathrm{L}}=10^5$.

Consequently, if the laser base of 5 GW-class output is available, the launcher has the capability of approximately 0.5 ton payload/launch. It is expected to deliver the 10^4 ton SPS through 24,000 launches at a quarter of the cost of an existing launcher. Moreover, if the mass-production effect can reduce C_{VP} to $100/kg, the launcher will cost about a tenth of the current mode.

4. Summary

Orbital launching from the ground to a GEO using the pulse-laser powered launcher is calculated using the performance modeled in the pulsejet, ramjet and rocket modes. Consequently, it can transfer 0.084 kg payload per 1 MW beam power to the GEO. The cost becomes a quarter of that of existing systems if one can divide a single launch into 24,000 multiple launches. Furthermore, if the payload is transferred through 10^5 launches, the launching system can reduce the cost to the order of $\$10^3$ per unit of payload mass.

5. References

Campbell, J. W. (1996). Project ORION: Orbital Debris Removal Using Ground-Based Sensors and Lasers, NASA TM-108522

Collins, P. (1993). The Promise of Electricity from Space for World Economic Development, *Proceedings of 5th International Energy Conference*, Vol. 3, pp. 50-59.

Humble, W. E. and Pierson, B. L. (1995). Maximum-Payload Trajectories for a Laser-Propelled Launch Vehicle, *Journal of Guidance, Control and Dynamics*, Vol. 18, No. 6, pp.1259–1266.

Jiang, Z., Takayama, K., Moosad, K. P. B., Onodera, O., and Sun, M. (1998). Numerical and Experimental Study of a micro-blast wave generated by pulsed-laser beam focusing, *Shock Waves*, Vol. 8, No. 6, pp.337-349.

Kare, J. T. (1986). Trajectory Simulation for Laser Launching, *Proceedings of the 1986 SDIO/DARPA Workshop on Laser Propulsion*, Vol. 2, pp.61-77.

Kare, T. J. (2004). Modular Laser Optics for HX Laser Launch, *Proceedings of the 3rd International Symposium on Beamed Energy Propulsion*, AIP Conference Proceedings, Vol.766, pp. 128-139.

Katsurayama, H., Komurasaki, K., Momozawa, A., and Arakawa, Y. (2003). Numerical and Engine Cycle Analyses of a Pulse Laser Ramjet Vehicle, *Transactions of the Japan Society for Aeronautical and Space Sciences, Space Technology Japan*, Vol. 1, pp. 9-16.

Katsurayama, H., Ushio, M., Komurasaki, K., and Arakawa, Y. (2004). Analytical Study on Flight Performance of a RP Laser Launcher, *Proceedings of 3rd International Symposium on Beamed Energy Propulsion*, AIP Conference Proceedings, Vol. 766, pp.117-127.

Katsurayama, H., Komurasaki, K., Hirooka, Y., Mori, K. and Arakawa, Y. (2008). Numerical Analyses of Exhaust and Refill Processes of a Laser Pulsejet, *Journal of Propulsion and Power*, Vol. 24, No. 5, pp.999-1006.

Katsurayama, H., Komurasaki, K., and Arakawa, Y. (2009). A Preliminary Study of Pulse-Laser Powered Orbital Launcher, *ACTA Astronautica*, Vol. 65, No. 7-8, pp.1032-1041.

Kompaneets, A. S. (1960). A point explosion in an inhomogeneous atmosphere, *Soviet Physics Doklady*, Vol. 5, No. 1, pp.46-48.

Komurasaki, K., Nakagawa, T., Ohmura, S., and Arakawa, Y. (2005). Energy Transmission in Space Using an Optical Phased Array, *Transactions of the Japan Society for Aeronautical and Space Sciences Space Technology Japan*, Vol. 3, pp. 7-11.

Liang, S.-M., Hsu, J.-L.,, and Wang, J.-S. (2001). Numerical Study of Cylindrical Blast-Wave Propagation and Reflection, generated by pulsed-laser beam focusing, *AIAA Journal*, Vol. 39, No. 6, pp.1152-1158.

Liang, S.-M., Wang, J.-S., and Chen, H. (2002). Numerical Study of Spherical Blast-Wave Propagation and Reflection, generated by pulsed-laser beam focusing," *Shock waves*, Vol. 12, No. 1, pp.59-68.

Libeau, M. A. and Myrabo, L. N. (2002). Combined Theoretical and Experimental Flight Dynamics Investigation of a Laser-Propelled Vehicle, AIAA Paper 2002-3781.

Mori, K., Komurasaki, K., and Arakawa, Y. (2002). Influence of the Focusing f Number on Heating Regime Transition in Laser Absorption Waves, *Journal of Applied Physics*, Vol.92, No.10, pp.5663-5667.

Mori, K., Komurasaki, K., and Arakawa, Y. (2004). Energy Transfer from a Laser Pulse to a Blast Wave in Reduced-Pressure Air Atmospheres, *Journal of Applied Physics*, Vol. 95, No. 11, pp. 5979-5983.

Mori, K., Komurasaki, K., and Arakawa, Y. (2004). Nozzle Scale Optimum for the Impulse Generation in a Laser Pulsejet, *Journal of Spacecraft and Rockets*, Vol. 41, No. 5, pp. 887-889.

Myrabo, L. N. (2001). World Record Flights of Beam-Riding Rocket Lightcraft: Demonstration of "Disruptive" Propulsion Technology, AIAA Paper 2001-3798.

Phipps, C. R., Reilly, P. R., and Campbell, J. W. (2000). Optimum Parameters for Laser Launching Objects into Low Earth Orbit, *Laser and Particle Beams*, Vol. 18, pp.661-695.

Phipps, C. R., Friedman, H., Gavel, D., Murray, J., Albrecht, G., George, E. V., Ho, C., Priedhorsky, W., Michaelis, M. M., and Reilly, J. P. (1996). ORION: Clearing Near-Earth Space Debris Using a 20-kW, 530-nm, Earth-Based, Repetitively Pulsed Laser, *Laser Particle Beams*, Vol. 14, pp. 1-4.

Steiner, H., Gretler, W., and Hirschler, T. (1998). Numerical Solution for Spherical Laser-Driven Shock Waves, *Shock Waves*, Vol. 8, No. 3, pp.139-147.

Toki, K. (1991). Conceptual Study of Laser Direct Launch, *The Journal of Space Technology and Science*, Vol. 8, No. 2, 1991, pp. 23-31.

Wang, T.-S., Chen, Y.-S., Liu, J., Myrabo, L. N., and Mead, F. B. Jr. (2002). Advanced Performance Modeling of Experimental Laser Lightcraft, *Journal of Propulsion and Power*, Vol. 18, No. 6, pp. 1129-1138.

Spencer, D. B. and Robert, D. C. (1995). Designing Continuous-Thrust Low-Earth-Orbit to Geosynchronous-Earth-Orbit Transfers, *Journal of Spacecraft and Rockets*, Vol. 32, No. 6, pp.1033-1038.

Kluever, C. A. and Oleson, S. R. (1998). Direct Approach for Computing Near-Optimal Low-Thrust Earth-Orbit Transfers, *Journal of Spacecraft and Rockets*, Vol. 35, No. 4, pp.509-515.

Sedov, L. I. (1993). *Similarity and Dimensional Methods in Mechanics, 10th Ed.*, CRC Press, Boca Raton.

Raizer, Y. P. (1977). *Laser-Induced Discharge Phenomena*, Consultants Bureau, New York and London.

Richard, J. C., Morales, C., and Myrabo, L. N. (1988) Transatmospheric Laser Propulsion of a 100 MW-Class Lightcraft Technology Demonstrator (LTD), AIAA Paper 88-2970.

Sanders, S., et al. (1994). High Power Coherent Two-Dimensional Semiconductor Laser Array, *Applied Physics Letters*,, Vol. 64, No. 12, pp.1478-1480.

Simons, G. A. and Pirri, A. N. (1977). The Fluid Mechanics of Pulsed Laser Propulsion, *AIAA Journal*, Vol. 15, No. 6, pp. 835-842.

Shirakawa, A., Saitou, T., Sekiguchi, T., and Ueda, K. (2002). Coherent Addition of Fiber Lasers by Use of a Fiber Coupler, *Optics Express*, Vol. 10, No. 21, pp.1167-1172.

Ushio, M., Komurasaki, K., Kawamura, K., and Arakawa, Y. (2008). Effect of Laser Supported Detonation Wave Confinement on Termination Conditions, *Shock Waves*, Vol. 18, No.1, pp.35-39.

2

"Impulsar": New Application for High Power High Repetition Rate Pulse-Periodic Lasers

V.V. Apollonov
Prokhorov General Physics Institute RAS, Vavilov str., 38, Moscow, 119991, Russia

1. Introduction

Since the 1970s, the possibility of using a laser engine to launch light satellites into orbit has been attracting the attention of researchers [1-9]. The solution of problems considered in [3] is still of current interest. This is an increase in the efficiency - the coupling coefficient J_r of using laser radiation (the ratio of the propulsion to the radiation power) by several times and the prevention of the shock damage of the apparatus, which appears when high-power repetitively pulsed laser radiation with low repetition rates / is used. For example, for J_r ~ 0.3 kN MW^{-1} (this value is typical for an air-jet laser engine of 1970th.), the mass of 200 kg, and the acceleration of 10g, the required laser power should be ~ 60 MW (the energy Q ~100 g in the TNT equivalent, f~ 100 Hz), and the power of a power supply should be 0.5-1 GW. However, it seems unlikely that such a laser will be created in the near future. In our experiments, J_r ~1 kN MW^{-1} (obtained experimentally) and 3-5kN MW^{-1} (estimate, special conditions), which allows us to reduce the laser power by a factor of 7-10. A power of 10-15 MW can be obtained already at present with the help of gas-dynamic lasers and HF/DF lasers by using the properties of repetitively pulsed lasing with high repetition rates and methods for power scaling of lasing [10, 11],checked already for CO2,HF and Nd YAG. .
To solve these problems, it was proposed to use repetitively pulsed radiation with f~ 100 kHz, the optical pulsating discharge (OPD), and the effect of merging of shock waves produced by the OPD [12-14]. The merging criteria were confirmed in experiments [15]. The OPD is laser sparks in the focus of repetitively pulsed radiation, which can be at rest or can move at high velocities [16-20]. The high-frequency repetitively pulsed regime is optimal for continuously-pumped Q-switched high-power lasers. In this case, the pulse energy is comparatively small and the stationary propulsion is possible.
The aim of our work is to verify experimentally the possibility of using pulse-periodic laser radiation with a high repetition rate to produce very effective stationary propulsion in a laser jet engine and to demonstrate the advantages of this technology for production of super long conductive canal for energy delivery from space.

2. Experimental

In the model considered in [12-14], the pulsed and stationary regimes are possible. Figure 1 explains the specific features of these regimes. An OPD is produced at the focus of a lens on the axis of a gas jet flowing from a high-pressure chamber or an air intake to a cylindrical reflector. The shock waves generated by the OPD merge to form a quasi-stationary wave - the high-pressure region between the OPD and reflector.

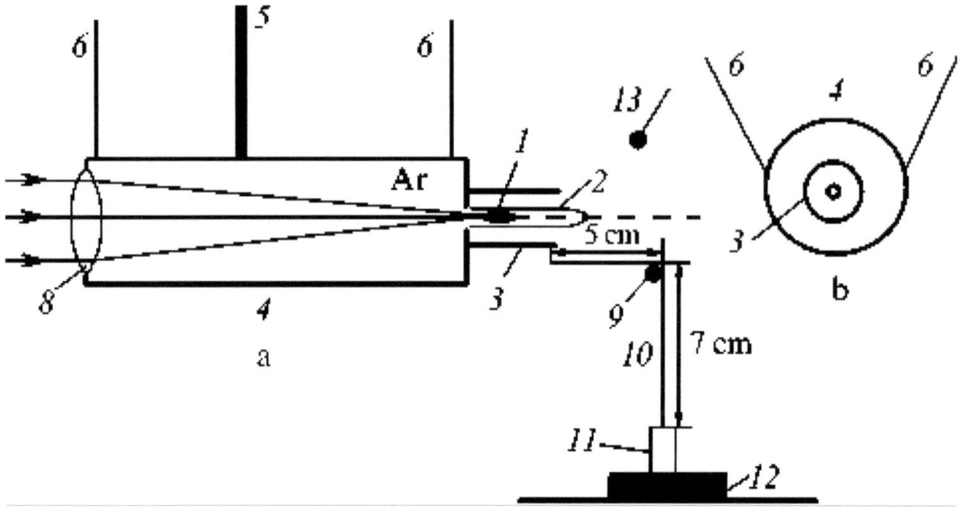

Fig. 1. Scheme of the experiment, side (a) and front (b) view: *(1)* OPD; *(2)* argon jet; *(3)* replaceable cylindrical attachment (reflector); *(4)* high-pressure chamber (rocket model); (5) elastic hose for argon supply; *(6)* model suspension wires; (7) laser radiation; *(8)* focusing lens; *(9)* block; *(10)* wire connecting cylinder *(3)* with weight *(11)*; *(12)* balance; *(13)* shock-wave pressure gauge.

As a result, the propulsion F_r appears. In a cylindrical reflector, the coupling coefficient is maximal, J_r=1.1NkW^{-1} [13], as for a plane explosion [21]. In the pulsed regime, the OPD is produced by trains of laser pulses. A narrow jet of diameter $D_j \sim 0.3R_d$ [13], which is smaller than the reflector diameter D_r, carries a plasma out from the OPD region, which is necessary for the efficient formation of shock waves. Here, R_d=2A5$(q/P_0)^{1/3}$ is the dynamic radius of a spark, q (in J) is the laser pulse energy absorbed in a spark, and P_0 (in atm) is the gas pressure. The propulsion acts during a pulse train, whose duration is limited by the air heating time. The hot atmospheric air is replaced by the cold air during the interval between pulses. In the stationary regime, gas continuously arrives to the reflector from the bottom, by forming a jet over the entire section. In experiments in this regime, we have $D_j \sim 2R_d \sim 3$ mm, which is comparable with the reflector diameter $D_r \sim 5$ mm.

The scheme of the experiment is shown in Figure 1. The OPD was produced by radiation from a pulsed CO_2 laser. The pulse duration was ~ 1 μs, the duration of the front peak was 0.2 μs. The pulse repetition rate was varied from 7 to 100 kHz, the pulse energy was 0.1-0.025 J. The peak power was 300-100 kW, the average power of repetitively pulsed radiation was W= 600 - 1700 W, and the absorbed power was $W_a = \eta W (\eta \approx 0.7)$. Figure 2 shows the shapes of the incident pulse and the pulse transmitted through the OPD region. Note that for a short pulse duration and high power, $\eta \sim 0.95$. Because the radiation intensity at the focus is lower than the optical breakdown threshold in air, the argon jet was used. The length l of sparks along the flow was ~ 0.5 cm.

The model of a rocket with a laser engine was a duralumin cylinder of diameter ~ 8 cm, length ~ 26 cm, and weight 1.1 kg, which was suspended on four thin wires of length 1.1 m and capable of moving only in the axial direction. A reflector (replaceable cylindrical

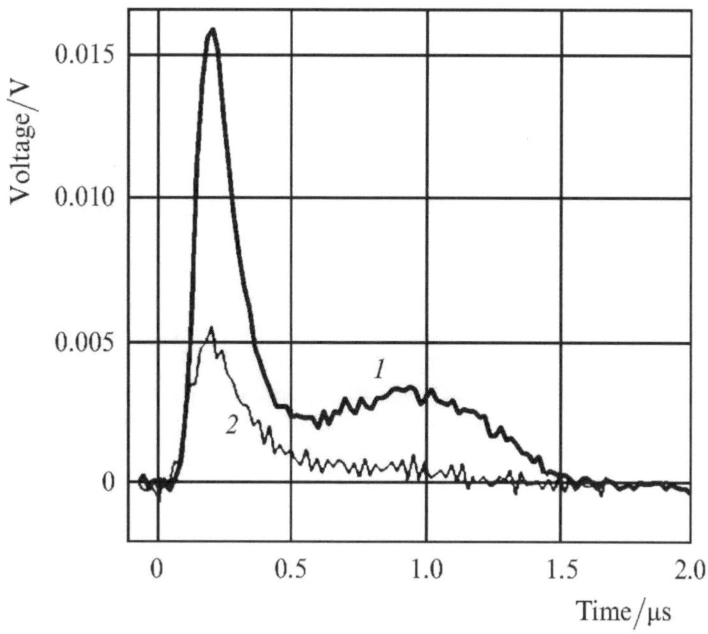

Fig. 2. Oscillograms of the laser pulse (*1*) and radiation pulse transmitted through the OPD (*2*) for f=50 kHz.

attachment) was mounted on the chamber end. Laser radiation was directed to the chamber through a lens with a focal distance of 17 cm. The argon jet was formed during flowing from a high-pressure chamber through a hole of diameter ~ 3 - 4 mm. The jet velocity V was controlled by the pressure of argon, which was delivered to the chamber through a flexible hose. The force produced by the jet and shock waves was imparted with the help of a thin (of diameter ~ 0.2 mm) molybdenum wire to a weight standing on a strain-gauge balance (accurate to 0.1 g). The wire length was 12 cm and the block diameter was 1 cm.

The sequence of operations in each experiment was as follows. A weight fixed on a wire was placed on a balance. The model was slightly deviated from the equilibrium position (in the block direction), which is necessary for producing the initial tension of the wire (~ 1 g). The reading F_m of the balance was fixed, then the jet was switched, and the reading of the balance decreased to F_1. This is explained by the fact that the rapid jet produces a reduced pressure (ejection effect) in the reflector. After the OPD switching, the reading of the balance became F_2. The propulsion F_r produced by the OPD is equal to F_1 - F_2. The pressure of shock waves was measured with a pressure gauge whose output signal was stored in a PC with a step of ~ 1 μs. The linearity band of the pressure gauge was ~ 100 kHz. The gauge was located at a distance of ~ 5 cm from the jet axis (see Figure 1) and was switched on after the OPD ignition (t = 0). The pressure was detected for 100 ms.

Let us estimate the possibility of shock-wave merging in the experiment and the expected values of F_r and J_r. The merging efficiency depends on the parameters $\omega = fR_d/C_0$ and $M_0 = V/C_0$ ($M_0 < 1$), where C_0 is the sound speed in gas. If the distance from the OPD region to the walls is much larger than R_d and sparks are spherical or their length l is smaller than R_d, then the frequencies characterizing the interaction of the OPD with gas are:

$$\omega_0 \approx 2.5M_0, \quad (1)$$

$$\omega_1 \approx 0.8(1 - M_0), \quad (2)$$

$$\omega_2 \approx 5.9(1 - M_0)^{1.5}. \quad (3)$$

For $\omega<\omega_1$, the shock waves do not interact with each other. In the range $\omega<\omega_1<\omega_2$, the compression phases of the adjacent waves begin to merge, this effect being enhanced as the value of ω approaches ω_2. In the region $\omega<\omega_2$, the shock waves form a quasi-stationary wave with the length greatly exceeding the length of the compression phase of the shock waves. For $\omega<\omega_0$, the OPD efficiently (up to ~ 30 %) transforms repetitively pulsed radiation to shock waves.

In the pulsed regime the value of M_0 in (1) corresponds to the jet velocity. Because shock waves merge in an immobile gas, $M_0 \approx 0$ in (2) and (3). The frequencies f=7-100 kHz correspond to R_d = 0.88 - 0.55 cm and ω = 0.2 - 1.7. Therefore, shock waves do not merge in this case. In trains, where the energy of the first pulses is greater by a factor of 1.5-2 than that of the next pulses ($\omega \approx 2$), the first shock waves can merge. The propulsion produced by pulse trains is $F_r = J_r \eta W$ = 0.3 N (~ 30 g), where-J_r = 1.1 N kW^{-1}, η = 0.6, and W~ 0.5 kW.

In the stationary regime for M_0 ~ 0.7, the shock wave merge because $\omega>\omega_2$ (ω = 1.8, $\omega_2 \approx 1.3$). A quasi-stationary wave is formed between the OPD and the cylinder bottom. The excess pressure on the bottom is $\delta P = P-P_0 = 0.54P_0(R_d/r)^{1.64} \approx$ 0.25- 0.5 atm, and the propulsion is $F_r \approx \pi(D_r^2 - D_j^2)\delta P/4$ = 0.03- 0.06kg.

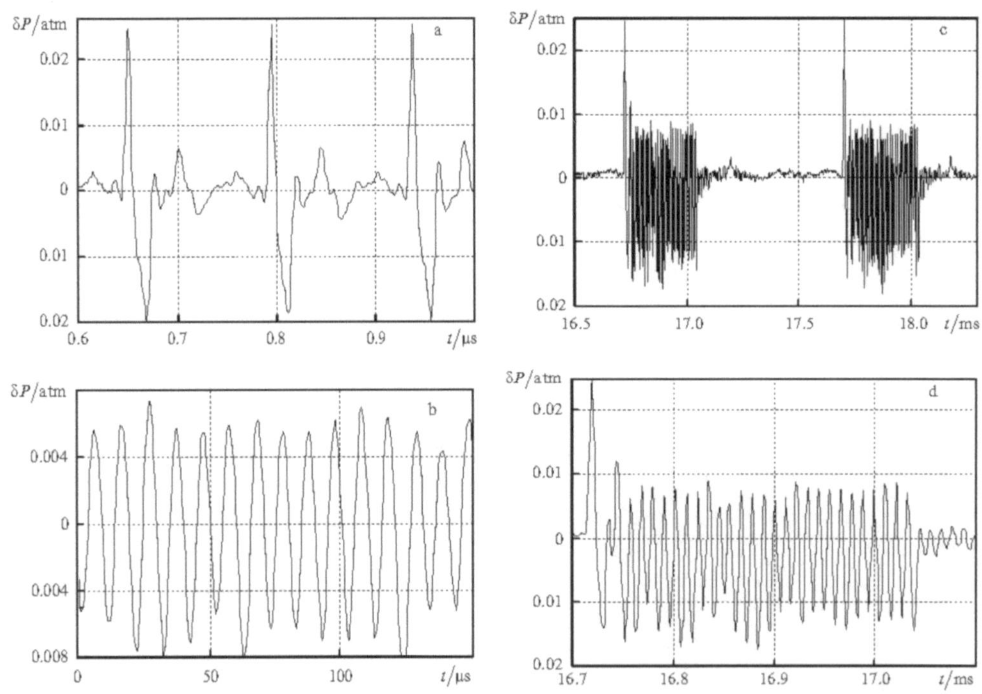

Fig. 3. Pressure pulsations produced by the OPD for V=300 ms^{-1} without reflector), f=7 kHz, W=690 W (a); f=100 kHz, W= 1700 W (b), and f=100 kHz, the train repetition rate φ=1 kHz, W=1000 W, the number of pulses in the train N=30 (c); the train of shock waves at a large scale, parameters are as in Fig. 3c (d).

3. Results of measurements

3.1 Control measurements

The jet propulsions F_j and F_r and the excess pulsation pressure $\delta P = P - P_0$ were measured for the model without the reflector. We considered the cases of the jet without and with the OPD. The jet velocity V and radiation parameters were varied. For V= 50, 100, and 300 m s^{-1}, the propulsion produced by the jet was Fj = 6, 28, and 200 g, respectively, and the amplitude of pulsations was $\delta P = 5\times10^{-6}$, 2×10^{-5} and 3×10^{-4} atm. The OPD burning in the jet did not change the reading of the balance. This is explained by the fact that the OPD is located at a distance of r from the bottom of a high-pressure chamber, which satisfies the inequality $r/R_d > 2$, when the momentum produced by shock waves is small [3, 22]. As follows from Fig. 3, pulsations $\delta P(t)$ produced by the OPD greatly exceed pressure fluctuations in the jet.

3.2 Stationary regime

The OPD was burning in a flow which was formed during the gas outflow from the chamber through a hole (D_j = 0.3 cm) to the reflector (D_r = 0.5 cm) (Figure 4). Because the excess pressure on the reflector bottom was ~ 0.5 atm (see above), to avoid the jet closing, the pressure used in the chamber was set equal to ~ 2 atm. The jet velocity without the ODP was V=300 and 400 ms^{-1}, F_j= 80 and 140 g. The OPD was produced by repetitively pulsed radiation with f= 50 and 100 kHz and the average power $W\approx1200$ W (the absorbed power was $W_a \approx 650$ W). Within several seconds after the OPD switching, the reflector was heated up to the temperature more than 100°C.
Figure 5 illustrates the time window for visualization of shock waves with the Schlieren system in the presence of plasma. Before 7 μs, the plasma is too bright relative to the LED source, and all information about the shock wave is lost. At 7 μs, the shock wave image could be discerned under very close examination. By 10 μs, the shock wave is clearly visible in the image; however, at this time the shock wave has nearly left the field of view. A technique was needed to resolve the shock waves at short timescales, when plasma was present.

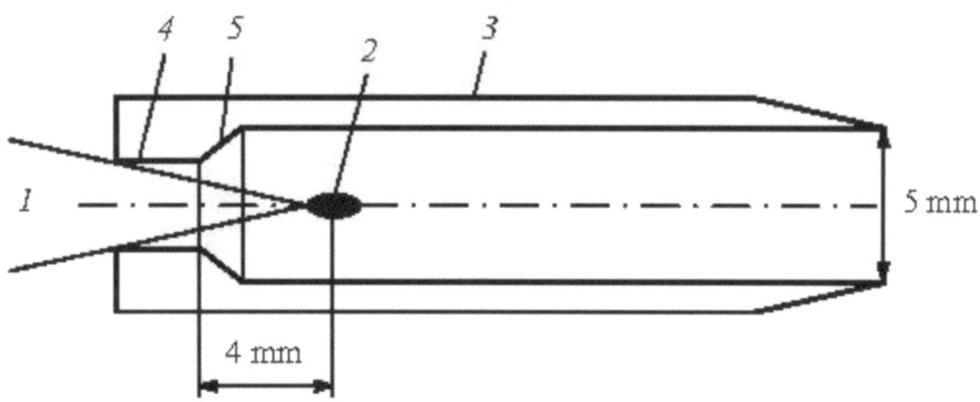

Fig. 4. Reflector of a stationary laser engine: (*1*) repetitively pulsed laser radiation with f=50 and 100 kHz, W=1200 W; (2) OPD; (*3*) reflector; (*4*) hole of diameter ~ 3 mm through which argon outflows from a high-pressure chamber (~ 2atm) to the reflector; (*5*) reflector bottom, the angle of inclination to the axis is ~ 30°.

For f= 50 kHz and V= 300 m s $^{-1}$, the propulsion is F_r = 40 g, and for V= 400 m s $^{-1}$ the propulsion is 69 g; the coupling coefficient is $J_r \approx 1.06$ N kW^{-1}. The propulsion F_r is stationary because the criteria for shock-wave merging in front of the OPD region are fulfilled. Downstream, the shock waves do not merge. One can see this from Figure 5 demonstrating pressure pulsations $\delta P(t)$ measured outside the reflector. They characterize the absorption of repetitively pulsed radiation in the OPD and, therefore, the propulsion. For f= 50 kHz, the instability is weak (±5 %) and for f= 100 kHz, the modulation $\delta P(t)$ is close to 100 %. The characteristic frequency of the amplitude modulation $f_a \approx 4$ kHz is close to $C_0/(2H)$, where H is the reflector length. The possible explanation is that at the high frequency f the plasma has no time to be removed from the OPD burning region, which reduces the generation efficiency of shock waves. The jet closing can also lead to the same result if the pressure in the quasi-stationary wave is comparable with that in the chamber. Thus, repetitively pulsed radiation can be used to produce the stationary propulsion in a laser engine.

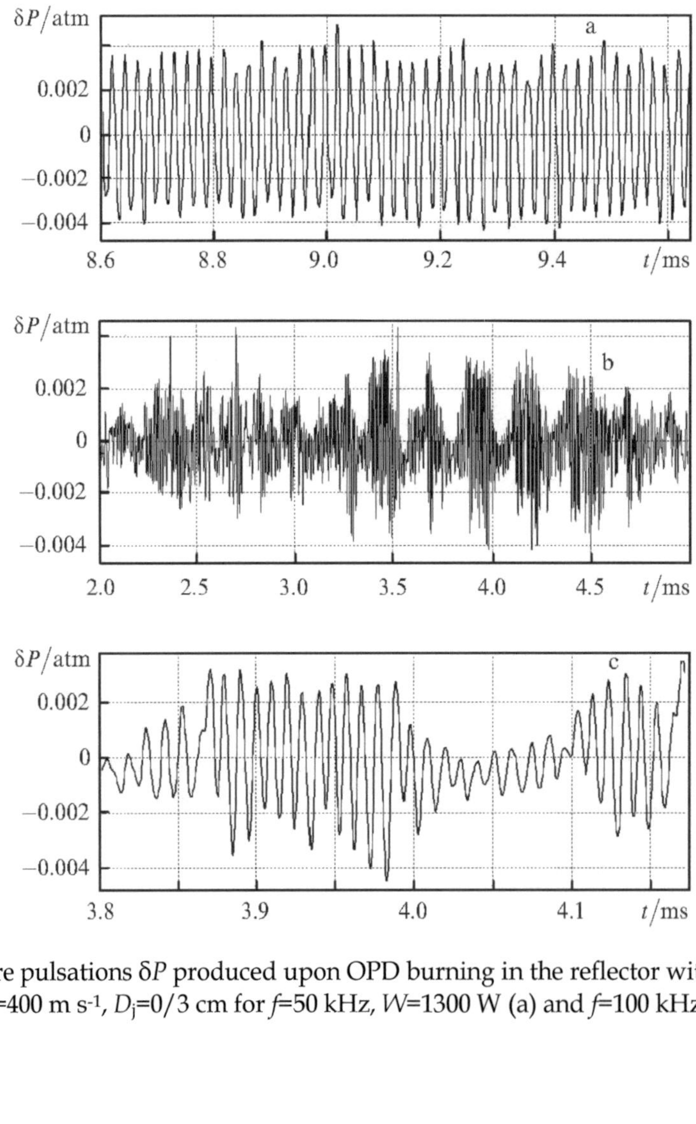

Fig. 5. Pressure pulsations δP produced upon OPD burning in the reflector with D_r=0.5 cm, H=4/6 cm, V=400 m s^{-1}, D_j=0/3 cm for f=50 kHz, W=1300 W (a) and f=100 kHz, W=1200 W (b, c).

3.3 Pulsed regime

To find the optimal parameters of the laser engine, we performed approximately 100 OPD starts. Some data are presented in Table 1. We varied the diameter and length of the reflector, radiation parameters, and the jet velocity (from 50 to 300 m s $^{-1}$). For V= 50 m s $^{-1}$ the ejection effect is small, for V= 300 m s $^{-1}$≈C_0, this effect is strong, while for V≈ 100 ms^{-1}, the transition regime takes place. In some cases, the cylinder was perforated along its circumference to reduce ejection. The OPD was produced by radiation pulse trains, and in some cases - by repetitively pulsed radiation. The structure and repetition rate of pulse trains was selected to provide the replacement of the heated OPD gas by the atmospheric air. The train duration was ~ 1/3 of its period, the number of pulses was N = 15 or 30, depending on the frequency f. The heating mechanism was the action of the thermal radiation of a plasma [23], the turbulent thermal diffusivity with the characteristic time ~ 300 μs [24] and shock waves.

The propulsion F_r was observed with decreasing the reflector diameter and increasing its length. The OPD burned at a distance of ~ 1 cm from the reflector bottom. One can see from Figure 6 that the shock waves produced by the first high-power pulses in trains merge. For f= 100 kHz, the pulse energy is low, which is manifested in the instability of pressure pulsations in trains. As the pulse energy was approximately doubled at the frequency f = 50 kHz, pulsations δP (t) were stabilized. The OPD burning in the reflector of a large diameter (D_r/R_d ≈ 4) at a distance from its bottom satisfying the relation r/R_d ≈ 3 did not produce the propulsion.

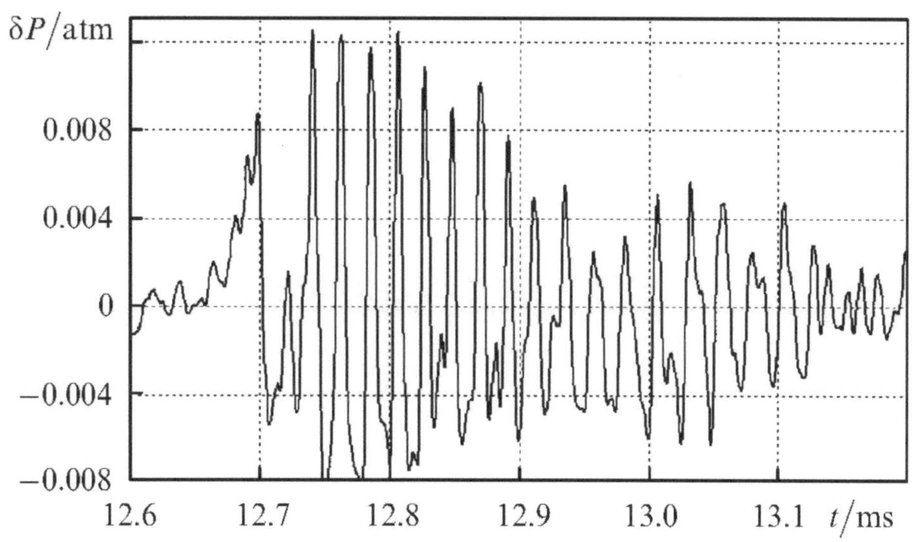

Fig. 6. Pressure pulsations δP in the OPD produced by pulse trains with φ=1.1 kHz, f=50 kHz, W=720 W, N=15, V=300 m s^{-1}, D_r=1.5 cm, H= 5 cm, D_j= 4 mm, and F= 4.5 g.

Table 1 presents some results of the measurements. One can see that the coupling coefficient J_r strongly depends on many parameters, achieving 1 N kW^{-1} in the stationary regime and 0.53 N kW^{-1} in the pulsed regime.

At present, the methods of power scaling of laser systems and laser engines, which are also used in laboratories, are being extensively developed [10, 25]. Let us demonstrate their application by examples. We observed the effect when the OPD produced the 'negative' propulsion F_t = -97 g (see Table 1), which correspond to the deceleration of a rocket. The value of J_r can be increased by approximately a factor of 1.5 by increasing the pulse energy and decreasing their duration down to ~ 0.2 μs. An important factor characterizing the operation of a laser engine at the high-altitude flying is the efficiency I_m of the used working gas. The value I_m = 0.005 kg $N^{-1}s^{-1}$ can be considerably reduced in experiments by using a higher-power radiation. The power of repetitively pulsed radiation should be no less than 10 kW. In this case, F_r will considerably exceed all the other forces. The gas-dynamic effects that influence the value of.F_r, for example, the bottom resistance at the flight velocity ~ 1 km s^{-1} should be taken into account.

f/kHz	φ/kHz	D_r/mm, [H/mm]	N	V/m s^{-1}	W/W	F_j/g	F_r/g	J_r/N kW^{-1}	Reflector material
45	RP	5, [46]	–	300	1300	80	40	0.61	duralumin
45	RP	5, [46]	–	400	1300	141	69	1.06	–"–
100	RP	5, [46]	–	400	1200	155	54	1.08	–"–
100	1	15, [50]	30	300	720	49	4	0.085	–"–
45	1	15, [50]	15	50	720	0.9	2.1	0.042	–"–
45	1	15, [50]	15	300	720	49.1	4.5	0.09	–"–
45	1	15, [50]	15	50	720	1.2	1.4	0.028	duralumin*
45	1	15, [50]	15	100	720	6.3	5.6	0.11	–"–
45	1	15, [50]	15	300	720	62.7	4	0.08	–"–
45	1	15, [50]	5	170	500	17.7	3.5	0.1	–"–
45	2	15, [50]	5	100	600	6.3	4.8	0.11	–"–
45	2	15, [50]	5	164	600	18.5	7.5	0.18	–"–
45	2	15, [50]	5	300	600	70	−4	0.095	–"–
12.5	RP	25, [35]	–	60	430	2.4	4	0.13	quartz
12.5	RP	25, [35]	–	100	430	5	7	0.23	–"–
12.5	RP	25, [35]	–	150	430	11	11	0.37	–"–
12.5	RP	25, [35]	–	300	430	51	16	0.53	–"–
12.5	RP	25, [35]	–	50	430	6	1	0.033	duralumin**
12.5	RP	25, [35]	–	100	430	12	7	0.23	–"–
12.5	RP	25, [35]	–	300	430	195	−97	−3.2	–"–

Table 1. Experimental conditions and results.

Note. Laser radiation was focused at a distance of 1 cm from the reflector bottom; * six holes of diameter 5 mm over the reflector perimeter at a distance of 7 mm from its exhaust; **six holes of diameter 5 mm over the reflector perimeter at a distance of 15 mm from its exhaust. Thus, our experiments have confirmed that repetitively pulsed laser radiation produces the stationary propulsion with the high coupling coefficient. The development of the scaling methods for laser systems, the increase in the output radiation power and optimization of the interaction of shock waves will result in a considerable increase in the laser-engine efficiency.

4. The impact of thermal action

A laser air-jet engine (LAJE) uses repetitively pulsed laser radiation and the atmospheric air as a working substance [1-3]. In the tail part of a rocket a reflector focusing radiation is located. The propulsion is produced due to the action of the periodic shock waves produced by laser sparks on the reflector. The laser air-jet engine is attractive due to its simplicity and high efficiency. It was pointed out in papers [26] that the LAJE can find applications for launching space crafts if ~ 100-kJ repetitively pulsed lasers with pulse repetition rates of hundreds hertz are developed and the damage of the optical reflector under the action of shock waves and laser plasma is eliminated. These problems can be solved by using high pulse repetition rates ($f\sim$ 100 kHz), an optical pulsed discharge, and the merging of shock waves [12, 13]. The efficiency of the use of laser radiation in the case of short pulses at high pulse repetition rates is considerably higher. It is shown in this paper that factors damaging the reflector and a triggered device cannot be eliminated at low pulse repetition rates and are of the resonance type.

Let us estimate the basic LAJE parameters: the forces acting on a rocket in the cases of pulsed and stationary acceleration, the wavelength of compression waves excited in the rocket body by shock waves, the radius R_k of the plasma region (cavern) formed upon the expansion of a laser spark. We use the expressions for shock-wave parameters obtained by us. A laser spark was treated as a spherical region of radius r_0 in which the absorption of energy for the time ~ 1 μs is accompanied by a pressure jump of the order of tens and hundreds of atmospheres. This is valid for the LAJE in which the focal distance and diameter of a beam on the reflector are comparable and the spark length is small. The reflector is a hemisphere of radius R_r. The frequency f is determined by the necessity of replacing hot air in the reflector by atmospheric air.

Let us estimate the excess of the peak value F_m of the repetitively pulsed propulsion over the stationary force F_s upon accelerating a rocket of mass M. It is obvious that $F_s = Ma$, where the acceleration $a = (10\text{-}20)g_0 \approx 100 - 200$ m s^{-2}. The peak value of the repetitively pulsed propulsion is achieved when the shock wave front arrives on the reflector. The excess pressure in the shock wave (with respect to the atmospheric pressure P_0) produces the propulsion $F_j(t)$ and acceleration a of a rocket of mass M. The momentum increment produced by the shock wave is:

$$\delta p_{\rm i} = \int_0^{1/f} F_{\rm i}(t)\mathrm{d}t \simeq F_{\rm a}t_{\rm a}\ [\mathrm{N\cdot s}]. \tag{4}$$

Here, F_a is the average value of the propulsion for the time t_a of the action of the compression phase of the shock wave on the reflector, and $F_m \approx 2F_a$. By equating δ_{Pi} to the momentum increment $\delta p_s = F/f = aM/f$ over the period under the action of the stationary propulsion F_s, we find:

$$\Delta = F_{\rm m}/F_{\rm s} = 2/(ft_{\rm a}). \tag{5}$$

The value of Δ, as shown below, depends on many parameters. The momentum increment per period can be expressed in terms of the coupling coefficient J as $\delta_{Pi} = JQ$, where Q [J] is the laser radiation energy absorbed in a spark. The condition $\delta_{Pi} = \delta p_s$ gives the relation:

$$W = aM/J \tag{6}$$

between the basic parameters of the problem: $W = Qf$ is the absorbed average power of repetitively pulsed radiation, and $J \approx 0.0001 - 0.0012$ N s J^{-1} [3, 13, 26].

The action time of the compression phase on the reflector is $t_a \sim R_c/V$, where $V \approx k_1C_0$ is the shock-wave velocity in front of the wall ($k_1 \sim 1.2$) and $C_0 \approx 3.4 \times 10^4$ cm s^{-1} is the sound speed in air. The length R_c of the shock wave compression phase can be found from the relation:

$$\frac{R_{\rm c}}{R_{\rm d}} = 0.26\left(\frac{h}{R_{\rm d}}\right)^{0.32}. \tag{7}$$

Here, h is the distance from the spark centre to the reflector surface and $R_d \approx 2.15(Q/P_0)^{1/3}$ is the dynamic radius of the spark (distance at which the pressure in the shock wave becomes close to the air pressure P_0). In this expression, R_d is measured in cm and P_0 in atm. The cavern radius can be found from the relation:

$$\frac{R_k}{R_d} = 0.6\left(\frac{r_0}{R_d}\right)^{0.29} = 0.22 - 0.3 \approx 0.25. \tag{8}$$

The final expression (8) corresponds to the inequality $r_0/R_d < 0.03 - 0.1$, which is typical for laser sparks (r_0 is their initial radius). Let us find the range of P_0 where the two conditions are fulfilled simultaneously: the plasma is not in contact with the reflector surface and the coupling coefficient J is close to its maximum [3, 22, 26]. This corresponds to the inequality $R_k<h<R_d$. By dividing both parts of this inequality by R_d, we obtain $R_k/R_d < h/R_d < 1$, or $0.25 < h/R_d < 1$. As the rocket gains height, the air pressure and, hence, h/R_d decrease. If we assume that at the start (P_0 = 1 atm) the ratio $h/R_d = 1$, where h and R_d are chosen according to (2), then the inequality $0.25 < h/R_d < 1$ is fulfilled for P_0 = 1 – 0.015 atm, which restricts the flight altitude of the rocket by the value 30 - 40 km (h = const).
The optimal distance h satisfies the relation $h/R_d \approx 0.25 b_i$ where $b_i \approx 4 - 5$. By substituting h/R_d into (7), we find the length of the shock-wave compression phase and the time of its action on the reflector:

$$\frac{R_c}{R_d} \approx 0.17 b_i^{0.32}, \tag{9}$$

$$t_a = \frac{0.17 b_i^{0.32} R_d}{k_1 C_0} = \frac{s_1 Q^{1/3}}{P_0^{1/3}} = \frac{s_1}{P_0^{1/3}}\left(\frac{aM}{Jf}\right)^{1/3}, \tag{10}$$

Where $s_1 = 0.37 b_i^{0.32}/(k_1 C_0) \approx 9\times10^{-6} b_i^{0.32}$. From this, by using the relation $\Delta = F_m/F_a = 2/(F t_a)$ we find:

$$\Delta = \frac{2P_0^{1/3}}{s_1 f^{2/3} W^{1/3}} = \frac{2P_0^{1/3} Q^{2/3} J}{s_1 aM} = \frac{2}{s_1 f^{2/3}}\left(\frac{P_0 J}{aM}\right)^{1/3}. \tag{11}$$

Of the three parameters Q, W, and f, two parameters are independent. The third parameter can be determined from expression (6). The conditions $1/f \sim t_a$ and $\Delta \approx 1 - 2$ correspond to the merging of shock waves [12].
The important parameters are the ratio of t_a to the propagation time $t_z = L/C_m$ of sound over the entire rocket length L (C_m is the sound speed in a metal) and the ratio of t_z to $1/f$. For steel and aluminum, C_m = 5.1 and 5.2 km s^{-1}, respectively. By using (10), we obtain:

$$U = \frac{t_a C_m}{L} = \frac{s_1 C_m}{L P_0^{1/3}} Q^{1/3}. \tag{12}$$

Here, L is measured in cm and C_m in cm s^{-1}. Expression (12) gives the energy:

$$Q = \frac{35.4P_0}{b_1^{0.96}}\left(U\frac{C_0}{C_m}\right)^3 L^3. \tag{13}$$

From the practical point of view, of the most interest is the case $U > 1$, when the uniform load is produced over the entire length L. If $U < 1$, the acceleration is not stationary and the wavelength of the wave excited in the rocket body is much smaller than L. If also $C_m/f < L$, then many compression waves fit the length L. The case $U \approx 1$ corresponds to the resonance excitation of the waves. Obviously, the case $U \leq 1$ is unacceptable from the point of view of the rocket strength.

By using the expressions obtained above, we estimate Δ, U, and R_k for laboratory experiments and a small-mass rocket. We assume that $b_i = 4$, $J = 5\times10^{-4}$ N s J^{-1}, and $s_1 = 1.4\times10^{-5}$. For the laboratory conditions, $M \approx 0.1$ kg, $R_r \approx 5$ cm, $L = 10$ cm, and $a = 100$ m s^{-2}. The average value of the repetitively pulsed propulsion F_{IP} is equal to the stationary propulsion, $F_{IP} = F_s = 10$ N; the average power of repetitively pulsed radiation is $W = F_{IP}/J = 20$ kW, and the pulse energy is $Q_p = W/f$. We estimate the frequency f and, hence, $Q_p \approx Q$ for the two limiting cases.

At the start, $P_0 \approx 1$ atm and the cavern radius R_k is considerably smaller than R_r. Here, as in the unbounded space, the laser plasma is cooled due to turbulent thermal mass transfer. For $Q_p <$ 20 J, the characteristic time of this process is 2-5 ms [8,9], which corresponds to $f = 500 - 200$ Hz. If $R_k \sim R_r$ ($P_0 < 0.1$ atm), the hot gas at temperature of a few thousands of degrees occupies the greater part of the reflector volume. The frequency f is determined by the necessity of replacing gas over the entire volume and is $\sim 0.5C_0/R_r$ -850 Hz. Let us assume for further estimates that $f = 200$ Hz, which gives $Q_p = 100$ J. We find from (7) and (8) that $\Delta = 74$ and $U =$ 3.5. This means that the maximum dynamic propulsion exceeds by many times the propulsion corresponding to the stationary acceleration. The action time of the shock wave is longer by a factor of 3.5 than the propagation time of the shock wave over the model length. For $P_0 = 1$ and 0.01 atm, the cavern radius is $R_k = 2.5$ and 11.6 cm, respectively.

5. The dynamic resonance loads

Let us make the estimate for a rocket by assuming that $M \approx 20$ kg, $R_r \approx 20$ cm, $L = 200$ cm, and $a = 100$ m s^{-2}. The average repetitively pulsed propulsion is $F_{IP} = F_S = 2000$ N, the average radiation power is W=4MW, for f= 200 Hz the pulse energy is $Q_p = 20$ kJ, $\Delta = 12.6$, U = 1, $R_k = 14.7$ and 68 cm ($P_0 = 1$ and 0.01 atm), and $F_m = 25.6$ kN = 2560 kg. One can see that the repetitively pulsed acceleration regime produces the dynamic loads on the rocket body which are an order of magnitude greater than F_s. They have the resonance nature because the condition $U \sim 1$ means that the compression wavelengths are comparable with the rocket length. In addition, as the rocket length is increased up to 4 m and the pulse repetition rate is increased up to 1 kHz, the oscillation eigenfrequency C_m/L of the rocket body is close to f (resonance).

Thus, our estimates have shown that at a low pulse repetition rate the thermal contact of the plasma with the reflector and strong dynamic loads are inevitable. The situation is aggravated by the excitation of resonance oscillations in the rocket body. These difficulties can be eliminated by using the method based on the merging of shock waves. Calculations and experiments [28] have confirmed the possibility of producing the stationary propulsion by using laser radiation with high laser pulse repetition rates. The method of scaling the output radiation power is presented in [10].

6. Matrix of reflectors

This matrix consists of N-element single reflectors, pulse-periodic radiation with a repetition rate of 100 kHz, pulse energy q and average power WC. All elements of the matrix are very similar (Figure 7), radiation comes with the same parameters: $qm = q / N$, $W_m = W_C / N$. The matrix of reflectors creates a matrix of OPD, each one is stabilized by gas flux with velocity - V_{Jm}. OPD's have no interactions in between. Elements structure of the matrix of reflectors could help find the solution for better conditions of gas flux penetration. In our case the number of reflectors was $N = 8$. Larger values of N are not reasonable.
The following estimations are valid for the boundary conditions: $W_C = 20$ MW ($W_m = 2.5$ MW), $f = 105$ Hz, $q = 200$J ($q_m = 25$J), $a_{rm} = 0.3$. Index 1 is for - P0 = 1 atm. (Start of "Impulsar") and index 2 for $P_0 = 0.1$ atm. (end of regime).
Radius of cylinder for each reflector

$$R_{rm} = \frac{0.2R_{dJm}}{\delta_m^{1/2}} = \frac{0.43\left(q_m/P_{J2m}\right)^{1/3}}{\delta_m^{1/2}} = \frac{R_r}{N^{1/3}} = 15.5cm$$

Focus of reflector ~ 5 cm. The size of matrix ~ 90 cm. Additional pressure is:

$$\delta P_{m1} = 1.56 \text{ atm. and } \delta P_{m2} = 0.55 \text{ atm.}$$

Force acting on matrix:

$$F_{am1} = 100 \cdot 103 \text{ N}, F_{am2} = 35.6 \cdot 103 \text{ N}.$$

Specific force for each element of the matrix (for MW of average power):

$$J_{m1} \approx 4000 \text{ N/MW}, J_{m2} \approx 1500 \text{ N/MW}.$$

The velocities of gas flux in the reflectors of matrix:

$$V_{J1} = 2520 \text{ m/s}, V_{J2} = 5440 \text{ m/s}$$

Flight control in this case can be done by thrust change for the different elements of the matrix of reflectors. At the same time, such a configuration could be very helpful in the realization of efficient gas exchange in the area of breakdown behind of the reflectors (Figure 7).
Thus, an OPD can be stationary or move at a high velocity in a gaseous medium. However, stable SW generation occurs only for a certain relation between the radiation intensity, laser pulse repetition rate, their filling factor, and the OPD velocity. The OPD generates a QSW in the surrounding space if it is stationary or moves at a subsonic velocity and its parameters satisfy the aforementioned conditions. The mechanism of SW merging operates in various media in a wide range of laser pulse energies. The results of investigations show that the efficiency of the high repetition rate pulse-periodic laser radiation can be increased substantially when a QSW is used for producing thrust in a laser engine [13, 14].

7. Super long conductive canal for energy delivery

Powerful lasers are capable to create the spending channels of the big length which are settling down on any distances from a radiator. At relatively small energies of single

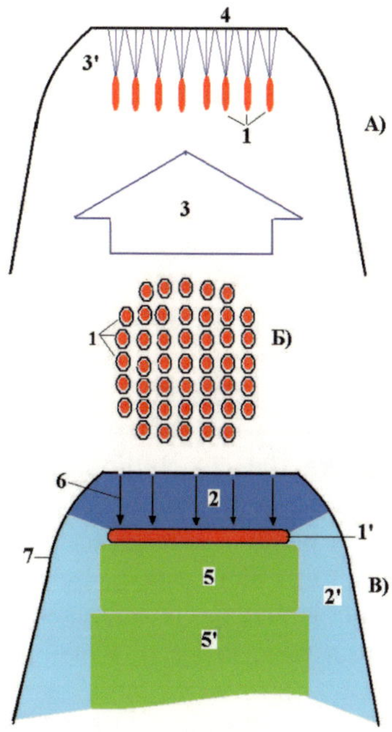

Fig. 7. "Impulsar" engine scheme based on QSW. A) Focusing system, Б) - OPD matrix, creating flat QSW; B) Plasma created inhomogeneities; 1) OPD elements; 1') Model of OPD (Distance from 1' to 4: less 10 cm); 2) Flat QSW, $(P - P0)/P0 \approx 0.5 - 3$; 2') Radial QSW, $(P - P0)/P0 < 0.1$; 3) Main beam, $q \sim 100$ J; 3' - focused beams $q \sim 3\text{-}5$J, creating QSW matrix; 4) Matrix of focusing elements and air injecting system; 5) OPD matrix of plasma decay; 5') OPD plasma turbulence ; 6) Gas flow; 7) Nozzle.

impulses the lengths of channels make about hundreds of meters. Since 1970 the successful attempts of their usage were undertaken for solution of problems of interception of lightning and blocking of overload waves on electric lines. The traditional lightning protection systems being used currently are not always in a position to ensure the desired level of efficiency. This stimulates the quest for new approaches to solve this problem. Laser protection against lightning is one of the most prospective trends that are being developed actively at present [29,30].

. While using this approach, it is assumed that the lightning discharge channel being developed is guided towards the conventional rod of the metal lightning rod along the plasma channel formed as a result of the laser induced breakdown of the atmosphere. This method is based on the concept of an active lightning rod, when a laser beam can be used for "triggering" and guiding a positive ascending leader from the tip of a lightning rod to a negatively charged thunderstorm cloud. It is expected that in contrast to the traditional approach, the use of laser spark will make it possible to control efficiently the very process of protection from lightning, ensure the selectivity of lightning capture, and provide safety of tall objects and large areas. Conductive canal in this case is about 10-15m long and main advantage of the approach is due to immediate appearance of laser produced prolongation of the lightning rod. But maximum length of the laser produced breakdown in the air was registered on the level of 100m and limited by optical method of laser energy delivery into the focal point. Where is the way to get conductive canal of much longer length?

The same goal to produce long conductive canal has ongoing French-German program "Teramobile", based on femto - second multi-photon lasers technology. But the goal is to get very long canal with very low level of electrical resistivity in comparison with canals produced by infrared laser radiation breakdown. The ionisation of air, produced by ultra-intense and ultra-short pulse can be put to use to channel bolts of lightning. As a "Teramobile" burst propagates it creates a sort of straight filament of ionised air, which should conduct electricity. If the laser were directed toward a dark and threatening thunderhead, it would trigger a lightning bolt that could be safely pushed away from doing harm. This capacity has already been demonstrated over a distance of a few meters only with a laboratory version of lightning, and tests on a more natural scale are limited by very high filaments resistivity. So what do we do with a mobile terawatt laser, if it is not good enough for the lightning control ? It can be used very effectively to study the propagation of intense laser light in the atmosphere, detect pollution, and control the parameters of fast objects in the space. Ultra - high intensity brings its own special qualities; it modifies significantly the index of refraction while it induces a focusing of the light beam along its path, the effect of the latter being to produce a self - guiding laser burst which can travel hundreds of meters. Another effect is that the luminous spectrum widens to yield a white laser whose light is composed of a wide range of wavelengths, which is important for a wide spectrum of applications.

There upon the well known program of creation of "Impulsar" represents a great interest, as this program in a combination with high-voltage high - frequency source can be useful in the solution of above mentioned problems. The principle of "Impulsar" operation can be shortly described as follows [31].

Jet draught of the offered device is made under influence of powerful high frequency pulse-periodic laser radiation. In the experiments the CO2 laser and solid - state Nd YAG laser systems were used. Active impulse appears thanks to air breakdown (<30km) or to the breakdown of vapour of low-ionizable material saturated by nano – particles (dust plasma), placed on the board in the vicinity of the focusing mirror - acceptor of breakdown waves. With each pulse of powerful laser the device rises up, leaving a bright and dense trace of products with high degree of ionization and metallization by nano - particles after ablation. The theoretical estimations and experimental tests show that with already experimentally demonstrated figures of specific thrust impulse the lower layers of the Ionosphere can be reached in several ten seconds that is enough to keep the high level of channel conductivity with the help of high frequency high voltage generator.

The space around globe represents a series of megavolt class condensers created by Earth surface, the cloudy cover, various layers of ionosphere and radiating belts. With the help of supported by high - voltage source of trajectory trace of "Impulsar" it is possible to create a conductive channel of required length and direction. In process of the optical vehicle lifting and conductive channel following it, the breakdown characteristics of the interval with decreasing for 5 orders of magnitude (90 km) density considerably reduce, than the process must be prolonged by the expanding of micro-discharges net and develop as self - supported process in the external field of all studied interval. It is important to notice, that presence of such an orbital scale channel allows us also to perform a number of important experiments from the Earth surface as well as from space. Ball and bead lightning investigation is the most interesting application for the long conductive canal technology based on "Impulsar" due to the intriguing possibility for investigator to set up the stationary laboratory with variable boundary conditions for effective tests. Most likely, their nature is

multiple. It would appear that natural ball lightning may be not one phenomenon but many, each with similar appearance but with different mechanisms of origin, different stability criteria, and somewhat different properties dependent upon the atmosphere and the environment present at the time of the event.
Consideration of a large set of available applications of high power high repetition rate pulse-periodic lasers give us strong confidence to open on that basis a new horizons of instrumental space science and wide spectrum of very new and important applications.

8. Acknowledgments

The author would like to acknowledge the valuable contributions made to the "Impulsar" program by N.P.Laverov, S.N.Bagaev, B.I. Katorgin, Yu.M.Baturin.V.N. Tishcenko, G.N. Grachev, V.V. Kijko, Yu.S. Vagin, and A.G. Suzdal'tsev.

9. References

[1] A. R. Kantrowitz *Astronautics and Aeronautics,* 10 (5), 74 (1972).
[2] A. Pirry, M. Monsler, R. Nebolsine *Raket. Tekh. Kosmonavt.,* 12 (9), 112 (1974).
[3] V. P. Ageev, A. I. Barchukov, F. V. Bunkin, V. I. Konov, A. M. Prokhorov, A. S. Silenok, N. I. Chapliev *Kvantovaya Elektron.,* 4, 2501 ,*Sov. J. Quantum Electron.,* 7, 1430 (1977).
[4] W. Schall *Proc. SPIE Int. Soc. Opt. Eng.,* 4065, 472 (2000).
[5] L. N. Myrabo, Yu. P. Raizer 2nd *Int. Symp. on Beamed Energy Propulsion* , Japan, p. 534.(2003)
[6] V. E. Sherstobitov, N. A. Kalitieevskiy, V. I. Kuprenyuk, A. Yu.Rodionov, N. A. Romanov, V. E. Semenov, L. N. Soms, N. V. Vysotina 2nd *Int. Symp. on Beamed Energy Propulsion* Japan, p. 296,(2003)
[7] V.Hasson, F. Mead, C. Larson *Ill Int. Symp. on Beamed Energy Propulsion* ,New York, p. 3
[8] K. Mori, L. Myrabo *Ill Int. Symp. on Beamed Energy Propulsion* (Troy, New York, 2004) p. 155,(2004)
[9] C. S. Hartley, T. W. Partwood, M. V. Filippelli, L. N. Myrabo, H. T. Nagamatsu, M. N. Shneider, Yu.P. Raizer *Ill Int. Symp. On Beamed Energy Propulsion* ,Troy, New York, p. 499,(2004)
[10] V. V. Apollonov, A. B. Egorov, V. V. Kiiko, V. I. Kislov,A. G. Suzdal'tsev .*Kvantovaya Elektron.,* 33, 753 ,*Quantum Elect ron.,* 33, 753 (2003).
[11] G. N. Grachev, A. G. Ponomarenko, A. L. Smirnov , V. B. Shulyat'ev *Proc. SPIE Int. Soc. Opt. Eng.,* 4165, 185 (2000)
[12] V. N. Tishchenko, V. V. Apollonov, G. N. Grachev, A. I. Gulidov, V. I. Zapryagaev, Yu.G. Men'shikov, A. L. Smirnov, A. V. Sobolev *Kvantovaya Elektron.,* 34, 941, *Quantum Electron.,* 34, 941 (2004).
[13] V. V. Apollonov, V. N. Tishchenko *Kvantovaya Elekton.,* 34, 1143 ,*Quantum Electron.,* 34, 1143 (2004)].
[14] V. V. Apollonov, V. N. Tishchenko *Kvantovaya Elekton.,* 36, 763,*Quantum Electron.,* 36, 763 (2006).
[15] G. N. Grachev, A. G. Ponomarenko,V. N. Tishchenko, A. L. Smirnov,S. I. Trashkeev, P. A. Statsenko, M. I. Zimin, A. A. Myakushina, V. I. Zapryagaev, A. I. Gulidov, V. M. Boiko, A. A. Pavlov, A. V. Sobolev *Kvantovaya Elektron.,* 36, 470 ,*Quantum Electron.,* 36, 470 (2006).

[16] U.Bielesch, M. Budde, B. Freisinger, F. Ruders, J. Schafer, J. Uhlenbusch *Proc. ICPIG XXI* (Arbeitsgemeinschaft, Plasmaphysik APP-RUB, p. 253.(1993)
[17] P. K. Tret'yakov, G. N. Grachev, A. I. Ivanchenko, V. I. Krainev, A. G. Ponomarenko, V. N. Tishchenko *Dokl. Akad. Nauk,* 336 (4), 466 (1994).
[18] L. N. Myrabo, Yu. P. Raizer *AIAA Paper,* No. 94-2451 (1994).
[19] V. Yu.Borzov, V. M. Mikhailov, I. V. Rybka, N. P. Savishchenko, A. S. Yur'ev *Inzh.-Fiz. Zh.,* 66 (5), 515 (1994).
[20] G. N. Grachev, A. G. Ponomarenko,A. L. Smirnov,P. A. Statsenko, V. N. Tishchenko, S. I. Trashkeev *Kvantovaya Elektron.,* 35, 973 ,*Quantum Electron.,* 35, 973 (2005).
[21] A. M. Prokhorov, V. I. Konov, I.Ursu, I. N. Mikheilesku *Vzaimodeistvie lazernogo izlucheniya s metallami* Interaction of Laser Radiation with Metals,Moscow: Nauka, (1988).
[22] V. P. Korobeinikov *Zadachi teorii tochechnogo vzryva* Problems of the Theory of Point Explosion,Moscow: Nauka, (1985).
[23] Yu.P. Raizer *Gas Discharge Physics* (Berlin: Springer, 1991; Moscow: Nauka, 1987).
[24] V. N. Tishchenko, V. M. Antonov,A. V. Melekhov ,S. A. Nikitin,V. G. Posukh,P. K. Tret'yakov, I. F. Shaikhislamov. *Pis'ma Zh. Tekh. Fiz.,* 22, 30 (1996).
[25] V. V. Apollonov, V. N. Tishchenko. *Kvantovaya Elektron.,* 37 (8), 798, *Quantum Electron.,* 37 (8), 798 (2007).
[26] F. V. Bunkin, A. M. Prokhorov .*Usp. Fiz. Nauk,* 119, 425 (1976).
[27] S. N. Kabanov, L. I. Maslova,T. I. Tarkhova,V. A. Trukhin, V. T. Yurov. *Zh. Tekh. Fiz.,* 60, 37 (1990).
[28] G. N. Grachev , V. N. Tishchenko , V. V. Apollonov , A. I. Gulidov , A. L. Smirnov, A. V. Sobolev, M. I. Zimin, *Kvantovaya Elektron.,* 37, 669,*Quantum Electron.,* 37, 669 (2007).
[29] Apollonov V. V. Optical engineering 44(1) 2005,
[30] Aleksandrov G.N., Ivanov V.L., Kadzov G.D., et al. Elektrichestvo (12), 47 (1980).
[31] V.V.Apollonov,"Super long conductive canal for energy delivery", Proceedings of GCL/HPL Symposium, Sofia -2010,SPIE 7751.
[32] V. V. Apollonov ,"To the space by laser light", Vestnik RANS 1, (2008);
[33] V.V.Apollonov, Patent RF "The conductive canal creation in nonconductive medium", № 2400005 от 20.05.09.

3

The Effect of the Time Structure of Laser Pulse on Temperature Distribution and Thermal Stresses in Homogeneous Body with Coating

Aleksander Yevtushenko[1] and Malgorzata Rozniakowska-Klosinska[2]
[1]*Bialystok University of Technology*
[2]*Technical University of Lodz*
Poland

1. Introduction

The need of increased precision and efficiency of thermal processing of modern construction materials with the aid of lasers, makes it especially timely to examine the problem of nonstationary temperature fields in non-homogeneous materials during the heating and cooling stages. This is the reason of continuous interest of many researchers (Kim et al., 1997; Loze & Wright, 1997; Said-Galiyev & Nikitin, 1993; Sheng & Chryssolouris, 1995).

Providing an example, such materials applied among other things in automotive and power industry, are these in the electrical steel - insulator's layer systems of which transformers' core or magnetic cores in electric engines are made. However, the core-loss occurs as a result of the overheating of these materials due to Joule's effect, this is the reason why the efficiency reduces in electrical devices. The induction currents, which are generated by changing magnetic field and connected with them magnetic structure domains has great influence on transformers' efficiency, too. It turns, that in order to decrease the transformer's core-loss, the size of magnetic structure domains should be decreased. This can be achieved by the application of pulsed laser heating of sheet steel (electrical steel - insulator's layer system), in such manner that homogeneous and stable stresses are made - it is a refinement method of magnetic structure domains. As a final result, a sheet steel with an energy lost of about 10% lower than for conventional sheet steel, is obtained. It should be underlined, that during the induce processing of stresses (setting the desired magnetic domains size) coating should not be damaged, the application of pulsed laser radiation satisfies this condition (Coutouly et al., 1999; Li et al., 1997).

Cleavage of the material in the process of thermal splitting results from tensile stresses when the sample is heated by the moving heat flux. When the stresses value exceeds the tensile strength of material then cracks arise on the surface of the processed sample and they follow the movements of the heat source. The cracks in the material are generated on condition that the temperature is higher than material's temperature corresponding to the thermal strength. But for the purpose of guaranteed destruction of the sample considerable temperature gradient must be produced by heating the smallest possible area. For this reason the heating should proceed quickly, in the pulsed mode and the maximum temperature should not exceed the temperature of melting for in such conditions the stress quickly disappears. Thermal splitting of brittle non-metals (like glass, ceramic materials, granite) is the easiest of all for they exhibit big difference between the melting temperature and the temperature of thermal strength. Low thermal conductivity of brittle materials is the cause why considerable thermal stresses are generated in thin subsurface layer in the initial moment. As a result, destruction of the sample takes place shortly after the heating process starts.

Laser treatment is one of the methods of improving the properties of coatings. It is applied when the heating of large areas is difficult or when heating should be limited to some specified parts of the product. Ceramic with zirconium dioxide ZrO_2 as a base constituent is one of the most promising materials for thermally insulating coatings (Dostanko et al., 2002). Functional parameters of such coatings can be improved by laser processing. As results the bigger density and smaller porosity are obtained. At the same time a specific structure of the processed surface is formed - many, nearly equidistant micro cracks arose on it. Usually the width of these micro cracks does not exceed 2 μm, the depth - 10 μm and the maximum length of this specific structure of the processed surface is smaller than 50 μm. When the surface with such a layer produced on it undergoes heating, the parts of the coating between micro cracks can slightly change their relative positions what retards the development of destructive macro cracks. This explains experimentally observed thermal strength increase of the zirconium dioxide ZrO_2 coatings after their laser processing (Dostanko et al., 2002).

Ornaments from granite in buildings like churches, theatres and hotels are often covered, to make them more attractive, with thin metallic foil (gold, platinum, copper). Remaining for years in polluted atmospheric air, sooner or later such ornaments require cleaning. Nowadays laser methods are used for this purpose. In the course of the process of cleaning the metallic foil can not be melted and must stay in thermal contact with the substrate. That is the cause why the problem of modelling temperature and stress fields generated by pulsed laser irradiation is so important for the system consisted of a bulk substrate of low thermal conductivity and a thin metallic coating deposited on it.

Analytical methods for calculation of the temperature fields generated by the pulse laser irradiation were developed mainly for homogeneous materials (Duley, 1976; Rykalin et al., 1985; Ready, 1971; Welch & Van Gemert, 1995; Hector & Hetnarski, 1996). On the other hand, the material to be split quite often has the form of a protective coating or thin film deposited on the homogeneous substrate. The mathematical model of controlled thermal splitting of homogeneous and piece-wise homogeneous bodies at assumption of uniform distribution of the heat flux intensity was considered earlier (Li et al., 1997; Yevtushenko et al., 2005). The reviews of different methods of analysis for the thermal phenomena connected with laser heating are presented in the relevant literature (Gureev, 1983; Rozniakowska & Yevtushenko, 2005).

Therefore, the aim of this chapter of the book is the analysis of temperature distribution and thermal stresses in the non-uniform body heated by the heat flow changing in time. For this purpose the analytical solution of the transient heat conduction problem and corresponding thermoelasticity quasi-static problem for the system, which consists of semi-infinite homogeneous substrate with coating and heated by the laser pulse with rectangular or triangular time shape, was used. The obtained solution determines the temperature and thermal stresses in the piecewise homogeneous body both in the heating phase at laser pulse irradiation and in the cooling phase, when the laser is switched off.

2. Temperature field

For small values of Fourier's numbers, which correspond to characteristic times of thermal splitting, the large part of the heat flux is directed into the body, perpendicularly to its surface. That makes it possible to consider the generation of temperature fields and thermal stresses as the one-dimensional non-stationary process. Let us consider the system of semi-infinite substrate with the coating of the thickness d (Fig. 1). Thermophysical properties of the substrate and the coating differ.

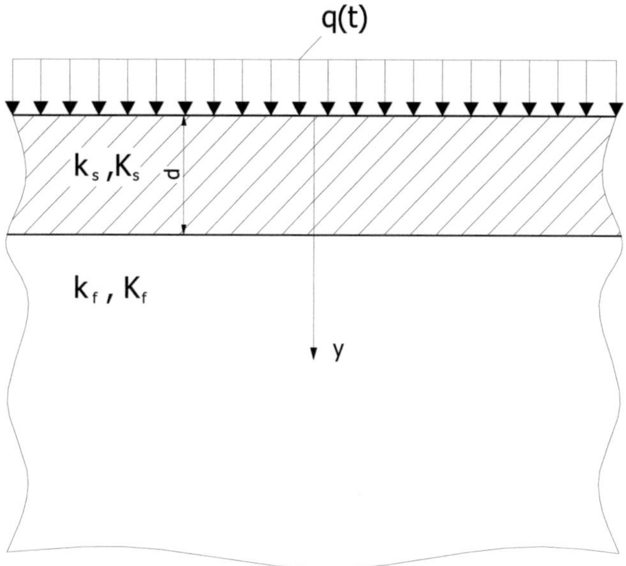

Fig. 1. Heating model of the homogeneous body with coating.

It is assumed, that the intensity of heat flux propagating in the coating material has the following form:

$$q(t) = Aq_0 q^*(t)\,, \tag{1}$$

where A is the absorption coefficient, q_0 is the characteristic value of heat flux intensity, t is the time. Usually one assumes a rectangular laser pulse shape

$$q^*(t) = \begin{cases} 1,\, 0 < t \le t_s, \\ 0,\, t > t_s, \end{cases} \tag{2}$$

or triangular one with respect to time

$$q^*(t) = \begin{cases} 2t / t_r,\, 0 < t \le t_r, \\ 2(t_s - t)/(t_s - t_r),\ t_r \le t \le t_s, \\ 0,\ t > t_s, \end{cases} \tag{3}$$

where t_r is the pulse rise time, t_s is the laser pulse duration. For comparative numerical analysis the parameters of functions (2) and (3) are chosen in such a manner that pulse duration and energy are the same in both cases. The perfect thermal contact between the substrate and the coating is assumed. All the values and parameters which refer to coating and the substrate in the further considerations will have bottom indexes *c* and *s* respectively. The dimensionless temperature distribution in the coating and in the substrate can be found from the solution of the following boundary-value problem of heat conduction:

$$\frac{\partial^2 T_c^*(\zeta,\tau)}{\partial \zeta^2} = \frac{\partial T_c^*(\zeta,\tau)}{\partial \tau}, \quad 0 < \zeta < 1, \tau > 0, \tag{4}$$

$$\frac{\partial^2 T_s^*(\zeta,\tau)}{\partial \zeta^2} = \frac{1}{k^*}\frac{\partial T_s^*(\zeta,\tau)}{\partial \tau}, 1 < \zeta < \infty, \tau > 0, \tag{5}$$

$$\left.\frac{\partial T_c^*}{\partial \zeta}\right|_{\zeta=0} = -q^*(\tau), \quad \tau > 0, \tag{6}$$

$$T_c^*(1,\tau) = T_s^*(1,\tau), \quad \tau > 0, \tag{7}$$

$$\left.\frac{\partial T_c^*}{\partial \zeta}\right|_{\zeta=1} = K^* \left.\frac{\partial T_s^*}{\partial \zeta}\right|_{\zeta=1}, \quad \tau > 0, \tag{8}$$

$$T_s^*(\zeta,\tau) \to 0, \ \zeta \to \infty, \tau > 0, \tag{9}$$

$$T_c^*(\zeta,0) = 0, \quad 0 \le \zeta \le 1, \tag{10}$$

$$T_s^*(\zeta,0) = 0, \quad 1 \le \zeta < \infty. \tag{11}$$

where

$$\zeta = \frac{z}{d}, \ \tau = \frac{k_c t}{d^2}, \ K^* = \frac{K_s}{K_c}, \ k^* = \frac{k_s}{k_c}, \ T_c^* = \frac{T_c}{T_0}, \ T_s^* = \frac{T_s}{T_0}, \ T_0 = A\frac{q_0 d}{K_c}, \tag{12}$$

K, k are coefficients of thermal conductivity and thermal diffusivity, respectively. Taking the equations (2) and (1) into account, the dimensionless temporal profile $q^*(\tau)$ of the laser pulse in the right side of the boundary condition (6) can be written in the form

$$q^*(\tau) = H(\tau_s - \tau), \tau > 0, \tag{13}$$

or

$$q^*(\tau) = \begin{cases} 2\tau / \tau_r, 0 < \tau \le \tau_r, \\ 2(\tau_s - \tau)/(\tau_s - \tau_r), \ \tau_r \le \tau \le \tau_s, \\ 0, \ \tau > \tau_s, \end{cases} \tag{14}$$

where

$$\tau_r = \frac{k_c t_r}{d^2}, \quad \tau_s = \frac{k_c t_s}{d^2}, \tag{15}$$

$H(\cdot)$ is the Heaviside's step function.

Solution of a boundary-value problem of heat conduction in friction (4)-(11) by applying the Laplace integral transform with respect to dimensionless time τ

$$L[T^*_{c,s}(\zeta,\tau);p] \equiv \overline{T}^*_{c,s}(\zeta,p) = \int_0^\infty T^*_{c,s}(\zeta,\tau)\exp(-p\tau)d\tau, \tag{16}$$

has form

$$\overline{T}^*_{c,s}(\zeta,p) = \overline{q}^*(p)\frac{\Delta_{c,s}(\zeta,p)}{\sqrt{p}\,\Delta(p)}, \tag{17}$$

where

$$\Delta_c(\zeta,p) = \mathrm{ch}[(1-\zeta)\sqrt{p}] + \varepsilon\,\mathrm{sh}[(1-\zeta)\sqrt{p}], \quad 0 \le \zeta \le 1, \tag{18}$$

$$\Delta_s(\zeta,p) = \exp\left[-(1-\zeta)\sqrt{\frac{p}{k^*}}\right], \quad \infty < \zeta \le 1, \tag{19}$$

$$\Delta(p) = \mathrm{sh}\sqrt{p} + \varepsilon\,\mathrm{ch}\sqrt{p}, \tag{20}$$

$\overline{q}^*(p)$ is the Laplace transform of dimensionless temporal profile $q^*(\tau)$ of the laser pulse, $\varepsilon = K^* / \sqrt{k^*}$ is known as the "thermal activity coefficient of the substrate in relation to the coating" (Luikov, 1986). Taking the expansion into account

$$\frac{1}{1-\lambda\exp(-2\sqrt{p})} = \sum_{n=0}^{\infty}\Lambda^n \exp(-2\sqrt{p}), \tag{21}$$

where

$$\Lambda^n = \begin{cases} (-1)^n |\lambda|^n, & -1 < \lambda \le 0, \\ \lambda^n, & 0 \le \lambda < 1, \end{cases} \quad \lambda = \frac{1-\varepsilon}{1+\varepsilon}, \tag{22}$$

the transforms (17)-(20) for the temperatures of the coating and substrate can be written as follows

$$\overline{T}^*_c(\zeta,p) = \frac{\overline{q}(p)}{\sqrt{p}}\left\{\sum_{n=0}^{\infty}\Lambda^n \exp[-(2n+\zeta)\sqrt{p}] + \sum_{n=1}^{\infty}\Lambda^n \exp[-(2n-\zeta)\sqrt{p}]\right\}, \quad 0 \le \zeta \le 1, \tag{23}$$

$$\overline{T}^*_s(\zeta,p) = \frac{2\overline{q}(p)}{(1+\varepsilon)\sqrt{p}}\sum_{n=0}^{\infty}\Lambda^n \exp\left\{-\left[(2n+1)\sqrt{p} + (\zeta-1)\sqrt{\frac{p}{k^*}}\right]\right\}, \quad \infty < \zeta \le 1. \tag{24}$$

The transforms (23) and (24) were obtained for arbitrary form of the function $q^*(\tau)$. Transition into the originals of integral Laplace transform will be at first considered for two special cases of laser pulses: with constant intensity $q_0^*(\tau)=1$, $\tau>0$ and linearly changing $q_1^*(\tau)=\tau$, $\tau>0$. Then, the corresponding Laplace transforms are (Luikov, 1986)

$$\bar{q}_i^*(p)=p^{-1-i},\, i=0,1. \tag{25}$$

Inversion of formulas (23) and (24) will be conducted independently for the transforms $\bar{q}_0^*(p)$ and $\bar{q}_1^*(p)$ (25) with the use of the convolution theorem for the integral Laplace transform (Luikov, 1986)

$$U_i(\tau)\equiv L^{-1}[\bar{q}_i^*(p)\bar{Q}(p);\tau]=\int_0^\tau q_i^*(s)Q(\tau-s)ds\,,\ \tau>0\,,\ i=0,1\,, \tag{26}$$

where

$$\bar{Q}(p)=\frac{\exp(-\sqrt{ap})}{\sqrt{p}}\,,\ Q(\tau)=\frac{1}{\sqrt{\pi\tau}}\exp\left(-\frac{a}{4\sqrt{\tau}}\right),\ \tau>0\,,\ a\geq 0\,. \tag{27}$$

Then, substitution of the functions $q_i^*(\tau)=1$, $i=0,1$ and $Q(\tau)$ (27) into right side of the formula (26) gives:

$$U_i(\tau)=\frac{1}{\sqrt{\pi}}\int_0^\tau \frac{s^i}{\sqrt{\tau-s}}\exp\left(-\frac{a}{4\sqrt{\tau-s}}\right)ds\,,\ \tau>0\,,\ i=0,1\,. \tag{28}$$

To calculate the integrals (28), a new variable

$$r=\frac{a}{4(\tau-s)}\,,\ w=\frac{1}{2}\sqrt{\frac{a}{\tau}}\,,\tau>0\,, \tag{29}$$

is introduced. Then we find

$$U_0(\tau)=\frac{1}{2}\sqrt{\frac{a}{\pi}}u_0(w)\,,\ U_1(\tau)=\frac{1}{2}\sqrt{\frac{a}{\pi}}\left[\tau u_0(w)-\frac{a}{4}u_1(w)\right], \tag{30}$$

where

$$u_i(w)=\int_{w^2}^\infty \frac{\exp(-r)}{r^{i+1}\sqrt{r}}dr,\, i=0,1\,. \tag{31}$$

Integration by parts of (31) gives

$$u_0(w)=2\sqrt{\pi}\,w^{-1}\mathrm{ierfc}(w)\,,\ u_1(w)=\frac{2}{3}\left[w^{-3}\exp(-w^2)-u_0(w)\right] \tag{32}$$

where

$$\mathrm{ierfc}(w)=\pi^{-1/2}\exp(-w^2)-w\,\mathrm{erfc}(w)\,,\ \mathrm{erfc}(w)=1-\mathrm{erf}(w)\,, \tag{33}$$

$\mathrm{erf}(x)$ is Gauss error function. After substitution of functions $u_i(w)$, $i=0,1$ (32) into formulas (30) is obtained finally

$$U_i(\tau)=2\tau^i\sqrt{\tau}F_i(w)\,,\ i=0,1\,, \tag{34}$$

where

$$F_0(x)=\mathrm{ierfc}(x),\quad F_1(x)=\frac{1}{3}[2(1+x^2)F^{(0)}(x)-x\,\mathrm{erfc}(x)]\,. \tag{35}$$

Taking into account the form of functions $U_i(\tau), i=0,1$ (34), from the formulas (23) and (24) the dimensionless temperatures $T^{(i)*}, i=0,1$ are found for the constant $q_0^*(\tau)=1$ and linearly changing with time $q_1^*(\tau)=\tau$, $\tau>0$ heat flux intensity, respectively

$$T_c^{(i)*}(\zeta,\tau)=\sum_{n=0}^{\infty}\Lambda^n T_{c,n}^{(i)*}(\zeta,\tau),\quad 0\le\zeta\le 1,\tau\ge 0,\quad i=0,1, \tag{36}$$

$$T_s^{(i)*}(\zeta,\tau)=\sum_{n=0}^{\infty}\Lambda^n T_{s,n}^{(i)*}(\zeta,\tau),\quad 1\le\zeta<\infty,\tau\ge 0,\quad i=0,1, \tag{37}$$

where

$$T_{c,0}^{(i)*}(\zeta,\tau)=2\tau^{i+1/2}F_i\left(\frac{\zeta}{2\sqrt{\tau}}\right),\ i=0,1, \tag{38}$$

$$T_{c,n}^{(i)*}(\zeta,\tau)=2\tau^{i+1/2}\left[F_i\left(\frac{2n+\zeta}{2\sqrt{\tau}}\right)+F_i\left(\frac{2n-\zeta}{2\sqrt{\tau}}\right)\right],\ i=0,1\,;\ n=1,2,\ldots, \tag{39}$$

$$T_{s,n}^{(i)*}(\zeta,\tau)=\frac{4\tau^{i+1/2}}{(1+\varepsilon)}F_i\left(\frac{2n+1}{2\sqrt{\tau}}+\frac{\zeta-1}{2\sqrt{k^*\tau}}\right),\ i=0,1\,;\ n=0,1,2,\ldots\,. \tag{40}$$

Subsequently, when the temperature fields $T_{c,s}^{(i)*}(\zeta,\tau)$, $i=0,1$ (36)-(40) are determined then temperature of the non-homogeneous body can be found from the expressions (Yevtushenko et al., 2007)

$$T_{c,s}^*(\zeta,\tau)=T_{c,s}^{(0)*}(\zeta,\tau)-T_{c,s}^{(0)*}(\zeta,\tau-\tau_s)H(\tau-\tau_s)\,,\ \zeta\ge 0,\ \tau\ge 0, \tag{41}$$

for the rectangular laser pulse (13) and

$$\begin{aligned}T_{c,s}^*(\zeta,\tau)=&\frac{2}{\tau_r}\left[T_{c,s}^{(1)*}(\zeta,\tau)-T_{c,s}^{(1)*}(\zeta,\tau-\tau_r)H(\tau-\tau_r)\right]-\\&-\frac{2}{\tau_s-\tau_r}\left[T_{c,s}^{(1)*}(\zeta,\tau-\tau_r)H(\tau-\tau_r)-T_{c,s}^{(1)*}(\zeta,\tau-\tau_s)H(\tau-\tau_s)\right],\zeta\ge 0,\ \tau\ge 0,\end{aligned} \tag{42}$$

for the triangular one (14).

It must be noted that equations (41) and (42) determine the temperature of the body (composed of coating and substrate) in a point beneath the heated surface ($\zeta\ge 0$), after time $\tau\ge 0$ from the beginning of laser irradiation to the moment, when the cooling of the body is completed. The solutions of the corresponding problems of heat conduction for homogeneous semi-space it is possible to obtained from the first component of the expressions (36) and (37) (for $n=0$) (Rozniakowska & Yevtushenko, 2005).

Taking the notation (12) into account, the temperature distributions in the coating and substrate write in the form

$$T_{c,s}(z,t) = T_0\, T^*_{c,s}(\zeta,\tau),\ \zeta \geq 0,\ \tau \geq 0\,, \tag{43}$$

where the dimensionless functions $T^*_{c,s}(\zeta,\tau)$ have the form (41) or (42).

3. Stresses and deformations

Experimental examinations of the controlled superficial splitting proved that from the three normal components of the stress tensor - longitudinal, lateral and in the direction of heating - only the lateral component σ_y is useful in thermal splitting (Dostanko et al., 2002). As the result of action of this component, thermal splitting proceeds in the direction of the heat flux movement trajectory. The longitudinal component σ_x is undesirable because at its enough greater values exceeding tensile strength of materials, the micro cracks oriented at various angles to the direction of splitting are created and divergence between the line of splitting and the direction of heat flow movement occurs. The normal component of stress tensor σ_z has no essential meaning in one-dimensional problem.

On the basis of these data quasi-static normal stresses σ_y in the coating induced by the non-stationary temperature field (36)-(40) can be determined from the equations, which describe thermal bending of thick plate of the thickness d with free ends (Timoshenko & Goodier, 1951):

$$\sigma_y(z,t) = \sigma_0\, \sigma^*_y(\zeta,\tau),\quad 0 \leq z \leq d, t \geq 0\,, \tag{44}$$

where

$$\sigma^*_y(\zeta,\tau) = \varepsilon^*_y(\zeta,\tau) - T^*_c(\zeta,\tau),\quad 0 \leq \zeta \leq 1, \tau \geq 0\,, \tag{45}$$

$$\varepsilon^*_y(\zeta,\tau) = \int_0^1 T^*_c(\zeta,\tau)d\zeta + 12(\zeta - 0{,}5)\int_0^1 (\zeta - 0{,}5)T^*_c(\zeta,\tau)d\zeta\,,\quad 0 \leq \zeta \leq 1, \tau \geq 0\,, \tag{46}$$

$\sigma_0 = \alpha_c E_c T_0 / (1 - \nu_c)$ is the stress scaling factor, α_c is the linear thermal expansion coefficient, E_c is the Young's modulus, ν_c is the Poisson's ratio of the coating material, $T^*_c(\zeta,\tau)$ is the dimensionless temperature field in the coating (41) or (42). When the heating of the plate's surface is realised with the uniform heat flux (2), (13) then the dimensionless lateral deformation ε^*_y (46) can be found from the equation (Yevtushenko et al., 2007):

$$\varepsilon^*_y(\zeta,\tau) = \varepsilon^{(0)*}_y(\zeta,\tau) - \varepsilon^{(0)*}_y(\zeta,\tau - \tau_s)H(\tau - \tau_s),\quad 0 \leq \zeta \leq 1, \tau \geq 0\,, \tag{47}$$

and for the triangular time shape of the heat pulse (3), (14) one has:

$$\begin{aligned}\varepsilon^*_y(\zeta,\tau) = \frac{2}{\tau_r}\left[\varepsilon^{(1)*}_y(\zeta,\tau) - \varepsilon^{(1)*}_y(\zeta,\tau - \tau_r)H(\tau - \tau_r)\right] - \\ - \frac{2}{\tau_s - \tau_r}\left[\varepsilon^{(1)*}_y(\zeta,\tau - \tau_r)H(\tau - \tau_r) - \varepsilon^{(1)*}_y(\zeta,\tau - \tau_s)H(\tau - \tau_s)\right], 0 \leq \zeta \leq 1, \tau \geq 0,\end{aligned} \tag{48}$$

where

$$\varepsilon_y^{(i)*}(\zeta,\tau)=\sum_{n=0}^{\infty}\Lambda^n\varepsilon_n^{(i)*}(\zeta,\tau),\ \ 0\le\zeta\le 1,\ \tau\ge 0,\ i=0,1, \tag{49}$$

$$\varepsilon_n^{(i)*}(\zeta,\tau)=Q_n^{(i)}(\tau)-\zeta R_n^{(i)}(\tau),\ i=0,1, \tag{50}$$

$$Q_n^{(i)}(\tau)=4I_n^{(i)}(\tau)-6J_n^{(i)}(\tau),\ \ R_n^{(i)}(\tau)=6I_n^{(i)}(\tau)-12J_n^{(i)}(\tau),\ i=0,1;\ n=0,1,2,\ldots, \tag{51}$$

$$I_n^{(i)}(\tau)=\int_0^1 T_{c,n}^{(i)*}(\zeta,\tau)d\zeta,\ \ J_n^{(i)}(\tau)=\int_0^1 \zeta T_{c,n}^{(i)*}(\zeta,\tau)d\zeta,\ \ i=0,1;\ n=0,1,2,\ldots, \tag{52}$$

and the dimensionless temperatures $T_c^{(i)*}(\zeta,\tau)$, $i=0,1$ has the form (36). Substituting the functions $T_{c,n}^{(i)*}(\zeta,\tau)$, $i=0,1;\ n=0,1,2,\ldots$, (38), (39) at the right side of equations (52) and taking into account the value of the integrals:

$$L_0(x)\equiv\int_0^x F_0(t)dt=\frac{1}{4}+\frac{x}{2\sqrt{\pi}}\exp(-x^2)-\frac{(1+2x^2)}{4}\operatorname{erfc}(x), \tag{53}$$

$$M_0(x)\equiv\int_0^x t\,F_0(t)dt=\frac{1}{6\sqrt{\pi}}-\frac{(1-2x^2)}{6\sqrt{\pi}}\exp(-x^2)-\frac{x^3}{3}\operatorname{erfc}(x), \tag{54}$$

$$L_1(x)\equiv\int_0^x F_1(t)dt=\frac{1}{8}+\frac{x(5+2x^2)}{12\sqrt{\pi}}\exp(-x^2)-\frac{(3+12x^2+4x^4)}{24}\operatorname{erfc}(x), \tag{55}$$

$$M_1(x)\equiv\int_0^x tF^{(1)}(t)dt=\frac{1}{15\sqrt{\pi}}-\frac{(1-4x^2-2x^4)}{15\sqrt{\pi}}\exp(-x^2)-\frac{x^3(5+2x^2)}{15}\operatorname{erfc}(x), \tag{56}$$

the following expressions were obtained:

$$I_0^{(i)}(\tau)=4\tau^{i+1}L_i\left(\frac{1}{2\sqrt{\tau}}\right),\ I_n^{(i)}(\tau)=4\tau^{i+1}\left[L_i\left(\frac{2n+1}{2\sqrt{\tau}}\right)-L_i\left(\frac{2n-1}{2\sqrt{\tau}}\right)\right],i=0,1;\ n=1,2,3\ldots, \tag{57}$$

$$J_0^{(i)}(\tau)=8\tau^{i+3/2}M_i\left(\frac{1}{2\sqrt{\tau}}\right),\ i=0,1, \tag{58}$$

$$\begin{aligned}J_n^{(i)}(\tau)=4\tau^{i+1}\Bigg\{&2\sqrt{\tau}\left[M_i\left(\frac{2n+1}{2\sqrt{\tau}}\right)-2M_i\left(\frac{n}{\sqrt{\tau}}\right)+M_i\left(\frac{2n-1}{2\sqrt{\tau}}\right)\right]-\\ &-2n\left[L_i\left(\frac{2n+1}{2\sqrt{\tau}}\right)-2L_i\left(\frac{n}{\sqrt{\tau}}\right)+L_i\left(\frac{2n-1}{2\sqrt{\tau}}\right)\right]\Bigg\},i=0,1;\ n=1,2,3\ldots\end{aligned} \tag{59}$$

It results from formulas (49) and (50) that the dimensionless lateral deformation ε_y^* is linearly dependent on the dimensionless distance ζ from the heated surface of the plate and that normal stresses σ_y (45) are proportional to the difference of that deformation and the dimensionless temperature $T_c^*(\zeta,\tau)$ (41) or (42).

4. Homogeneous material

The solutions of the corresponding problems of heat conduction for homogeneous semi-space it is possible to obtained from the first component of the expressions (36) and (37) (for $n=0$). Assuming the same thermophysical properties of the foundation and coating ($K_c = K_s = K$, $k_c = k_s = k$) from formulas (22) leads that $\varepsilon = 1$, $\lambda = 0$, $\Lambda = 0$ and then from the expressions (36)-(40) one obtains

$$T^{(0)*}(\zeta,\tau) = 2\sqrt{\tau}\,\mathrm{ierfc}\left(\frac{\zeta}{2\sqrt{\tau}}\right),\ 0 \le \zeta < \infty,\ \tau \ge 0, \tag{60}$$

$$T^{(1)*}(\zeta,\tau) = \frac{2}{3}\tau\sqrt{\tau}\left\{2\left[1+\left(\frac{\zeta}{2\sqrt{\tau}}\right)^2\right]\mathrm{ierfc}\left(\frac{\zeta}{2\sqrt{\tau}}\right)-\left(\frac{\zeta}{2\sqrt{\tau}}\right)\mathrm{erfc}\left(\frac{\zeta}{2\sqrt{\tau}}\right)\right\},\ 0 \le \zeta < \infty,\ \tau \ge 0. \tag{61}$$

In the present case, the parameter d in formulas (12) can be, for example, the radius of the irradiated zone. Substitution of these solutions (60) and (61) into equations (41) and (42) for $\tau = \tau_s$, gives as a result the dimensionless temperature at the moment when laser is being turned off for the rectangular pulse

$$T^*(\zeta,\tau_s) = T^{(0)*}(\zeta,\tau_s),\ 0 \le \zeta < \infty, \tag{62}$$

and for triangular one

$$T^*(\zeta,\tau_s) = \frac{2}{\tau_r}\left[T^{(1)*}(\zeta,\tau_s) - \frac{\tau_s}{\tau_s-\tau_r}T^{(1)*}(\zeta,\tau_s-\tau_r)\right],\ 0 \le \zeta < \infty. \tag{63}$$

The final results for the irradiated surface, $\zeta = 0$ in these two cases, have the following known form (Gureev, 1983), respectively

$$T^*(0,\tau_s) = 2\sqrt{\frac{\tau_s}{\pi}}, \tag{64}$$

and

$$T^*(0,\tau_s) = \frac{8}{3}\sqrt{\frac{\tau_s}{\pi}}\left(\frac{\tau_s}{\tau_r}\right)\left(1-\sqrt{1-\frac{\tau_r}{\tau_s}}\right). \tag{65}$$

The values of dimensionless lateral stresses $\overset{*}{\sigma}_y$ corresponding to the temperature fields $T^*(\zeta,\tau)$ (41) or (42) at functions $T^{(i)*}(\zeta,\tau)$, $i=0,1$ (60) and (61) can be obtained from equations (45)-(51) for $n=0$. Taking into account the formulas (53)-(58) the equations (51) leads:

$$Q_0^{(0)}(\tau) = 4\tau\left[1-2\sqrt{\frac{\tau}{\pi}}+2\sqrt{\tau}\,\mathrm{ierfc}\left(\frac{1}{2\sqrt{\tau}}\right)\right], \tag{66}$$

$$R_0^{(0)}(\tau) = 4\tau\left[\frac{3}{2} - 4\sqrt{\frac{\tau}{\pi}} + \sqrt{\frac{\tau}{\pi}}\exp\left(-\frac{1}{4\tau}\right) + \left(3\sqrt{\tau} - \frac{1}{2\sqrt{\tau}}\right)\operatorname{ierfc}\left(\frac{1}{2\sqrt{\tau}}\right)\right], \quad (67)$$

$$Q_0^{(1)}(\tau) = 4\tau^2\left[\frac{1}{2} - \frac{4}{5}\sqrt{\frac{\tau}{\pi}} - \frac{1}{5\sqrt{\pi}}\left(\sqrt{\tau} - \frac{1}{6\sqrt{\tau}}\right)\exp\left(-\frac{1}{4\tau}\right) + \left(\sqrt{\tau} - \frac{1}{60\tau\sqrt{\tau}}\right)\operatorname{ierfc}\left(\frac{1}{2\sqrt{\tau}}\right)\right], \quad (68)$$

$$\begin{aligned} R_0^{(1)}(\tau) = 4\tau^2\Bigg[&\frac{3}{4} - \frac{8}{5}\sqrt{\frac{\tau}{\pi}} + \frac{1}{10\sqrt{\pi}}\left(\sqrt{\tau} + \frac{3}{2\sqrt{\tau}}\right)\exp\left(-\frac{1}{4\tau}\right) + \\ &+\left(\frac{3}{2}\sqrt{\tau} - \frac{1}{2\sqrt{\tau}} - \frac{3}{40\tau\sqrt{\tau}}\right)\operatorname{ierfc}\left(\frac{1}{2\sqrt{\tau}}\right)\Bigg]. \end{aligned} \quad (69)$$

5. Numerical analysis

The dimensionless input parameters of the calculations are: spatial coordinate ζ, time (Fourier's criterion) τ, pulse time rise τ_r, duration of the pulse (time of heating) τ_s, ratio of the coefficients of thermal conductivity and thermal diffusivity of the substrate and coating $K^* = K_s / K_c$ and $k^* = k_s / k_c$. Isolines for the dimensionless temperatures $T^* = T / T_0$ and normal stresses $\sigma_y^* = \sigma_y / \sigma_0$ were drawn in the coordinates (ζ, τ) for different temporal profile of the heat pulse. All calculations were conducted for the pulses with dimensionless duration $\tau_s = 0.15$, which is characteristic for irradiation done by CO_2 laser, which emits light at wavelength 10.6 μm (Rykalin et al., 1985).

Firstly, the temperature and stress distributions for the case when the coating and the substrate have the same thermophysical and mechanical properties (homogeneous semi-space) will be analysed. Isotherms of the dimensionless temperature T^* at heating by rectangular and triangular, with different rise times, pulses are presented in Figure 2a-d.

In case of the rectangular pulse the maximum temperature on the surface of the irradiated bulk sample is achieved at the end of the pulse $\tau_s = 0.15$ and its value is $T_{\max}^* = 0.429$ (Fig. 2a). For triangular pulse with different rise times the evolution of temperature proceeds differently - with the increase of back front steepness the moment when the highest temperature is achieved moves closer from the middle of the pulse duration interval (for small values of τ_r) to the moment when laser is switched off τ_s (Fig. 2b-d).

So, for the three considered triangular laser pulses with $\tau_r = 0.001; 0.075; 0.149$ the maximum values of dimensionless temperature are equal $T_{\max}^* = 0.412; 0.475; 0.566$ and are achieved in the moments $\tau = 0.08; 0.10$ and 0.149, respectively. With increase of the maximal temperature, the effective depth of heating (the depth where the temperature decreases to 5% of its maximum value on the surface) also increases. The presented analysis indicates that the greater value of this depth can be obtained when the laser pulse has the gentle fore front and steep back front.

Isolines for the dimensionless lateral stresses σ_y^* are presented in Figure 3a-d. For heating with the rectangular pulse, in the time interval $0 < \tau \le 0.15$ the regions of compressive lateral stress ($\sigma_y^* < 0$) occur near the border surfaces $\zeta = 0$ i $\zeta = 1$ (Fig. 3a). Inside this layer tensile stresses are generated ($\sigma_y^* > 0$).

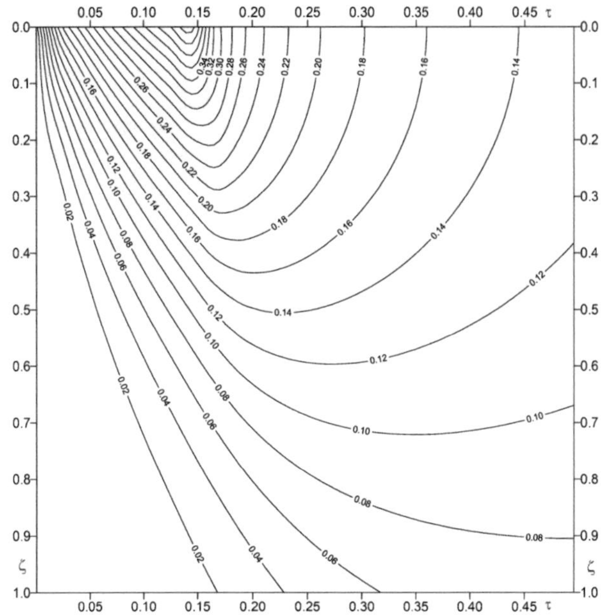

(a)

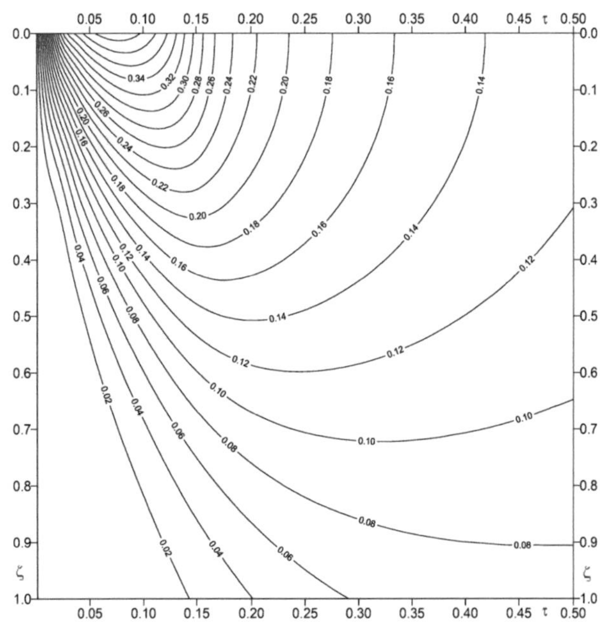

(b)

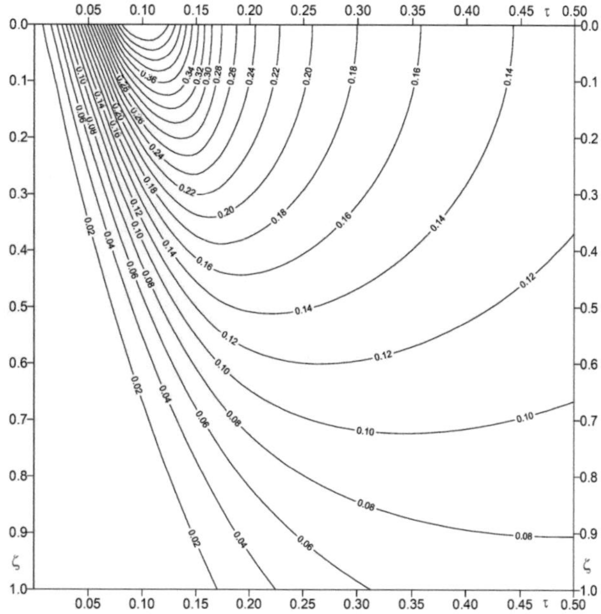

(c)

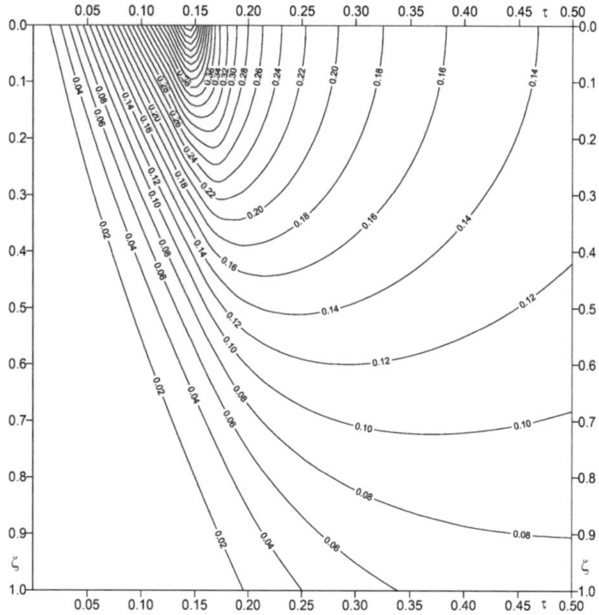

(d)

Fig. 2. Isotherms of dimensionless temperature T^* for the dimensionless laser pulse duration $\tau_s = 0.15$: a) rectangular laser pulse; b) c), d) triangular laser pulses for the dimensionless rise time $\tau_r = 0.001$; $\tau_r = 0.075$ and $\tau_r = 0.149$, respectively (Yevtushenko et al., 2007).

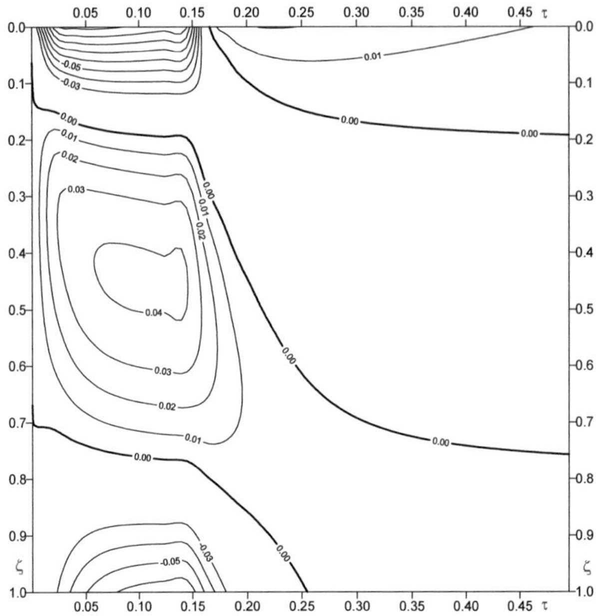
0.05 0.10 0.15 0.20 0.25 0.30 0.35 0.40 0.45 τ
0.0 0.1 0.2 0.3 0.4 0.5 0.6 0.7 0.8 0.9 ζ 1.0
-0.05
-0.03
0.00
0.01
0.02
0.03
0.04
-0.05

(a)

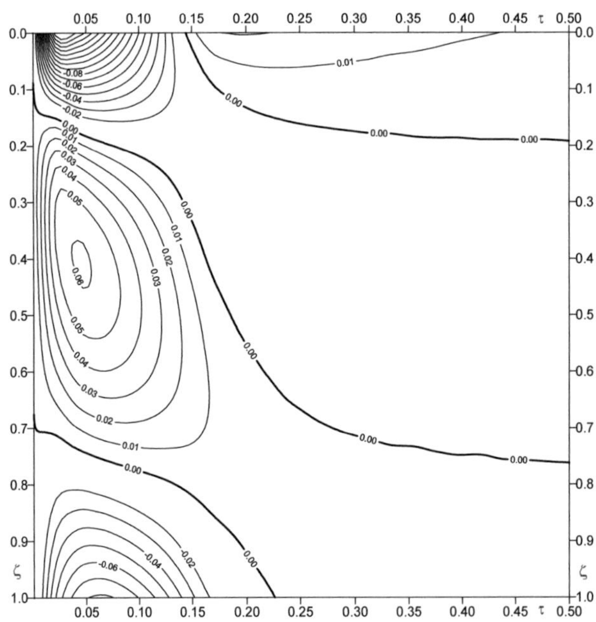
0.05 0.10 0.15 0.20 0.25 0.30 0.35 0.40 0.45 τ 0.50
0.0 0.1 0.2 0.3 0.4 0.5 0.6 0.7 0.8 0.9 ζ 1.0
-0.08
-0.06
-0.04
-0.02
0.00
0.01
0.02
0.03
0.04
0.05
0.06
-0.06

(b)

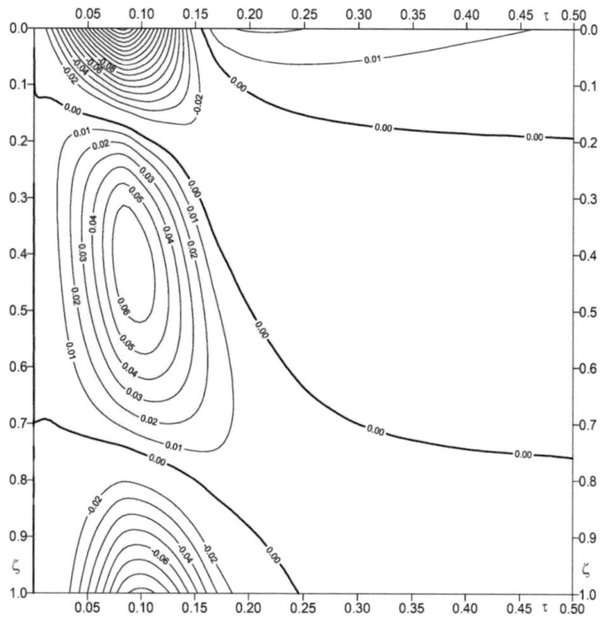

(c)

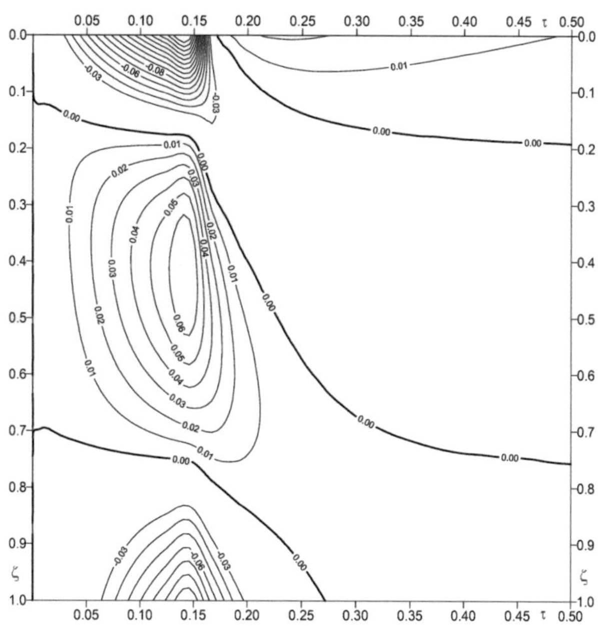

(d)

Fig. 3. Isolines of dimensionless lateral stress σ_y^* for the dimensionless laser pulse duration $\tau_s = 0.15$: a) rectangular laser pulse; b) c), d) triangular laser pulses for the dimensionless rise time $\tau_r = 0.001$; $\tau_r = 0.075$ and $\tau_r = 0.149$, respectively (Yevtushenko et al., 2007).

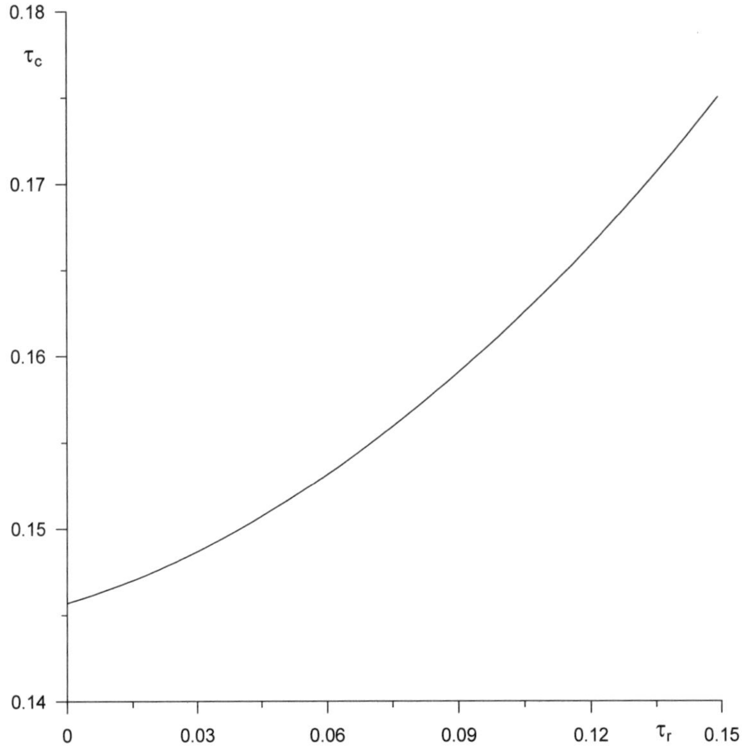

Fig. 4. Dimensionless time τ_c of the sign change of lateral stress σ_y^* on the irradiated surface $\zeta = 0$ versus pulse rise time τ_r for dimensionless time of heating $\tau_s = 0.15$ (Yevtushenko et al., 2007).

Between the regions with compressive and tensile stresses there are two isolines with zero stresses. The third line of zero stresses, which "descends" from the surface $\zeta = 0$, appears when the heating is over ($\tau > 0.15$). It means that during relaxation phase, when there is no more heating, the sign of stresses changes and the region of tensile stresses expands further from the heated surface - the line of zero stresses moves into the plate with the time increase (Fig. 3a).

Somewhat different distribution of dimensionless lateral stress σ_y^* is observed when irradiation is realised with the triangular laser pulses (Fig. 3b-d). In the heating phase ($0 < \tau < 0.15$) compressive superficial stresses occur and the isotherms are much denser near the points of the greatest intensity of the heat flux. For the pulse with steep fore front ($\tau_r = 0.001$) it appears at the very beginning of heating and for the pulse with the steep back front ($\tau_r = 0.149$) - nearly in the moment when the heat source is switched off. The change of stress sign does not necessarily take place after the heat source is off, what was the case for uniform heating.

For the heating with triangular pulse with steep fore front the isoline of zero stresses occurs on the irradiated surface already during the heating phase, before its end (Fig. 3b). The magnitude of tensile stresses during the relaxation phase is more or less the same for the laser pulses of rectangular and triangular shape.

For greater pulse rise times τ_r the dimensionless time τ_c, connected with the change of stresses type from compressive to tensile one also increases (Fig. 4). This dependence is approximately described by the equation $\tau_c = 0.8173\tau_r^2 + 0.075\tau_r + 0.1457$.
In thermal processing of brittle materials, it is the sign change of superficial stresses what plays key role in controlled thermal splitting (Dostanko et al., 2002). The beginning of superficial cracks generation is accompanied with the monotonic increase of tensile lateral stresses and makes the controlled evolution of the crevices possible. By equating the relation (45) to 0 at $\tau = \tau_c > \tau_s$, $\zeta = 0$ and taking into account the equations (42), (48), (49), one obtains:

$$Q_0^{(0)}(\tau_c) - Q_0^{(0)}(\tau_c - \tau_s) = \frac{2}{\sqrt{\pi}}(\sqrt{\tau_c} - \sqrt{\tau_c - \tau_s}), \tag{70}$$

where the function $Q_0^{(0)}(\tau)$ has the form (66). With the absolute inaccuracy, smaller than 3%, the solution of nonlinear equation (70) can be approximated by the function $\tau_c = -1.133\tau_s^{\ 3} + 1.172\tau_s^{\ 2} + 0.604\tau_s + 0.052$ (Fig. 5).

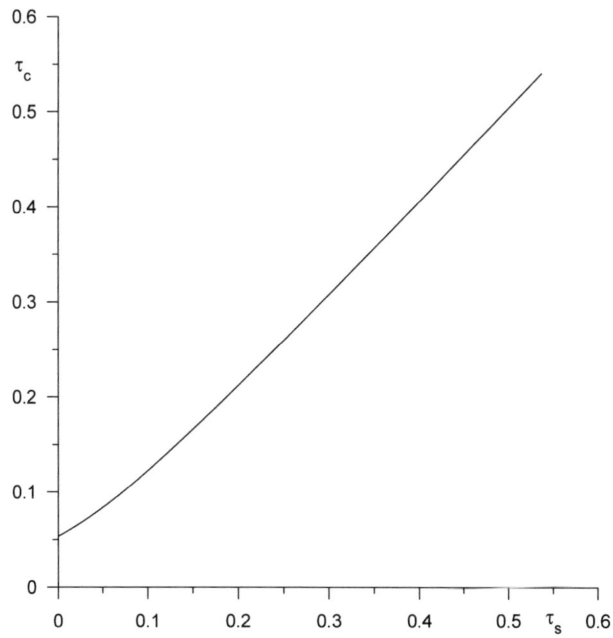

Fig. 5. Dimensionless time τ_c of the sign change of lateral stress σ_y^* on the irradiated surface $\zeta = 0$ versus dimensionless duration τ_s of the rectangular laser pulse (Yevtushenko et al., 2007).

The forced cooling of the surface in the moment time $t = t_c$ would cause the jump of temperature $\Delta T = T(0,t_c - 0) - T(0,t_c + 0)$ in thin superficial layer. From the equations (45) and (46) it follows, that the dimensionless lateral deformation of the plate ε_y^* is determined by the integral characteristic of temperature only. For this reason, the rapid cooling of the thin film, practically, does not change the surface deformation $\varepsilon_y(0,t_c + 0) = \alpha_c T(0,t_c - 0)$ but at the same moment it produces the increase of the normal stresses $\sigma_y(0,t_c+0)=\alpha_c\Delta T$. Finally, the development of the superficial crack can be described as a series of the following phases:

1. due to local short heating a surface of the sample in it the field of normal lateral stresses is formed;
2. tensile lateral stresses occur near the subsurface cooled region and are proportional to the temperature jump observed before and after the cooling agent is applied;
3. when the stresses exceeds the tensile strength of the material, the surface undergoes tear;
4. development of the crack into the material is limited by the regions of lateral compressive stress, which occur beneath the cooled surface.

As mentioned in the introduction, there is considerable interest (for scientific and practical reasons) in thermal processing of ceramic coatings from zirconium dioxide ZrO_2. Authors presented the numerical examinations of thermal stresses distribution for the system consisting of ZrO_2 ceramic coating ($K_c=2.0\ \mathrm{W/(mK)}$, $k_c=0.8\cdot10^{-6}\mathrm{m^2/s}$), deposited on the 40H steel substrate ($K_s=41.9\ \mathrm{W/(mK)}$, $k_s=10.2\cdot10^{-6}\mathrm{m^2/s}$) (Fig. 6). The coefficient of thermal activity for this system is equal $\varepsilon=5.866$ and the parameter λ found from the equation (22), has the value $\lambda=-0.708$. Thermal diffusivity of zirconium dioxide is small when compared with the value for steel. That difference is the cause of high temperatures on the processed surface and considerably higher than for the homogeneous half-space (one order of magnitude) lateral tensile stresses generated in the superficial layer when the heating is finished. So, the thermal processing of the coating from zirconium dioxide leads to the generation of superficial cracks, which divide the surface into smaller fragments. Of course the distribution of cracks at different depths depends on the heat flux intensity, the diameter of the laser beam, pulse duration and other parameters of the laser system.

But when using dimensionless variables and parameters the results can be compared and the conclusion is that for the heating duration $\tau_s=0.15$, penetration depth of cracks for coating-substrate system (ZrO_2–40H steel) is, more than two times greater than for the homogeneous material (one can compare Figs. 3a and 6).

The opposite, to the discussed above, combination of thermo-physical properties of the coating and the substrate is represented by the copper–granite system, often used in ornaments decorating interiors of the buildings like theatres and churches. For the copper coating $K_c=402\ \mathrm{W/(mK)}$, $k_c=125\cdot10^{-6}\mathrm{m^2/s}$, while for the granite substrate $K_s=1.4\ \mathrm{W/(mK)}$, $k_s=0.505\cdot10^{-6}\mathrm{m^2/s}$, what means that the substrate is practically thermal insulator and the coating has good thermal conductivity (see Figs. 7 and 9). The distribution of lateral thermal stresses for copper–granite system is presented in Fig. 8. In this situation, when the thickness of the coating increases, the temperature on the copper surface decreases. Therefore the effective depth of heat penetration into the coating is greater for the better conducting copper than for thermally insulating zirconium dioxide (ZrO_2) (see Figs. 6 and 8). We note that near to the heated surface $\zeta=0$ lateral stresses σ_y are compressive not only in the heating phase $0<\tau<0.15$ but also during relaxation time, when the heat source is off. Considerable lateral tensile stresses occur during the cooling phase close to the interface of the substrate and the coating, $\zeta=1$. This region of the tensile stresses on the copper-granite interface can destroy their contact and in effect the copper coating exfoliation can result.

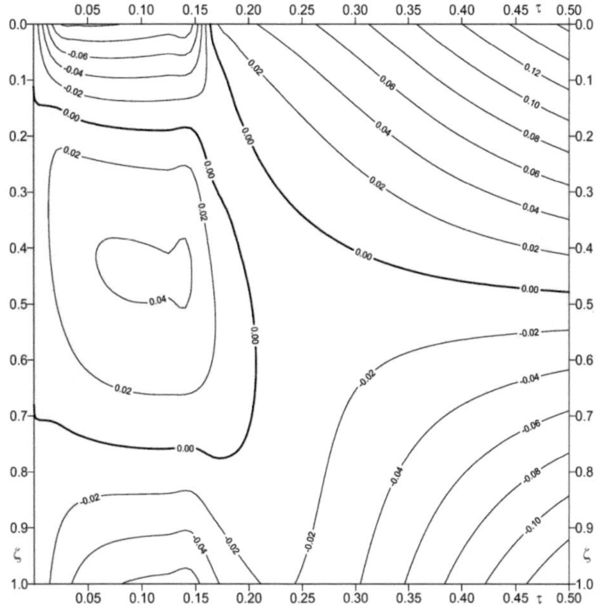

Fig. 6. Isolines of dimensionless lateral stress σ_y^* for ZrO_2 ceramic coating and 40H steel substrate at rectangular laser pulse duration $\tau_s = 0.15$ (Yevtushenko et al., 2007).

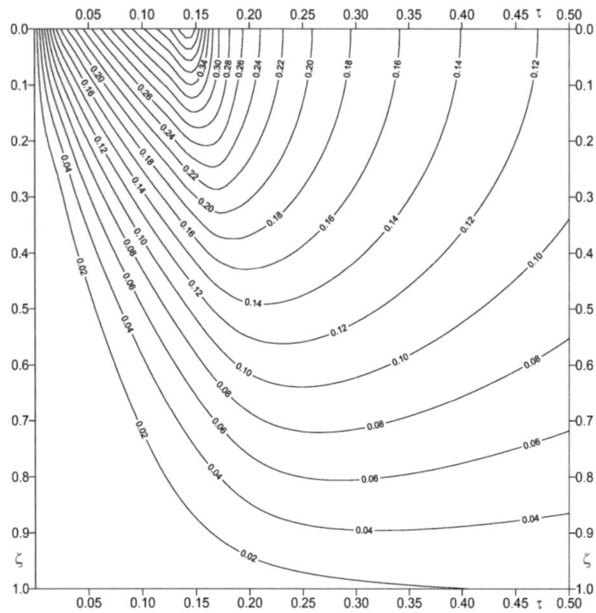

Fig. 7. Isotherms of dimensionless temperature T^* for ZrO_2 ceramic coating and 40H steel substrate at rectangular laser pulse duration $\tau_s = 0.15$ (Yevtushenko et al., 2007).

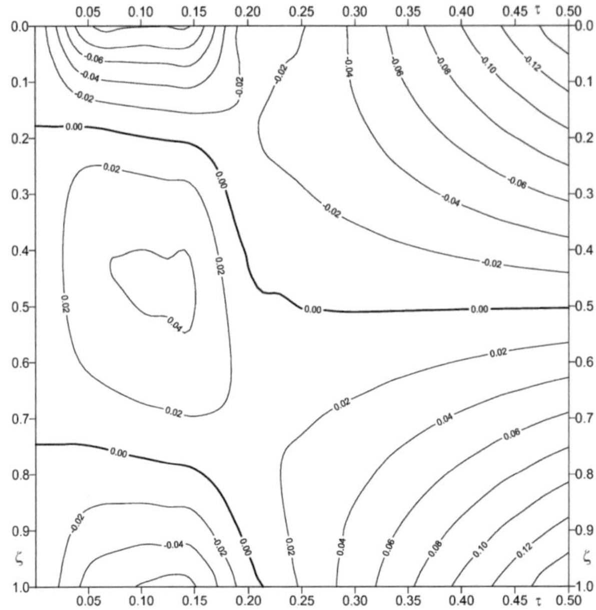

Fig. 8. Isolines of dimensionless lateral stress σ_y^* for copper coating and granite substrate at rectangular laser pulse duration $\tau_s = 0.15$ (Yevtushenko et al., 2007).

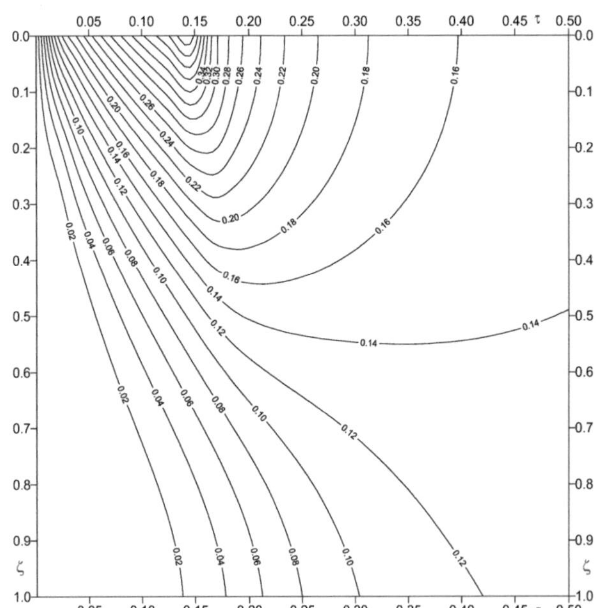

Fig. 9. Isotherms of dimensionless temperature T^* for copper coating and granite substrate at rectangular laser pulse duration $\tau_s = 0.15$ (Yevtushenko et al., 2007).

6. Effective absorption coefficient during laser irradiation

The effective absorption coefficient A in the formula (1) and (12) is defined as the ratio of laser irradiation energy absorbed on the metal's surface and the energy of the incident beam (Rozniakowski, 2001). This dimensionless parameter applies to the absorption on the metal's surface, on the very sample surface (so called "skin effect"). The absorption coefficient A can be found in book (Sala, 1986) or obtained on the basis of calorimetric measurements (Ujihara, 1972). The mixed method of effective absorption coefficient determination for some metals and alloys was presented by Yevtushenko et al., 2005. This method is based on the solution of axisymmetric boundary-value heat conduction problem for semi-space with circular shape line of division in the boundary conditions and on the metallographic measurements of dimensions of laser induced structural changes in metals. The calculations in this method are very complex because, in particular, the numerical calculation of the Hankel's integrals has to be done. Therefore, we shall try to use with this purpose obtained above the analytical solution of the transient one-dimensional heat conduction problem for homogeneous semi-space in the form

$$T(z,t) = AT_0' T^*(\zeta,\tau), \quad z \geq 0, t \geq 0, \tag{71}$$

where, taking the formula (12) into account, the coefficient $T_0' = T_0 / A$ and the dimensionless temperature $T^*(\zeta,\tau)$ is defined by formulae (41) and (60). It should be noticed that the temperature on the irradiated surface has maximum value at the moment of laser switching off, for $t = t_s$ ($\tau = \tau_s$), and in the superficial layers the maximum is reached for $t \equiv t_h = t_s + \Delta t$ (in dimensionless units, for $\tau_h = \tau_s + \Delta\tau$, $\Delta\tau = k\Delta t / d^2$, d is the radius of the irradiated zone). The parameter Δt ($\Delta\tau$) is known as "the retardation time" (Rozniakowska & Yevtushenko, 2005). The time interval, when the temperature T reaches its maximum in the point $z = z_h$ beneath the heated surface, can be found from the condition:

$$\frac{\partial T(z_h,t)}{\partial t} = 0, \; t > t_s > 0. \tag{72}$$

By taking into account the solutions (71), (41) and (60), the equation (72) can be rewritten as:

$$\frac{\partial T^*(\zeta_h,\tau)}{\partial \tau} = \frac{1}{\sqrt{\pi\tau}} \exp\left(-\frac{\zeta_h^2}{4\tau}\right) - \frac{1}{\sqrt{\pi(\tau-\tau_s)}} \exp\left[-\frac{\zeta_h^2}{4(\tau-\tau_s)}\right] = 0, \; \tau > \tau_s > 0, \tag{73}$$

where $\zeta_h = z_h / d$. After substituting $\tau \equiv \tau_h = \tau_s + \Delta\tau$ in equation (73), one gets

$$\sqrt{\frac{\Delta\tau}{\tau_s + \Delta\tau}} = \exp\left[-\frac{\zeta_h^2 \tau_s}{4\Delta\tau(\tau_s + \Delta\tau)}\right]. \tag{74}$$

From the equation (74) for the known dimensionless hardened layer depth ζ_h and the pulse duration τ_s, the dimensionless retardation time $\Delta\tau$ can be found. On the other hand, at known $\Delta\tau$ from equation (74) we find the dimensionless hardened layer depth ζ_h of maximum temperature can be found:

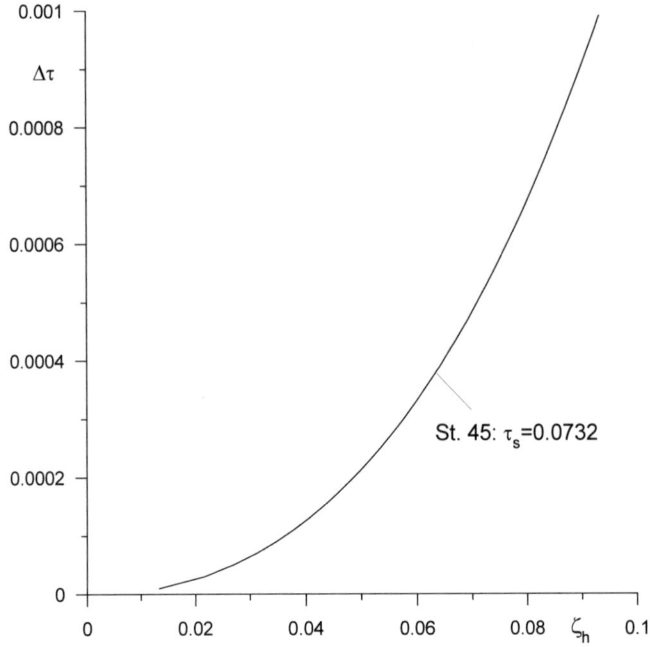

(a)

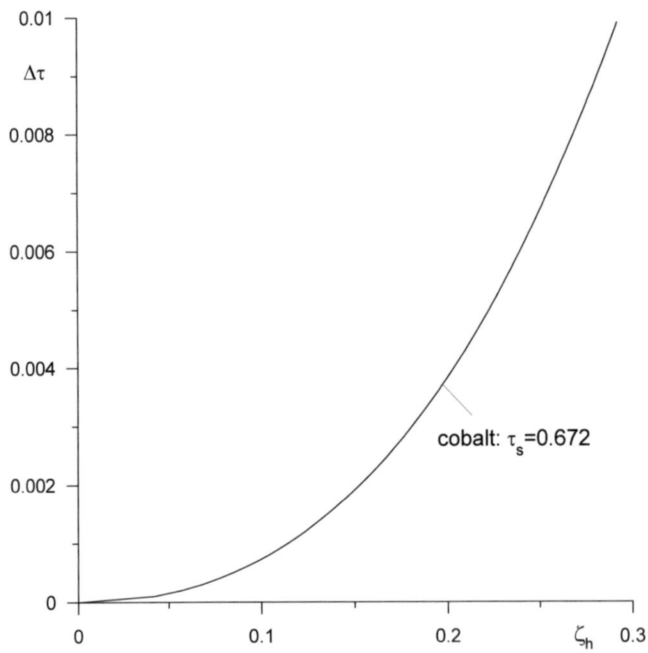

(b)

Fig. 10. Dimensionless hardened layer depth ζ_h from the heated surface versus dimensionless retardation time $\Delta\tau$, for the dimensionless laser pulse duration: a) $\tau_s = 0.0732$ (St.45 steel sample); b) $\tau_s = 0.672$ (Co monocrystal sample) (Yevtushenko et al., 2005).

parameter / metal	laser type	d mm	t_s ms	$q_0 \times 10^{-9}$ W/m^2	K Wm^{-1}K^{-1}	$k \times 10^5$ m^{2}s^{-1}	T_h K	z_h μm
St. 45	Nd:YAG	0.64	2	0.58	33.5	1.5	1123	40
Co	QUANTUM15	0.35	4.5	4.62	70.9	1.83	693	100

Table 1. Input data needed for the calculations of the effective absorption coefficients for the St.45 steel and Co monocrystal samples.

$$\zeta_h = \sqrt{2\Delta\tau\left(1+\frac{\Delta\tau}{\tau_s}\right)\ln\left(1+\frac{\tau_s}{\Delta\tau}\right)}. \tag{75}$$

The curves which represent the dependency of ζ_h on the parameter $\Delta\tau$, for fixed values $\tau_s = 0.0732$ and $\tau_s = 0.672$ are shown in Figs. 10a and 10b, respectively. The dimensionless retardation time quickly increases with the distance increase from the heated surface. The dimensionless pulse durations τ_s were calculated from equation (12) with the use of material constants characteristic for St. 45 steel and Co monocrystals (Table 1), which were presented by Yevtushenko et al., 2005.

Assuming the temperature T_h of the structural phase transition, characteristic for the material, is achieved to a depth z_h from the heated surface at the moment t_h. It should be noticed that for steel the region of structural phase transitions is just the hardened layer, while for cobalt – it is the region where, as a result of laser irradiation, no open domains of Kittel's type are observed.

It can be assumed that the thickness of these layers z_h, is known – it can be found in the way described by Rozniakowski, 2001. Then, from the condition

$$T(z_h, t_h) = T_h, \tag{76}$$

the following formula, which can be used for the determination of the effective absorption coefficient, is obtained:

$$A = \frac{T_h}{T_0'}\left[T^*(\zeta_h, \tau_h)\right]^{-1}, \tag{77}$$

where dimensionless temperature T^* is expressed by equations (41) and (60), the dimensionless retardation time $\Delta\tau$ can be found from the equation (75) and the constant $T_0' = q_0 d / K$. The input data needed for the calculations by formula (77) are included in Table 1. Experimental results obtained by Rozniakowski, 1991, 2001 as well as the solutions for the axisymmetric (Yevtushenko et al., 2005) and one-dimensional model are presented in

parameter / metal	$T_0' \times 10^{-5}$, K^{-1}	ζ_h, -	τ_s, -	$\Delta\tau \times 10^3$ -	A exp.	A axisym.	A one-dimen.
St. 45	0.111	0.0625	0.0732	0.36	$0.3 \div 0.5$	0.42	0.41
Co	0.228	0.286	0.672	9.8	0.1	0.112	0.045

Table 2. Values of the effective absorption coefficient for the St.45 steel and Co monocrystal samples.

Values of the effective absorption coefficient for the St.45 steel sample irradiated with pulses of short duration $\tau_s = 0.0732$, found on the basis of the solutions for axisymmetric ($A = 0.42$) and one-dimensional ($A = 0.41$) transient heat conduction problem are nearly the same, and correspond to the middle of the experimentally obtained values range $A = 0.3 \div 0.5$ (Table 2). The cobalt monocrystal samples were irradiated with pulses of much longer duration $\tau_s = 0.672$. In this case, there is more than twofold difference of A values found on the basis of the solutions for axisymmetric ($A = 0.112$) and one-dimensional ($A = 0.045$) transient heat conduction problem. Moreover, only the value of effective absorption coefficient obtained from the axisymmetric solution of transient heat conduction problem corresponds to the experimental value $A = 0.1$. In that manner, it was proved that the solution of one-dimensional boundary heat conduction problem of parabolic type for the semi-space can be successfully applied in calculations of the effective absorption coefficient only for laser pulses of dimensionless short duration $\tau_s << 1$. Otherwise, the solution of axisymmetric heat conduction problem must be used.

7. Conclusions

The analytical solution of transient boundary-value heat conduction problem of parabolic type was obtained for the non-homogeneous body consisting of bulk substrate and a thin coating of different material deposited on its surface. The heating of the outer surface of this coating was realised with laser pulses of the rectangular or triangular time structure.

The dependence of temperature distribution in such body on the time parameters of the pulses was examined. It was proved that the most effective, from the point of view of the minimal energy losses in reaching the maximal temperature, is irradiation by pulses of the triangular form with flat forward and abrupt back front.

Analysis of the evolution of stresses in the homogeneous plate proves that when it is heated, considerable lateral compressive stresses occur near the outer surface. The value of this stresses decreases when the heating is stopped and after some time the sign changes - what means that the tensile stresses takes place. The time when it happens increases monotonously with increase of a thermal pulse duration (for rectangular laser pulses) or with increase of rise time (for triangular laser pulses). When the lateral tensile stresses exceed the strength of the material then a crack on the surface can arise. The region of lateral compressive stresses, which occur beneath the surface, limits their development into the material.

The presence of the coating (for example, ZrO_2) with thermal conductivity lower than for the substrate results in considerably higher than for the homogeneous material, lateral tensile stresses in the subsurface after the termination of heating. The depth of thermal splitting is also increased in this case. When the material of the coating (for example, copper) has greater conductivity than the substrate (granite) then the stresses have compressive character all the time. The coating of this kind can protect from thermal splitting. The region that is vulnerable for damage in this case is close to the interface of the substrate and the coating, where considerable tensile stresses occur during the cooling phase.
The method for calculation of the effective absorption coefficient during high-power laser irradiation based on the solution of one-dimensional boundary problem of heat conduction for semi-space, when heating is realised with short pulses, was proposed, too.

8. References

Coutouly, J. F. et al. (1999). Laser diode processing for reducing core-loss of gain-oriented silicon steels, *Lasers Eng.*, Vol. 8, pp. 145-157.

Dostanko, A. P. et al. (2002). *Technology and technique of precise laser modification of solid-state structures*, Tiechnoprint, Minsk (in Russian).

Duley, W. W. (1976). *CO_2 Lasers: Effects and Applications*, Acad. Press, New York.

Gureev, D. M. (1983). Influence of laser pulse shape on hardened coating depth, *Kvant Elektron* (in Russian), Vol. 13, No. 8, pp. 1716-1718.

Hector, L. G. & Hetnarski, R. B. (1996). Thermal stresses in materials due to laser heating, In: R. B. Hetnarski, *Thermal Stresses* IV, pp. 1-79, North–Holland.

Kim, W. S. et al. (1997). Thermoelastic stresses in a bonded coating due to repetitively pulsed laser radiation, *Acta Mech.*, Vol. 125, pp. 107-128.

Li, J. et al. (1997). Decreasing the core loss of grain-oriented silicon steel by laser processing, *J. Mater. Process. Tech.*, Vol. 69, pp. 180-185.

Loze, M. K. & Wright, C. D. (1997). Temperature distributions in a semi-infinite and finite-thickness media as a result of absorption of laser light, *Appl. Opt.*, Vol. 36, pp. 494-507.

Luikov, A. V. (1986). *Analytical Heat Diffusion Theory*, Academic Press, New York.

Ready, J. F. (1971). *Effects of high-power laser radiation*, Academic Press, New York.

Rozniakowska, M. & Yevtushenko, A. A. (2005). Influence of laser pulse shape both on temperature profile and hardened coating depth, *Heat Mass Trans.*, Vol. 42, pp. 64-70.

Rozniakowski, K. (1991). Laser-excited magnetic change in cobalt monocrystal, *J. Materials Science*, Vol. 26, pp. 5811-5814.

Rozniakowski, K. (2001). *Application of laser radiation for examination and modification of building materials properties*, (in Polish), BIGRAF, Warsaw.

Rykalin, N. N. et al. (1985). *Laser and electron-radiation processing of materials*, (in Russian), Mashinostroenie, Moscow.

Said-Galiyev, E. E. & Nikitin, L. N. (1993). Possibilities of Modifying the Surface of Polymeric Composites by Laser Irradiation, *Mech. Comp. Mater.*, Vol. 29, pp. 259-266.

Sala, A. (1986). *Radiant properties of materials*, Elsevier, Amsterdam.

Sheng, P. & Chryssolouris, G. (1995). Theoretical Model of Laser Grooving for Composite Materials, *J. Comp. Mater.*, Vol. 29, pp. 96-112.

Timoshenko, S. P. & Goodier, J. N. (1951). *Theory of Elasticity*, McGraw-Hill, New York.

Ujihara, K. (1972). Reflectivity of metals at high temperatures, *J. Appl. Phys.*, Vol. 43, pp. 2376-2383.

Welch, A. J. & Van Gemert, M. J. C. (1995). *Optical-thermal response of laser-irradiated tissue*, Plenum Press, New York and London.

Yevtushenko, A. A. et al. (2005). Evaluation of effective absorption coefficient during laser irradiation using of metals martensite transformation, *Heat Mass Trans.*, Vol. 41, pp. 338-346.

Yevtushenko, A.A. et al. (2007). Laser-induced thermal splitting in homogeneous body with coating, *Numerical Heat Transfer, Part A.*, Vol. 52, pp. 357-375.

4

High-order Harmonic Generation

Krzysztof Jakubczak
Institute of Physics, Academy of Sciences of the Czech Republic
Czech Republic

1. Introduction

X-rays were observed for the first time by Wilhelm Conrad Röntgen in 1895 (Röntgen, 1895). During the first century since that great event X-rays were benefiting mostly from their spatial resolution capability. However, recently it was possible to take advantage also from their temporal resolution due to novel sources providing ultrashort bursts of short-wavelength radiation (i.e. wavelength $\lambda < 100$ nm) and to get an inside view of physical processes in molecules and atoms. One possibility of how to obtain ultrashort bursts of coherent extreme ultraviolet (abbreviated XUV or EUV; wavelength spectral range between 10-100 nm), soft X-ray (1-10 nm) and/or X-ray radiation (< 1 nm) is by high-order harmonic generation (HHG) process. It involves interaction of laser light at a given frequency during which it is being converted into integer multiples of the fundamental frequency through a highly nonlinear interaction with a conversion medium (typically a noble gas; Brabec & Krausz, 2000). Laser-driven HHG uses acceleration of electrons on time-scales that are of the order of an optical cycle of the laser field. Currently this technique gives rise to the shortest flashes of light ever generated in a laboratory which are typically of the order of a few hundreds of attoseconds (1 as = 10^{-18} s; Paul et al., 2001; Kienberger et al., 2004; Schultze et al., 2007). When laser field of intensity of about 10^{14} - 10^{15} W/cm^2 and time durations in

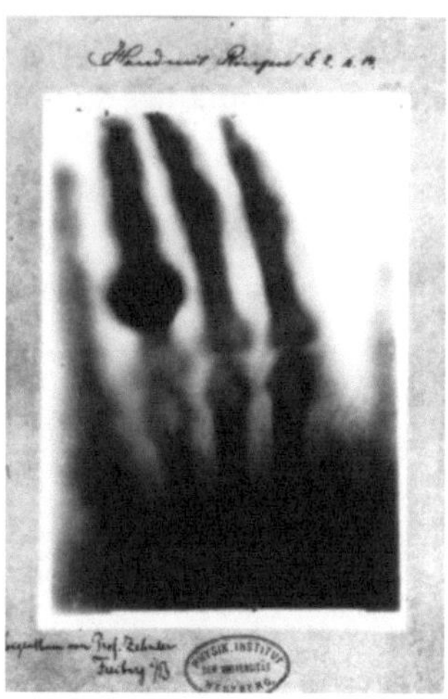

Fig. 1. Ms. Röntgen's hand. First medical imaging with X-rays (December 22, 1985; source: wikipedia.org).

range of femtoseconds (1 fs = 10^{-15} s) is applied to the gas, a plateau of equally intense harmonics of very high order can be observed. The atom is ionized when the absolute electric field of the laser is close to its crest during an optical cycle and is pulled away from the parent ion. Since the laser electric field changes its sign about a quarter of a period later, the electron will slow down, stop at a position far from the ion and start to accelerate back towards it (Corkum, 1993). When it returns to the ion, it can possess significant amount of kinetic energy, much larger than the photon energy but being its multiple. This energy plus the ionization potential is transferred into emitted photon energy as soon as the electron recombines with its parent ion, which gives rise to very high harmonic orders observed in the experiments (Macklin, 1993). Thus HHG represents a source of coherent X-rays bursts of ultrashort time duration. Additionally, the HHG source features spectral tunability from UV to hard X-rays. Moreover, advantage of particular importance is a very high repetition rate of HHG which is given by the repetition rate of the driving laser only and can be easily as much as few kHz (Schultze et al., 2007)!

It has been shown that high-order harmonic pulse comprises train of attosecond pulses (Papadogiannis et al., 1999). This great advantage constitutes a stimulus for further development of high-order harmonic sources, especially of the techniques leading to generation of single attosecond pulses. Nowadays, well explored and most frequently deployed are:

- usage of very short IR laser pulses (< 5 fs) (Christov et al., 1997; Baltuska et al., 2003),
- a technique called *polarization gating* (Sola et al., 2006).

The details of the aforementioned techniques will not be discussed in detail here; however, it is worth noting that the intension of improvement of high-order harmonic sources has become a boost for laser technology progress leading to development of laser systems emitting pulses with duration in the range of single optical cycle (~ 3.3 fs at ~810 nm central wavelength) and shifting the laser pulse central wavelength to the mid-infrared spectral range (MIR) in around 2-3 µm. Besides, the lasers' repetition rates have been significantly increased typically to a few kHz (and energy ~mJ per pulse; e.g. Schultze et al., 2007). Another recent achievement of particular interest is carrier-envelope absolute phase stabilization (CEP).

State-of-the-art HHG sources require not only development of the high-harmonic source itself but also sophisticated metrology techniques and methods for characterization of femtosecond and attosecond pulses (Véniard et al., 1996; Drescher et al., 2002; Kienberger et al., 2002; Mairesse et al., 2005; Itatani et al., 2002; Sansone et al., 2008).

Due to unusual combination of all properties that high-order harmonics feature, they immediately found vast number of unprecedented applications. For example, a number of

Fig. 2. Typical spectrum of high-order harmonics (conversion medium: argon; Jakubczak a)).

experimental results have been recently published related to time-resolved investigation of atomic processes. For instance manipulation of drift energy of photoelectron wave packets (so called "steering of wave packets") and their imaging (e.g. Kienberger et al., 2007), measurement of relaxation and lifetime dynamics in an atom by the direct measurement in time domain with attosecond resolution (e.g. Baltuska et al., 2003; Kienberger et al., 2002) in contrary to thus far frequency-domain measurements of transition linewidths (Becker & Shirley, 1996), spectroscopy of bound electron dynamics in atoms and molecules (Hentschel et al., 2001), observation of interference of coherent electron wave packets (Remetter et al., 2006), probing molecular dynamics (Niikura et al., 2002) and real-time tomography of molecular orbitals (Itatani et al., 2004).
Moreover, novel and very promising schemes for HHG have been recently demonstrated, e.g., generation of harmonics during reflection of super intense ultrashort IR laser pulses ($I > 10^{17}$ W/cm^2) from plasma mirror oscillating at relativistic velocities on the surface of a solid state target (Quéré et al., 2006), or generation of HHG from interaction of IR femtosecond laser pulses with molecules (N_2, H^{2+}; Lorin et al., 2008).

2. Physical mechanisms of high-order harmonic generation

If material is subjected to a strong electric field, nonlinear polarization of the material is induced. The magnitude of the arisen polarization strongly depends on the intensity of the incident radiation. At moderate and low intensity values the external electric field does not influence significantly the electronic structure of the irradiated atoms. The potential barriers can be just slightly modified and Stark effect can be observed. To great probability the atoms remain in their ground state and extension of their ground state wave function is of the order of Bohr radius ($5.2917 \cdot 10^{-11}$ m). All nonlinear phenomena taking place in this regime are well described by the perturbation theory. Thus it is referred as *the perturbative regime* of nonlinear optics. Comprehensive discussion on phenomena and related theory in the perturbative regime can be found e.g. in Boyd, 2003. Some of nonlinear optical phenomena in this regime are:

- harmonic radiation generation (second, third, etc.),
- optical parametric amplification,
- optical rectification,
- stimulated Raman scattering,
- self-phase modulation,
- self-focusing.

However, when the electric field strength of the incident radiation is comparable to (or higher than) atomic electric field strength ($5.142 \cdot 10^{11}$ V/m ; Brabec & Krausz, 2000) then the potential barriers are strongly modified. With high probability the electrons from the most-outer atomic shells may be liberated either through the tunnel ionization or the above-barrier ionization (depending on the external field strength; see Fig. 3 and Fig. 4). Subsequently, if the field is linearly polarized electron wave packets will start oscillatory motion. The amplitude of oscillations exceeds Bohr radius and cycle-averaged kinetic energy of electron wave packet surpasses binding energy (Brabec & Krausz, 2000).
Range of intensities implying these phenomena defines *the strong field nonlinear optics regime*. In contrary to the perturbative regime, here, the nonlinear response of the polarization of the medium is affected by the ionization process. The nonlinear treatment can be only applied

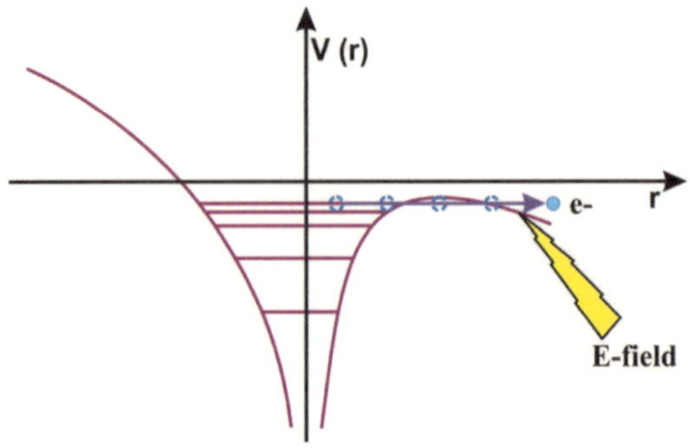

Fig. 3. Tunnel ionization. The atomic potential affected by the external electric field whose the field strength is comparable to the atomic fields. It is plausible that the electrons from the most-outer atomic will be unbound. This transition is often referred as optical field ionization (OFI).

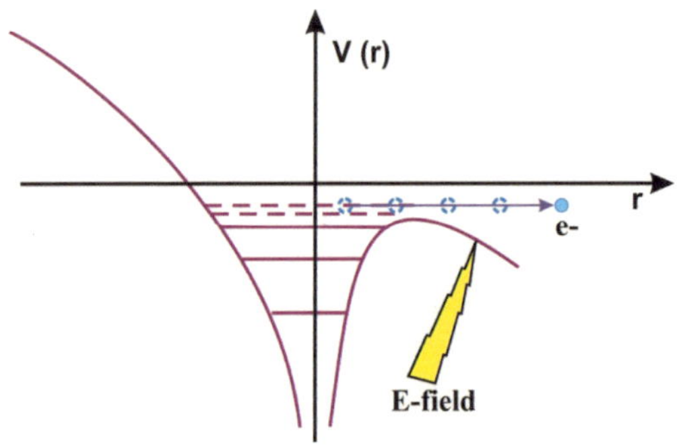

Fig. 4. In this case, the applied electric field is higher than the atomic field strength. The atomic potential barrier is suppressed and electrons from most-outer shells are liberated through above barrier ionization.

to an electron which is in very close vicinity of a parent ion. As soon as it is released by optical field its response is linear to the electric field and may be treated by classical laws of motion (Corkum, 1989; Corkum, 1993).

Very interesting phenomena are present in the intermediate range of parameters, in the so called *intermediate regime*, i.e. between the perturbative and the strong field regimes. They include long-distance self-channeling when nonlinear Kerr effect causes beam focusing, on the one hand, and free electrons cause its defocusing, on the other. This interplay leads to the channeling of the propagating intense pulse (even at distances as long as a few meters). Another interesting phenomenon in this regime is multiphoton ionization, where the total amount of absorbed energy exceeds the ionization potential (Fig. 5).

When electric field strengths are even higher, the nonlinearities become stronger. Electric field is able to optically liberate electrons from inner shells of the atom and the wiggling energy of an electron is comparable with its rest energy mc^2. This is a launch of *relativistic regime.*

Publications of crucial importance related to the intermediate to strong-field nonlinear optics regimes were made by Keldysh (Keldysh, 1965) and Ammonsov, Delone and Krainov (Ammosov et al., 1986). Keldysh defined a parameter, which was later named after him that allows determining whether tunneling or multiphoton process is dominant for particular experimental conditions. It reads:

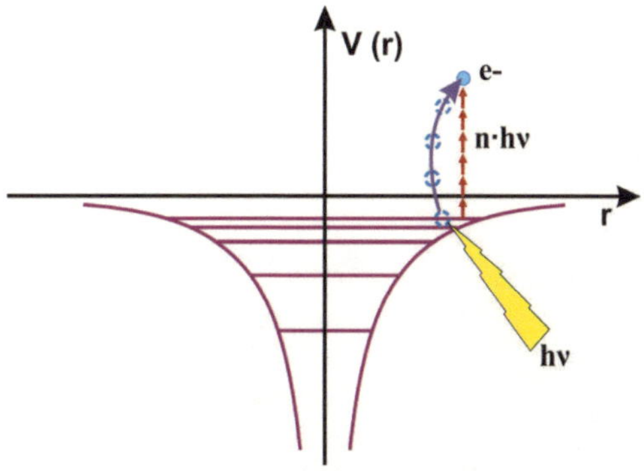

Fig. 5. Multiphoton ionization process: n-photons are absorbed. The total energy of absorbed photons (n*hν; n - number of absorbed photons, h - Planck's constant, ν - light frequency) exceeds ionization potential.

$$\gamma = \sqrt{\frac{I_p}{2 \cdot U_p}} \tag{1}$$

Where:
I_p - is ionization potential of a nonlinear medium,
U_p - is ponderomotive potential, which is cycle-averaged quivering energy of an electron in the external laser field. It is defined as:

$$U_p = \frac{e^2 \cdot E_0{}^2}{4m_e \cdot \omega^2} \tag{2}$$

Where:
e - stands for charge of electron,
m_e - is mass of electron,
E_0 - external field amplitude oscillating at frequency ω.
Substitution of constants leads to simplified relation:

$$U_p[eV] = 0.97 \cdot 10^{-13} I[W / cm^2] \lambda^2 [\mu m] \tag{3}$$

The laser field amplitude can be estimated from relation:

$$E_0{}^2[V / cm] = \frac{I[W / cm^2]}{\frac{1}{2} Z_0} \tag{4}$$

Where:
I - is laser intensity [V/cm^2],
Z_0 - is vacuum impedance.

$$Z_0 = \sqrt{\mu_0 \varepsilon_0} = 377[V / A] \tag{5}$$

Where:
μ_0 - is vacuum permeability, $\mu_0 = 1.26 \cdot 10^{-6}$ [H/m],

ε_0 - is vacuum permittivity, ε_0=8.85·10^{-12} [F/m].
If $\gamma >> 1$ multiphoton ionization dominates. However, if $\gamma << 1$ tunneling ionization takes over. By these simple formulas it is possible to divide regimes of nonlinear optics in intensity domain as depicted in Fig. 6.

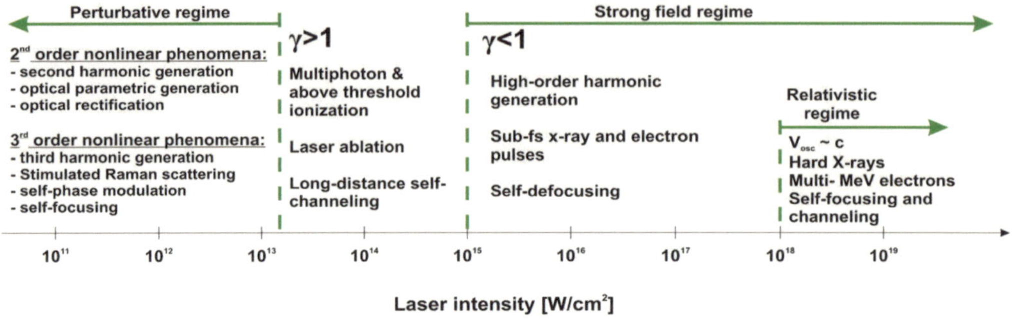

Fig. 6. Regimes of nonlinear optics.

Ionization rates are of core importance when discussing interaction of the intense laser pulses with matter. Their estimations were performed within quaststatic field approximation. The first approach was proposed by Keldysh. The second was developed by Ammonsov, Delone and Krainov (also known as ADK theory named after the acronyms of the names). Additionally, the ionization rate calculation could be performed by an exact numerical solution of the time-dependent Schrödinger equation. The Keldysh theory possesses a source of a discrepancy between the other theories which comes from the fact that it neglects Coulomb potential in an atom. The difference leads to lower ionization rates compared to other approaches (e.g. in case of He and H by about 1-2 orders of magnitude) (Brabec & Krausz, 2000). The discrepancies between theories increase with the electric field strength. For example, for He there is no difference between ionization rates obtained from ADK theory and solution of the time-dependent Schrödinger equation as long as the field strength does not exceed 0.2 atomic unit.

3. High-order harmonic generation in gaseous media

3.1 Microscopic analysis

High-order harmonic generation process takes place when linearly polarized ultrashort laser pulses of intensity of a few times 10^{13} W/cm^2 to < 10^{16} W/cm^2, and time duration from picoseconds to a few femtoseconds (Pfeifer et al., 2006), are applied to a nonlinear medium (atoms, atom clusters, molecules and plasmas).
HHG process can be understood using semi-classical *three-step model* (Corkum, 1993):
Step I - Ionization. When an atom is exposed to external electric field, the potential of the atom is modified by a factor of $e\vec{E}(t)\vec{r}$. Then, the resulting potential is equal to:

$$V(\vec{r},t) = -\frac{e^2}{4\pi\varepsilon_0 r} + e\vec{E}(t)\vec{r} \tag{6}$$

With increasing strength of the external field $\vec{E}(t)$ the probability of tunnel-ionization by the low-frequency laser field of most-outer-shell electrons increases significantly (the natural potential of atom is being cancelled).

Step II - Propagation. When an electron wave function undergoes tunnel-ionization from its parent atom the free electron wave packet is affected primarily by the external electric field (and not by the field of the parent atom) and is accelerated by this field. When the laser field reverses its sign, the electrons slow down and are re-accelerated back towards the atom. The free electron motion can be described by laws of classical physics (Pfeifer et al., 2006):

$$v(t) = \int_0^t -\frac{e}{m}E(t')dt' + v_0 = -\frac{E_0 e}{m\omega}\sin(\omega t) + v_0 \quad (7)$$

where: v(t) - is instant velocity of an electron, v_0 - is electron drift velocity.
If we consider an initially bound electron (x=0), with zero drift velocity, its velocity can be described by (Pfeifer et al., 2006):

$$v(t) = \int_0^t -\frac{e}{m}E(t')dt' + v_0 = -\frac{E_0 e}{m\omega}[\sin(\omega t + \varphi) - \sin(\varphi)] \quad (8)$$

and its position (Pfeifer et al., 2006):

$$x(t) = \int_0^t v(t')dt' = \frac{E_0 e}{m\omega^2}[\cos(\omega t + \varphi) - \cos(\varphi)] + \sin(\varphi)t \quad (9)$$

Where: φ is phase of electric field at which atom is ionized (often referred as: "electron is born").
Dependence of the electrons paths as a function of time, where the electric field phase is a parameter, reveals that just a fraction of the electrons are probable to return to the parent ion and re-collide (contributing to emission of radiation). The problem is addressed in more detail in Pfeifer et al., 2006. This is the reason of the optimization of the electric field phase.
Step III - Recombination. After re-acceleration of the electron wave packet towards the ion it is plausible that electron will collide with the ion and recombine. The excess of kinetic energy is transferred to the momentum of the emitted photon. The electrons which recombine with ions will emit harmonic radiation with energy of spectral lines defined as follows:

$$\hbar\nu = E_{kin} + I_p \quad (10)$$

Where E_{kin} - is kinetic energy of an electron acquired by absorption of n-photons of the driving field, I_p - is ionization potential corresponding to the shell from which the electron has been ripped off by the field.
There exists a limit on the maximum emitted energy. It is given by maximum kinetic energy that electron gains during acceleration. It was claimed that the path and kinetic energy of the electron is controlled by the phase of the electric field (Eq. 7 – Eq. 9). If phase is ~18° the kinetic energy of electron is maximized and its value is $\sim 3.17 \cdot Up$ (Pfeifer et al., 2006)[1]. Thus the energy of the highest harmonic order is given by:

$$\hbar\nu = 3.17U_p + I_p \quad (11)$$

[1] 18° = ~314 mrad.

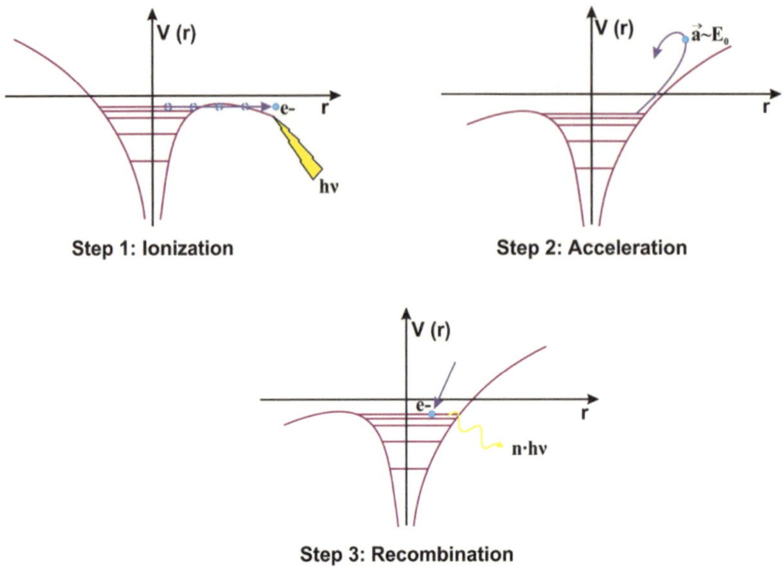

Fig. 7. Summary of the 3-step model of HHG. The first step is tunnel-ionization of an atom. Next, an electron is driven away from its parent ion in the external electric field of an intense laser pulse. When the oscillating laser field reverses its sign the particle is re-accelerated back towards the atom and, finally, recombines. The last step leads to emission of a photon.

The HHG process can be also explained in the formalism of quantum mechanics (Itatani et al., 2004). The returning electron wave packet overlaps remaining portion of initial wave function. The coherent addition of the two wave functions induces a dipole as asymmetric displacement of the electron wave packet. The dipole oscillates as continuum wave function (of the free electron packet) propagates. The oscillating dipole is a source of harmonic radiation and harmonic spectrum is given by Fourier transform of dipole acceleration. Instantaneous frequency of the dipole is determined by kinetic energy of the recombining electrons (i.e. electron wave packet and emitted photons are linked by the energy conservation: $\hbar \nu = E_k$, where the ionization potential is omitted due to the fact that E_k of electrons is "seen" by bound-state electron wave function, compare to Eq. 11). It is worth noting that the electron wave packet and the emitted photons are mutually coherent.

It is important to point the influence of tunnel-ionization process on HHG. The ionization rate increases with the amplitude of the electric field (Keldysh, 1965; Ammosov et al., 1986) leading to generation of free electrons what results in their increasingly stronger contribution to HHG. Most energetic electrons are produced at 18°-phase of the electric field. On the other hand, multiphoton ionization produces constant number of electrons depending only on the intensity of the laser pulses and not on the field phase. From cut-off law (Eq. 11) we also know that energy of photons may be increased by increasing the ponderomotive potential U_p (when $\gamma << 1$), thus shifting more into the strong-field regime.

3.2 Macroscopic analysis - phase-matching considerations

In most generic sense HHG process analysis is divided into two logical constituents: micro- and macroscopic. The three-step model deals only with atomic-scale phenomena leading to emission of a photon, thus, this part of the process is assigned to microscopic analysis of

HHG. However, there are also macroscopic issues to be concerned. The crucial problem in this sense for HHG process is a phase-matching of the two propagating beams: XUV from HHG and of the driving laser. If we consider mismatch between m^{th} harmonic order and driving laser field oscillating at fundamental frequency ω_f the wave vector of the mismatch could be written as (Pfeifer et al., 2006):

$$\Delta k = mk(\omega_f) - k(m\omega_f) \tag{12}$$

In general there are three major components of total Δk that may be written as:

$$\Delta k = \Delta k_{natural} + \Delta k_{plasma} + \Delta k_{foc} \tag{13}$$

Refractive index is a function of frequency. Since the phase velocity of radiation at given wavelength depends on refractive index, in general, radiation of different wavelengths propagates with different velocities what leads to de-phasing of particular spectral components. This input to total phase mismatch is called natural dispersion.

Because only little fraction of free electrons that had been generated by the laser field recombines with parent ion free-electron clouds are created. These free electrons give raise to additional component of refractive index (Brabec & Krausz, 2000):

$$n_{plasma} = \sqrt{1 - \frac{N_e}{N_c(\omega_f)}} \tag{14}$$

$$N_c(\omega) = \frac{\varepsilon_0 m_e \omega^2}{e^2} \tag{15}$$

Where:

N_e - is free-electron density,

N_c - is critical plasma density.

This leads to phase mismatch wave vector component (Pfeifer et al., 2006):

$$\Delta k_{plasma} = mk_{plasma}(\omega_f) - k_{plasma}(m\omega_f) = \frac{\omega_p{}^2(1-m^2)}{2mc\omega_f} \tag{16}$$

HHG requires peak intensities in range of 10^{13} - 10^{16} W/cm^2. In order to obtain such high intensities one has to focus driving laser beam. However, focusing involves phase-shift of the driving field along beam propagation direction 'z' (so called *Gouy phase shift*; Jaeglé, 2006):

$$\varphi_{geo}(z) = -\arctan(\frac{2z}{b}) \tag{17}$$

Where "b" is a confocal parameter defined as follows (Jaeglé, 2006):

$$b = \frac{2\pi a^2}{\lambda} \tag{18}$$

Where "a" is beam radius in the focal spot.

Gouy phase shift leads to the phase mismatch wave vector component (Pfeifer et al., 2006):

$$\Delta k_{foc} = mk_{foc}(\omega_f) - k_{foc}(m\omega_f) = \frac{2(m-1)}{b} \tag{19}$$

If the HHG process takes place in a hollow fiber the Δk_{foc} component is replaced by:

$$\Delta k_{cap} = mk_{cap}(\omega_f) - k_{cap}(m\omega_f) = \frac{u_{nl}{}^2 c(1-m^2)}{2md^2\omega_f} \tag{20}$$

Where:
u_{nl} - is l^th^ zero of Bessel function, and
d - is capillary inner radius.
There is also a component to phase mismatch originating from the fact that when the driving field pulse propagates in a gas it is defocused by the free electrons density gradient what leads to drop of intensity. Since the dipole moment is roughly linearly proportional to $-U_p / \omega_f$ the intensity drop implies de-phasing. However, this contribution is negligible compared to the de-phasing due to focusing and free-electrons generated index of refraction.
Phase-matching condition can be fulfilled by:

- tuning gas density (modification of gas density leads to modification of the index of refraction),
- changing position of the focus with respect to the gas resulting in minimization of Gouy phase-shift influence (gradient of function defined by Eq. 17 is highest at the beam waist),
- in the hollow-fiber geometry modification of the fiber parameters can develop perfect phase-matching,
- free electron density may be controlled by intensity and time duration of laser pulses.

3.3 Generic properties of high-order harmonics spectrum

An important feature of the harmonic spectrum is its universal shape. As already mentioned there is a spectral region of roughly equally intense spectral lines, so called *plateau*. The plateau is preceded by increase of spectral lines intensity in longer wavelengths which is subsequently followed by abrupt intensity drop-off. The short-wavelength part of spectrum extends to the limit defined by cut-off law. From this formula (Eq. 11) one can infer how to extend the wavelength range of HHG, e.g., by increasing ionization potential (proper choose of conversion medium or by working with ions for subsequent shells have higher ionization potential). Another way of doing so is by raising intensity or increasing wavelength of driving laser field (revoke Eq. 2 or Eq. 3).
It has been shown that intensity of mth harmonic order is proportional to square of phase-matching factor ($m^{th} \propto |Fq|^2$; Jaeglé, 2006). This factor rapidly decreases with *m* leading to drop of intensity of spectral lines in the long-wavelength part of harmonic spectrum.
Additionally, in the plateau region the scaling law has been observed (Jaeglé, 2006):

$$I_m \propto b^3 \cdot \Delta t \tag{21}$$

Where, I_m - is intensity of mth harmonic order, b is confocal parameter, Δt is driving laser pulse duration. The dependence on b^3 comes from the fact that total number of photons is

spatially integrated. The dependence on duration of the pulse comes from temporal integration of the HHG process.
In most common scheme of HHG when only one-wavelength driving laser field is used (in gaseous media) the HHG spectra contain only odd harmonic orders. This could be understood by formalism presented, e.g., in Boyd, 2003. This approach originates from perturbative theory of description of nonlinear optical phenomena and is followed here after Boyd.
When low-strength external field $\vec{E}(t)$ (<< ~10^9 W/cm^2) interacts with matter the material polarization $\vec{P}(t)$ (or dipole moment per unit volume) is responding to excitation in linear fashion to the applied field. This relation is linked by linear susceptibility and reads:

$$\vec{P}(t) = \chi^{(1)}\vec{E}(t) \tag{22}$$

However, if the field increases the nonlinear response of the harmonic oscillator appears. If the field is not too strong (for intensities smaller than ~10^{14} W/cm^2) $\vec{P}(t)$ can be expanded in Taylor series:

$$\vec{P}(t)(t) = \chi^{(1)}\vec{E}(t) + \chi^{(2)}\vec{E}^2(t) + \chi^{(3)}\vec{E}^3(t) + \chi^{(4)}\vec{E}^4(t) + ... \tag{23}$$

If any arbitrary medium features e.g. third order nonlinearity ($\chi^{(3)}$ does not vanish) the medium polarization is capable of being a source of nonlinearities of the third order (e.g. third-harmonic generation, nonlinear index of refraction and secondary phenomena having their origin in the dependence of $n = f(I)$, etc.). In general, there is a strong dependence between type of symmetries of media and their nonlinear properties. For instance, if a medium is centrosymmetric all its even-order susceptibilities vanish and thus those media are not able to give rise to any of even-order nonlinear phenomena. This is also the case of gasses (as well as e.g. liquids and amorphous solids). Since gases display inversion symmetry it is possible to obtain odd-order HHG only.

3.4 HHG geometries involving gaseous media

The most common experimental setup of HHG comprises (apart from laser system and diagnostic apparatus) a *gas puff* target (L'Huillier & Balcou, 1993). Such a target is basically a gas valve injecting a portion of a gas at desired pressure. The valve is repetitively open and laser pulses interact in proximity of valve exhaust. Typical repetition rate of the gas puffs is < 100 Hz. The valves may have either circular symmetry or could be elongated. Elongated valves provide higher XUV beam outputs but limiting factor is re-absorption in a gas thus long valves are used for wavelengths > ~20 nm (also because of the longer coherence length for longer wavelengths). On the other hand, circular (e.g. 0.5 mm diameter) gas puff valves are used in shorter-wavelength HHG.
High-order harmonics are also generated in gas cells. A gas cell is a simple container filled with a noble gas at moderate pressure (few tens of mbar). Arrangements with gas cells are very comfortable to work with because, compared to the gas puff targets, there are fewer parameters to optimize to obtain phase matching. In this case the phase matching is obtained by tuning only longitudinal position of a cell and gas pressure in it. For gas cells

can be as long as desired (due to technical ease of construction compared to gas puff valves) this setup is also favorable when maximization of interaction length is wanted.

Another possible geometry of HHG involves hollow fibers filled with a conversion gas (Rundquist et al., 1998). In such a fiber laser pulses are propagating even meters long resulting in efficient transfer of driving field energy into XUV beam. This geometry is also popular due to ease of control of phase-matching by the fiber parameters (Eq. 20).

4. High-order harmonic generation by molecules

Essentially, the physical mechanism of HHG in molecules is the same as in atomic gases (see section 3) and can be understood by the same 3-step model. However, orbital structure of molecules and thus the description of the mechanism of HHG in this case is significantly different from situation when atomic conversion media are involved.

There have been many molecules proven to be capable of HHG:

- N_2 (Itatani et al., 2004; Sakai & Miyazaki, 1995),
- CS_2, hexane, N_2 (Velotta et al., 2001),
- O_2, CO_2 and N_2 (Kanai et al., 2007).

A very interesting feature of HHG in molecules is that signal yield of particular harmonic order depends on laser light polarization ellipticity and its orientation with the respect to the direction of molecular axis. In general, the signal yield is highest for linearly polarized light perpendicularly oriented to the axis of the molecule (Kanai et al., 2007).

The fact that yield of HHG depends on orientation of the molecule with the respect to light polarization direction suggests that the HHG process (especially in molecules) is strongly affected by the shape of orbitals of the molecule. This idea has been motivation to Itatani et al., 2004. Finding the relation between orientation of the molecule and spectral intensities of high-order harmonics they succeeded to perform inverse calculation and obtained tomographic reconstruction of the most-outer orbital of N_2 molecule. Since these orbitals are responsible for chemical properties of the molecules the results have great impact on the state of our knowledge. Such direct measurement of orbitals is a first step to the "molecular movie" showing, e.g., time-resolved process of creation of chemical bonds.

Another very interesting feature of HHG from molecules was unveiled during experiments in which the influence of molecular structure complexity on HHG efficiency was investigated. It turned out that the increasing complexity of the molecule is unfavorable for efficient HHG. When dissociating pre-pulse was applied the HHG yields were higher compared to case with unaffected molecules (Velotta et al., 2001; Hay et al., 2002). It has been inferred (Jaeglé, 2006) that the origin of the higher conversion efficiency in atomic media relies in fact that dipole phase depends on the angle between molecular axis and the pump laser polarization. An additional de-phasing mechanism would exist between the emitters in randomly aligned molecular media and it could imply worse phase matching compared with the monomers (Hay et al., 2002).

5. High-order harmonic generation from solid targets

HHG process can also take place at the interface between vacuum and solid targets. The physical process leading to HHG is different here and is explained in terms of resonant absorption. Let us now introduce reference system presented in Fig. 8.

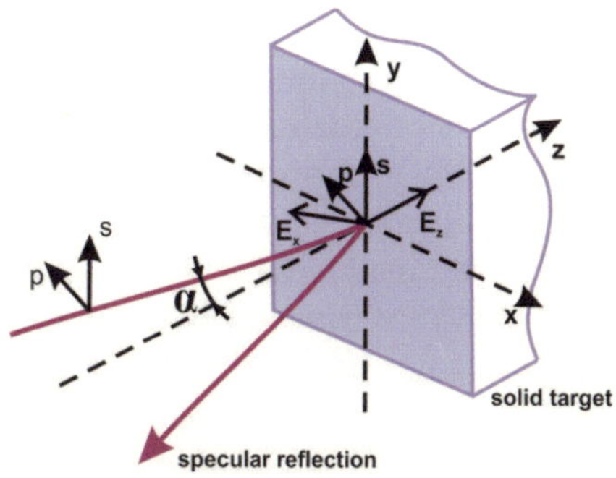

Fig. 8. HHG from solid target surface geometry.

From electromagnetic analysis of reflection of radiation at an interface between two media it is known that in case of specular reflection at oblique angles ($\alpha > 0°$) "s" polarization is reflected according to Snell's law, or, alternatively, this polarization, does not propagate in z direction (due to the boundary conditions). On the other hand, "p" polarization can propagate in z direction into the medium. When a very intense laser pulse impinges upon the target the plasma is created. In such a plasma light propagates to the layer of free-electron density equal to plasma critical density (Eq. 24). The laser radiation cannot propagate any farther, but, instead, the free-electron plasma wave (plasmon) is induced by "p" polarized light at frequency equal light frequency (*resonant absorption*) and directionality of propagation along z axis.

$$N_c = \frac{4\pi^2 c^2 \varepsilon_0 m_e}{e^2 \lambda_L{}^2} \approx \frac{1.1 \cdot 10^{21}}{\lambda_L{}^2[\mu m]}[cm^{-3}] \tag{24}$$

When laser light at frequency ω_L mixes with the induced plasmon at frequency ω_p, the light at frequency of $2\omega_L$ is generated. The $2\omega_L$-electromagnetic wave may propagate further into plasma since plasma critical density for $2\omega_L$ is higher ($4N_c$). Then, light at $2\omega_L$ mixes with that at ω_p generating $3\omega_L$ component. This frequency radiation can propagate until it reaches density of $9N_c$, and so on (up to some upper-limit density N_u). This cascaded phenomenon leads to generation of harmonics of the order limited by plasma frequency and its maximum value is given by (Jaeglé, 2006):

$$q_{\max} \approx \frac{N_u}{N_c} \tag{25}$$

The approximate relation for N_u is given by (Jaeglé, 2006):

$$N_u = \frac{E_0{}^2}{8\pi T_b} \tag{26}$$

T_b is background temperature (Carman et al., 1981).

Very important parameter is $I \cdot \lambda^2[\mu m^2 W / cm^2]$ product. For λ=1 μm and intensities below 10^{18} W/cm^2 Lorentz force can be approximated only by component coming from electric field, i.e. by $e \cdot E_z$. However, if intensity becomes comparable to 10^{18} W/cm^2 the relativistic effects bring increasingly significant input to Lorentz force from magnetic field component.

For laser pulse intensities below the mentioned value (or for $I\lambda^2 < 10^{18}$ μm² W/cm²) the only polarization that may induce harmonic generation is "p" polarization. However, above this value HHG may take place due to nonlinear mixing between longitudinal and transverse oscillations resulting in possibility of HHG from "s" polarized light (Jaeglé, 2006).

Another way of describing of HHG from solid-vacuum interface involves an oscillating plasma mirror. It could be shown by mathematical analysis that light reflected at critical surface oscillating at relativistic velocities contains harmonics of the incident beam (Jaeglé, 2006). This approach; however, will not be discussed in more detail here. Instead, some general properties of HHG from solids shall be discussed. Probably the most interesting feature of HHG from solids is that all-order harmonic radiation is obtained. Additionally, it has been observed that the emission solid angle changes with the intensity of driving laser field (Jaeglé, 2006). In general the emission cone is larger compared to the laser beam. When $I\lambda^2$ reaches values of 10^{15} - 10^{16} μm²W/cm² the emission solid angle strongly increases; for values above 10^{17} μm²W/cm² it is found to be isotropic (no angular distribution of HHG has been found - Jaeglé, 2006). Moreover, HHG efficiency and signal yield from solids drop above 10^{16} μm²W/cm². The two phenomena are attributed to the transition from specular reflection regime of HHG to diffusion reflection regime.

The key parameters of HHG from solid-vacuum interfaces sources are:

- Intensity of the driving field (or more exactly the product of $I\lambda^2$), which is basically the only limiting factor to the order of generated harmonics,
- angle of incidence of the driving field upon the target,
- contrast of the driving laser pulse.

Nowadays, this type of HHG source is drawing lots of attention in the scientific world. This is due to huge potential capabilities of this type of source (e.g. odd and even harmonic orders generation and their number limited only by available intensity of the driving laser fields, conversion efficiency increasing with $I\lambda^2$). For more detailed discussion on HHG from solids see, e.g., (Carman et al., 1981; Bezzerides et al., 1982; Dromey et al., 2006; Balcou et al., 2006; Tarasevitch et al., 2007; Quéré et al., 2008; Thaury et al., 2007).

6. Applications of high-order harmonic sources

High-order harmonics, due to their high repetition rate operation, tunability and high coherence degree, have already found a number of interesting applications. For instance, they are used in material sciences, life sciences and detection technology. As an example high-order harmonics together with femtosecond NIR beam were used to efficiently modify tribological properties of materials. The results from these experiments are presented in the sub-section 6.1. High-order harmonics are also very practical in metrology of multilayer (ML) optics (sub-section 6.2).

6.1 Materials surface processing

Recently, a new method for materials processing suitable for efficient machining of transparent materials has been demonstrated (Mocek et al., 2009). The technique utilizes simultaneous interaction of NIR femtosecond laser pulses generated by Ti: Al_2O_3 laser system (Δt=32 fs, E=2.8 mJ, λ_c=820 nm) and the second harmonic, combined with extreme ultraviolet (XUV) high-order harmonics with the strongest spectral line at 21.6 nm.

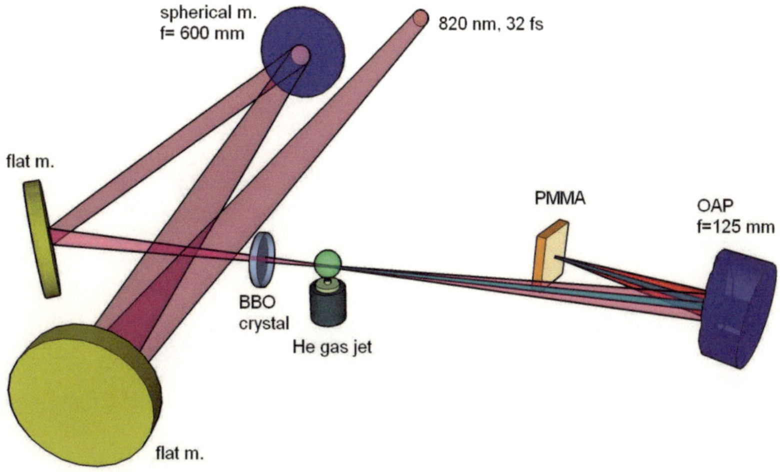

Fig. 9. Schematic of the experimental setup for surface modification by dual action of XUV and Vis-NIR ultrashort pulses (Mocek et al., 2009).

The experimental setup is shown in Fig. 9. For strong HHG a two-color laser field, consisting of fundamental and second harmonic (SH) of a femtosecond laser pulse, was applied to a gas jet of He (Kim et al., 2005). Femtosecond laser pulses at 820 nm with an energy of 2.8 mJ and pulse duration of 32 fs were focused by a spherical mirror (f =600 mm) into a He gas jet. For SH generation, a 200-µm-thick beta-barium borate (BBO) crystal was placed between the focusing mirror and gas jet so that, after the BBO crystal, the laser field consisted of both the SH and the residual fundamental laser fields. For the optimum SH conversion the BBO crystal was placed ~400 mm from the focusing mirror and the energy conversion efficiency was about 27 %. A gas jet with a slit nozzle of 0.5 mm width and length of 6 mm was used (Kim et al., 2008). The gas pressure in the interaction region was 150 Torr (~0.2 bar). Generated HHG were first characterized using a flat-field soft X-ray spectrometer equipped with a back-illuminated X-ray charge coupled device. Optimization of the two-color HHG source was performed by selecting the gas jet position while controlling the relative phase between the two fields. The strongest harmonic at the 38th order (at 21.6 nm) reached energy of ~50 nJ.

Subsequently, the spectrometer was replaced with a 1-inch diameter off-axis paraboloidal mirror (OAP, f = 125 mm at 13°) with a Mo:Si multilayer coating (R = 30 % at 21 nm) placed 245 mm from the HHG source. The sample (500-nm thin layer of PMMA spin-coated on a 315 µm thick silicon substrate) was positioned 125 mm from the OAP, perpendicularly to the incident beam. The measured reflectivity of the OAP in the optical region was 37 %. The measured diameter (FWHM) of the HHG beam incident on the OAP was 280 µm while the diameter of the fundamental and SH laser beams was ~4 mm. The morphology of irradiated target surface was first investigated by Nomarski differential interference contrast optical microscope, and then with an atomic force microscope (AFM) operated in the tapping mode to preserve high resolution.

The estimated fluence on the surface of the PMMA was 97 µJ/cm^2 at 21.6 nm, 14.7 mJ/cm^2 at 820 nm, and 6.3 mJ/cm^2 at 410 nm per shot, respectively. As all these values lie far below the ablation threshold for PMMA by infrared (2.6 J/cm^2 for single-shot and 0.6 J/cm^2 for 100 shots - Baudach et al., 2000) as well as by XUV (2 mJ/cm^2 - Chalupsky et al., 2007) radiation, no damage of target surface was expected.

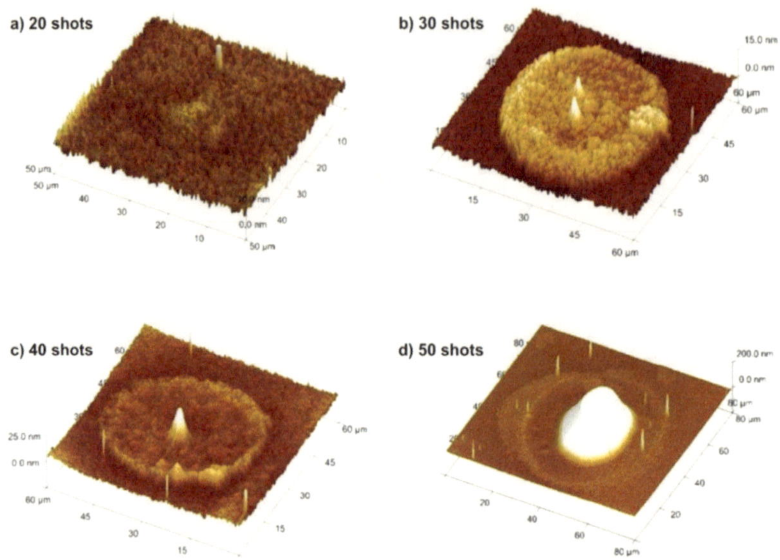

Fig. 10. AFM images of the PMMA surface simultaneously irradiated with XUV and Vis-NIR ultrashort pulses (Mocek et al., 2009).

The PMMA target was irradiated under the following conditions: a) Vis-NIR beams only (no gas jet in operation), b) XUV beam only (0.4 µm Al filter placed in front of the OAP), and c) mixed XUV/Vis-NIR field (21.6 nm + 820 nm + SH). In cases a) and b) we have not observed any signs of surface damage after irradiation by ~3000 shots, although the accumulated dose was significantly higher than in case c). The target surface remained significantly unaffected, preserving its original quality and roughness. In striking contrast, the application of the mixed XUV/Vis-NIR field resulted in clearly visible, irreversible surface modification after irradiation with only a few shots. The damage is characteristic due to material expansion, quite different from ablation craters observed in experiments with pure XUV pulses (Chalupsky et al., 2007; Juha et al., 2005; Vaschenko et al., 2006; Krzywinski et al., 2007; Mocek et al., 2006). A very interesting feature in Fig. 10 is the formation of nano-scale spike(s) in the center of a uniform flat pedestal. The size of pedestal increases with the number of shots applied to the target (~20 µm for 10 shots, ~60 µm for 50 shots) while the spikes have a radius of the order of few microns down to a few tens of nanometers in length (~15 nm for 30 shots, ~25 nm for 40 shots). The target was also analyzed by Fourier-transform infrared spectroscopy, which revealed that the expanded material is exclusively PMMA, not the silicon from substrate.

In another experiment (the same experimental setup) 890 nm thick amorphous carbon (a-C) sample was simultaneously irradiated by beams: 15 mJ/cm^2 at 820 nm, 6 mJ/cm^2 at 410 nm, and 0.1 mJ/cm^2 at 21.6 nm. The sample irradiated solely with NIR-VIS light exhibited only negligible surface changes, whilst in the case of combined fields the area of the modified surface dramatically increases. The damage occurred in a single pulse exposure. RMS microroughness of NIR-VIS and XUV/NIR-VIS illuminated samples was estimated to be 4.7 nm and 176.5 nm, respectively (evaluated only in the interaction region), and 5.7 nm for the unexposed sample (Horcas et al., 2007). The efficiency of the dual action is much higher than in the previous experiment due to larger fluencies of the beams. The difference in the

surface changes visible in AFM scans suggests that the structures shapes are fluence-dependent.
Even stronger enhancement of surface processing by the dual action was observed on a-C sample. Laser induced periodic surface structures (LIPSS, or "ripples") with spatial period of ~550 nm were created during 10-shot exposure in both cases; however, a clear difference in modulation depth was observed: the peak-to-valley depth was ~130 nm for NIR-VIS irradiated sample, while ~200 nm for combined XUV/NIR-VIS. Moreover, despite its random distribution, the frequency of occurrence of LIPSS is much higher in the case of the mixed fields. RMS microroughness was 13.5 nm for NIR-VIS illumination, 46.2 nm for XUV/NIR-VIS, and 8.3 nm for an unexposed sample, respectively.
Detailed discussion on experimental results and physical processes involved in the dual action can be found in Jakubczak et al., b) and Jakubczak et al., c).

6.2 Multilayer optics metrology

The high-order harmonic source was also used for reflectivity measurements of multilayer mirror (ML) versus the incidence angle. The experimental setup is presented in Fig. 11. The driving laser field featured the parameters: λ_c=820 nm, Δt=30 fs, E=4.5 mJ. For efficient HHG generation the beam was frequency doubled with BBO crystal with conversion efficiency of ~27 %. Subsequently, the two beams were focused by a spherical mirror (f=125 mm) into a gas cloud created by a helium gas jet at 3 bar backing pressure. The gas jet was 9 mm long. To filter out IR laser light 200 nm Al filter was used. The high-order harmonic source was optimized for the strongest harmonic line at 21.6 nm. The mirror comprised molybdenum-silicon multilayer structure deposited on BK7 substrate. It was designed for operation at 45° and 21.6 nm wavelength. The mirror was placed on the rotational stage together with the XUV sensitive absolutely calibrated photodiode allowing change of the incidence angle β under vacuum. The photodiode is additionally covered with 150 nm Al filter. The signal was acquired with an oscilloscope.

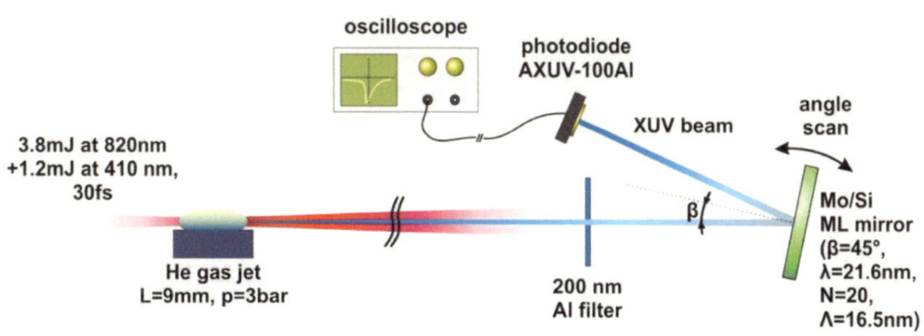

Fig. 11. Experimental setup of measurement of reflectivity of multilayer mirror versus incidence angle (Jakubczak c)).

Experimental results are presented in Graph 1. The measured and simulated reflectivity for the mirror are presented. The measured reflectivity was 31.5 % (fitted value) and simulated ~29 %. The insignificant difference could be due to the imperfections of manufacturing process as well as because of the possible additional errors (e.g. source instability or estimation error when taking into account additional input from neighboring harmonics).

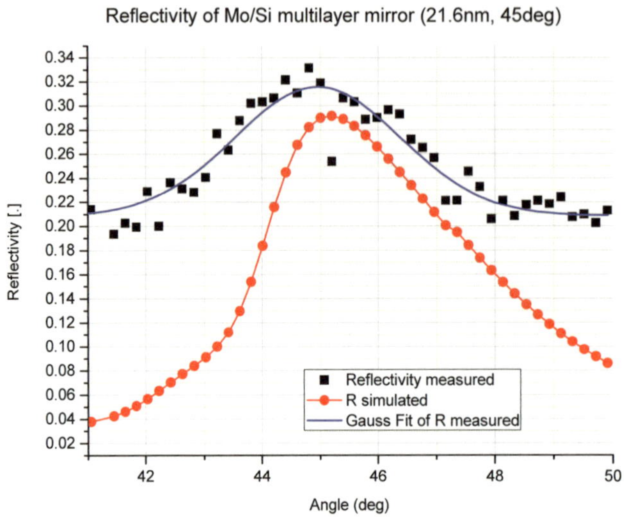

Graph 1. Reflectivity measurement of Mo/Si flat multilayer mirror designed for 21.6 nm radiation incident at 45° angle.

7. Conclusion

High-order harmonics from their first observation have been extensively investigated resulting in a good understanding of physical processes involved in their generation. The main physical processes can now be well understood on the basis of the three step model proposed by P. B. Corkum (Corkum, 1993). Due to the great advent in the ultrafast lasers technology, impressive progress has been recently observed in the efficiency of HHG, cut-off extension to shorter-wavelength region (even to hard X-rays) and generation of attosecond pulses. This also allowed increase of energy of XUV pulses obtained from HHG by means of two-color HHG as well as amplification of HHG in XRL plasma amplifiers. Having at hand sub-μJ pulses of coherent pico- and femtosecond XUV pulses it became feasible to find practical applications of the source. Some of them cover investigation of ultrafast processes in atoms, others; however, are of very practical use in, e.g., structuring of surface of the materials or multilayer optics metrology. Future perspectives for HHG sources are very promising. Additionally, there is lots of activity observed in the generation of harmonics from vacuum-solid interface. Both techniques require; however, further improvements in femtosecond laser technology - especially in terms of pulse intensity and pulse contrast. For instance, efficient HHG from plasma at the interface of the vacuum and solid target intensities higher than 10^{17} W/cm^2 are required with the pulse contrast greater than 10^6. Moreover, generation of single attosecond pulses requires carrier-envelope stabilization of the driving laser field. Finally, in order to extend cut-off in HHG from gases there are built femtosecond laser systems in the mid-infrared spectral region what found its motivation from the fact that cut-off spectral position is proportional to $I \cdot \lambda^2$ (Eq. 3).

The HHG source is assessed to be a very promising for needs of the future ultra compact sources of coherent radiation in XUV, soft X-ray and even X-ray spectral regions. It should be stressed that at the moment there is no other source of short-wavelength coherent radiation and of ultrashort pulse duration at the same time. The source would be also available at laboratory scale providing an unprecedented tool for investigation of the ultrashort physical processes. It should be stated; however, that high-order harmonics cannot substitute every kind of XUV radiation source. As an example, this type of the source is not of practical use in plasma backlighting.

8. Acknowledgments

This work was partially supported by the Czech Science Foundation (grants 202/07/J008, 202/08/1734), by the Czech Ministry of Education, Youth and Sports (project 7E09113), and by the Academy of Sciences of the Czech Republic (projects KAN300100702, M100100911).

9. References

Ammosov, M. V., Delone, N. B., Krainov, V. P., (1986). *Sov. Phys. - JETP*, Vol. 64, 1191.

Balcou, P., Haroutunian, R., Sebban, S., Grillon, G., Rousse, A., Mullot, G., Chambaret, J.-P., Rey, G., Antonetti, A., Hulin, D., Roos, L., Descamps, D., Gaarde, M. B., L'Huillier, A., Constant, E., Mevel, E., von der Linde, D., Orisch, A., Tarasevitch, A., Teubner, U., Klöpfel, D., and Theobald, W., (2002). High-order-harmonic generation: towards laser-induced phase-matching control and relativistic effects, *Appl. Phys. B*, Vol. 74, pp. 509-515.

Baltuska, A., Udem, Th., Uiberacker, M., Hentschel, M., Goulielmakis, E., Gohle, Ch., Holzwarth, R., Yakovlev, V. S., Scrinzi, A., Hänsch, T. W., & Krausz, F. (2003). Attosecond control of electronic processes by intense light fields", *Nature*, Vol. 421, pp. 611-615 .

Baudach, S., Bonse, J., Krüger, J., Kautek, W., (2000). Ultashort pulse laser ablation of polycarbonate and polymethylmethacrylate, *Appl. Surf. Sci.*, Vol. 154-155, pp. 555-560.

Becker, U., and Shirley, D. A., (1996). VUV and Soft X-Ray Photoionization, *Plenum*, New York.

Bezzerides, B., Jones, R. D., and Forslund, D. W., (1982). Plasma Mechanism for Ultraviolet Harmonic Radiation Due to Intense CO_2 Light, *Phys. Rev. Lett.*, Vol. 49, No. 3, pp. 202 - 205 (1982).

Boyd, R. W., (2003). *Nonlinear Optics*, Second Edition, Academic Press.

Brabec, T. and Krausz, F., (2000). Intense few-cycle laser fields: Frontiers of nonlinear optics, *Rev. Mod. Phys.*, Vol. 72, No. 2, pp. 545-591.

Carman, R. L., Forslund, D. W., and Kindel, J. M., (1981). Visible Harmonic Emission as a Way of Measuring Profile Steepening, *Phys. Rev. Lett.*, Vol. 46, No. 1, pp. 29 - 32.

Chalupsky, J., Juha, L., Kuba, J., Cihelka, J., Hakova, V., Koptyaev, S., Krasa, J., Velyhan, A., Bergh, M., Caleman, C., Hajdu, J., Bionta, R. M., Chapman, H., Hau-Riege, S. P., London, R. A., Jurek, M., Krzywinski, J., Nietubyc, R., Pelka, J. B., Sobierajski, R., Meyer-ter-Vehn, J., Krenz-Tronnier, A., Sokolowski-Tinten, K., Stojanovic, N., Tiedtke, K., Toleikis, S., Tschentscher, T., Wabnitz, H., and Zastrau, U., (2007). Characteristics of focused soft X-ray free-electron laser beam determined by ablation of organic molecular solids, *Opt. Express*, Vol. 15, 6036.

Christov, P., Murnane, M. M., and Kapteyn, H. C., (1997). High-Harmonic Generation of Attosecond Pulses in the "Single-Cycle" Regime, *Phys. Rev. Lett.*, Vol. 78, No. 7, pp. 1251 - 1254.

Corkum, P. B., Burnett, N. H., and Brunel, F., (1989). Above-threshold ionization in the long-wavelength limit, *Phys. Rev. Lett.*, Vol. 62, No. 11, pp. 1259 - 1262 (1989).

Corkum, P. B., (1993). Plasma perspective on strong field multiphoton ionization, *Phys. Rev. Lett.*, Vol. 71, No. 13, 1994-1997.

Drescher, M., Hentschel, M., Kienberger, R., Uiberacker, M., Yakovlev, V., Scrinzi, A., Westerwalbesloh, Th., Kleineberg, U., Heinzmann, U., & Krausz, F., (2002). Time-resolved atomic inner-shell spectroscopy, *Nature*, Vol. 419, pp. 803-807.

Dromey, B., Zepf, M., Gopal, A., Lancaster, K., Wei, M. S., Krushelnick, K., Tatarakis, M., Vakakis, N., Moustaizis, S., Kodama, R., Tampo, M., Stoeckl, C., Clarke, R., Habara, H., Neely, D., Karsch, S., and Norreys, P., (2006). High harmonic generation in the relativistic limit, *Nature Physics*, Vol. 2, pp. 456 - 459.

Hay, N., Velotta, R., Lein, M., de Nalda, R., Heesel, E., Castillejo, M., and Marangos, J. P., (2002). High-order harmonic generation in laser-aligned molecules, *Phys. Rev. A*, Vol. 65, 053805.

Hentschel, M., Kienberger, R., Spielmann, Ch., Reider, G. A., Milosevic, N., Brabec, T., Corkum, P., Heinzmann, U., Drescher, M., Krausz, F., (2001). Attosecond metrology, *Nature*, Vol. 414, pp. 509-513.

Horcas, I., Fernandez, R., Gomez-Rodriguez, J. M., Colchero, J., Gomez-Herrero, J., and Baro, A. M., (2007). WSxM Software, *Rev. Sci. Instrum.*, Vol. 78 , 013705.

Itatani, J., Quéré, F., Yudin, G. L., Ivanov, M. Yu., Krausz, F., and Corkum, P. B., (2002). Attosecond Streak Camera, *Phys. Rev. Lett.*, Vol. 88, No. 17, 173903-1.

Itatani, J., Levesque, J., Zeidler, D., Niikura, H., Pépin, H., Kieffer, J. C., Corkum, P. B., and Villeneuve, D. M., (2004). Tomographic imaging of molecular orbitals, *Nature*, Vol. 432, pp. 867-871.

Jaeglé, P., (2006). *Coherent Sources of XUV Radiation. Soft X-Ray Lasers and High-Order Harmonic Generation*, Springer.

Jakubczak, K., a)- Doctoral Thesis.

Jakubczak, K., Mocek, T., Chalupsky, J., Lee, G. H., Kim, T. K., Park, S. B., Nam, Ch. H., Hajkova, V., Juha, L., and Rus, B., b). Enhanced surface structuring by XUV/NIR ultrafast dual action - in preparation.

Jakubczak, K., Mocek, T., Rus, B., Polan, J., Hrebicek, J., Sawicka, M., Sobota, J., Fort, T., Pina, L., c). Beam properties of fully optimized, table-top, coherent source at 30 nm - Optoelectronics Review - accepted.

Juha, L., Bittner, M., Chvostova, D., Krasa, J., Kozlova, M., Pfeifer, M., Polan, J., Präg, A. R., Rus, B., Stupka, M., Feldhaus, J., Letal, V., Otcenasek, Z., Krzywinski, J., Nietubyc, R., Pelka, J. B., Andrejczuk, A., Sobierajski, R., Ryc, L., Boody, F. P., Fiedorowicz, H., Bartnik, A., Mikolajczyk, J., Rakowski, R., Kubat, P., Pina, L., Horvath, M., Grisham, M. E., Vaschenko, G. O., Menoni, C. S., and Rocca, J. J., (2005). Short-wavelength ablation of molecular solids: pulse duration and wavelength effects, *J. Microlith. Microfab. Microsyst.*, Vol. 4, 033007.

Kanai, T., Minemoto, S., and Sakai, H., (2007). Ellipticity Dependence of High-Order Harmonic Generation from Aligned Molecules, *Phys. Rev. Lett.* Vol. 98, 053002.

Keldysh, L. V., (1965). *Sov. Phys. - JETP*, Vol. 20, 1307.

Kienberger, R., Hentschel, M., Uiberacker, M., Spielmann, Ch., Kitzler, M., Scrinzi, A., Wieland, M., Westerwalbesloh, Th., Kleineberg, U., Heinzmann, U., Drescher, M., Krausz, F., (2002). Steering Attosecond Electron Wave Packets with Light, *Science,* Vol. 297, pp. 1144-1148 (2002).

Kienberger, R., Goulielmakis, E., Uiberacker, M., Baltuska, A., Yakovlev, V., Bammer, F., Scrinzi, A., Westerwalbesloh, Th., Kleineberg, U., Heinzmann, U., Drescher M. & Krausz F., (2004). Atomic transient recorder, *Nature,* Vol. 427, pp. 817-821.

Kienberger, R., Uiberacker, M., Kling, M. F., Krausz, F., (2007). Attosecond physics comes of age: from tracing to steering electrons at sub-atomic scales, *J. Modern Optics,* Vol. 54, No. 13-15, pp. 1985-1998.

Kim, I J., Kim, C. M., Kim, H. T., Lee, G. H., Lee, Y. S., Park, J. Y., Cho, D. J., and Nam, Ch. H., (2005). Highly Efficient High-Harmonic Generation in an Orthogonally Polarized Two-Color Laser Field, *Phys. Rev. Lett.*, Vol. 94, 243901.

Kim, I J., Lee, G. H., Park, S. B., Lee, Y. S., Kim, T. K., Nam, Ch. H., Mocek, T., and Jakubczak, K., (2008). Generation of submicrojoule high harmonics using a long gas jet in a two-color laser field, *Appl. Phys. Lett.*, Vol. 92, 021125.

Krzywinski, J., Sobierajski, R., Jurek, M., Nietubyc, R., Pelka, J. B., Juha, L., Bittner, M., Letal, V., Vorlicek, V., Andrejczuk, A., Feldhaus, J., Keitel, B., Saldin, E., Schneidmiller, E., Treusch, R., and Yurkov, M., (2007). Conductors, semiconductors, and insulators irradiated with short-wavelength free-electron laser, *J. Appl. Phys.*, Vol. 101, 043107.

L'Huillier, A., and Balcou, Ph., (1993). High-order harmonic generation in rare gases with a 1-ps 1053-nm laser, *Phys. Rev. Lett.,* Vol. 70, No. 6, pp. 774 - 777.

Lorin, E., Chelkowski, S., and Bandrauk, A. D., (2008). Attosecond pulse generation from aligned molecues - dynamics and propagation in H^{2+}, *New Journal of Physics,* Vol. 10, 025033.

Macklin, J. J., Kmetec, J. D., Gordon, C. L., (1993). High-order harmonic generation using intense femtosecond pulses, *Phys. Rev. Lett.*, Vol. 70, No. 6, pp. 766-769.

Mairesse, Y., and Quéré, F., (2005). Frequency-resolved optical gating for complete reconstruction of attosecond bursts, *Phys Rev. A,* Vol. 71, pp. 1-4, 011401.

Mocek, T., Rus, B., Kozlova, M., Stupka, M., Präg, A. R., Polan, J., Bittner, M., Sobierajski, R., and Juha, L., (2006). Focusing a multimillijoule soft x-ray laser at 21 nm, *Appl. Phys. Lett.*, Vol. 89, 051501.

Mocek, T., Polan, J., Homer, P., Jakubczak, K., Rus, B., Kim, I J., Kim, C. M., Lee, G. H., Nam, Ch. H., Hajkova, V., Chalupsky, J., and Juha, L., (2009). Surface modification of organic polymer by dual action of extreme ultraviolet/visible-near infrared ultrashort laser pulses, *J. Appl. Phys.*, Vol. 105, 026105.

Niikura, H., Légaré, F., Hasbani, R., Bandrauk, A. D., Ivanov, M. Yu., Villeneuve, D. M., and Corkum, P. B., (2002). Sub-laser-cycle electron pulses for probing molecular dynamics, *Nature*, Vol. 417, pp. 917-922.

Papadogiannis, N. A., Witzel, B., Kalpouzos, C., and Charalambidis, D., (1999). Observation of Attosecond Light Localization in Higher Order Harmonic Generation, *Phys. Rev. Lett.*, Vol. 83, No. 21, pp. 4289 - 4292.

Paul, P. M., Toma, E. S., Breger, P., Mullot, G., Augé, F., Balcou, Ph., Muller, H. G., Agostini P., (2001). Observation of a Train of Attosecond Pulses from High Harmonic Generation, *Science,* Vol. 292, pp. 1689-1692.

Pfeifer, T., Spielmann, C., and Gerber, G., (2006). Femtosecond x-ray science, *Rep. Prog. Phys.*, Vol. 69, pp. 443-505.

Quéré, F., Thaury, C., Monot, P., Dobosz, S., and Martin, Ph., (2006). Coherent Wake Emission of High-Order Harmonics from Overdense Plasmas, *Phys. Rev. Lett.*, Vol. 96, 125004-1.

Quéré, F., Thaury, C., Geindre, J-P., and Martin, Ph., (2008). Comment on "Transition to the Relativistic Regime in High Order Harmonic Generation", *Phys. Rev. Lett.* Vol. 100, 089401.

Remetter, T., Johnsson, P., Mauritsson, J., Varjú, K., Ni, Y., Lépine, F., Gustafsson, E., Kling, M., Khan, J., López-Martens, R., Schafer, K. J., Vrakking, M. J. J. and L'Huillier, A., (2006). Attosecond electron wave packet interferometry, *Nature Physics*, Vol. 2, pp. 323 - 326.

Roentgen, W. C., (1896). On a new kind of rays, *Nature*, Vol. 53, pp. 274-276.

Rundquist, A., Durfee, Ch. G., Chang, Z., Herne, C., Backus, S., Murnane, M. M., Kapteyn, H. C. (1998). Phase-Matched Generation of Coherent Soft X-rays, *Science*, Vol. 280, pp. 1412 - 1415.

Sakai, H., Miyazaki, K., (1995). High-order harmonic generation in nitrogen molecules with subpicosecond visible dye-laser pulses, *Appl. Phys. B*, Vol. 61, pp. 493-498 (1995).

Sansone, G., Benetti, E., Vozzi, C., Stagira, S., and Nisoli, M., (2008). Attosecond metrology in the few-optical-cycle regime, *New Journal of Physics*, Vol. 10, 025006.

Schenkel, B., Biegert, J., Keller, U., Vozzi, C., Nisoli, M., Sansone, G., Stagira, S., De Silvestri, S., and Svelto, O., (2003). Generation of 3.8-fs pulses from adaptive compression of a cascaded hollow fiber supercontinuum, *Opt. Lett.*, Vol. 28, No. 20, pp. 1987-1989 (2003).

Schultze, M., Goulielmakis, E., Uiberacker, M., Hofstetter, M., Kim, J., Kim, D., (2007). Powerful 170-attosecond XUV pulses generated with few-cycle laser pulses, *New Journal of Physics*, Vol. 9, Vol. 243, pp. 1-11.

Sola, J., Mével, E., Elouga, L., Constant, E., Strelkov, V., Poletto, L., Villoresi, P., Benedetti, E., Caumes, J.-P., Stagira, S., Vozzi, C., Sansone, G., and Nisoli M., (2006). Controlling attosecond electron dynamics by phase-stabilized polarization gating, *Nature Physics*, Vol. 2, pp. 319 - 322.

Tarasevitch, A., Lobov, K., Wünsche, C., and von der Linde, D., (2007). Transition to the Relativistic Regime in High Order Harmonic Generation, *Phys. Rev. Lett.*, Vol. 98, 103902.

Thaury, C., Quéré, F., Geindre, J.-P., Levy, A., Ceccotti, T., Monot, P., Bougeard, M., Réau, F., d'Oliveira, P., Audebert, P., Marjoribanks, R., & Martin, Ph., (2007). Plasma mirrors for ultrahigh-intensity optics, *Nature Physics*, Vol. 3, pp. 424 - 429.

Vaschenko, G., Etxarri, A. G., Menoni, C. S., Rocca, J. J., Hemberg, O., Bloom, S., Chao, W., Anderson, E. H., Attwood, D. T., Lu, Y., and Parkinson, B., (2006). Nanometer-scale ablation with a table-top soft x-ray laser, *Opt. Lett.*, Vol. 31, 3615.

Velotta, R., Hay, N., Mason, M. B., Castillejo, M., and Marangos, J. P., (2001). High-Order Harmonic Generation in Aligned Molecules, *Phys. Rev. Lett.*, Vol. 87, 183901.

Véniard, V., Taïeb, R., and Maquet, A., (1996). Phase dependence of (N+1)-color (N > 1) ir-uv photoionization of atoms with higher harmonics, *Phys. Rev. A*, Vol. 54, No. 1, pp. 721-728 (1996).

5

Burst Mode Q:switched Laser Pulses for Plasma Excitation in LIBS Analysis

Luis Ponce[1], Lesther Moreira[2], Eduardo de Posada[1], Miguel Arronte[1], Teresa Flores[1] and Eugenio Rodriguez[1]
[1]*CICATA-Altamira km 14,5 Carretera Tampico Puerto industrial, Altamira 89600, Tamps,*
[2]*IMRE-Havana University, Vedado 10400, C. Habana,*
[1]*México*
[2]*Cuba*

1. Introduction

The Laser Induced Breakdown Spectroscopy (LIBS), is a technique that has been firmly established for the rapid determination of the elemental composition (Cremers, 2006). It relies on material ablation by using a short duration laser pulse with high density energy enough to produce plasma. By analyzing the light emitted by the plasma, it is possible to determine the elemental composition of practically any material. This technique has significant advantages over other conventional analytical techniques (Cremers, 2007; Rusak et al., 1998; Song et al., 1997). For example, requires no sample preparation, may be performed in solid, liquid or gas phase in samples with any shape or dimension. Moreover, it allows an in-depth study in order to characterize the composition of multi-component material (Adamson et al., 2007; Radziemski et al., 1983). It is especially suitable for field work by offering the possibility of real-time analysis with high portability. Thanks the above mentioned advantages, the LIBS technique has experienced strong growth, which is reflected in a large and growing number of publications.

LIBS is not a new technique: firsts laser-induced breakdown studies go back to the early 1960s and important application studies date from the 1980s with the work of Radziemski (Tognoni et al., 2002). A comprehensive review of LIBS development and applications through the mid-1990s was produced by Rusak *et al.(Rusak et al., 1997).*

The technique has many attributes that make it an attractive tool for chemical analysis, particularly as regards its potential as a field-portable sensor for geochemical analysis. LIBS is relatively simple and straightforward, so skilled analysts are not required. Little to no sample preparation is required, which eliminates the possibility of adulteration of the sample through improper handling or storage or cross-contamination during sample preparation. LIBS provides a real-time response and simultaneous multi-element detection and analysis. The laser plasma is formed over a very limited spatial area, so that only a very small amount of sample (picograms to nanograms) is engaged in each laser micro-plasma event. All components of the instrument can be made small and rugged for field use and LIBS sensors can be operated either as a point sensor or in a standoff detection mode. The

detection limits of LIBS are in the low hundreds to tens of ppm range for most common elements, so field-portable LIBS should be capable of field surveying and screening for the geochemical exploration and environmental remediation applications envisaged.
The existing equipment on the market used predominantly solid-state lasers based on Nd:YAG crystals. This is because this type of lasers, in addition to the advantages of robustness and compactness; allow obtaining very short pulses with high power density. To ensure the appropriate parameters, it works in the so-called Q: Switch regime. In this mode of operation, is introduced into the laser cavity an optical shutter based on a crystal whose transmittance is electrically switched, allowing the production of single-pulse of several nanoseconds in duration and very high intensity. In the last years several works have reported an improvement of plasma intensity, and therefore the limits and quality of detection, using systems that generate two consecutive pulses, delivered from one or two synchronized lasers (Tognoni et al., 2002). Recently, it was reported a system with dual crystal modulator by using two LiF crystals (Galbacs et al., 2005). While this system improved the signal-noise relation and the intensity of the plasma emission, has the disadvantage associated with the gradual degradation of these crystals (Zabello et al., 2003).
In our study we propose a new low cost LIBS system, which uses passively Q:Switched, capable to produce pulse trains that enhances the plasma intensity. The employment of this regime is illustrated by LIBS characterization of art object. On the other hand, as an example of on-line monitoring of industrial process, and also the possibilities for different pulse regimes, we present the applications of free-running regime in the de-thorning process of cactus, a food Mexican product.

2. A low cost and portable LIBS device description

Figure 1 shows an outline of the experimental setup. A Nd:YAG laser emitting at 1064 nm, produces pulses with energy adjustable up to 0.9 J. The active element is a 6 x 60 mm Nd:YAG rod, pumped by a xenon lamp. It uses a Q:Switch consisting of a Cr:YAG crystal with 6 mm in diameter and 3 mm in thick and initial transmittance of 14%. The light emitted by laser is focused through a 5 mm focal length lens. The distance between the lens and the surface of the sample was 8.5 cm to achieve an area of 0.19 cm^2. For the experimental conditions used, this means a yield of 4.7 J/cm^2. The sample is placed on metallic base ensuring that the sample surface is located at the focal point.
An optical system coupled with the entry of optical quartz fiber, whose entry can be moved respect to the sample surface. By this way it is possible to capture the emission from a specific region of the plasma.
The spectrometer used in our configuration was an USB4000 Ocean Optics, with 350 nm to 900 nm spectral range and 0.2 nm resolution. The spectra were processed in computer by using Spectra Suite software.

2.1 Nd:YAG laser with passive Q:Switch

In the laser head of our device the pump radiation is produced by xenon pulse lamp. The light from this lamp is concentrated on the active medium of Nd:YAG through an special reflector made from monolithic quartz with external metallic coating. The quartz block is doped with atomic Ce at 1 % doping level. This feature avoids the optical damage of the active medium due the ultraviolet light emitted by xenon lamp. On the other hand, the

conversion of UV to visible radiation contributes to enhance the absorption of pump light in laser crystal and, in consequence, the laser efficiency.
For obtaining a compact device, the resonator mirrors of 99,8% and 50 % reflectance respectively, were placed at 1.5 cm form the rod ends. The rod dimensions are 5 x 50 mm. As a Q:Switch element a YAG:Cr crystal with dimensions 6 mm in diameter and 3 mm thick was used. The initial transmittance of Q:Switch was 21 % for 1064 nm wavelength.

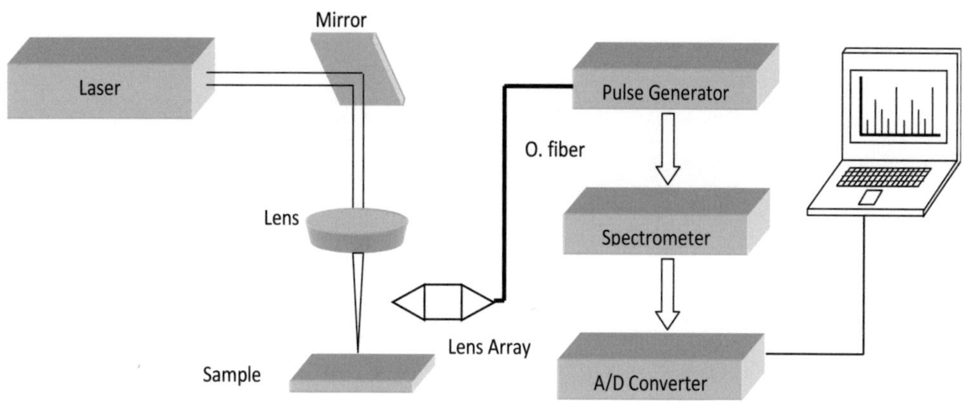

Fig. 1. Schematic of experimental setup

For this configuration, the laser can produce a train of pulses with duration for each pulse of about 20 ns, separated by 10 µs interval, with a total duration of the pulse train about 700 µs. In Figure 2 the acquired signal of laser pulse for passive Q:Switched laser is showed.

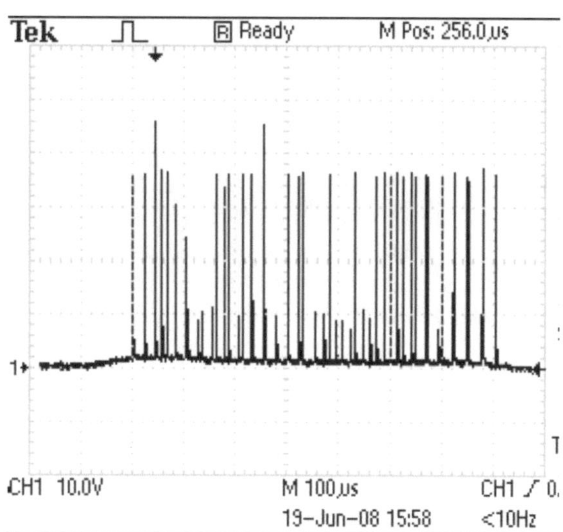

Fig. 2. Oscilogramme of pulse train obtained with passive Q:Switch.

All system elements are mounted on a stereo-microscope in order to obtain better facility for positioning and observation of samples as well as to follow the results of irradiation. The dimensions and weight of LIBS device is 40 x 20 x 15 cm and 5 kg respectively. Figure 3 shows a photograph of device, called MicroLIBS.

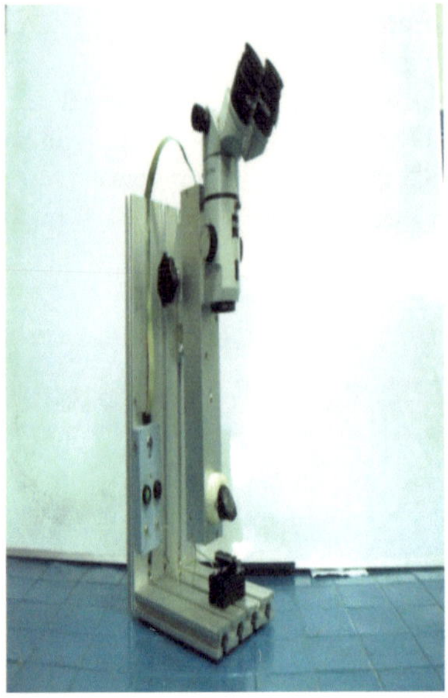

Fig. 3. Photograph of MicroLIBS system.

3. Results and discussion

The use of LIBS as an analytical tool depends on three underlying assumptions: (i) that material ablation is stoichiometric so that the composition of the plasma generated is fully representative of the sample composition; (ii) that an optically thin plasma is generated so that emission from the central high-temperature portion of the plasma plume is not reabsorbed in the colder plasma boundary region; and (iii) the plasma is in local thermodynamic equilibrium. These conditions have been verified by Chan & Russo (Chan & Russo, 1991) and Corsi *et al.* (Corsi et al., 2000) and it is understood how to realize these conditions analytically.
The use of LIBS for elemental detection is simple and readily accomplished via peak matching against a spectral library constructed in advance for a specific application. For this type of application, which can be accomplished through statistical analysis of LIBS spectra (Anzano et al., 2000), it is only relative peak intensities and overall spectral shape that is important, not absolute peak intensities. Quantitative LIBS analysis of specific elements in natural materials is significantly more difficult because of the broad issue known as "matrix effects". This is an observed dependence of the intensity of the LIBS emission response to the nature of the material analysed (Eppler et al., 1996), which manifests as variations in laser-target coupling and resultant plasma intensity variations. These two facets of the sample, which are generally lumped together as "matrix effects"(Aguilera et al., 2009; Corsi et al., 2000; Ctvrtnickova et al.; Lu et al., 2010; Ma et al.; Mohamed, 2008; Shen et al., 2009; Windom & Hahn, 2009), will determine LIBS signal reproducibility, namely sample composition and sample surface character.

3.1 Multi-pulse excitation

An important approach for enhance the detection limits and the signal-noise relation was reported early (Ponce et al., 2008), and consist in the use of passively Q:Switched multi-pulse scheme in order to perform the plasma formation-excitation process. In order to

establish a comparison, it was used an electro-optically Q:Switched Nd:YAG laser with similar parameters with the passively Q:Switched laboratory-made device. In the first case, the laser pulse energy of 0.14 J was focused on an area of 1 mm diameter, thus obtaining a fluency of 4.6 J/cm^2, similar to that used in the passively Q:Switched laser. The only experimental difference: Pulse duration of 20 ns in single pulse configuration and 700 µs total duration for the multi-pulse configuration. In Figure 4 are observed an example of spectra captured for both lasers. The spectra chosen to illustrate the comparison were collected form Prickly Pears spines. As can be seen, are clearly identified several peaks associated with the CaII, MgII and CI. These peaks, like the rest of the spectrum, have higher intensity and better signal-noise relation for a train of pulses configuration.
These results can be explained through a several steps process (Zabello et al., 2003). Initially, the absorption occurs in the first laser pulse at the sample surface, with the consequent overheating above the melting point. This causes the explosive material ablation, his rupture and rapid heating of the surrounding atmosphere, and a strong electronic emission. The steam flow expands and produces a shock wave, which in its motion dragging behind an area of low pressure. The initial pressure is restored after about 100 µs, as estimated experimentally (Kabashin et al., 1990).

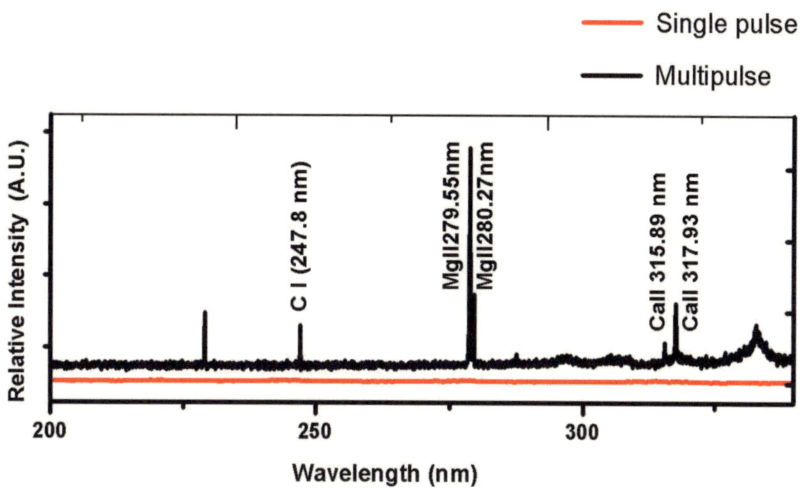

Fig. 4. LIBS spectra captured by using multipulse (above) and single-pulse (below) configurations.

By influencing subsequent pulses, separated from each other by a few tens of microseconds, the conditions for these interactions are the same as the first except for the presence of the above mentioned area of low pressure and a broadcast electronics (noise), much lower, because at that time the flow of electrons has disappeared. Moreover, the highest line intensity for multipulse regime can be explained by the additional excitation pulses produced in the steam of initially ejected material. Thus, the use of pulse trains by using passive Q:Switch, produce higher emission intensities and betters signal noise rate.

3.2 Free-running excitation

Typically, the LIBS technique is performed using Q:Switched lasers with high power pulses (megawatt range), produced with time durations in nanosecond range. As far as we know, no LIBS experiments were performed by using free-running regime until the recent reports of LIBS in Prickly Pear (Flores et al., 2009), because the low energy density for this mode of laser operation. However, in the mentioned report, due to the strong absorption in areoles, an intense plasma emission takes place even for microsecond range in pulse duration with kilowatts/cm^2 range of laser fluency. As we demonstrated, this allows to use in experiments not only a Q:Switched laser but also a free running low cost Nd:YAG laser regime. The experiments in free-running regime have a great practical interest because at this regime is possible to obtain a higher productivity in spines elimination. On the other hand, the laser operating at this regime is less expensive and the operation cost is lower.

At present, our research group is working of high energy and long pulse LIBS excitation, performed at millisecond pulse time duration. The special interest of this exotic conditions in plasma generation, is explained by two reasons: The possibility of simulation of some natural events like interstellar jets or atmospheric thunders from one side, and, the on-line monitoring of several industrial applications in millisecond range as hole perforation or pulse welding from the other side.

In principle, as lower pulse duration them better resolution can be obtained, if we taking in to account the fact that typically with shorter pulse duration avoid higher power densities and better plasma ionization level. I also important the fact that the plasma life is extended no more than several microseconds, and, in consequence, if the pulse duration is longer, them the interaction with subsequent excitation must be taken in consideration in the LIBS spectra interpretation. As an example of line-width behaviour for different pulse duration, LIBS spectra captured on metal sample are showed in figure 5.

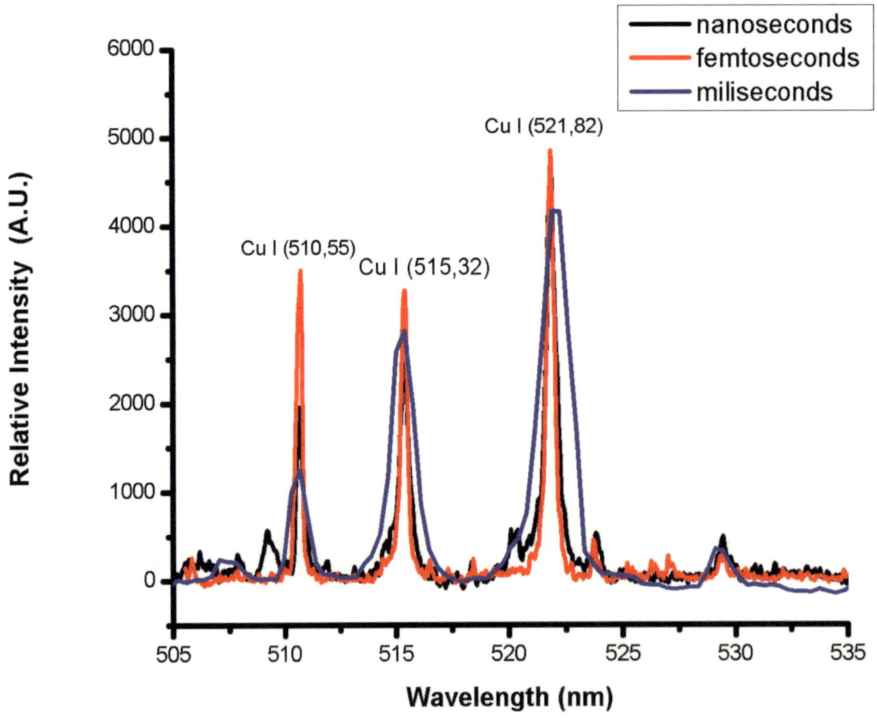

Fig. 5. LIBS spectra obtained from Cu sample.

3.3 Field applications: Free-running excitation for Cactus de-thorning

An exciting example of our LIBS device is the on-line elemental determination for laser de-thorning process. In this new application a Nd:YAG (Flores et al., 2009) laser is used for selective ablation of thorns of cactus called opuntia. In figure 6, is showed a deep-profile LIBS measurement of thorn-cortex systems illustrating the behavior of Na(I) 589.5 nm line and C-H band.

These curves shows that LIBS technique contribute to control the spine elimination process: For C-H line the relative intensity reach the maximal value after two or three pulses and decrease gradually until the full thorn elimination reaching in this moment a fixed value that correspond to lower percent of C-H concentration in cortex. In opposite direction, the Na line intensity grows gradually reaching a higher level simultaneously with C-H decrease. The high concentration of Na in cortex is reached after spine elimination precisely when the inner part of cactus is to be ablated.

On the other hand, it is another important difference between the two curves showed in figure 6. While the Na pike dependence versus number of pulses growth continuously, the C-H dependence shows periodical changes. It's remarkable that similar variations were detected in previous work (Arronte et al., 2010), by photo-acoustic technique when the signal intensity versus number of pulses was measured. In principle, these changes can be attributed to periodical character of combustion process in laser de-thorning, consisting in

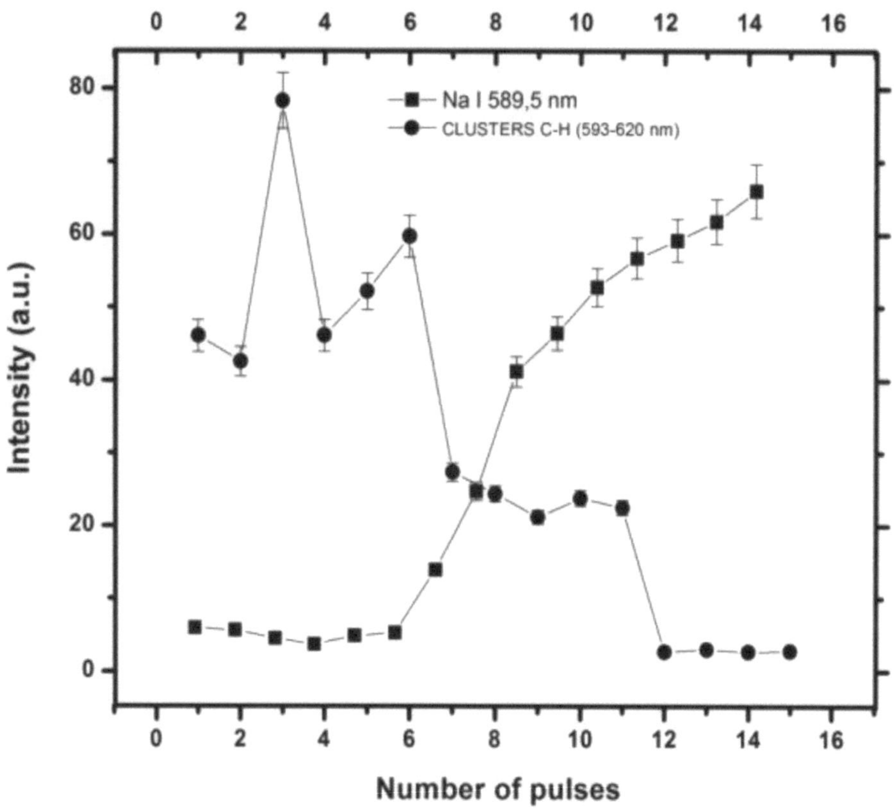

Fig. 6. Intensity profile curves Na I (589.5 nm) y emission band of the C-H clusters

several stages combustion-ablation: While the first pulse prepare the surface making it darker by combustion, the second pulse irradiate a surface with strong absorption and ablate it at very high rate.

3.4 Field applications: Bust mode Q:Switched excitation for LIBS analysis of art metal object

LIBS offers a potential alternative to other techniques used in art conservation, works of art, archaeology, etc., being a rapid elemental analysis technique, performed in situ and virtually non-destructive (Fotakis 2006). A LIBS's particularly interesting application is the analysis of metal objects with different compositional layers, particularly when trying to get in-depth composition profiles to distinguish the different elements present in a metal object and can develop a proper restoration process, among other uses, see Fig. 7a. Thus, the determination of the elemental composition, so far unknown, of the different layers present in a 20th Century Metal Jug, of Japanese origin, belonging to a private collection and the only one of its kind (see Figure 7b), is performed using a prototype LIBS equipment. For this purpose, successive burst mode laser pulses are applied to a well-localized area of the sample. The determination of the components avoid to the better selection of the restoration method to be used.

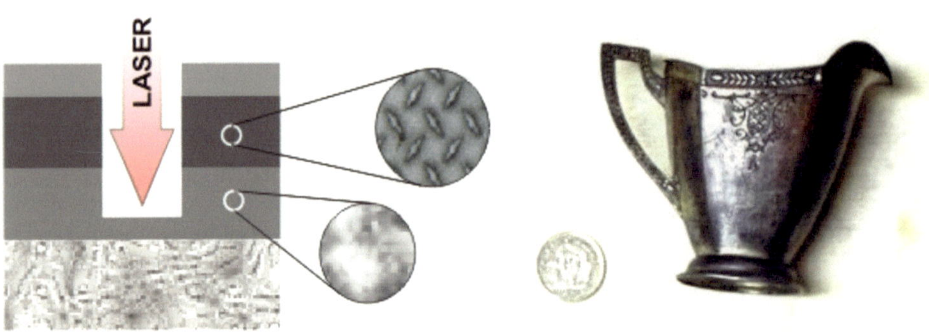

(a) (b)

Fig. 7. (a) An schematic of in-depth laser microanalysis performed by LIBS technique, (b) The picture of Japanese Jug.

A wide spectral window was selected, covering approximately 450 nm and that was centered at 575 nm, so that any elements contained in the metal object, including those elements that can arise by the influence of environmental factors such as Ca, Mg, Si, Al , Na and K, may be measured simultaneously.

The Jug was analyzed superficially in four different and distant points to determine the surface composition and ensure the minimum possible sampling and thereby minimize damage as a result of ablation. The normalized characteristic spectrum for each of the samples is shown in the Figure 8. As shown, certain elements are represented in each of the measurements performed. The Cu constitutes the fundamental component and in turn the one that shows major proportion in relation to the other materials. Other elements were found to a lesser extent such as Ca, Si, Na and K, including marked relative differences between their intensities, showing changes in their proportions. These elements are attributed to deposition of dirt and impurities on the surface of the sample. Another element determined is the Pb which, similarly to the other aforementioned elements, has lower intensity values as compared to Cu. Finally, O and N were determined essentially using the same behavior in all measurements, and their presence is due to mainly the combustion process during the laser-matter interaction with the atmosphere, among others. Each time a laser pulse is applied on the surface of the Jug, a portion of the material that makes up the surface is removed and plasma is generated. If the laser is applied all the time in one place, it will penetrate the material (see Fig. 7a). In order to obtain the composition profile and identify different elements and layers present in the Jug, a multispectral analysis of the emission of the plasma generated by each laser shot was performed. The procedure to determine each layer consisted of recording pulse-to-pulse variations in the intensity of emission lines characteristics of maximum amplitude of the obtained constituent elements of the surface: Ca (II) 392.83 nm, Pb (I) 405.56 nm, Cu (I) 521.24 nm, Si (II) 546.04 nm, Na (I) 588.67 nm and K (I) 765.93 nm. The intensity of these elements depending on the number of pulses is shown in the Fig. 9.

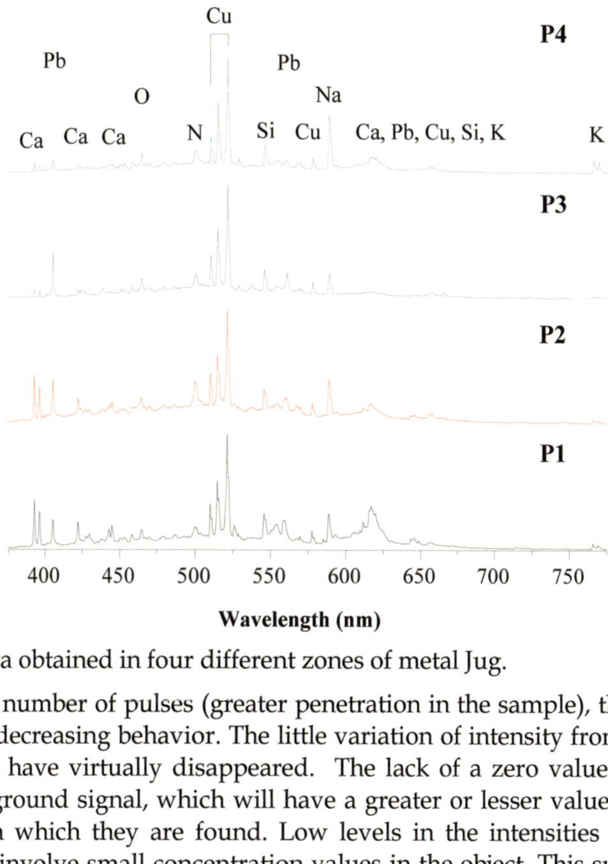

Fig. 8. LIBS spectra obtained in four different zones of metal Jug.

By increasing the number of pulses (greater penetration in the sample), the elements Ca, Si, Na, and K had a decreasing behavior. The little variation of intensity from pulse 4 indicates that the elements have virtually disappeared. The lack of a zero value is justified by the presence of background signal, which will have a greater or lesser value depending on the spectral region in which they are found. Low levels in the intensities of these elements, compared to Cu, involve small concentration values in the object. This supports the idea of the presence of elements resulting from the deposition of dirt and impurities on the Jug's surface. Metal objects often show heterogeneity and surface oxidation caused by heat treatment during construction and environmental degradation. These factors can cause variation in the morphology and elemental composition of ancient metal alloys, which results in pollution of surface and global object.

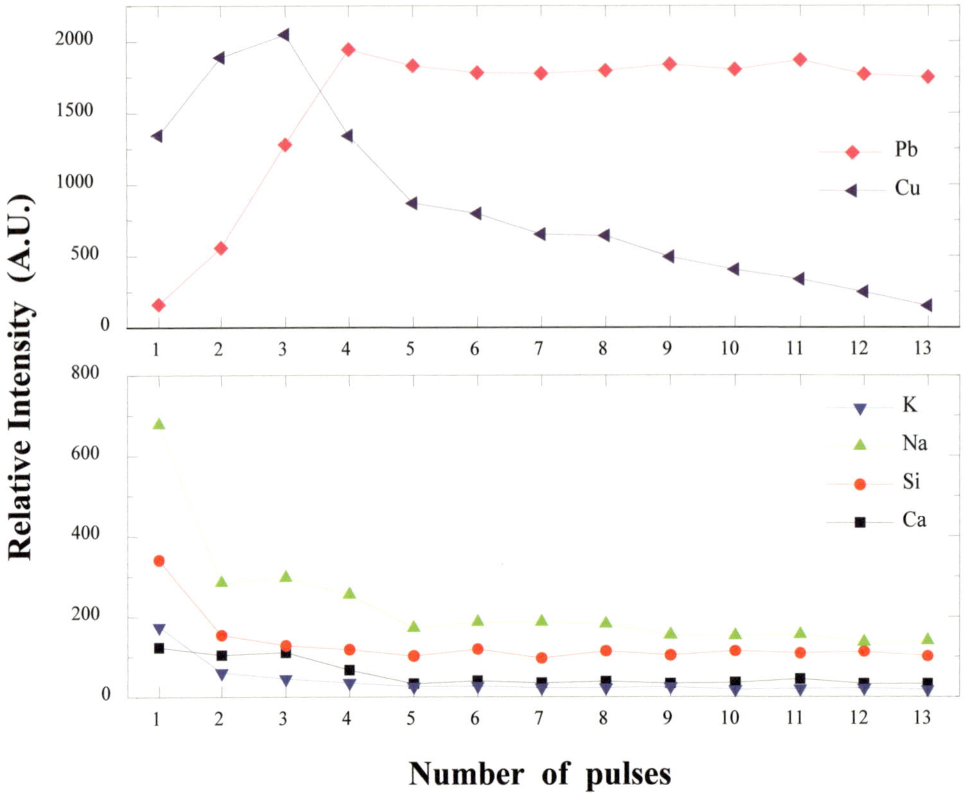

Fig. 9. In-depth profiles for Ca(II) 392.83 nm, Pb(I) 405.56 nm, Cu(I) 521.24 nm, Si(II) 546.04 nm, Na(I) 588.67 nm and K(I) 765.93 nm as function of number of pulses.

The Cu, during the first pulses (pulse 1 - pulse 3), shows increasingly higher intensity values, constituting the main element on the jug's surface (pulse 1) and, turn on, a small interior portion of the jug. The Pb, initially (pulse 1) at very small intensity levels, has a linear increase until coming to its maximum value after pulse 4. After this pulse, the intensity of Cu starts to decrease until reaching a practically zero value. On the other hand, the Pb maintains a stable behavior during the following pulses. After pulse 13, virtually all elements have disappeared being the Pb the majority element in the jug's composition. Figure 10 shows the superposition of the Jug's spectrum, after pulse 13, with a spectrum pattern of Pb showing a perfect match between these spectrums. It is, therefore, evident the presence of two phases or layers in the Jug: phase 1 with Cu as the majority element, and phase 2 for Pb.

On the other hand, the Cu-Pb interface, through the behavior between the intensity ratios of Cu/Pb and Pb, precisely on the pulse 4 is observed clearly in the Figure 11.

Precisely, the Pb reaches its maximum value and the ratio Cu/Pb becomes lesser than 1 which, from the following pulses, tends linearly to zero. While the existence of two layers in the depth profile of the Jug was determined, from Figure 11 can be seen that the Cu layer, in addition to Ca, Si, Na and K, there is the presence of Pb. This result comes from the thickness of Cu layer and the laser energy. While energy was selected in an optimal work area; nonetheless, given the thickness of Cu layer, this energy level remains relatively

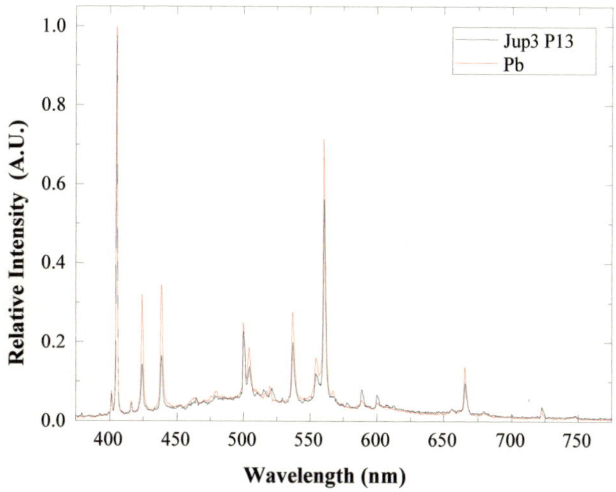

Fig. 10. Comparison between the spectra of metal Jug and Pb standard.

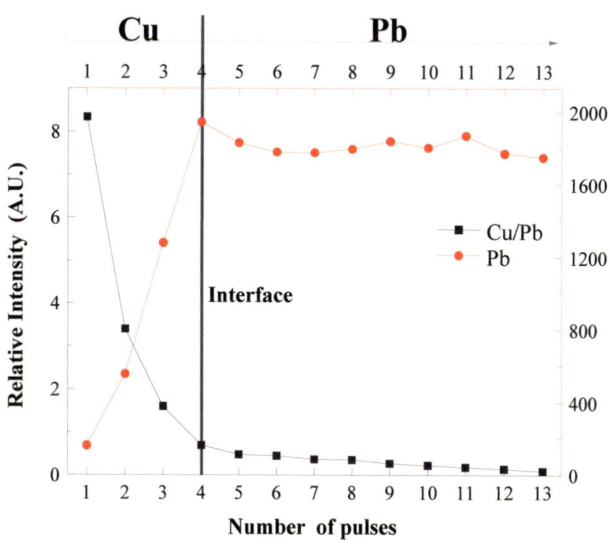

Fig. 11. In-depth behavior of Cu/Pb rate and Pb relative intensity.

sufficient to interact with the next layer (Pb) and generate also laser ablation. Similarly, the presence of Cu in the Pb layer can be seen in the Figure 9. In this case, the ratio is given in the form of how laser energy is distributed. The distribution of energy of the pulse laser is not exactly homogeneous (at the edges it is lower than in the center); so the amount of material removed from the irradiated region will be lower in edges than in the center, so that debris of the previous layer may be always remained as it moves in depth.[20]

From the crater produced by 13 laser pulses in interaction with the jug, the crater's wall can be seen, see Figure 12.

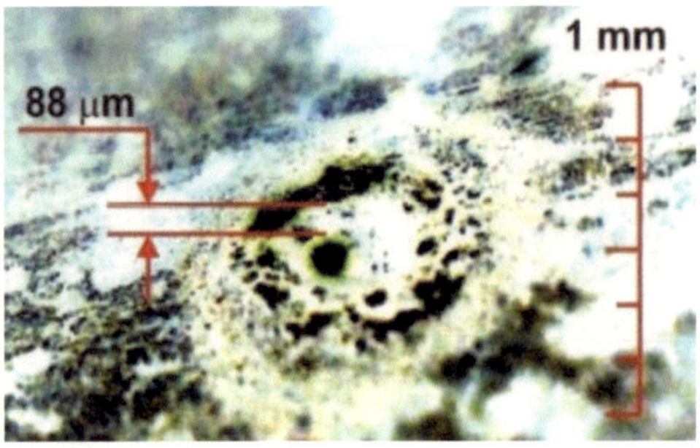

Fig. 12. Contrast image of laser treated zone captured by optical microscopy (magnification 60x).

Referencing the image scale, the approximate depth of the crater can be determined. Therefore, knowing the depth of the crater and the number of pulses, it is possible to determine how much material is extracted in every interaction of laser with the Jug and, in turn, determine the thickness of Cu layer. Making the necessary calculations, we find that the thickness of the crater's wall is approximately 88 μm, which implies, for a total of 13 laser pulses, a penetration factor of 6.8 μm/pulse. Therefore, as estimated, the thickness of Cu layer is 27 μm knowing that it ends after pulse 4.

7. Conclusions

Recent technological advances are leading to the development of fully field-portable LIBS systems as show MicroLIBS, the compact, portable and low cost LIBS device developed by IMRE-Havana University. The use of Cr:YAG crystal as Q:Switch element allows simplifying the design and reducing costs as compared to electro-optical Q:Switch and also to guarantee more duration that LiF elements. The application of pulse trains as an excitation source helps to achieve a higher intensity of plasma emission and a substantial improvement in the signal-to-noise ratio compared with single-pulse systems.

The advances in modification of laser parameters open new applications no limited to elemental determination but also covering the laser material processing on-line monitoring or the simulation of specific plasma events in wide range of energies or pulse durations.

Regarding the fields applications, two examples were given: On one side, the real-time monitoring of de-thorning process for one food product, and, on the other side, the compositional characterization of heritage object.

For the first field application, the burst-mode excited LIBS technique is shown to be a suitable method for new industrial processes. Additionally, the approach of complex band emission pattern recognition can be used for determining of contamination problems in such vegetable food.

For the case of metal Jug analysis, the multi-pulse regime shows the possibilities of the technique for heritage objects characterization. The presence of key elements (Cu and Pb) was determined. They are distributed in layers, being the Cu the surface element with an estimated thickness of 27 um. The elements Ca, Si, Na and K are impurities caused by dirt and the environment that influence the object's surface.

In resume, the multi-pulse excited LIBS technique serves as an in situ, low-cost and virtually non-destructive analytical tool, in the identification of elements and environmental monitoring or characterizing a wide spectrum of process and objects.

8. References

Adamson, M., Padmanabhan, A., Godfrey, G.J., Rehse, S.J., 2007, Laser-induced breakdown spectroscopy at a water/gas interface: A study of bath gas-dependent molecular species. Spectrochimica Acta Part B: Atomic Spectroscopy 62, 1348-1360.

Aguilera, J.A., Aragón, C., Madurga, V., Manrique, J., 2009, Study of matrix effects in laser induced breakdown spectroscopy on metallic samples using plasma characterization by emission spectroscopy. Spectrochimica Acta - Part B Atomic Spectroscopy 64, 993-998.

Anzano, J.M., Gornushkin, I.B., Smith, B.W., Winefordner, J.D., 2000, Laser-induced plasma spectroscopy for plastic identification. Polymer Engineering and Science 40, 2423-2429.

Arronte, M., Ortega, E., Ponce, L., de Posada, E., Rodriguez, E., Flores, T., 2010. Real-time monitoring of de-thorning process in Opunctia Nopalea by using a PILA technique. Acoustic Technique 1, 1-10.

Corsi, M., Palleschi, V., Salvetti, A., Tognoni, E., 2000, Making LIBS quantitative: A critical review of the current approaches to the problem. Research Advances in Applied Spectroscopy 1, 41-46.

Cremers, D.A., 2007, Remote Analysis by LIBS: Application to Space Exploration, In: Jagdish, P.S., Surya, N.T. (Eds.) Laser-Induced Breakdown Spectroscopy. Elsevier, Amsterdam, pp. 353-379.

Cremers, L.J.R.a.D.A., 2006, Laser-Induced Breakdown Spectroscopy (LIBS): Fundamentals and Applications.

Ctvrtnickova, T., Mateo, M.P., Yañez, A., Nicolas, G., Laser Induced Breakdown Spectroscopy application for ash characterisation for a coal fired power plant. Spectrochimica Acta - Part B Atomic Spectroscopy.

Chan, W.T., Russo, R.E., 1991, Study of laser-material interactions using inductively coupled plasma-atomic emission spectrometry. Spectrochimica Acta Part B: Atomic Spectroscopy 46, 1471-1486.

Eppler, A.S., Cremers, D.A., Hickmott, D.D., Ferris, M.J., Koskelo, A.C., 1996, Matrix effects in the detection of Pb and Ba in soils using laser-induced breakdown spectroscopy. Applied Spectroscopy 50, 1175-1181.

Flores, T., Ponce, L., Arronte, M., de Posada, E., 2009, Free-running and Q:Switched LIBS measurements during the laser ablation of Prickle Pears spines. Optics and Lasers in Engineering 47, 578-583.

Galbacs, G., Budavan, V., Geretovszky, Z., 2005, Multi-pulse laser-induced plasma spectroscopy using a single laser source and a compact spectrometer. Journal of Analytical Atomic Spectrometry 20, 974-980.

Kabashin, A.V., Konov, V.I., Nikitin, P.I., Prokhorov, A.M., Konjević, N., Vikor, L., 1990, Laser-plasma generation of currents along a conductive target. Journal of Applied Physics 68, 3140-3146.

Lu, Y.F., Shen, X.K., Ling, H., 2010. Laser-induced breakdown spectroscopy combined with spatial confinement of plasmas and laser-induced fluorescence for trace-materials detection. In, Shanghai, pp. 697-704.

Ma, Q.L., Motto-Ros, V., Lei, W.Q., Boueri, M., Zheng, L.J., Zeng, H.P., Bar-Matthews, M., Ayalon, A., Panczer, G., Yu, J., Multi-elemental mapping of a speleothem using laser-induced breakdown spectroscopy. Spectrochimica Acta - Part B Atomic Spectroscopy.

Mohamed, W.T.Y., 2008, Improved LIBS limit of detection of Be, Mg, Si, Mn, Fe and Cu in aluminum alloy samples using a portable Echelle spectrometer with ICCD camera. Optics and Laser Technology 40, 30-38.

Ponce, L., Flores, T., Arronte, M., Hernandez, L.C., Bilmes, G.M., Alvira, F., 2008, Laser Induced Breakdown Spectroscopy widh multi-pulse excitation. Revista Cubana de Física, 25, N2A, pp. 85-87.

Radziemski, L.J., Loree, T.R., Cremers, D.A., Hoffman, N.M., 1983, Time-resolved laser-induced breakdown spectrometry of aerosols. Analytical Chemistry 55, 1246-1252.

Rusak, D.A., Castle, B.C., Smith, B.W., Winefordner, J.D., 1997, Excitational, vibrational, and rotational temperatures in Nd:YAG and XeCl Laser-Induced plasmas. Spectrochimica Acta Part B: Atomic Spectroscopy 52, 1929-1935.

Rusak, D.A., Castle, B.C., Smith, B.W., Winefordner, J.D., 1998, Recent trends and the future of laser-induced plasma spectroscopy. TrAC - Trends in Analytical Chemistry 17, 453-461.

Shen, X.K., Ling, H., Lu, Y.F., 2009. Laser-induced breakdown spectroscopy with high detection sensitivity. In, San Jose, CA.

Song, K., Lee, Y.I., Sneddon, J., 1997, Applications of laser-induced breakdown spectrometry. Applied Spectroscopy Reviews 32, 183-235.

Tognoni, E., Palleschi, V., Corsi, M., Cristoforetti, G., 2002, Quantitative micro-analysis by laser-induced breakdown spectroscopy: A review of the experimental approaches. Spectrochimica Acta - Part B Atomic Spectroscopy 57, 1115-1130.

Windom, B.C., Hahn, D.W., 2009, Laser ablation - Laser induced breakdown spectroscopy (LA-LIBS): A means for overcoming matrix effects leading to improved analyte response. Journal of Analytical Atomic Spectrometry 24, 1665-1675.

Zabello, E., Syaber, V., Khizhnyak, A., 2003. Spectral-analytical characteristics of laser plasma under multi-pulse excitation regime. In, pp. 220-222.

Fotakis, C., Lasers in the Preservation of Cultural Heritage; Principles and applications, Institute of Physics Publishing , (New York, 2006).

6

Numerical Simulations of Temperature-dependence on Distributed Bragg Reflector (DBR) and Performance Analyses for Proton-Implant/Oxide Confined VCSEL: Comparison with Transmission Matrix, Matrix Calculating Methods and Macleod Model

Tzu-Chiang Chen
Chung Cheng Institute of Technology, National Defense University
Taiwan, Republic of China

1. Introduction

This chapter mainly focuses on the simulation for temperature-dependent Distributed Bragg Reflector (DBR) of 850nm vertical cavity surface emitting laser (VCSEL) with Transmission Matrix (TMM), Matrix Calculating Methods (MCM) and Macleod Model and performance for comparison with proton-implant/oxide confined process on VCSEL. Using well-developed temperature-dependent DBR-reflectivity solver with Mathcad simulator, we have successfully compared the Macleod Model simulator with theoretical self-developed solution based on the Transmission Matrix (TMM), Matrix Calculating Methods (MCM) and find very good agreement with previous results while accounting for influences of conjugated part of refractive index and graded Al compositions of DBR materials. Moreover, optoelectronic performance of Proton-Implant/Oxide Confined 850 nm VCSEL have been demonstrated on this chapter using temperature-dependent power output, voltage/injection current, transverse operating wavelengths, optical spectral characteristics, slope efficiency and transverse optical modes with an approximated Marcatili's method extracted and measurement from systematically measuring experiments. Through adequate and precise LD device design and processes, we have proposed the high performance single-mode proton implanted in contrast to the oxide confined 850 nm VCSEL. Under nominal temperature-variety and keeping operating temperature of 30 °C, the maximum power output of 10 micro meter aperture proton implanted VCSEL exceeds 5 mW while injecting current of 10 mA, and the threshold voltage, injecting current, peak-wavelength, differential resistance are 1.8 V, 3.2 mA, 851 nm and 36.8 ohm, respectively.

2. Research focuses

Vertical cavity surface emitting laser has a number of inherent advantages including low divergence of the circular beam, which is emitted from the top surface of the laser diode in order to efficiently couple into fiber for communication. One of the most unique characteristic operation of a VCSEL is that it operates in single longitudinal mode due to its natively short effective cavity length. And many researchers also try to achieve high efficiency, high power, single transverse mode operation and insensitive temperature feature in VCSEL for high-speed light-wave communication and higher information capacity of the communication-linking network. By far, the common compound semiconductor material used for fabricating the reflector mirrors of VCSEL is the AlGaAs system because of the large refractive index difference between the lattice matched GaAs and AlAs binaries. J. Talghader et.al. have investigated the thermal dependence of the refractive index of GaAs and AlAs measured using semiconductor multiplayer optical cavities (J. Talghader et.al., 2003). The VCSEL-type passive optical cavities exhibiting substantially different longitudinal optical mode shifts with temperature has been demonstrated. For conventional edge-emitting semiconductor lasers, the far-field and beam characteristics depend on laser structure. However, the detailed study of far-field modes and beam features for VCSEL with different window diameters (ω) and active-layer apertures (s) as function of injected current has been proposed at a certain temperature (Iga, K. et.al., 1984). In our study, the temperature-dependent reflectivity spectra of linear graded Al-composition DBR in 850 nm-VCSEL were simulated and analyzed using two-kinds of transfer matrix methods, which compared with the results from the multi-layer films evolution software of essential Macleod. To our knowledge, little work has yet investigated the thermal effect on spatial distributions of far-fields patterns. So, the influences of injected current and temperature-varying have been displayed in this work.

3. Theory and experimental setups

3.1 Description of VCSEL structure

The VCSEL studied are 1λ cavity length, GaAs quantum well (QW)-based structure with 15 μm-circular selectively oxided cavity apertures for current confinement. As seen in Fig.1, the DBR mirrors were formed with alternating layers of top graded mirrors of 20 pairs of $Al_xGa_{1-x}As/Al_yGa_{1-y}As$ ($x= 0\sim0.9$, $y= 0\sim0.12$) quarter-wave stacks and a bottom mirror of 34 pairs of $Al_yGa_{1-y}As/Al_xGa_{1-x}As$ quarter-wave stacks. The graded Al composition in DBR mirrors is to built electric field uniformly in the VCSEL structure which avoid sharp energy discontinuity between DBR and active region. Three periods of quantum wells are sandwiched with double DBRs and centered in active region; lasing occurs in the 850 nm

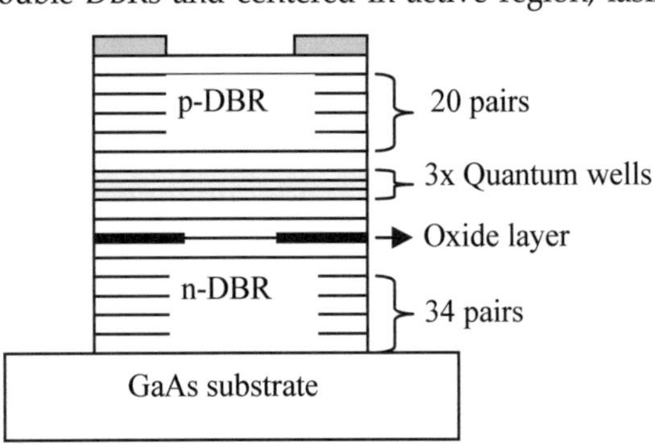

Fig. 1. Structure of the selectively oxidized VCSEL with circular aperture of 10μm in diameter and with active region of quantum well as the gain medium

range for the fundamental mode. The excess two thin layers of $Al_{0.98}Ga_{0.02}As$ are oxidized to efficiently form a circular aperture. The top and bottom DBR mirrors were doped moderately for decreasing the series resistance of VCSEL.

3.2 Calculation of refractive index of $Al_xGa_{1-x}As$

We use a semi-empirical method for calculating the room temperature refractive index of $Al_xGa_{1-x}As$ at energies below the direct band edge and this quantity is important in the design of GaAs heterostructure lasers. M. A. Afromowitz has used the interpolation scheme and compare the results of the calculation of refractive index for the $Al_xGa_{1-x}As$ with experimental data (M. A. Afromowitz, 1974). From this model, these equation of n(x,λ) reproduces as yielding,

$$n(x,\lambda)=\sqrt{\left[\frac{E_d(x)}{E_o(x)}+\frac{E_d(x)\times E_p^2(\lambda)}{E_o^3(x)}+\frac{E_d(x)\times E_p^4(\lambda)}{2\times E_o^3(x)\times\left(E_o^2(x)-E_\Gamma^2(x)\right)}\times\ln\left(\frac{2\times E_o^2(x)-E_\Gamma^2(x)-E_p^2(\lambda)}{E_\Gamma^2(x)-E_p^2(\lambda)}\right)\right]+1} \quad (1)$$

where n denotes the refractive of $Al_xGa_{1-x}As$, x : Al composition, λ: optical light wavelength, Ep: incident optical energy, E_0: effective oscillated energy, E_d: dispersion energy, and E_Γ: band gap energy. The equations for E_0, E_d , and E_Γ as the function of alloy composition yield

$$E_p(\lambda)=\frac{1.239852066\times10^3}{\lambda} \quad (2)$$

$$E_o(x)=3.65+0.871\times x+0.179\times x^2 \quad (3)$$

$$E_d(x)=36.1-2.45\times x \quad (4)$$

$$E_\Gamma(x)=1.424+1.266\times x+0.26\times x^2 \quad (5)$$

In this section, using two kinds of described methods for transfer matrix method simulate DBR mirrors (C. Chen, 2002). The detailed deduction of the models as follows:

3.3 Transmission matrix method, TMM (Furman, Sh.et.al., 2002)

Consider the optical waves are propagated along with z-direction as shown in Fig. 2 And the electric fields display as the next two equations:

$$A_j(z_2)=e^{-j\beta_i(z_2-z_1)}A_i(z_2) \quad (6)$$

$$B_j(z_2)=e^{j\beta_i(z_2-z_1)}B_i(z_2) \quad (7)$$

$$\beta_i=k_o n_i=k_o\left(n_{i,re}+n_{i,im}\right)\ k_o=\frac{2\pi}{\lambda} \quad (8)$$

where β_i is propagation constant, k_o : wave number, n_i : complex refraction index, $n_{i,re}$: refraction index, $n_{i,im}$: extinction coefficient

For the purpose of convenience, the equations (6) and (7) can be achieved as matrix type：

$$\begin{bmatrix} A_j(z_2) \\ B_j(z_2) \end{bmatrix} = \begin{bmatrix} e^{-j\beta_i L} & 0 \\ 0 & e^{j\beta_i L} \end{bmatrix} \begin{bmatrix} A_i(z_1) \\ A_i(z_1) \end{bmatrix} = P_i \begin{bmatrix} A_i(z_1) \\ A_i(z_2) \end{bmatrix} \tag{9}$$

where P_i is propagation matrix, $L = Z_2 - Z_1$,

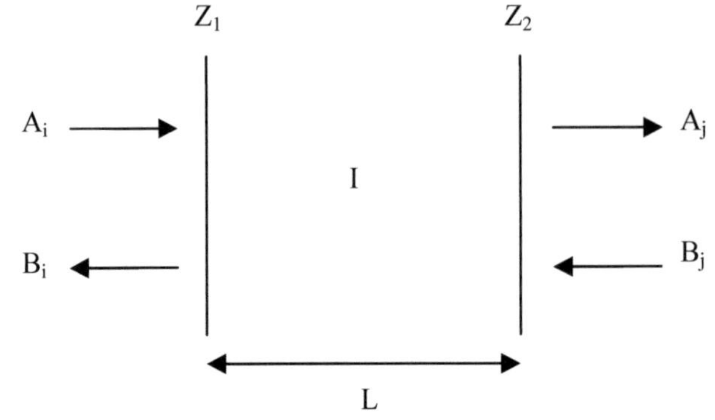

Fig. 2. Schematic diagram of wave propagation in region I

The optical waves are transmitted through the different interface regions as plotted in Fig.3, using the below boundary conditions:

$$A_j + B_j = A_i + B_i \tag{10}$$

$$-n_j A_j + n_j B_j = -n_i A_i + n_i + B_i \tag{11}$$

then let equation (10) and (11) exhibit as 2-dimensional matrix type,

$$\begin{bmatrix} A_j \\ B_j \end{bmatrix} = \begin{bmatrix} \dfrac{n_j + n_i}{2n_j} & \dfrac{n_j - n_i}{2n_j} \\ \dfrac{n_j - n_i}{2n_j} & \dfrac{n_j + n_i}{2n_j} \end{bmatrix} \begin{bmatrix} A_i \\ B_i \end{bmatrix} = T_{j,i} \begin{bmatrix} A_i \\ B_i \end{bmatrix} \tag{12}$$

in where $T_{j,i}$ is propagation between matrix medium i and medium j.

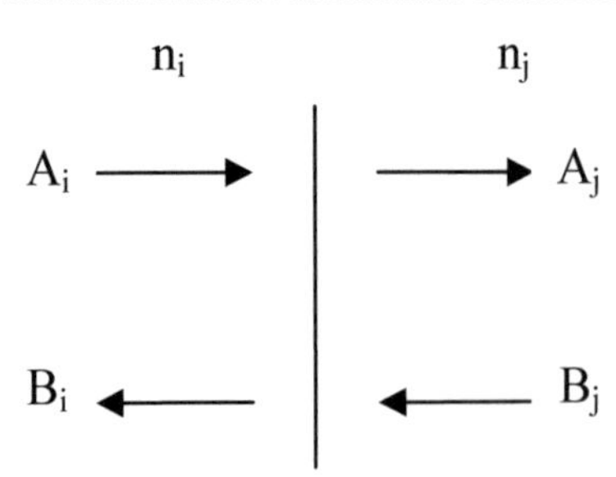

Fig. 3. Schematic diagram of wave propagation in interface between n_i and n_j

Now, the structure with multi-layer stacks is displayed in Fig. 4, then the scheme of wave propagation can be written with propagation matrix form,

$$\begin{bmatrix} A_2 \\ B_2 \end{bmatrix} = T_P \begin{bmatrix} A_1 \\ B_1 \end{bmatrix} \tag{13}$$

$$T_P = T_{1,2} P_2 T_{2,1} P_1 = \begin{bmatrix} t_{11} & t_{12} \\ t_{21} & t_{22} \end{bmatrix} \tag{14}$$

Assume A2= 0, the reflective coefficient can be written as:

$$R_1 = \frac{A_1}{B_1} = -\frac{t_{12}}{t_{11}} \tag{15}$$

then the reflective ratio is shown as

$$R = |R_1|^2 \tag{16}$$

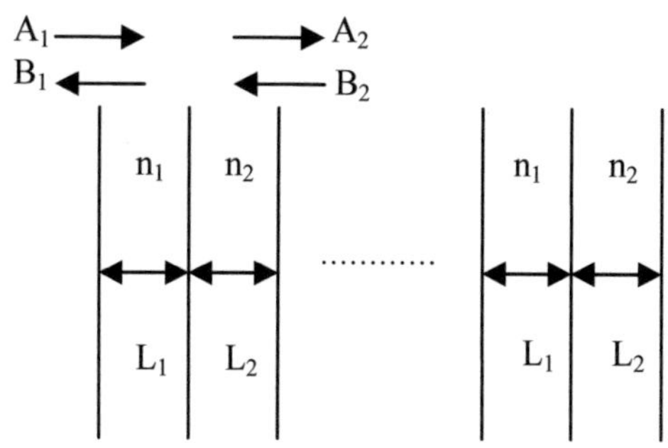

Fig. 4. The multiplayer structure consists of $n_1$and n_2 mediums

3.4 Matrix calculating method, MCM (Furman, Sh.et.al., 2002)

Consider the multiplayer structure as plotted in Fig. 5, the electric field u(z) in layer j and magnetic field v(z) are located in contiguous interface. While applying the boundary conditions of transverse electric field, the overall transfer matrix from the first to the M layer can be exhibited in the below matrix:

$$\begin{pmatrix} u \\ v \end{pmatrix}_{z=z_j} = \begin{pmatrix} \cos\varphi_j & \left(i/q_j \right) \sin\varphi_j \\ iq_j \sin\varphi_j & \cos\varphi_j \end{pmatrix} \begin{pmatrix} u \\ v \end{pmatrix}_{z=z_j-1} \tag{17}$$

where $\varphi_j = kn_j d_j$, $k = \dfrac{2\pi}{\lambda}$, $d_j = z_j - z_{j-1}$, $q_j = n_j$, let

$$M_j = \begin{pmatrix} \cos\varphi_j & \left(i/q_j \right) \sin\varphi_j \\ iq_j \sin\varphi_j & \cos\varphi_j \end{pmatrix}$$

For a series of multi-layer structure, the overall transfer is the product of individual transfer, i.e.,

$$\begin{pmatrix} u \\ v \end{pmatrix}_{z=z_a} = M_m M_{m-1} \ldots M_1 \begin{pmatrix} u \\ v \end{pmatrix}_{z=0} \tag{18}$$

$$M = M_m M_{m-1} \ldots M_1 = \begin{pmatrix} m_{11} & m_{12} \\ m_{21} & m_{22} \end{pmatrix} \tag{19}$$

According to overall transfer matrix, the reflective coefficient can be written as:

$$r = \frac{n_a m_{11} - n_s m_{22} + n_a n_s m_{12} - m_{21}}{n_a m_{11} + n_s m_{22} + n_a n_s m_{12} + m_{21}} \tag{20}$$

so, the reflective ratio is described as

$$R = |r|^2 \tag{21}$$

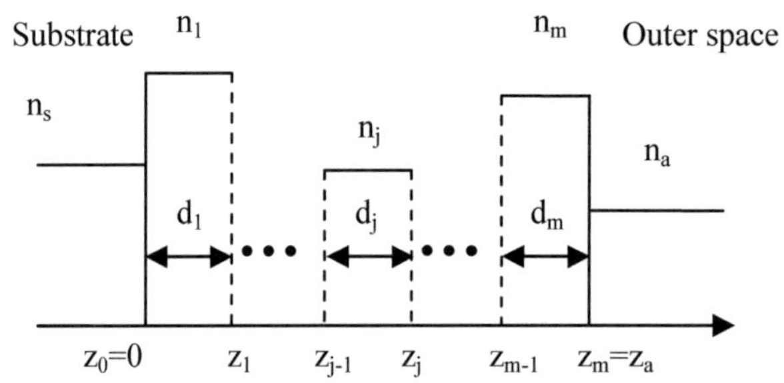

Fig. 5. The schematic diagram of multilayers consisting of substrate and outer space

For further verifying the precision of reflectivity spectra using TMM and MCM method, the multi-layer films evolution software of essential Macleod is utilized to simulate for the same 850nm-VCSEL structure which includes the database built-up refractive index with respect to linearly graded Al composition from equation (1) to (5). The schematic diagrams of unit period in top and bottom DBR mirrors are shown as Fig. 6(a) and (b), respectively. The unit period of bottom DBR mirror consisting of 4 layers of AlxGa1-xAs(x: 0.12→0.9), $Al_{0.9}Ga_{0.1}As$, $Al_xGa_{1-x}As$(x: 0.9→0.12) and $Al_{0.12}Ga_{0.88}As$ is quasi-symmetric to top one for $Al_xGa_{1-x}As$ (x: 0.9→0.12), $Al_{0.12}Ga_{0.88}As$, $Al_xGa_{1-x}As$ (x: 0.12→0.9) and $Al_{0.9}Ga_{0.1}As$. However top and bottom DBR mirrors have 20 and 34 periods, respectively from theoretical calculations.

From simulated results as shown in Fig. 7 and 8, it is notable that the reflectivity spectra achieved from TMM and MCM method are well agreed with the ones from using Macleod software at room temperature. The maximum reflectivity of top and bottom DBR mirrors are >96.4% and 99.98%. The peak wavelengths are located in 840 nm. It can overall satisfy the specification of high DBR performance.

(a)	(b)
$Al_{0.9}Ga_{0.1}As$	$Al_{0.12}Ga_{0.88}As$
$Al_xGa_{1-x}As$ (x:0.12→0.9)	$Al_xGa_{1-x}As$ (x:0.9→0.12)
$Al_{0.12}Ga_{0.88}As$	$Al_{0.9}Ga_{0.1}As$
$Al_xGa_{1-x}As$ (x:0.9→0.12)	$Al_xGa_{1-x}As$ (x:0.12→0.9)
GaAs substrate	GaAs substrate

Fig. 6. Unit periods of (a) top and (b) bottom DBR mirror

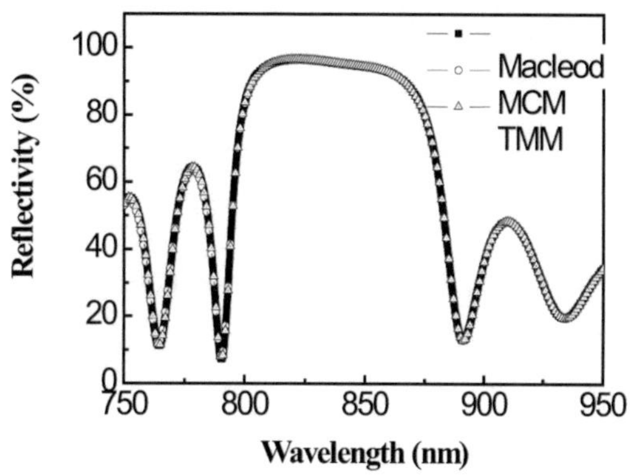

Fig. 7. The spectra of top-DBR mirror

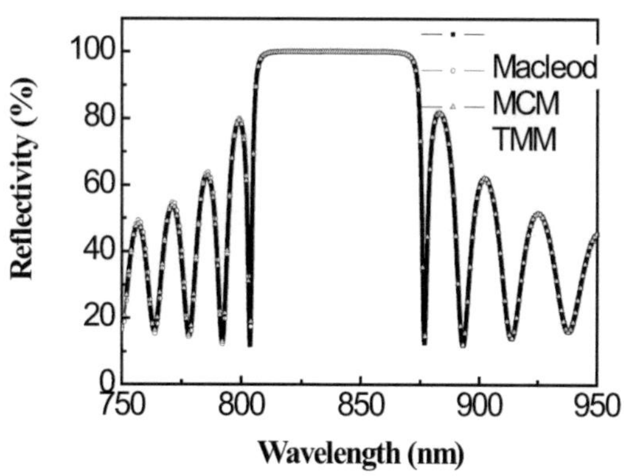

Fig. 8. The spectra of bottom-DBR mirror

Next, considering the temperature-dependent refractivity of DBR mirrors, it is due to the changes of refractive indexes in linearly graded $Al_xGa1\text{-}xAs$ and GaAs. (J. Talghader et.al., 2003) as follows:

$$\left(\frac{dn}{dT}\right)_{GaAs} = (2.67 \pm 0.07) \times 10^{-4} \, / \, °C, \tag{22}$$

$$\left({dn}/{dT} \right)_{AlAs} = (1.43 \pm 0.07) \times 10^{-4} \ /\ °C \quad (23)$$

Estimate the temperature influences on reflectivity spectra from adding the equation (22) and (23) into the previous TMM, MCM method and material database in Macleod software. From Fig. 9 and 10, it is only 0.5Å/°C of spectra red-shift in DBR mirrors from 30 to 80°C, which are comparable with the proposed results from J. L. Shen et.al. (S. Chuang, 1995)

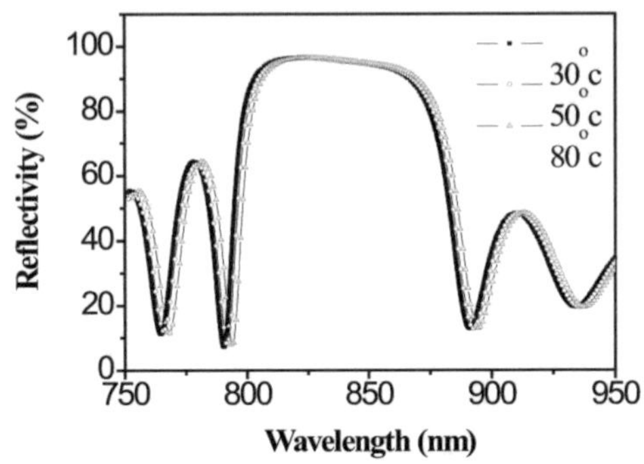

Fig. 9. The temperature dependent spectra of top-DBR mirror

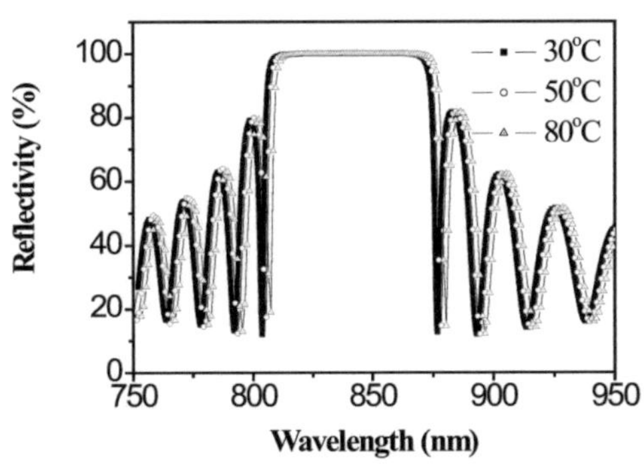

Fig. 10. The temperature dependent spectra of top-DBR mirror

4. Theory of optical dielectric slab waveguides (S. Chuang, 1995)

The dielectric waveguide theory is very useful studying heterojunction semiconductor lasers and the far-field patterns of VCSEL can be explained with the optical waveguide theory in our experiments. At first, let us consider a slab waveguide as shown in Fig. 11, where the width w >>thickness d, and the field dependence on y is negligible, i.e., $\partial / \partial y \equiv 0$. From the wave equation

$$\left(\nabla^2 + \omega^2 \mu \varepsilon \right) E = 0 \quad (24)$$

We shall find the solutions for the fields everywhere. We assume that the waveguide is symmetric, that is, the permittivity and the permeability are ε and μ, respectively, for

|x|≧d/2, and ε1 and μ1 for |x|<d/2. The origin has been chosen to be at the center of the guide, since the waveguide is symmetric; therefore, we have even-mode and odd-mode solutions. Wee separate the solutions of the fields into two classes: TE polarization and TM polarization. Here, a waveguide mode or a normal mode is defined as the wave solution to Maxwell's equations with all of the boundary conditions satisfied; the transverse spatial profile and its polarization remain unchanged while propagating down the waveguide.

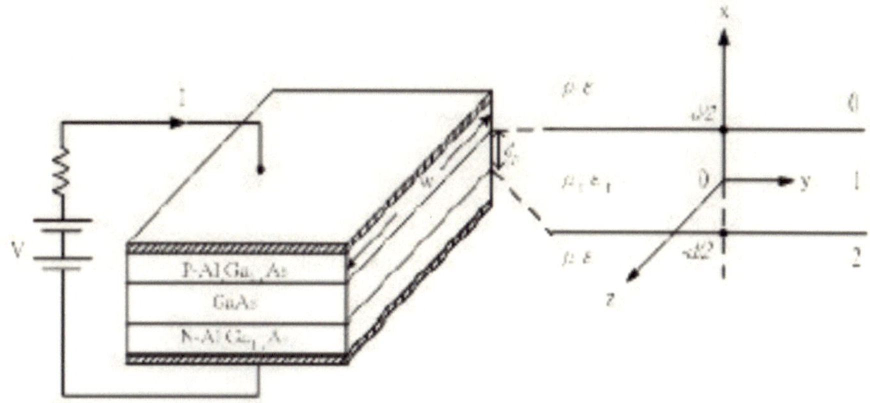

Fig. 11. A simplified heterojection diode structure for waveguide analysis

5. Experimental setups for L-I-V and far-field patterns

The capturing the electronic and optical output characteristics of 850 nm VCSEL were achieved with setup diagram as shown in Fig. 12. At first, the device is mounted on 3-D axis holder consisting of the copper plate hybridized with TE cooler and temperature sensor assembly controlled by temperature controller. The temperature range can be tuned from 30 to 100 °C. Then the electro-probe is linked between R6243 LD Driver (S. T. Su, et.al., 2006) and the device. The near-IR detector is collected with Q8221 Optical Power Meter

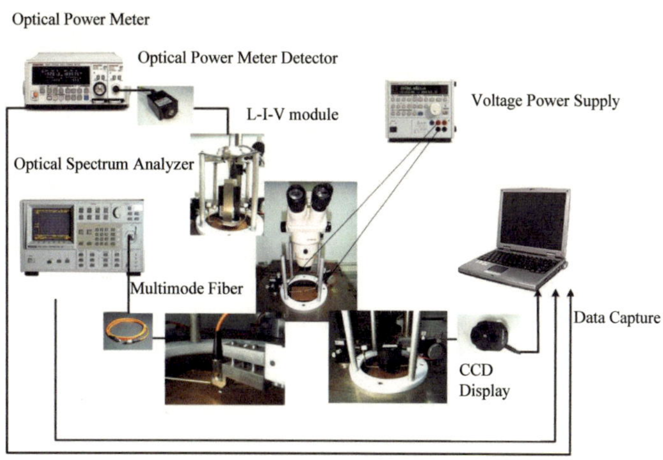

Fig. 12. The electronic and optical measurements consisting of L-I-V, spectra and far-field pattern

(Advantest Corp., 2002) for L-I measurement or Advantest Q8383 Optic Spectrum System (Advantest Corp., 1994) for LD spectra. All the equipments are controlled by the computer through the NI-GPIB interfaces. The Newport Beam Profile Acquisition module is separated from the above equipments excluding the temperature varying system for far-field pattern measurement.
In addition, measures the characteristic which temperature-dependent. Install the device which temperature changing on n 3-D axis holder consisting of the copper plate hybridized. Designed temperature-dependent measuring system is demonstrated in Fig. 13. Variable temperature unit include thermoelectric cooler, temperature controller, heat sink and temperature Sensor. The temperature controller controls the power that the cooler device outputs. Then reach temperature the temperature controller from temperature sensing device that purpose to achieve temperature control.

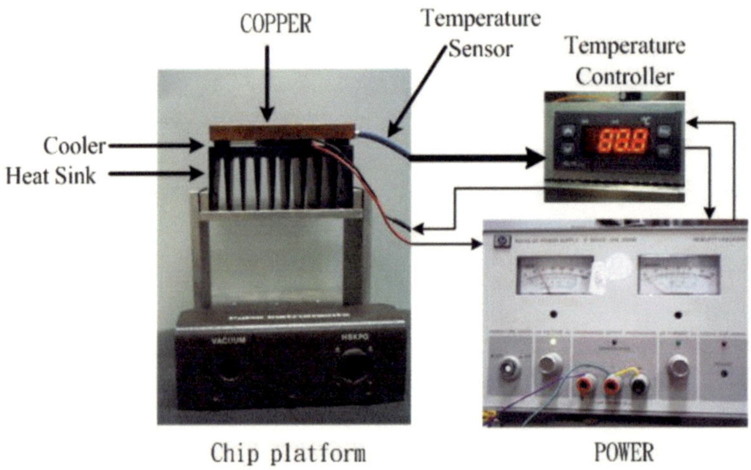

Fig. 13. Variable temperature measuring system

6. Results and discussions

6.1 Oxide confined 850 nm VCSEL

Using best empirical and selectively oxidized parameters of device process, the temperature-dependent I-V measurement of the optimized VCSEL sample are shown in Fig.14. From 30 to 80 °C of I-V measurement, temperature is the higher and the higher slope of IV-curve is presented. While showing the environment temperature rises, energy gap and energy barrier reduce of material. So, diffused current is easy to cross energy gap of surface. On the other hand, the surface defect of junction will be reduced to catch electrons with improvement of temperature, too. The differential resistances of the optimized VCSEL sample are shown in Fig. 15. From 30 to 80 °C and oxidized times of 70 min, the averagely differential resistance of Oxide confined 850 nm VCSEL is 25 Ω.
The temperature-dependent spectra of the optimized VCSEL sample are shown in Fig. 16. From 30 to 80 °C and under the driving current of 30mA and oxidized times of 70 min, the

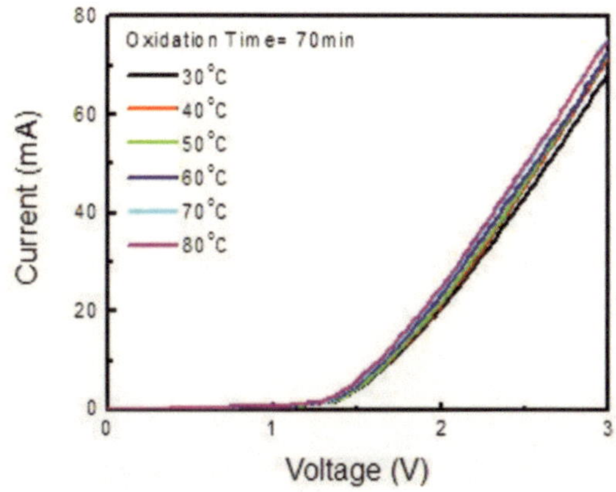

Fig. 14. I-V curve of oxide confined 850nm VCSEL devices

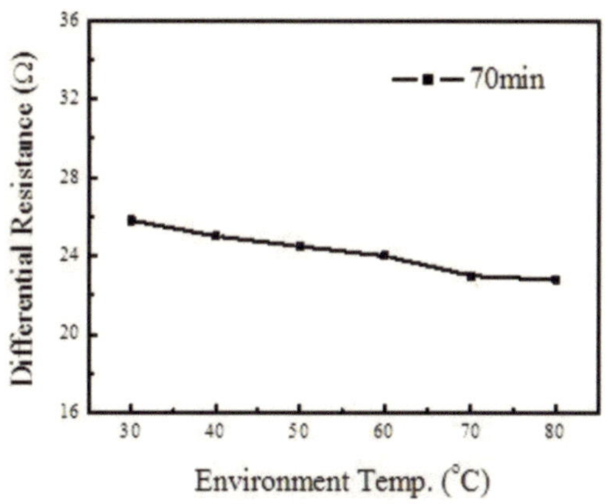

Fig. 15. The temperature-dependent Differential Resistance of Oxide confined 850 nm VCSEL

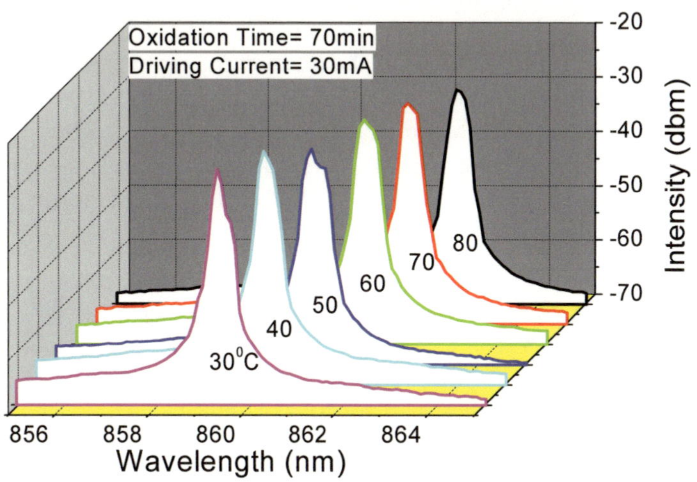

Fig. 16. The temperature-dependent spectra of Oxide confined 850 nm VCSEL

spectra exhibit the quiasi-single transverse modes successfully excluding the 50°C spectrum due to mode hopping. The shifts of FWHM and peak-wavelength red-shift are 0.71±0.05 and 0.06 nm/°C, respectively. In the analysis on far-field patterns (Advantest Corp., 1993), 30°C laser beam profiles are shown in Fig. 17 after selectively oxidized times of 60 min. From the diagram, it is clear to note that the HE_{10} or EH_{10} lasing modes occur suddenly from spontaneous mode once the injected currents increase to 24 mA. While the injected current increases continuously to 26 mA, the transverse modes are transited to HE_{11}. Many modes happen when the injected current is higher than 27 mA.

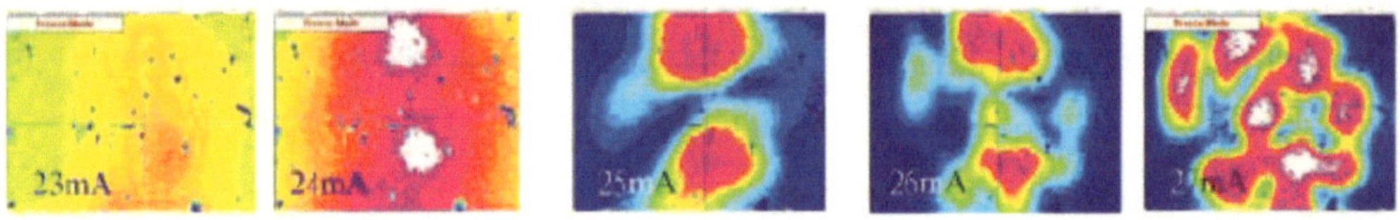

Fig. 17. The far-field pattern of 850 nm VCSEL as function of injected currents, oxidized times of 60 min

Similarly, 30°C laser beam profiles after selectively oxidized times of 70 min are shown in Fig. 18. From the profiles, it is found that the HE_{00} fundamental lasing mode occurs suddenly from spontaneous mode once the injected currents increase to 17 mA. While the injected current increases continuously to 20 mA, the transverse modes are transited to HE_{10} or EH_{10}. Higher modes like HE_{20} or EH_{20} modes happen when the injected current is higher to 29 mA.

Fig. 18. The far-field pattern of 850 nm VCSEL as function of injected currents, oxidized times of 70 min

Differently, 30°C laser beam profiles are shown in Fig. 19 after selectively oxidized times of 80 min. From the profiles, it is very strange that the higher-order lasing modes like HE_{20} or EH_{20} modes occurs directly from spontaneous mode once the injected currents increase to 15mA. While the injected current increases continuously to 19 mA, the transverse modes are transited to lower-order HE_{10} or EH_{10}. Mode-merging phenomenon happen when the injected current is higher to 30 mA

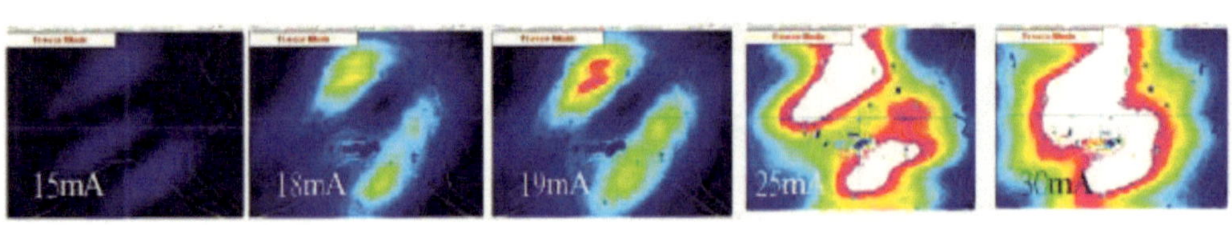

Fig. 19. The far-field pattern of 850 nm VCSEL as function of injected currents, oxidized times of 80 min

Compared with the Fig. 17 and 19, the spontaneous mode is transited to lasing mode with respect to aperture size and injected current values. The main reason possibly is that the temperature increasing of the hetero-junction for VCSEL attribute to increasing injected current. This can result in the changes of refractive index in the cavity of VCSEL so that the k would be subsequently changed to be a large influence on the far-field mode-transition. The parameter of aperture size is also key factor for current density, which cause mode-transition. The aperture size of oxidized times of 60 min for VCSEL is larger than one of oxidized times approaching to 80 min. It may result in the divergent distribution and obscure profile for the mode patterns as displayed in the pattern transition increasing injected current from 23 to 24 mA shown in Fig. 17. In contrast with Fig. 19, the smaller aperture and threshold current density lead to constrictive mode patterns and clear profiles. Similarly, the formation of feature on mode-pattern transition has been explained as the mentioned before in Fig. 17 to 19. The asymmetric aperture window due to non-uniformity of selectively oxidized process may be lead to the initial laser mode appear with higher order mode. It also results in non-uniformity and asymmetry for pattern distribution.

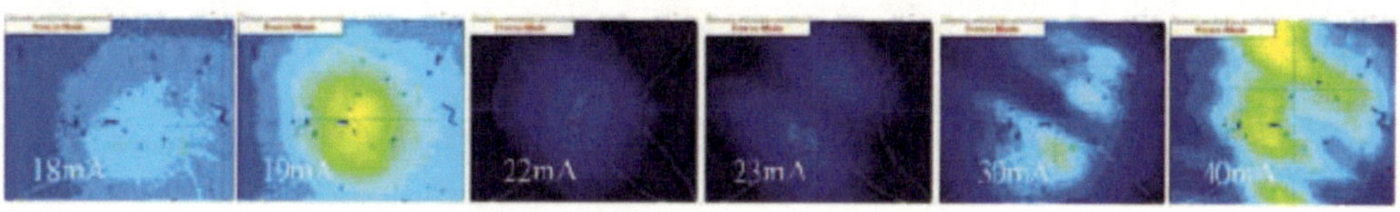

Fig. 20. The far-field pattern of 850 nm VCSEL as function of injected currents, oxidized times of 70 min

Far-field patterns, 80 °C laser beam profiles are shown in Fig. 20 after selectively oxidized times of 70 min. From this Figure, it shows that the HE_{00} lasing modes appear while injected current with 19 mA. Increasing injected current to 23 mA, it transits to HE_{10} or EH_{10}, then to higher order mode of HE_{20} or EH_{20} while further increasing current to 40 mA. In compared with Fig. 18 and 20, two mode pattern distributions are very similar excluding mode intensities. It can be to depict that insensitive operation temperature for far-field pattern distributions. The thermal effect like electron-phonon interaction mechanism leads to enhancement of non-radiation process which may be key factor for intensity degradation.

6.2 Proton-Implant 850nm VCSEL (D. Burak. et.al., 1998),

Contrast to the above selectively oxidized VCSEL performance, the best empirical and proton implanted parameters of device process are treated in the VCSEL sample. The temperature-dependent I-V measurements of the optimized proton-implanted VCSEL sample are shown in Fig. 21. Similarly, from 30 to 80 °C of I-V measurement, the higher temperature is, the larger of slope of IV-curves are. (G. T. Dang et.al., 2003) The depicted phenomenon is previously mentioned. The differential resistances of the optimized VCSEL sample are shown in Fig. 22. From 30 to 80 °C, the averagely differential resistance of proton-implanted 850 nm VCSEL are 36.8 Ω. The differential resistances and turn-on voltages of proton-implanted are higher than those of oxidized confined ones, respectively. It could be explained that the ion-implant process treated in the VCSEL causes some

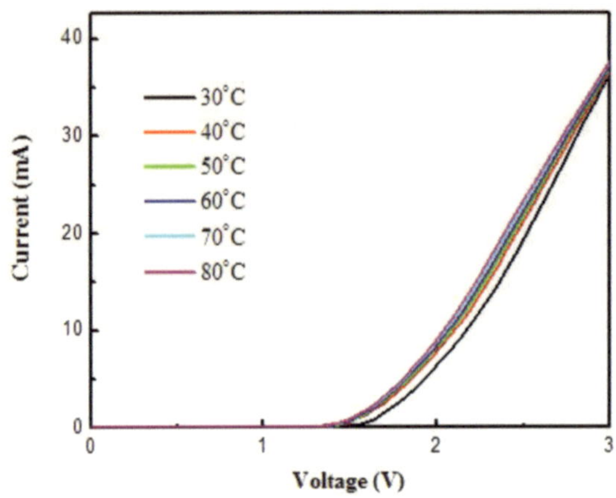

Fig. 21. I-V curve of proton-implant 850 nm VCSEL devices

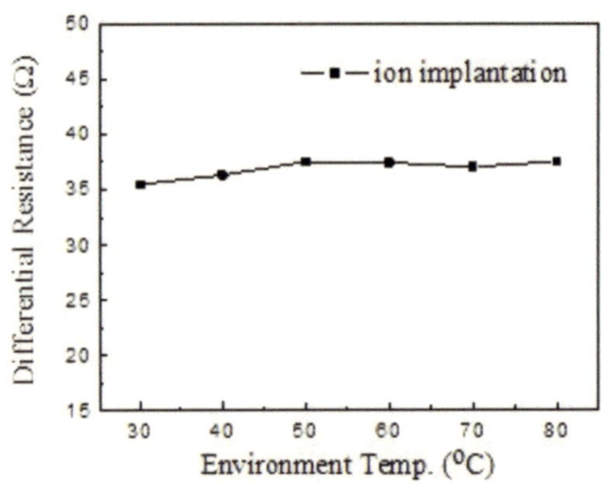

Fig. 22. The temperature-dependent differential resistance of proton-implant 850 nm VCSEL

disorder and damage in contact region even though the high-resistance regions are designed to be fabricated far from the top contact layer. (E.W. Young et.al., 2003)
The temperature-dependent spectra of the optimistically ion-implanted VCSEL sample are shown in Fig. 23. From 30 to 80°C and under the driving current of 10 mA. The shifts of FWHM and red-shifts of peak-wavelength are 0.12 and 0.07 nm/°C, respectively. It is worthy to note that the shifts of FWHM in the proton-implanted VCSEL is smaller with compared to that of above oxidized confined VESEL under the same effective dimension of the optical cavity. And the comparable red-shifts of peak-wavelength in the two different confined types. In the analysis on far-field patterns (Advantest Corp, 1993), 30°C laser beam profiles of the VCSEL sample after ion-implanted treatment are shown in Fig. 24. From the

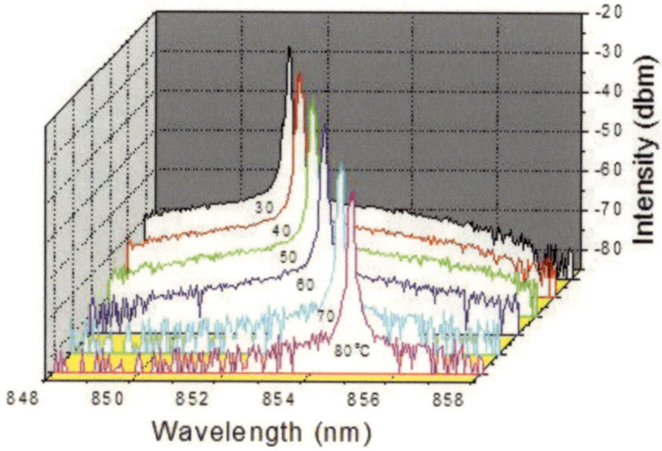

Fig. 23. The temperature-dependent spectra of Proton-Implant 850 nm VCSEL

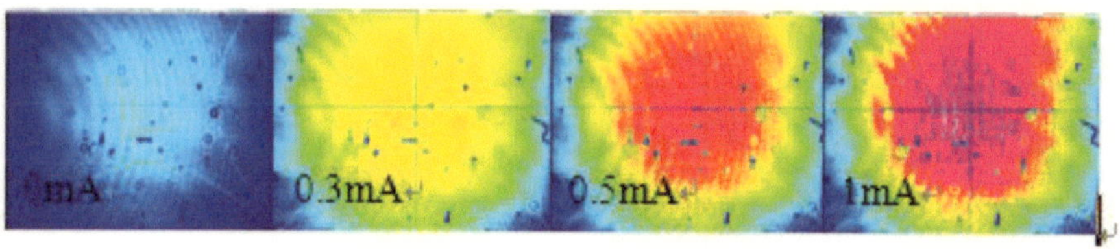

Fig. 24. The far-field pattern of Proton-Implant 850 nm as function of injected currents, temperature of 30°C

diagram, it is clear to obtain that the single lasing modes occur suddenly from spontaneous mode once the injected currents increase to 0.3 mA. While the injected current increases continuously to 1 mA, lasing intensities eventually become stronger and the profile remain the same shape. The transverse modes always are single modes. And the threshold current of proton-implanted VCSEL is much smaller than that of the sample treated in oxidized confined process. (P. K. Kondratko et.al., 2003)
Similarly, 80°C laser beam profiles after ion-implanted treatment are shown in Fig. 25. From the profiles, it is found that the single fundamental lasing mode occurs suddenly from spontaneous mode once the injected currents increase to 1.2 mA. And the profiles of far-field pattern remain the quasi-circular shape. While the injected current increases continuously to 1.3 mA, the transverse modes still are single modes. Only lasing intensity becomes stronger like that of the low temperature operation. It can be depicted that the proton-implanted VCSEL have a good performance to be operated in higher circumambient temperature.
The experimental results in summary of the oxidized and proton-implanted confined VCSEL as sown in Table I, as well as the assistance of using theoretical DBR simulated with transfer matrix method (TMM), matrix calculating method (MCM), Marcatili's method. Table I demonstrates the superior performance of VCSEL treated in ion-implanted process

contrast to the oxidized confined VCSEL. However, the low-differential resistance and lower-cost process with high-temperature oxidized treatment in VCSEL has some benefits for the specific optical-communication application as short-distance data transmission.

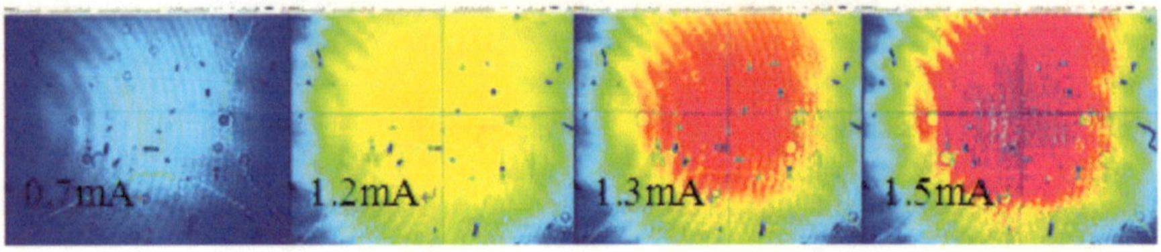

Fig. 25. the far-field pattern of Proton-Implant 850 nm as function of injected currents, temperature of 80°C

VCSEL	Oxidized		Ion Implanted	
Peak wavelength	859 nm	IF= 30 mA	851 nm	IF=10 mA
Threshold current	16.9 mA	CW	3.2 mA	CW
Threshold voltage	1.9 V	IF=30 mA	1.8 V	IF=10 mA
Series differential Resistance	25.8 Ω	IF=30 mA	36.8 Ω	IF=10 mA
Lasing mode	Multi mode		Single mode	

Table 1. The comparison sheets of oxide confined 850 nm VCSEL and Proton-Implant 850 nm VCSEL

7. Conclusions

In the theoretical simulation, the optical TMM and MCM method as well as multi-layer films evolution software of essential Macleod have been proposed to verify the model validity. Besides, the operation temperature leading changes of material refractive index is considered for reflectivity spectra on graded $Al_xGa_{1-x}As$/ GaAs DBR mirrors. For oxidized confined 850 nm VCSEL, under injected current of 30 mA and the operation temperature increasing from 30 to 80°C, the FWHM shifts and peak-wavelength red-shifts are 0.71±0.05 and 0.06 nm/°C. It can be concluded that the aperture size, hetero junction temperature changes and uniformity of selectively oxidized process have very critical influences on the far-field mode pattern distributions, mode numbers, mode transitions. For proton-implanted 850 nm VCSEL, under smaller injected current of 10mA and the operation temperature increasing in the same temperature region of the above oxidized confined VCSEL, the FWHM shifts and peak-wavelength red-shifts are 0.12 and 0.07 nm/°C, respectively. The summary of our experimental results as well as the assistance of the DBR simulation using the TMM, MCM and Macleod's models can be concluded that the optimized 850 nm VCSEL has been proposed in the promising application for high efficient and low-cost optical fiber and free space data communications in the future.

8. References

Talghader; J & Smith, J. S. (1995). Thermal dependence of the refractive index of GaAs and AlAs measured using semiconductor multilayer optical cavities, *Appl. Phys. Lett.*, vol. 66, pp. 335-337

Iga, K.; Ishikawa, S.; Ohkouchi, S. & Nishimura, T. (1984). Room Temperature Pulsed Oscillation of GaAlAs/GaAs Surface-Emitting Injection Laser, *Appl. Phys. Lett.*, vol. 45, pp. 348-350

Afromowitz M. A. (1974). Refractive Index of Ga1-xAlxAs, *Solid State Communications*, vol. 15, pp. 59-63

C. Chen. (2002).Vertical-cavity surface-emitting laser with stable single transverse mode and stable polarization, *SPIE* .pp. 14-16, Taipei, Taiwan

Furman, Sh. & Tikhonravov, A.V. (1992). Basics of optics of multilayer systems, pp. 21-26, ADAGP, Frontiers, France

S. Chuang. (1995). Physics of optoelectronic devices, pp. 242-278, *John Wiley*, New York

S. T. Su; S. F. Tang; T. C. Chen; C. D. Chiang; S. T. Yang & W. K. Su. (2006). Temperature dependent VCSEL optical characteristics based on graded AlxGa1-xAs/GaAs distributed Bragg reflectors: reflectivity and beam profile analyses, *SPIE*. Vertical-Cavity Surface-Emitting Lasers X, Vol. 6132, pp. 0L01-0L10

Advantest Corp. (2002). R6243/44 DC voltage current source/monitor operation manual, chapter 3-4, *Advantest Corp.*,Tokyo, Japan,

Advantest Corp. (1994). Q8221 optical multi-power meter operation manual, chapter 4, *Advantest Corp.*, Tokyo, Japan

Advantest Corp. (1993). Q8381A/8383 optical spectrum analyzer operation manual, chapter 4, *Advantest Corp.*, Tokyo, Japan

D. Burak; S. A. Kemme; R. K. Kostuk & R. Binder. (1998). Spectral identification of transverse lasing modes of multimode index-guided vertical-cavity surface-emitting lasers, *Appl. Phys. Lett.*, vol. 73, pp. 3501-3503,

G. T. Dang; R. Mehandru; B. Luo; F. Ren; W. S. Hobson; J. Lopata; M. Tayahi; S. N. G. Chu; S. J. Pearton; W. Chang & H. Shen. (2003). Fabrication and Characteristics of High-Speed Implant-Confined Index-Guided Lateral-Current 850-nm Vertical Cavity Surface-Emitting Lasers, *Journal of Lightwave Technology*, vol. 21, NO. 4, APRIL

E. W. Young; Kent D. Choquette; Jean-François P. Seurin; Shun Lien Chuang; K. M. Geib & Andrew A. Allerman. (2003). Comparison of Wavelength Splitting for Selectively Oxidized, Ion Implanted, and Hybrid Vertical-Cavity Surface-Emitting Lasers, *IEEE Journal of Quantum Electronics*, vol. 39, NO. 5, MAY

P. K. Kondratko; E. W. Young; Jean-Fran; cois Seurin; Shun Lien Chuang & K. D. Choquette. (2002). Performance of Proton-Implant/Oxide Aperture VCSELs and Comparison with Vector Optical Model, *SPIE.* Vertical-Cavity Surface-Emitting Lasers VI, vol. 4649, pp. 71-76

Part 2

Laser Diagnostics

7

Various Ambiguities in Generating and Reconstructing Laser Pulse Parameters

Chandrasekhar Roychoudhuri
University of Connecticut
USA

1. Introduction

The pulse generation out of a laser cavity is a collaborative and evolving interaction process between EM fields (first spontaneous and then stimulated), and the intracavity device meant to introduce *mode locking* process. When we carry out the actual *mode lock* analysis, we do take into account of the interpaly between all the tempral dynamics of the cavity gain medium, cavity round trip time and the evolution of the termporal behavior of the *mode locking* element (a saturable absorber or a Kerr cell). It is this mode locking *element* that facilitates the enforcement of *locking* the phases of the cavity spontaneous emissions towrads in-phase stimulated emissions with its own temporal gating characteristics. On the observational level, this representation of the *mode locking* process has been serving us well (Milonni & Eberly, 2010; Krausz & Ivanov, 2009) and hence we have stopped questioning whether we have learned everything there are to learn about generating ultra short leaser pulses (Roychoudhuri & Prasad, 2009; Roychoudhuri & Prasad, 2006; Roychoudhuri et al 2006). Consider the paradox discussed further in the next section. Homegeneoulsly broadened gain media, like Ti-Sapphire laser, when succeed in generating *transform limited pulses*, mathematically it is equivalent to $a(t)\exp[i2\pi\nu_0 t]$, an E-vector oscillating with a unique frequency ν_0 under the envelope function $a(t)$. A recent measurement does show such a uniqe E-vector undulation under a few fs pulse (see Fig.1b). What happened to all the longitudinal modes? Have they all interacted with each other and synthesized themselves into a single carrier frequency as is implied by the time-frequency Fourier theorem (TF-FT)? Section-2 will show experimental results underscoring several *ambiguous* interpretations of measured data that we have been maintaining in the literature on mode lock physics. In Section-3 we will develop the methodology of thinking, *Interaction Process Mapping Epistemology* (IPM-E), which will help us discover the universal NIW-principle, Non-Interaction of Waves (Roychoudhuri, 2010), valid for all propagating waves within the linear regime. In Section-4 we will implement this IPM-E and the NIW-principle to show that all the case examples of ambiguities underscored in Section-2 can be resolved satisfactorily.

The purpose of this article is two-fold. The first purpose is to convince the readers that it is not the *phase locking* and *field-field interaction* between the *longitudinal modes* that re-groups the laser field energy into temporal pulses, rather it is the fast time-gating properties of the intra-cavity devices that are most important factors in advancing the field of ultra short pulse laser technologies. We believe that proper understanding of the deeper physical processes behind light-matter interaction processes will clear out our minds from the clutter of ambiguities and then we can emulate nature's actual processes to advance the field at a rate faster than that has been taking place in the past. The second purpose is to draw attention to the need of articulating our methodology of thinking (epistemology) that goes behind gathering and organizing information related to a natural phenomenon, which then give rise to a working theory. Then the next generation of physicists, empowered by their newer and more precision measurement tools along with newer matheamatical tools, can re-evaluate the foundational hypotheses behind the *working* theories for further advancement of physics. We have not yet reached the stage where we can safely assume that the basic edifice of physics has already been constructed; as if we just need to discover the pieces of stones of right shape and size to fit into the existing edifice.

2. Recognizing the fundamental ambiguities

All of our experimental data about any laser pulse parameter are gathered from quantitative measurements of some physical transformations experienced by some material medium, like a detector, after absorbing energy from one or multiple superposed light beams incident on them. Before we get into a better method of understanding of such processes, we need to establish that there does exist ambiguities behind the very concept of *mode lock* theory.

2.1 Can superposed modes create a new mean frequency?

Current literature (Milonni & Eberly, 2010; Krausz & Ivanov, 2009; Siegman, 1986) has accepted that *mode locked* laser pulses are generated by the summation process that take place between the monochromatic beams of electromagnetic waves with carrier frequencies having a periodic separation of $\delta\nu = c / 2L$. Eq.1 is set up for N longitudinal modes, all in phase with equal unit amplitude having a cavity round trip time as the inverse of mode spacing, $\tau = 1 / \delta\nu$:

$$E_{sum}(\nu_0, t) = \sum_{n=0}^{N-1} e^{i2\pi(\nu_0 + n\delta\nu)t} \approx \frac{\sin N\pi\delta\nu t}{\sin \pi t\delta\nu} e^{i2\pi\nu_0 t} \equiv a(t / \tau)e^{i2\pi\nu_0 t} \quad (1)$$

The normalized intensity envelope for the pulse train, in two different mathematical forms, is given by:

$$I(t) = (1 / N^2)\left|E_{sum}(\nu_0, t)\right|^2 = \frac{1}{N^2}\frac{\sin^2 N\pi\delta\nu t}{\sin^2 \pi\delta\nu t} \equiv \frac{1}{N} + \frac{2}{N^2}\sum_{p=1}^{N-1}(N-p)\cos[2\pi p\delta\nu t] \quad (2)$$

The operational implications of Eq.1 and 2 are that the superposed continuous longitudinal cavity modes interact with each other by themselves and re-arrange their temporal energies into a new train of *mode locked* pulses and convert the periodic mode frequencies into a new single mean frequency ν_0 (see Fig.1a). Surprisingly, a novel measurement process does reveal that the electric vector oscillate in a single carrier frequency (see Fig.1b) if the laser is stabilized with great care. This clipped out 4.5 fs pulse was generated by a *mode locked* Ti-Sapphire laser, a *homogeneously broadened* gain medium (Krausz & Ivanov, 2009).
Now, a question to the reader. Can collinearly superposed propagating EM waves in the linear domain generate a new E-vector frequency without the aid of any interaction with some material medium? Can the laser gain medium itself carry out this summation? But

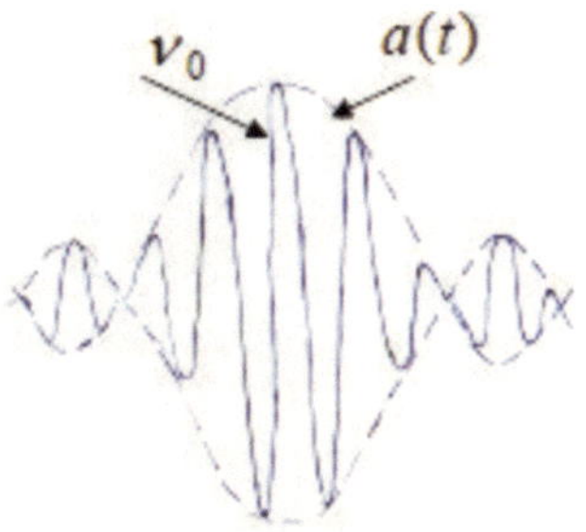

(a)

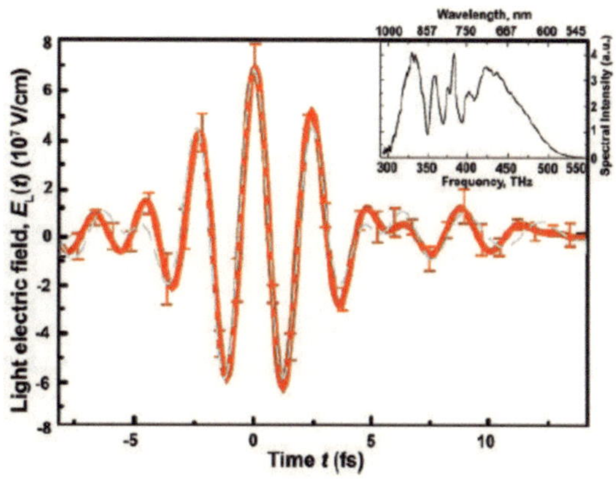

(b)

Fig. 1. (a): A mathematical envelope function (dashed curve) implied by Eq.1 is sketched that defines the time varying amplitude for a single E-vector oscillating at a unique frequency ν_0. Only a single major pulse out of an infinite train has been shown. (b): Demonstration of the existence of a single carrier frequency in a 4.5fs pulse by directly measuring the harmonically undulating E-vector strength (taken from Fig.12 in Krausz & Ivanov, 2009).

then, why do we need high intensity laser beams propagating through some special *nonlinear material medium* with preferred orientation to generate sum or difference frequencies?

Contradictions and paradoxes abound in this field. A He-Ne laser with *inhomogeneously broadened gain medium,* when *mode locked,* its longitudinal modes do not get converted into a single central carrier frequency (see Fig.2b), even though the pulse width (Fig.2a) and the intrinsic line width of the individual longitudinal modes (Fig.2c) corroborate extreme phase stability between the modes needed for the required *mode locking* condition (Allen et al, 1969). If Eq.1 does represent the real physical process behind *mode locking,* and if that is corroborated by the result of Fig.1a and c, then we should conclude that Allen et al did not really achieve *mode locking* in spite of *locked* phases between the modes!

2.2 Can a homogeneously broadened gain medium oscillate in all the allowed cavity modes?

Now, another question for the reader. An excellent Ti-Sapphire crystal, in a CW laser cavity, runs normally at a single longitudinal mode determined by the gain-line center where the gain is highest because the Ti-atoms are embedded in a homogeneously broadened gain medium. Can the spectral behavior of Ti-atoms become inhomogeneously broadened under mode locked conditions, allowing all the potential cavity modes to oscillate, as allowed by *inhomogeneous* Ne-atoms in a He-Ne gas laser? If *mode locking* field-field interaction is the cause behind obtaining ultra short pulses from a Ti-Sapphire laser, then the gain medium needs to become functionally inhomogeneously broadened! The alternate explanation is that it is the periodic Fourier side band frequencies, matched with the cavity modes, which provide seeds for multi frequency oscillation (Milonni & Eberly, 2010) even though the gain medium always remains homogeneously broadened. Then the question arises as to which physical process carries out the Fourier decomposition of a pulse envelope to generate the

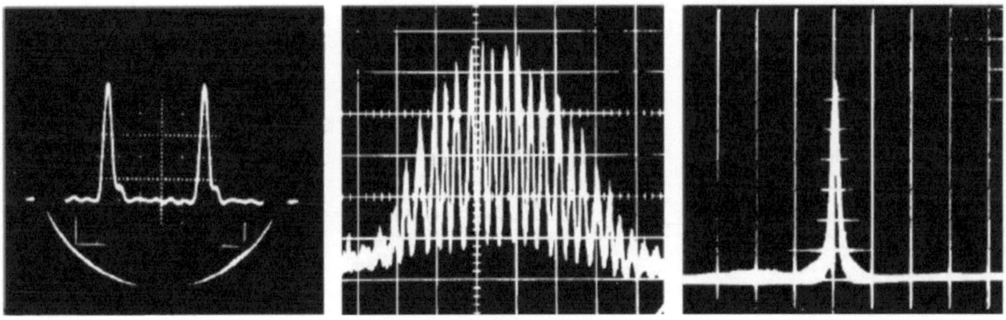

(a) Mode locked pulse train (b) Longitudinal mode spectrum (c) 10KHz intrinsic line width

Fig. 2. Experimental data from a mode-locked He-Ne laser showing intensity vs. time for a mode-locked pulse train in (a), longitudinal mode spectrum in (b), and the heterodyne high resolution line width, 10KHz, of one individual laser mode in (c) (from Allen et al, 1969).

longitudinal mode seeds? Note that any device that can carry out the Fourier transformation process, must posses some memory to be able to first *read* the shape of the *entire amplitude envelope* of the pulse and then *carry out* the mathematical Fourier *decomposition process* to generate *real physical Fourier frequencies* to promote stimulated emissions at these *side band* frequencies! However, we know that the response time of excited atoms to stimulating radiations are well below pico seconds, if not femto seconds, or even less.

2.3 Can the time-frequency Fourier theorem (TF-FT) be a principle of nature?

For the Fourier side bands to exactly match the cavity allowed mode frequencies, the oscillating amplitude envelope and its periodicity must already correspond to the final *mode locked* envelope and the pulse train. The possibility exists that the spontaneous emissions accidentally gets in phase and opens up the stuarable absorber gate and a pulse starts to reverberate through the cavity while iteratively perfecting itself towards the ideal mode locked envelope, and at the same time, the Fourier decomposition process of the amplitude envelope (generation of the side band frequencies) also continues to evolve into a perfectly matching frequency set with the cavity modes. For this temporal evolution to work in favor of our current hypothesis, the time frequency Fourier theorem (TF-FT) must be a physical principle of nature. In other words, the pulsed light amplitude, even when the carrier E-vector is oscillating in a unique single frequency, must have inherent affinity to re-represent themselves as the summation of periodic *Fourier frequencies* as is demanded by the TF-FT. Then it is possible that the Ti-atoms will be literally stimulated by all the allowed cavity mode frequencies, as per TF-FT. Mathematical logic wise it is plausible. Can this be the physical reality? Then the evolving weaker Fourier side band frequencies must be able to compete with the stronger gain line center. Further, if the TF-FT is a physical principle of nature, then superposed coherent light beams must be able to interact with each other and re-distribute their energy in time and space to create energy pulses without the need of mediation of any material medium. In other words, inhomogeneously broadend lasers, like He-Ne, with very high-Q cavity (narrow mode width and high coherence time), should show spontaneous break up into random pulsations, which is not observed in reality.

2.4 Why regular CW He-Ne lasers show mode-lock-like pulsations?

However, a fast detector stimulated by a regular CW He-Ne laser does show oscillatory current exactly mimicking *mode locking* behavior! Consider the measurements carried out in our lab using an ordinary commercial CW He-Ne laser as shown in Fig.3 (Lee, 2004; Roychoudhuri et al, 2006). Fig3a shows the spectrally resolved longitudinal modes of the laser as displayed by an optical spectrum analyzer (OSA). This OSA was a very slowly scanning Fabry-Perot spectrometer and the output spectral lines never showed any pulsations. The result implies that the laser is running CW with three dominant modes and two weak modes under the usual 1.5 GHz Ne-gain envelope. However, Fig.3b shows the intensity envelope that implies the laser is *mode locked*, even though it is not. The data was recorded through a combination of a high speed detector and a high speed sampling scope. Fig.3c is a computer simulation of the resultant amplitude, which is the sum of the five modes shown in Fig.3a. When the output of the high speed detector was analyzed by an electronic spectrum analyzer (ESA), one can identify the self-heterodyne signals (beat between all the individual modes), as shown in Fig.3d. This clearly corroborates the result of Fig.3a that the optical frequencies (modes) are oscillating independent of each other and have not merged into a mean frequency as predicted by Eq.1, or the intensity trace of Fig.2b may imply.

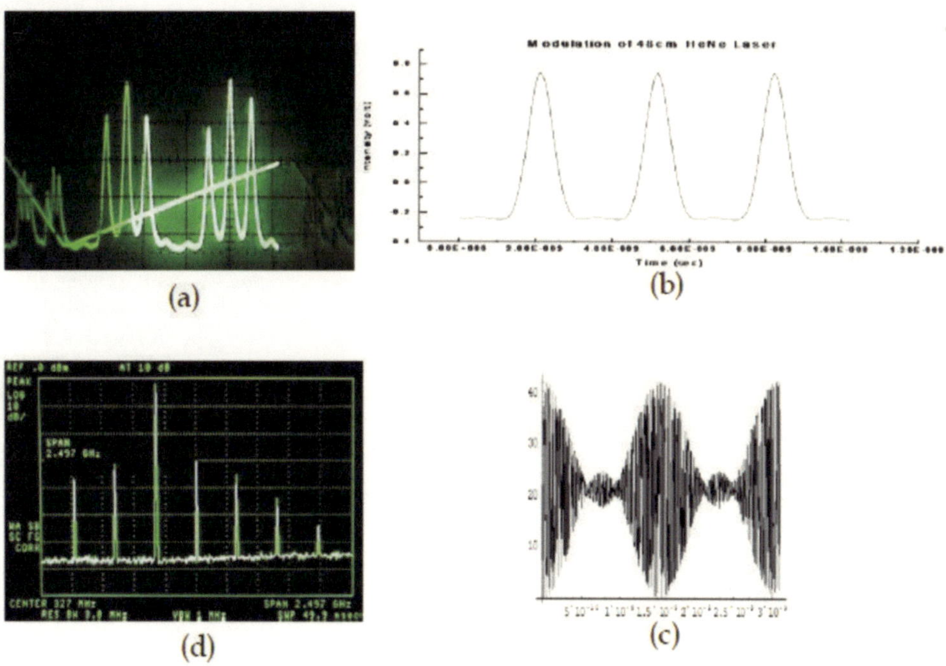

Fig. 3. Is this He-Ne laser mode locked? Data gathered from an ordinary He-Ne laser. (a) Longitudinal modes resolved and displayed through a very slow scanning Fabry-Perot spectrometer. (b) Laser intensity trace recorded by a 40 GHz sampling oscilloscope as detected by a 25 GHz detector. (c) Computer model of the amplitude envelope for the sum of all the modes displayed in (a) as if they are in same phase (mode locked), which corroborates the measured intensity envelope in (b). (d) Display of the 25 GHz detector output as analyzed by an electronic spectrum analyzer. [from Roychoudhuri et al, 2006].

2.5 Is synthetic mode locking possible?

Next we present another experiment to test whether simple superposition of a set of periodically spaced frequencies with steady mutual phase coherence, can automatically generate *mode lock* pulses. Fig.4a shows the schematic diagram of the experimental set up. With the help of an acoustooptic modulator, a single frequency (ν_0) beam from an external cavity stabilized diode laser is converted into three coherent beams of three periodic frequencies (ν_0 & $\nu_0 \pm \delta\nu$) and then superposed into a single collinear beam. The beam was then analyzed to check whether it became *mode locked* and pulsing with a single carrier frequency. The intensity envelope as registered by a 25GHz detector and then displayed by a 40GHz high speed sampling scope is shown in Fig.4c; the trace does correspond to a pulse train that would be generated by a *three-mode-locked* laser. However, a simultaneous spectral analysis of a sample of the same beam through a slowly scanning Fabry-Perot spectrometer displayed the presence of the original three frequencies (ν_0 & $\nu_0 \pm \delta\nu$). If they were *mode locked* then, as per TF-FT, we should have registered only the mean (central) frequency ν_0. Analysis of the current from the high speed detector by an electronic spectrum analyser (as in Fig.3d, but not shown here), had displayed two mode-beating lines at $\delta\nu$ and $2\delta\nu$, corroborating that Fourier synthesis did not take place even though the high speed detector current implies *mode locking*! Clearly some apparently successful mathematical modeling of data can mislead us to wrong conclusions. Light-matter interaction processes behind all measurements must be investigated thoroughly before convincing ourselves about any specific properties of light.

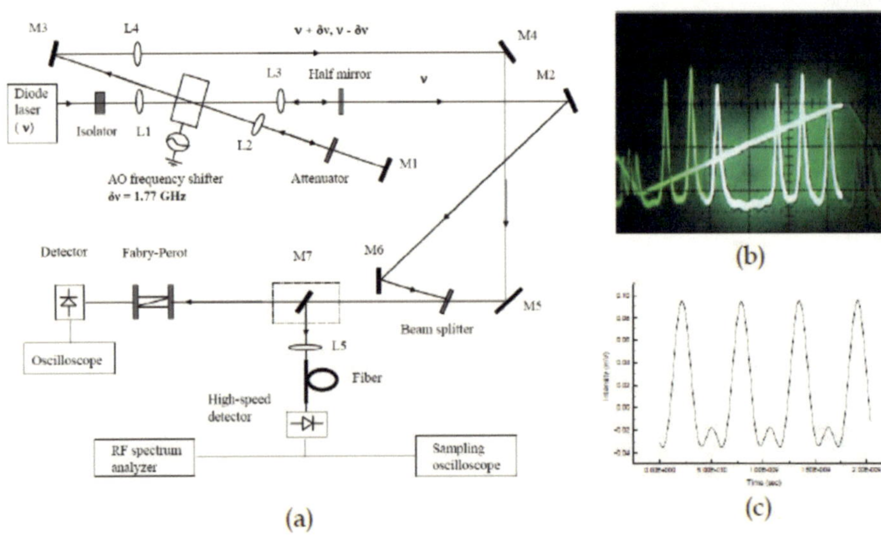

Fig. 4. (a) shows experimental arrangement to generate three periodically spaced coherent frequencies (modes) from an external-cavity-stabilized single frequency diode laser using an acoustooptic modulator. The beams are then collinearly superposed and analyzed for possible mode-lock behavior. (b) shows the spectral display of the three independently oscillating frequencies through a slowly scanning high resolution Fabry-Perot spectrometer. (c) displays the photo-electric current on a high speed sampling scope generated by a high speed photo-detector. (from Lee, 2004; Roychoudhuri et al 2006).

2.6 Can autocorrelation data unambiguously determine the existence of ultrashort pulses?

Next we present experimental results to demonstrate that a measured train of autocorrelation spikes, which may imply the existence of a train of ultra short pulses in a laser beam, may not necessarily represent the actual physical reality! The data shown in Fig.5 were generated using a Q-switched diode laser with a saturable absorber facet (Roychoudhuri et al, 2006), which was generating a steady train of 12ps pulses at about one millisecond interval. Fig.5a shows the time averaged spectrum generated by a high resolution grating spectrometer. There are some 32 modes present and the spacing is about 0.4839nm or $\delta\nu \approx 200$ GHz (199.74 GHz) at $\lambda = 852.5$ nm. The experimental resolving power from the graph is clearly narrower than 100GHz. This is also supported by computation using the TF-FT corollary, $\delta\nu\delta t \approx 1$. The pulse width of $\delta t = 12$ps, derived by Lorentzian fitting from the autocorrelation trace of Fig.5b, implies that the individual spectral fringe width should be about $\delta\nu = 83.3$ GHz, which is clearly smaller than 100GHz, as observed above. The cavity round trip time is 5ps (1/200GHz), which is less than half the Q-switched pulse width. So, the Q-switch pulse width had time to carry out a couple of reverberations and establish cavity longitudinal modes through stimulated emissions.

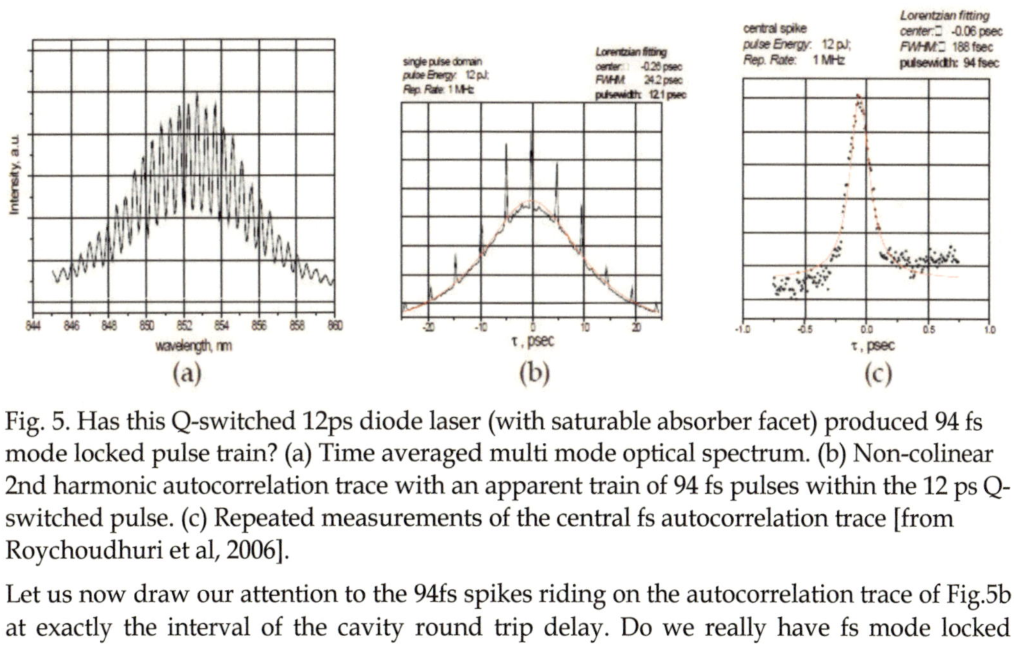

Fig. 5. Has this Q-switched 12ps diode laser (with saturable absorber facet) produced 94 fs mode locked pulse train? (a) Time averaged multi mode optical spectrum. (b) Non-colinear 2nd harmonic autocorrelation trace with an apparent train of 94 fs pulses within the 12 ps Q-switched pulse. (c) Repeated measurements of the central fs autocorrelation trace [from Roychoudhuri et al, 2006].

Let us now draw our attention to the 94fs spikes riding on the autocorrelation trace of Fig.5b at exactly the interval of the cavity round trip delay. Do we really have fs mode locked pulses within each 12ps Q-switched pulses? As per $\delta\nu\delta t \approx 1$, the spectral line width corresponding to 94fs pulses should be more than 10,000GHz. But the half width of the spectrum is less than 3000GHz. Of course, one may argue that the pedestal (lower envelope of the spectrum) of Fig.5a shows the spectral broadening due to the fs spikes and it is not the spontaneous emission background. It is a difficult proposition because in a 5ps cavity a 12ps pulse does not have enough time to over-ride the dominance of spontaneous emissions when the diode is pumped by current pulses of nano second duration and kilo amperes peak value repeated at KHz.

3. Discovering the principle that resolves the ambiguities

Why do we need to discuss the methodology of thinking (epistemology) in a hard-core scientific paper? Since this is not the normal custom, some readers should feel free to skip this section and jump to Section-4 and find the resolutions of the ambiguities raised in the last section. Then they can come back to read and appreciate the utility of this section on epistemology. Here we develop an epistemology, we call the Interaction Process Mapping Epistemology (IPM-E) whose objective is to visualize the invisible interaction processes that give rise to the measurable data. Current physics stops once we have successfully modeled the measured data, which we call Measurable Data Modeling Epistemology (MDM-E). When IPM-E is applied systematically to light-matter interaction processes behind registration of optical superposition effects, one can discover that, in reality, superposed light beams do not interact (interfere) by themselves. We call this NIW-principle since Non-Interaction of Waves (NIW) is a general principle of nature in the linear regime, which has remained unrecognized due to our consistent epistemology of ignoring what is not directly measurable or observable.

3.1 Introducing the Interaction Process Mapping Epistemology (IPM-E)

A careful analysis of the methodology of our thinking behind the development of theories (information gathering and organizing) is a vitally important task because it will allow us to critically and objectively evaluate the various steps that went behind existing working theories and then modify/correct them as our technologies for all measurements keep on dramatically improving. We know that all human constructed theories are necessarily incomplete as they have been organized based on insufficient information about the universe; and everything in the universe is interconnected, sometimes overtly and other times subtly. We still do not know what an electron is. And yet, our current knowledge of the universe has exploded during the last few hundred years through several punctuated *revolutions* as claimed by Kuhn [Kuhn, 1996] in modeling observable information. Over the centuries, we have clearly experienced that all of our theories have been iteratively corrected, improved and/or replaced as our sensor (measurement) technologies have been enhancing with time. But this dynamism in physics has steadily *slowed down* over the last several decades as we have remained focused on maximizing the *utilities* of current *working* theories, instead of iteratively improving upon their foundational hypotheses. This *slow down* can be appreciated from the list of recently published books by many authors, some of whom are very well known (Silverman, 2010; Woit, 2007; Laughlin, 2006; Smolin, 2006; Penrose, 2005). In contrast to complex epistemologies by these authors, we define a very simple and pragmatic epistemology that, beside solving the ambiguities encountered in the field of pulsed lasers, also solves many other paradoxes encountered in both classical and quantum optics presented elsewhere (Roychoudhuri, 2010; 2009a; 2009b; 2009c; 2008; 2007a). As mentioned earlier, this is because the core epistemology of physics has remained basically same for several centuries: *MDM-E*. While measurable data had been, are, and will remain as the key validation approach for all of our theories, we need to graduate to the next deeper level of epistemology so we can understand and visualize the invisible interaction processes that give rise to the measurable data. We have named this epistemology: *IPM-E*. In reality, inventors of new technologies have always tended to use IPM-E without articulating as such. They have always appreciated nature as a creative system engineer. They think like *reverse engineers* and visualize the invisible interaction processes in nature using their imaginative faculty and then emulate different natural processes to invent and innovate useful new technologies. This has been the most vitally important practical step behind our successful evolution. It has been the unusually rapid rate of technology inventions that has helped humans to become the most dominant species of the biosphere. We have gone so far ahead of other specie using our technologies that we have started to ignore that we are just another species and we cannot and must not try to defy the laws behind the biospheric and the cosmospheric evolutions.

Our MDM-E guided mathematical models seem to be working as the measured data are modeled reasonably well and can predict new measurements quite accurately. And yet, we have all these confusing ambiguities, paradoxes and contradictions, identified in the last section. Clearly MDM-E is somehow falling short of helping us visualize the interaction processes behind laser *mode locking*! We must be missing something fundamental behind our assumption of summing the various light beams (mode locking). We do not accept a generic curve fitting polynomial as a proper theory for a phenomenon under consideration because it contains too many free parameters. However, when it is a compact and elegant mathematical relation like that of Planck's Blackbody Radiation, we are elated because it also leads to further prediction of new phenomena that we have never measured before. However, Planck's relation still does not help us visualize the physical processes behind radiation absorption and emission. Otherwise quantum mechanics would have been invented by early 1900. Planck proposed that only the energy exchange process is quantized as $\Delta E_{mn} = h\nu_{mn}$, but once the released quantum of energy ΔE_{mn} evolves into an EM wave packet, it propagates diffractively as a classical wave. But, within five years of Plack's discovery, in 1905 Einstein discovered some *quantumness* in the photoelectric data and assigned this *quantumness* to the wave packets of light (spontaneous emissions), rather than to the binding energies of the electrons, and declared light to be *indivisible quanta* and missed the opportunity to discover quantum mechanics himself. It was Bohr who formally proposed that the electron binding energy was quantized in atoms in 1913, which was evident from Ritz-Rydberg formula for atomic spectral lines. But the formulation of formal quantum mechanics (QM) had to wait until 1925. The dominant interpretation of this QM categorically instructs us not to waste time in trying to visualize the details of interaction processes between electrons, protons and neutrons that have build the entire observable material universe! One should also note that this QM *does not have a rigid hypothesis that only quantum entities can exchange energy with each other*. Otherwise, a classically accelerated electron in a He-Ne discharge tube could not have shared a fraction of its *kinetic energy* in raising the quantized Ne-atoms from their ground to an upper excited level! And before the end of the decade of *quantum revolution*, Dirac assigned *self-interference* property to these *indivisible quanta*, we now call photons (Dirac, 1974). And, now, over the last couple of decades, we have been claiming to successfully carry out quantum communication, computation and encryption exploiting this unique *self-interference* property of single *indivisible* but *nonlocal* photons.

Remarkably, even though our instruments, interferometers and detectors, are very well *localized* (physically finite) in space, *single photons* are unlocalized in a coherent CW beam as if it is like a Fourier monochromatic mode. We assume that they are equally well *unlocalized* within a 0.3 micron long (1fs) pulse since the pulse is apparently built out of many infinitely extended Fourier monochromatic modes (Fourier transform of the pulse envelope). The point is MDM-E guided successful theories are not guiding us to discover unambiguous *pictures* as to how nature really carries out its interaction processes. We need something better! So, the author has initiated publications and an international conference series to promote deeper investigation on the nature of light (OSA 2003; SPIE 2005, 07, 09). Readers are very welcomed to join us to accelerate the growth of optical science and engineering with a deeper foundation; the 4th biannual conference is set for August 2011 during the SPIE Annual Conference.

The seed for seeking a better epistemology is already built into MDM-E. We want to understand all the cosmic logics (laws) behind all the interaction processes that are behind the dynamically evolving cosmic system. But we do not have any direct access to these logics. However, our successful theories organized as mathematical equations represent strict relationships between cause and effect and match data to model potential cosmic laws. And the long list of staggering successes of our endeavor clearly imply that the laws of nature must be very logical, causal and hence *invariant*. Otherwise we could not have achieved so much successes in understanding the processes behind the evolution of the cosmic system. It is then safe to conclude that the interaction processes guided by these cosmic laws must also be *invariant*, albeit invisible. If we want to extract reality out of nature, we must anchor our epistemology with something in nature which is accessible to us through our measurement and theorizing processes and yet invariantly connected to the cosmic logics more deeply than the measurable data alone. It is then logically safe for us to shift our epistemology one-step deeper to construct and refine theories based upon our attempts to visualize the interaction processes, rather than carrying out only *curve fitting* of measured data. Curve-fitting MDM-E provides neither the guidance, nor the incentives, and nor the challenges necessary to try to *iteratively enhance* our already *working* theories, but IPM-E does. The interaction processes, being invisible and elusive, pose constant mental challenges to us. And yet they represent nature's invariant referent source for gathering feedback information and keep on refining them iteratively for as long as it takes! All knowledge must be advanced and refined through iterative feedback loop. Successes with MDM-E brings complacency, while construction of process map achieved with IPM-E keeps on bringing perpetual challenges. A map never becomes the actual terrain! Our sustained and successful evolution will be assured by such an epistemology. However, IPM-E does not replace MDM-E. IPM-E coopts MDM-E as its foundational tool but empowers it with the iterative debating tool to challenge all *working theories* and force them to evolve or make room for new theories.

Since invention of new technologies fundamentally rely upon our capability of emulating various interaction processes in nature and the concomitant cosmic logics (laws), it will be more productive for us to think like reverse engineers, as far as nature is concerned. We must stay focused on becoming discoverers of the laws of nature rather than trying to invent them and then tell her how she should function! All human scientific logics (epistemology) must keep on evolving to refine our grasp of unknown cosmic logics that enforce continuous cosmic evolution including our minds.

To further appreciate the utility of IPM-E, we need to dissect the epistemological steps behind creating theories The two key steps are (i) information gathering/generating challenge (IGC) and (ii) information organizing challenge (IOC). The first step is also known as the *Measurement Problem* in quantum philosophy and the second step may be identified as the eternal *Theorizing Problem*! This is because an analysis of the first problem makes us aware that we are forever challenged by nature in gathering complete information about even the simplest entity in nature we try to study. We may call it the Incomplete Information Challenge (IIC)! For example, the electron was discovered in 1897 by J. J.

Thompson and is being analyzed by tools of quantum mechanics since 1925. It is being utilized in engineering machines like free-electron lasers and electron microscopes for many decades. Yet, even today, we still do not really know its real structure and from where the electron gets its charge and mass. Accepting wave-particle duality as the final answer that physics can provide is akin to accepting a moving bulge under a carpet as the final limit of our knowledge for fear of lifting the proper carpet corner and discovering a running mouse! We must not *suppress* or *moth-ball* deeper questions related to *interaction processes* by imposing further layers of sophisticated mathematical models. We cannot solve the *Measurement Problem* and the IIC driven *Theorizing Problems* by inventing more and more mathematical constructs alone. Mathematical tools are our key guide to help us extract many new information, but tools cannot be used as substitutes for the information we seek.

3.1.1 Information generating and gathering challenge (Measurement Problem).

The founders of Quantum Mechanics correctly recognized that there is a "Measurement Problem" in QM. However, the author believes that the problem is more generic than has been assumed. The problem pervades all scientific endeavors as we try to generate and gather information about anything we try to study in nature. This information challenge or measurement problem cannot be overcome by any clever mathematical theorem! It will be obvious once we delineate the simple steps behind any experiment or information generating/gathering process.

1. **Measurable Transformation:** We can measure only some physical transformations in our sensors triggered by physical transformations experienced by the interactants under study. [No need to invoke God or human consciousness!]
2. **Energy Exchange:** All physical transformations require some energy exchange between the interactants under study and the sensor used to register the transformation.
3. **Force of Interaction:** Energy exchange, and consequent transformations, must be guided by an allowed force, or forces, of interaction(s) between the interactants and our sensors. "Classical" interactions allow continuously variable energy exchange dictated by the experimental context. QM interactions in the linear domain (transition between pre-existing QM levels) allow only discrete amount of "resonant" energy exchange.
4. **Physical Superposition (Sphere of Influence) or "Locality" of All Interactions:** Interactants must be within each other's sphere of influence to be able to interact under the guidance of an allowed force to transfer energy and then undergo transformations. Thus "locality" can be of galactic distance for galaxies under gravitational force, or it can be 10<-15> meter for strong nuclear force between nucleons.

Corollary 1: Impossibility of Interaction-free Transformation (IIT): It is self evident from the items described in 1 to 4 to above.

Corollary 2: Incomplete Information Challenge (IIC): We can never gather all the information about anything through any experiment since the details of none of the interaction processes and those of the interactants are completely known to us. Further, almost all measuring instruments require amplification of the original information created by original transformation through new interaction step(s) in the detecting system. And we know that we can rarely transmit (or, transfer for recoding as data) the original but unknown information with perfect fidelity through complex systems that have their own unknowns. We are after Cosmic-Logics (rules of interactions)! They are not directly accessible through Data-Math model alone.

3.1.2 Information organizing challenge (theorizing problem)

Purpose of physics is to discover the unknown *cosmic logics* (laws) and the underlying invisible *interaction processes* that we can emulate to create technologies to assure our *sustainable evolution*. We are challenged for ever! Through millennia of observations, we found that nature behaves in logical patterns. So, we have refined the faculty of *human logics* to *frame logical questions* that help us discover *conceptual continuity* between diverse observations. Then we bring some *cause-effect logical congruence* between them by inventing well-articulated *hypotheses* to compensate for the missing information. For a quantitative best fit (curve fitting), we have then invented a new language of more refined logics, we call mathematics or *mathematical logics*. In other words, *human logics* frame questions and hypotheses about *cosmic logics* to match the organized observations and then create a quantitative fit to the data using *mathematical logics*. The use of a generic n-th degree polynomial could have been the simplest choice. However, we have found that simpler and more elegant a mathematical expression (theory) is, more success it shows towards predicting newer phenomena, leading us to believe that we are getting closer to the actual cosmic logics (laws). Continuous endeavor to iteratively refine the map of invisible interaction processes will take us closer and closer to the real cosmic logics. However, we know that the construction of a very realistic map can never be the substitute for traveling in the real terrain!

3.1.3 Importance of identifying and articulating the epistemology

In the last section on information gathering, we have identified the perpetual IIC imposed on us by nature. It is the IIC that forces us to hypothesize, or invent (dream up) information that was not directly available through measured data. This is where the genius minds of Galileo, Newton, Fraday, Maxwell, Einstein, etc., appear to us so impressive and worthy of reverence. Importance of the epistemology now becomes apparent. *Framing the question determines the answer one can extract out of nature's interaction processes.* And framing the question is dictated by the personal epistemology, or the mode of thinking, which is a complex product of individual genome and how that genome absorbs, nurtures and cultivates the individual mind towards his/her personal epistemology. A small subset of a very complex but organized system can be modeled by many different sets of rules, none of which may exactly coincide with the original set of rules that has organized the whole system (Johnson, 2009). A simple illustration could be obtained by giving a very intelligent child to solve the puzzle of the world map, but with all the pieces inverted and randomized. The child most likely will quickly succeed in assembling Australia, Madagascar, England, Italy and some other parts of the world having unique boundaries. Inverting the partially *solved* puzzle will demonstrate that there are many wrong pieces perfectly fitting inside some of the *solved* countries! This is because the puzzle pieces are built out of only a few set patterns (except the boundaries). Based on our current state of knowledge, the cosmic evolution is being guided by only a few cosmic logics (laws), perhaps constituting only four forces as theorized now. So, as our scientific epistemology evolves further, we should not be surprised to discover that some of our mathematical logics (theories) that have fitted the data very well, are *happy coincidences,* like segments of the solved world map-puzzle. We should remain humble enough to hold our judgments from declaring that we have already found the *God's Equation* (Aczel, 1999).

Our attempts to iteratively try to refine our imagined map, personally or collectively, to depict a particular interaction process will be greatly benefited if we explicitly lay out our personal epistemologies as to how we have framed our questions, what hypotheses we have constructed to fill up the missing information (IIC), etc. In this context, it is rather surprising to recognize the 2500 year old advice by Gautam Budhha of India: To visualize the invisible cosmic elephant, it is better to learn to think the way the blind people do and then collaborate to draw a preliminary sketch (map) of the perceived elephant and then keep on improving upon it iteratively through perpetual debates. That our personal perceptions are dictated by our survival needs, and hence far from our capability of objective analysis of actual reality, was brought to our understanding by Buddha by *framing a brilliantly simple question*: How can blind people *observe* and *model* an elephant? Objectively speaking, we are blind! Our neural network literally *imagines* images, smells, tastes, pains and pleasures out of the various sensorial inputs (registered *transformations* in our bodily measuring instruments!) for the key purpose of survival and successful evolution. We must now learn to use our free brain to go beyond survival and start consciously constructing road-maps for our purposeful evolution.

3.2 Applying IPM-E to discover the NIW-principle (Non-Interaction of Waves)

Detection of the presence of electromagnetic waves (energy) is always carried out by some material based detector that has the intrinsic capability to respond to the incident oscillatory electric field and then absorb some energy from the field and consequently undergo some measurable transformation. We never *see* any electromagnetic waves, *visible or invisible*. We can only indirectly interpret the presence of some EM wave(s) from the various transformations (*meter readings*) presented to us by our measuring instruments based on initial transformations experienced by the key sensors (detectors). In fact, EM waves do not possess any real physical attribute that we can call color. Frequency is the real physical parameter. Three types of retinal molecules in our eyes, sensitive to *red, green* and *blue* frequencies, communicate with our visual cortex to generate the interpretation of colors, which have evolved for the necessity of our successful evolution. Successful evolution-congruent neural networks, which are at the root of our personal epistemologies, are not the most reliable thinking machines to discover cosmic logics in a straight forward manner. Their job is to successfully adapt for survival to the changing environment generated through the interplay of cosmic logics within the evolving cosmic system. Development of scientific epistemology, to see the world as is, not as required for survival, requires extra awareness to systematically cultivate objective thinking. We need to learn to *observe* what are happening *inside-the-boxes*, first inside our own active brains, then within the biosphere and finally in the cosmo-sphere, while *standing outside-the-box* using our evolved imaginative *free-brain*. This is where IPM-E can play the role of an important thinking tool for consciously constructing our *purposeful* evolution. Darwin's *Natural Selection* have been helping us to evolve as a species. But the evolution of our free-brain is being both accelerated and strongly guided by our *Cultural Selection* (Roychoudhuri, 2010a). Our conscious awareness of this *Cultural Selection,* e.g. IPM-E in preference to MDM-E, and eventually, something superior, is needed to assure the successful evolution of human scientific minds. We can then take the responsible ownership of the biosphere and then eventually graduate to become *cosmo-zens* (cosmic citizens)!

Let us come back to the application of IPM-E to understand the interaction processes behind *interference of light*. What we call *interference of light* is, in reality, a resultant transformation experienced by some detector in response to simultaneous stimulations induced by multiple superposed beams due to their simultaneous presence on the detecting molecule. If its intrinsic properties do not allow it to experience simultaenous stimulations, it can not report any superposition effects; and there is no interference! Responses (or measured transformations) of all light detectors are further complicated by the follow-on steps through which we amplify and register the final transformations as recorded data. In every step, we may loose some information about the original transformation we intend to measure due to further interaction steps during these amplification stages (recall IIC).

The reader may now raise the *interaction process* based question as to whether (i) the superposed EM waves first create a new resultant EM wave and then impose the effective stimulation on the detector, or (ii) the detecting molecule separately but simultaneously responds to all the oscillating E-fields. The mathematical model and experimental data are in favor of the latter proposition. The resolution was carried out by using a two-beam interferometer with variable polarizers in its two arms (Roychoudhuri & Barootkoob, 2008). Coherent or not, EM waves cross through each other, or propagate coilinearly, without creating any transformations in their individual physical parameters. This property of Non-Interaction of Waves (NIW) is a generic principle of nature. No propagating waves interact with each other in the linear domain. They pass through each other unperturbed without any permanent re-arrangement in their phase or energy distribution, either in the space or in the time domain. This is the universal NIW-principle (Roychoudhuri, 2010b). Otherwise the fidelity of our daily vision (due to light waves) and hearing (due to sound waves), critical to our successful evolution, would not have worked! If the light beams interacted with each other to create new energy distributions, our Radios would not have been able to extract clear music out of multiple carrier frequencies and our internet signal, utilizing WDM (Wavelength Domain Multiplexing) technologies, would have created undecipherable heterodyne noise instead of clear data for each channel.

In reality, Hyathem around 1080, clearly noted in his book that light beams belonging to different sources do not interact with each other (Ronchi, 1970). He simply observed the images of a set of candles formed through the pinhole of a *Camera Obscura,* which remained unaltered even when he lit or unlit different candles. This observation should not be explained away assuming *incoherence* of light (Roychoudhuri, 2006b). Surprisingly, Dirac during late 1920's, while quantizing the electromagnetic field, and based on already developed Bose-Einstein statistics, clearly recognized that photons are non-interacting Bosons. And yet, to accommodate the apparently successful classical model of *interference of light*, declared, "...each photon then interferes only with itself" (Dirac, 1974) as if self-interference (appearance and disappearance) of a stable elementary particle is logically self-consistent with our mathematical logics and *cause-effect* driven cosmic logics! In fact, computation of direct photon-photon interaction, or scattering cross section in the *material-free vacuum,* has been found to be unmesurably small (Tomasini et al, 2008).

We conclude this section on epistemology by raising the following questions to the readers. Could, treating the hypotheses proposed & developed successfully by Newton, Einstein, Heisenberg, etc., as final and inviolable theories, just because they have been *working*, threaten the further evolution of human scientific minds (logics), and hence, the evolution of our sciences and technologies, and hence, eventually our very existence? Are we all suffering from the *Messiah Complex* so deep that we are irrecoverably believing that the final foundation for our *scientific edifice has already been firmly established*? Should we just stay focused on discovering only the right size *stones* to fit into the already established scientific edifice?

4. Applying IPM-E and the NIW-Principle to resolve the ambiguities

Once we have accepted the NIW-principle, then IPM-E dictates us to hold our final conclusions regarding the phenomenon we call *interference of light* before we have thoroughly explored light-matter interaction processes and the impact of the quantum properties of detecting molecules during the process of photo electron emission (or electron transfer from valence to the conduction band). Obviously, we cannot simply sum the optical fields directly as we have been doing for centuries (Roychoudhuri 2010b, Roychoudhuri 2009a), albeit the fact that TF-FT, based on summation of fields, has been *working* and Maxwell's wave equation *accepts* any *linear combination* (sum) of harmonic waves as its solution, as $\vec{E}_{total}(t)$ shown below:

$$\vec{E}_{total}(t) = \vec{E}_1(t)\exp(i2\pi\nu_1 t) + \vec{E}_2(t)\exp(i2\pi\nu_2 t) + \ldots + \vec{E}_n(t)\exp(i2\pi\nu_n t) \qquad (3)$$

While the above equation may appear to be a mathematically correct representation of the physically superposed n-waves $\vec{E}_n(t)$, it is a logically wrong representation of physics because the mathematical operation sign "+" in physics represents an interaction mediated by some force of interaction; and there are no force of interaction between waves in the linear domain. We may create the convention of representing a physically superposed but non-interacting set of waves by using the semicolon sign as separator of different waves Roychoudhuri 2006a):

$$\vec{E}_{set}(t) \subset [\vec{E}_1(t)\exp(i2\pi\nu_1 t);\ \vec{E}_2(t)\exp(i2\pi\nu_2 t);\ \ldots\ ;\ \vec{E}_n(t)\exp(i2\pi\nu_n t)] \qquad (4)$$

We can now give a better operational interpretation of the mathematical observation that Maxwell's wave equation can accept any linear combination of harmonic waves: Same volume of the vacuum or a noninteracting medium in the linear domain, can allow simultaneous presence and co-propagation of multiple EM waves. Let us now assume that we have introduced a photo detector within the physical volume of the superposed waves. Then each one of the E-vectors try to polarize the detecting molecules (or their clusters) and make them oscillate to the induced frequencies provided the intrinsic quantum properties of the molecules allow that. Let us use the symbol for this linear polarizability as ${}^1\chi(\nu)$, then the various stimulations can be represented as:

$$\vec{\psi}_n(t) \equiv \vec{d}_n(t) = [{}^1\chi(\nu)][\vec{E}_n] \Rightarrow {}^1\tilde{\chi}(\nu)E_n \qquad (5)$$

The summation sign we use for superposition effect implies a process. Something has to carry out this process to keep the system causal. So the resultant stimulation can now be represented by Eq.6, while the summation, or the superposition effect is now carried out by the same dipole molecule (or their cluster) as they respond simultaneously and absorb energies from all the *quantum compatible* fields. A similar interaction relation can also be assumed for intra-cavity saturable absorbers:

$$\vec{\Psi}_{total}(t) = \Sigma_n \vec{\psi}_n = \vec{d}_1(t)\exp(i2\pi\nu_1 t) + \vec{d}_2(t)\exp(i2\pi\nu_2 t) + \ \ldots\ + \vec{d}_n(t)\exp(i2\pi\nu_n t) \qquad (6)$$

If the quantum transition bands of the detecting dipoles can respond simultaneously to only a subset of the frequencies from ν_p to ν_{p+r} , determined, say, by the band widths and the band gap of a solid state detector (or, a saturable absorber), then the effect of only these superposed waves will be experienced by the detector. Normally, the rest of the waves will remain unrecognized by the detector. The terms within the curly bracket below corresponds to the total resultant stimulations experienced by the detector belonging to the *quantum compatible* frequencies. To underscore the NIW-principle, we are showing the other superposed waves, outside the curly bracket and without the summation signs, since there are no interaction between these waves.

$$\vec{\Psi}_{total}(t) = \vec{E}_1(\nu_1,t)e^{i2\pi\nu_1 t};...\left\{\vec{d}_p(v_p,t)e^{i2\pi\nu_p t} + \vec{d}_{p+1}(v_{p+1},t)e^{i2\pi\nu_{p+1} t} + ... + \vec{d}_{p+r}(v_{p+r},t)e^{i2\pi\nu_{p+r} t}\right\}; \quad (7)$$
$$... \ \vec{E}_n(\nu_n,t)e^{i2\pi\nu_n t}.$$

For example, a silicon detector will not respond to either the high frequency (energy) X-ray or low frequency (energy) infra-red photons. The transformation experienced by the detector, or transfer of photo electrons from the valence to the conduction band due to energy transfer from all the *quantum compatible* fields to the photodetector, is given by the quantum mechanical recipe, the ensemble average of the square modulus of the effective complex amplitude stimulation:

$$\mathrm{D(t)} = \left\langle \vec{\Psi}^*(\mathrm{t}) \cdot \vec{\Psi}(\mathrm{t}) \right\rangle \quad (8)$$

In reality, all frequencies, even those outside the *quantum compatible* band, stimulate the detecting dipoles accessing their nonlinear polarizabilities, however, normally the strengths of these nonlinear stimulations are very weak compared to the linear stimulations. The stimulations due to higher order polarizability terms are never zero in the real world:

$$\vec{\Psi}_{total}(t) = \Sigma_n {}^1\chi(\nu)\vec{E}_n(t)e^{i2\pi\nu_n t} + \Sigma_n {}^2\chi(\nu)\vec{E}_n^2(t)e^{i2\pi\nu_n t} + \ ... \ + \Sigma_n {}^m\chi(\nu)\vec{E}_n^m(t)e^{i2\pi\nu_n t} \quad (9)$$

With these background, we can now approach to resolve the ambiguities, contradictions and paradoxes mentioned in Section 1.

4.1 Can superposed modes create a new mean frequency?

Readers are now requested to recall Section 2.1 and Fig.1 and Fig.2. IPM-E and the consequent NIW-principle tell us that the derivation of a *mean frequency* by mathematical summation of in-phase periodic frequencies in the linear domain is not a physical process that is allowed in nature. A saturable absorber placed inside a He-Ne laser, having the inhomeogeneously broadened gain medium, must always run at all the allowed longitudinal modes (Allen et al 1969). By virtue of the NIW-principle, these frequencies remain independent of each other even when their phases are locked with the help of cavity reverberation through the intracavity saturable absorber. The intensity absorbed by the saturable absorber, $I_{absrb.}$, is given by:

$$I_{sat.}(t) = \left|\Sigma_{n=0}^{N-1} {}^1\chi a e^{i2\pi(\nu_0+n\delta\nu)t}\right|^2 = {}^1\chi^2 \mathrm{N}^2 a^2 \frac{\sin^2 N\pi\delta v t}{\sin^2 \pi t\delta\nu} \quad (10)$$

The absorber simply behaves as a fast temporal gate and, of course, the absorption efficiency rapidly enhances as the various modes become phase locked, which, in turn, makes the width of the individual modes much narrower than those in a free running CW laser (see Fig.3c). However, for a hohmogeneously broadened gain medium like Ti-Sapphire, there is no *locking of modes* since the gain medium tends to run at the gain line center under steady pulsing conditions (see Fig.1b). Accidental in-phase spontaneous emissions around the gain line center triggers the saturable absorber to start functioning as the required time-gating device. The readers should note that except for a constant factor ${}^1\chi^2$ representing the linear polarizability of the saturable absorber molecules, this Eq.10 looks identical to the Eq.2, which we have been using to justify the *mode lock* hypothesis, as summation of modes, to be a correct one! Working mathematics may confuse us, but a systematic application of of IPM-E will direct us in the right direction.

4.2 Can a homogeneously broadened gain medium oscillate in all the allowed cavity modes?

Readers are requested to recall the discussion in Section 2.2. In general, a steadily oscillating laser with homogeneously broadened gain medium will always run at the gain line center (Siegman 1986), unless an intra-cavity device flattens the gain line envelop. Note that broader the gain bandwidth, shorter is the atomic life time and hence faster the lasing atom can recycle itself to contribute energy by stimulated emission in the next pulse. Only during the transient oscillation periods, as the laser is repetitively Q-switched, can one observe the presence of multiple longitudinal modes due to transient competitions enhanced by the presence of strong spontaneous emissions at all allowed frequencies. Of course, if the medium develops many defect centers, effectively creating *inhomogeneous* gain regions, one can also observe the presence of multiple longitudinal modes. And an excellent example is the case of a Q-switched diode laser, essentially homogeneously broadened, which we are analysing in Section-2.6 and 4.6 (see Fig.5a). But a well stabilized and periodically pulsing laser will always select the frequency at the highest gain point, which can be tuned by inserting an intra-cavity frequency tuner that introduces preferential losses over the entire gain line except around the desired frequency. This is why the *heroic* measurement of the carrier frequency of a *mode locked* Ti-Sapphire laser shows a unique carrier frequency under the pulse envelope (Fig.1b).

4.3 Can the time-frequency Fourier theorem be a principle of nature?

Readers are requested to recall the discussion in Section 2.3. The time-frequency Fourier theorem (TF-FT) is a very useful and a self-consistent mathematical theorem, however, by virtue of the NIW-principle, it cannot represent superposition of EM waves unambiguously in the most general sense. Further, it certainly does not model the physical process of light-matter interaction processes in the mathematical form generally presented in books. However, we have been effectively using the TF-FT as if it is a principle of nature as it is capable of predicting the correct measured data, except for a *constant*, the polarizability of the detecting dipole by the incident electric vector. While the TF-FT has been a very successful mathematical tool in most of the branches of physics, it incorrectly implies field-field interaction:

$$a(t) = \int_0^\infty \tilde{a}(f)\exp[i2\pi ft]df; \quad \tilde{a}(f) = \int_0^\infty a(t)\exp[-i2\pi ft]dt \tag{11}$$

Let us take a particular example to illustrate our point. Consider a two-modes, CW He-Ne laser beam of equal amplitude, passing through a Michelson interferometer (Michelson, 1962; the discoverer of the Fourier transform spectroscopy). Each mode will send two beams with a fixed phase delay on the detector that will experience the simultaneous presence of four beams (Roychoudhuri 2006b):

$$\begin{aligned} D(\tau) &= \left| {}^1\chi e^{i2\pi\nu_1 t} + {}^1\chi e^{i2\pi\nu_2 t} + {}^1\chi e^{i2\pi\nu_1 (t+\tau)} + {}^1\chi e^{i2\pi\nu_2 (t+\tau)} \right|^2 \\ &= {}^1\chi^2 \left| e^{i2\pi\nu_1 t} + e^{i2\pi\nu_1 (t+\tau)} \right|^2 + {}^1\chi^2 \left| e^{i2\pi\nu_2 t} + e^{i2\pi\nu_2 (t+\tau)} \right|^2 \\ &= {}^1\chi^2 [4 + 2(\cos 2\pi\nu_1\tau + \cos 2\pi\nu_2\tau)] \end{aligned} \tag{12}$$

As per Michelson's assumption, light beams corresponding to different optical frequencies are *incoherent* and hence they do not *interfere*. Accordingly, in the second step of Eq.12, we have applied the square modulus operation on each pair of terms separately corresponding to the two different frequencies. Note that the application of this correct but *arbitrary* mathematical logic depends upon the *incoherency* hypothesis. As per Michelson, one then extracts the oscillatory part of the data, after discarding the *DC* part. Application of the mathematical Fourier transform operation then yields the extraction of the source *spectrum*:

$$\begin{aligned} &D_{Osc.}(\tau) = \cos 2\pi\nu_1\tau + \cos 2\pi\nu_2\tau \\ &\text{Or, } \tilde{D}(\nu) = \delta(\nu - \nu_1) + \delta(\nu - \nu_2) \end{aligned} \tag{13}$$

Michelson's assumption of *incoherence* between frequencies were observationally correct as he could only use very slow optical detectors, slow eyes and photographic plates available in those days. He also incorrectly assumed light beams interfere by themselves to create energy re-distribution (fringes). Scrutinizing Eq.12 reveals why his summing the fields and our summing the dipole stimulations give identical results, except for a constant multiplying factor. This is because the mathematical logic allows us to take the common constant in an equation out of all terms and place it as a common multiplier. This is good enough for MDM-E driven science, but because of the application of this mathematical logic, we have lost sight of the real interaction process for centuries. Further, since the discovery of fast optical detectors, we know that different optical beams corresponding to different frequencies are really not *incoherent* and they give rise to heterodyne difference (beat) frequencies. We now have a new tool, called heterodyne spectroscopy:

$$\begin{aligned} D_{osc}(t,\tau) &= {}^1\chi^2 \left| e^{i2\pi\nu_1 t} + e^{i2\pi\nu_2 t} + e^{i2\pi\nu_1(t+\tau)} + e^{i2\pi\nu_2(t+\tau)} \right|^2 \\ &= {}^1\chi^2[4 + 2\cos 2\pi(\nu_1 - \nu_2)t + 2[\cos 2\pi(\nu_1 - \nu_2)(t+\tau) \\ &\quad + \cos 2\pi\{(\nu_1 - \nu_2)t + \nu_1\tau\} + \cos 2\pi\{(\nu_1 - \nu_2)t - \nu_2\tau\}] \\ &\quad + 2\,{}^1\chi^2[\cos 2\pi\nu_1\tau + \cos 2\pi\nu_2\tau]. \end{aligned} \tag{14}$$

Note that in the above Eq.14, if we have a slow detector, or a an electric circuit with a long LCR time-constant, the time dependent terms will average out to zero leaving behind the DC term of Michelson and we will be left with what Michelson's photographic plate would have registered, which is same as the final result of Eq.12. The readers should now appreciate that application of IPM-E and the NIW-principle are critical tools for developing working engineering principles behind optical instruments.

It is customary to expect new predictions from new principles. The prediction of the NIW-principle is that neither Fourier synthesis, nor Fourier decomposition can take place in the domain of optical frequencies where the detection is carried out by quantized atoms, molecules or their assemblies, as in solid state detectors. We have validated these predictions through two separate experiments. In the first experiment (Lee & Roychoudhuri, 2003; Lee, 2004), we have superposed two phase-stable optical beams with two different optical frequencies 2GHz apart and exactly bisecting one of the fine-structure Rb-resonance line. The Rb-atoms could not respond to these collinear beams since both the frequencies were outside their quantum stimulation boundary, which proves that Fourier synthesis to generate a new average frequency did not take place. When we tuned either one of these two frequencies matching the actual Rb-resonance line, strong resonance fluorescence was clearly observable.

In the second experiment (Roychoudhuri and Tayahi 2006), we demonstrated that optical detectors cannot respond to Fourier decomposition frequencies corresponding to the envelope of a pulse. We carried out a heterodyne experiment by mixing the beam from a stable external-cavity CW diode laser beam with a steady pulse train derived from a separate Bragg-grating-stabilized diode laser but after the beam was amplitude modulated by an external LiNbO3 modulator. The main heterodyne line remained unaltered whether the second laser beam was modulated or kept CW, proving that photodetectors cannot respond to Fourier frequencies corresponding to the pulse envelope. Of course, when the second diode beam was modulated, an oscillatory current corresponding to the modulation frequency was separately observable, besides the unaltered heterodyne beat frequency. The broader implications of using TF-FT as a principle of nature in the field of classical and quantum optics has been summarized in (Roychoudhuri, 2009a; 2007b); and earlier realization of the concept can be found in (Roychoudhuri, 1976).

4.4 Why regular CW He-Ne lasers show mode-lock-like pulsations?

Readers are now requested to recall the discussion in Section 2.4 and Fig.3 where an ordinary He-Ne laser appears to behave like a *mode locked* laser. Such a deceptive behavior from a He-Ne laser is all the more reason to apply IPM-E to detection processes and then appreciate that the conventional mode locking assumption behind the physical pulsation of a laser is incorrect. In a CW He-Ne laser, as per the NIW-principle, each mode oscillate independent of the others, but naturally with phases that remain steady relative to each other for periods of miliseconds or longer due to very high-Q cavity (high reflective cavity mirrors). So, when a multimode He-Ne beam stimulates a fast detector, its resultant *coherent amplitude* stimulation can be described for intervals of milliseconds as:

$$d(t) = \Sigma_{n=0}^{N-1} {}^{1}\chi e^{i2\pi(\nu_0+n\delta\nu)t} \approx {}^{1}\chi \frac{\sin N\pi\delta\nu t}{\sin \pi\delta\nu t} e^{i2\pi\nu_0 t} \tag{15}$$

Eq.15 is identical to Eq.1 except for the multiplicative constant ${}^{1}\chi$ representing the linear polarizability of the detecting molecules. The temporal variation in the detector current, when normalized, is given by:

$$D(t) = |d(t)|^2 = ({}^{1}\chi^2 / N^2)[\sin^2 N\pi\delta\nu t / \sin^2 \pi\delta\nu t] \tag{16}$$

With a fast oscilloscope trace, *using internal trigger signal,* as is the case for Fig.3b, one will naturally observe *mode-lock-like* oscillation in the detector current. For a quantitative computational match, one needs to set N=5 in Eq.16 and adjust the intensities of the five modes as per the spectrum in Fig.3a.

The self-heterodyne lines of Fig.3d can also be appreciated by re-deriving the detector current in a different trigonometric form, as shown as in Eq.17, and then isolate the oscillatory term from the DC term, as in Eq.18:

$$D(t) = d^*(t)d(t) = \frac{{}^{1}\chi^2}{N} + \frac{2\,{}^{1}\chi^2}{N^2}\sum_{p=1}^{N-1}(N-p)\cos[2\pi p\delta\nu t] \tag{17}$$

$$D_{osc}(t) = \frac{2^1 \chi^2}{N^2} \sum_{p=1}^{N-1} (N-p)\cos[2\pi p \delta \nu t] \tag{18}$$

It is the electronic design of an ESA to display the oscillatory current in terms of sum of harmonic terms, as in Eq.18. Again, the *internal sampling time constant* must be set at about a millisecond or less (He-Ne laser coherence time). For N=5, we should have four harmonic lines with difference frequencies at $\delta\nu$, $2\delta\nu$, $3\delta\nu$ and $4\delta\nu$, as can be seen in Fig.3c around the central *zero frequency* line (Lee 2004; Roychoudhuri et al 2006).

4.5 Is synthetic mode locking possible?

Readers are now requested to recall discussions in Section 2.5 and Fig.4. Again, by virtue of the NIW-principle, it is not possible to generate oscillatory light energy simply by superposing phase-stable CW light beams with a periodic frequency difference in free space. However, based on the observation and analysis of section 4.4, it is clear that one can always generate *pulsating response out of a material medium* whose energy absorbing dipoles can respond to all the frequencies simultaneously and can recycle their measurable transformation with the required high speed (Lee 2004; Roychoudhuri et al 2006). If the phase locked multi-frequency source has a pulse envelope a(t), the measured width will be narrower than a(t) squared. *If the detector response time is slower than the periodicity* $\tau = (1/\delta\nu)$ *implied by the Eq.16, the observation will not corroborate a mode-lock-like behavior (periodic pulsation of energy).*

4.6 Can autocorrelation data unambiguously determine the existence of ultrashort pulses?

Readers are now requested to recall the discussion in Section 2.6 and Fig.5. There we have shown that a Q-switched diode, with a built-in saturable absorber was able to generate 12ps pulse for each Q-switching current pulse. But the autocorrelation trace (Fig.5b and c) showed 94fs pulse train within each 12ps Q-switched pulse. Are these fs spikes artifacts of measurement process, or the laser is really partially *mode locked*? The author believes that they are artifacts of measurement and product of very strong nonlinear second harmonic energy conversion due to high-visibility fringes as the path delay changes by the cavity round-trip delay in the non-colinear second harmonic-generation interferometer. In such an interferometer the laser beam is replicated into two beams and then superposed with a variable path delay.

People who are familiar with two-beam interferometry or holography using a CW multimode laser to achieve unit visibility fringes, they must set the relative path delay between the two beam exactly as $n\tau$ (integral multiple of cavity round trip time). Fig.6 below shows the computer plot of two-beam fringe visibility, or the modulus of the autocorrelation function, using the longitudinal mode spectrum (without the background pedestal) shown in Fig.5a.

Amplitude of the second harmonic generation is proportional to the intensity of the stimulating signal and hence the recorded intensity in the non-colinear 2nd harmonic signal (autocorrelation spikes of Fig.5b & c) is proportional to the square of the fringe visibility function shown in Fig.6, and hence they are even narrower than these visibility peaks. As mentioned in Section 2.6, the existence of a true ~100fs pulse would imply a registered spectral broadening of 10,000 GHz. But our spectrum consists of about 32 individual lines of half-width 83.3 GHz wide over a pedestal of approximately 1,300GHz half-width, not 10,000GHz! We believe that the pedestal represents the broad spontaneous emission, real carrier frequencies across the entire gain bandwidth, rather than an instrumental spectral broadening due to a 100fs laser pulse.

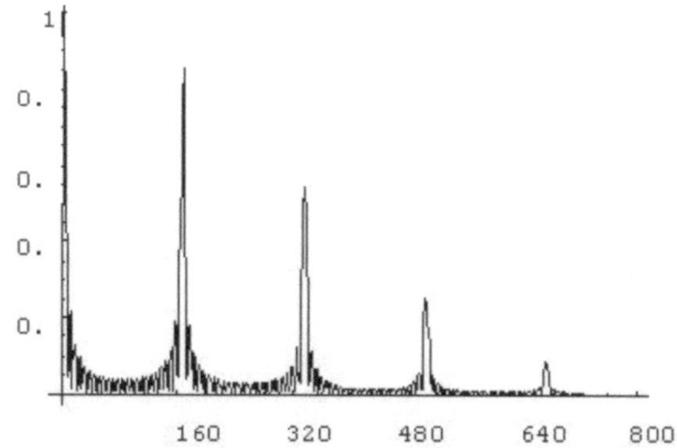

Fig. 6. Computer simulation of sharp but decaying fringe visibility due to superposition of two replicated beams produced from a beam containing 32 mode frequencies of Fig5a.

To further appreciate the arguments presented here, we need to re-formulate foundational mathematical relations for spectrometry. Attempts to visualize the processes behind recorded data (measured transformations in detectors), will reveal further roots behind some of the misconceptions and ambiguities discussed above.

4.6.1 Generalized formulation of spectral fringe broadening.

We will now derive formulations to distinguish between apparent spectral fringe broadening due to a pulse carrying a single carrier frequency and multiple carrier frequencies under the same temporal envelope. But, let us first justify this attempt by underscoring the paradox built into the classical formulation of spectrometry. Classical spectrometry is formulated based on propagating a CW *monochromatic* beam $\exp[-i2\pi\nu_0 t]$ that has a constant amplitude and a unique carrier frequency existing over all time, which cannot exit in the real world as it would contain infinite energy and violate the law of conservation of energy. For a grating with N-slits and a step delay τ, the classical approach derives a spectral fringe of a finite width, $\delta\nu = (1/N\tau)$ (Born & Wolf, 1980), instead of a *delta function* $\delta(\nu-\nu_0)$:

$$I_{cw}(\nu,\tau) = \left|\Sigma_{n=0}^{N-1}(1/N)e^{-i2\pi\nu\tau}\right|^2 = \frac{1}{N^2}\frac{\sin^2 N\pi\nu\tau}{\sin^2 \pi\nu\tau} \equiv \frac{1}{N} + \frac{2}{N^2}\Sigma_{p=1}^{N-1}(N-p)\cos[2\pi p\nu_0\tau] \quad (19)$$

Then it appropriately identifies that this fringe width $\delta\nu$ does not correspond to the physical presence of any new carrier frequencies, rather it is an instrumental artifact, defined as the *instrumental CW response function,* which we must deconvolve from experimental spectral fringes to extract actual *spectrum*. However, when a pulsed light beam generates a spectral fringe, we automatically assume that it represents the real pulse spectrum (or, real frequency content) and the broadening is due to Fourier spectrum that a pulse *inherently* contains. This conjecture is based upon the assumption that the TF-FT is a functional principle in nature, which we have argued against in this and the last sections. In the next few paragraphs, we will derive the most general spectrometer response function due to a time-finite pulse, $a(t)\exp[-i2\pi\nu_0 t]$, that has a unique carrier frequency. We will propagate this carrier frequency and the envelope function through spectrometer directly, instead of the Fourier spectrum of the pulse envelope existing over all time.

Let us now apply our IPM-E and the NIW-principle to optical spectrometers. From the standpoint of IPM-E, spectrometers like gratings and Fabry-Perots are *linear* pulse replicators; they function as wavefront dividers (as in an N-slit grating) and amplitude dividers (as in a Fabry-Perot interferometer of finesses ~N), respectively. These *linear* beam replicating devices cannot generate new optical frequencies out of the incident beam, irrespective of the length of the pulse (fs or infinite CW). Further, according to the NIW-principle, these replicated beams *do not interfere* or interact with each other by themselves to create new energy distribution (fringes). The optical components associated with the spectrometers physically superposes the replicated beam-train with a periodic time delay on a detector, which then registers a transformation according to its absorbed energy distribution determined by the actual carrier frequencies. Each spectral line will have the same instrumental line broadening determined by the common incident pulse envelope. We will consider here the case of a grating only. For details of derivations and other implications in the field of optics, the interested readers should consult the following references (Roychoudhuri & Tayahi 2006; Roychoudhuri 2004; Roychoudhuri et al 2003). For the purpose of easy illustration, we are considering only a transmission blazed grating with four steps. It replicates the incident pulse into N periodically delayed pulses with a step delay of τ as shown in Fig.7.

We define the total time delay between the first and the N-th replicated pulse as the spectrometer time constant, τ_0, which connects with the classical definition of resolving power R=mN as shown below in Eq.20. For a grating working in the m-th order, we can write:

$$\tau_0 = N\tau = N.(m\lambda / c) = R\lambda / c = R / \nu \tag{20}$$

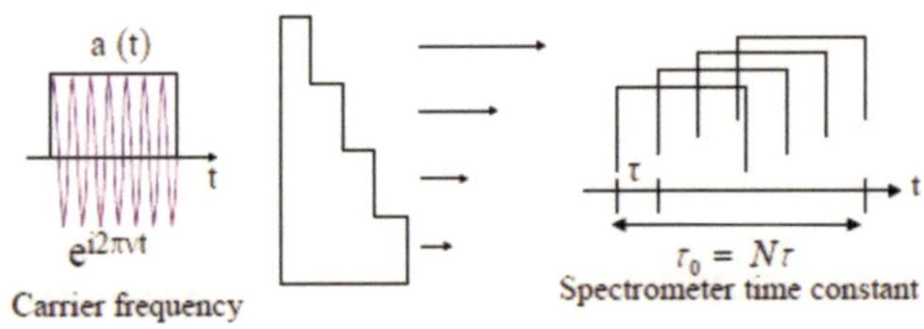

Fig. 7. Generation of a train of N pulses from a single incident pulse by an N-slit grating with a periodic delay of τ. In a spectrometer this train of partially overlapped pulses are superposed on a detector whose response generate the spectrometer fringes (from Roychoudhuri et al, 2003).

From the view point of *mapping the process behind spectrometer fringe evolution*, τ_0 plays a critical role in spectrometry, to be apparent later, which was never explicitly appreciated in classical spectrometry. The normalized dipolar amplitude stimulation of the detector at the output of the spectrometer can be given by:

$$d_{\text{det}}(t,\nu_0) = \Sigma_{n=0}^{N-1}(^1\chi / N)a(t-n\tau)\cdot\exp[i2\pi\nu_0(t-n\tau)] \tag{21}$$

The detector then generates fringes by absorbing energies dictated by the realtive delays in the superposed pulse train determined by $n\nu_0\tau$:

$$D_{\text{det}}(t,\nu_0) \equiv \left| d_{\text{det}}(t,\nu_0)\right|^2 = \left|\Sigma_{n=0}^{N-1}(^1\chi / N)a(t-n\tau)\cdot\exp[i2\pi\nu_0(t-n\tau)]\right|^2 \tag{22}$$

A ps streak camera will display an apparent spectral fringe whose width will vary in time as per Eq.22. We certainly should not interpret this time varying fringe width as time evolving spectrum! Note that the case of temporally chirping spectrum is not modeled by Eq.22. If the detector energy is integrated over a duration longer than that of the entire pulse train, which ordinary spectrometers do, then we have:

$$I_{pls}(\nu_0,\tau) = \int D_{\det}(\nu_0,t)dt = \int dt\left|\Sigma_{n=0}^{N-1}({}^1\chi / N)a(t-n\tau)\cdot\exp[i2\pi\nu_0(t-n\tau)]\right|^2 \qquad (23)$$

Now we get time integrated fringe pattern as the *pulse-response-function* for a single carrier frequency,

$$I_{pls}(\nu_0,\tau) = \frac{{}^1\chi^2}{N} + \frac{2\,{}^1\chi^2}{N^2}\Sigma_{p=1}^{N-1}(N-p)\gamma(p\tau)\cos[2\pi p\nu_0\tau] \qquad (24)$$

Where, $p \equiv |n-m|$ and the autocorrelation factor $\gamma(p\tau)$ is defined by:

$$\gamma(p\tau) = \int d(t-n\tau)d(t-m\tau)\,dt \Big/ \int d^2(t)\,dt \qquad (25)$$

If there are many carrier frequencies with a normalized intensity function $b(\nu)$ under the same pulse envelope $a(t)$, as is the case discussed for a Q-switched diode laser in Fig.6, then the time integrated fringe spectrum should be represented by the convolution:

$$I_{mult.freq.}(\nu,\tau) = b(\nu)\otimes I_{pls}(\nu,\tau) \qquad (26)$$

Let us now bring connections to classical spectrometric results. When the pulse width δt is much greater than the spectrometer time constant τ_0, say, $\delta t \geq 10\tau_0$, then the overlap between all the replicated pulses will be approximately complete and hence all the autocorrelation factor $\gamma(p\tau)$ will be close to unity for all values of p. Then using Eq.24 and 25, we can write:

$$\underset{\substack{\gamma\to 1\\ \delta t>10\tau_0}}{Lt.}\; I_{pls}(\nu_0,\tau) = \frac{{}^1\chi^2}{N} + \frac{2\,{}^1\chi^2}{N^2}\sum_{p=1}^{N-1}(N-p)\cos[2\pi p\nu_0\tau] \equiv \frac{{}^1\chi^2}{N^2}\frac{\sin^2\pi N\nu_0\tau}{\sin^2\pi\nu_0\tau} = {}^1\chi^2 I_{cw}(\nu_0,\tau) \qquad (27)$$

Thus, for a pulse of width δt that is much longer than the time constant of a spectrometer $\tau_0 = R\lambda / c$, our time-domain formulation converges identically to the classical result, except for the characteristic detector polarizability constant ${}^1\chi^2$. After recovering the classical *monochromatic-beam-result*, we would like to underscore that our *pulse-respnse-function* $I_{pls}(\nu,\tau)$ of Eq.24 can also be expressed, using Parseval's energy conservation theorem (Roychoudhuri et al 2003), as a convolution between the classical *CW-response-function* and the normalized Fourier intensity *spectrum* $\tilde{A}(\nu)$, the normalized square modulus of the Fourier transform of $a(t)$:

$$I_{pls}(\nu,\tau) \approx \int_{-\infty}^{\infty} |d_{\det}(t,\nu)|^2 dt = I_{cw}(\nu)\otimes\tilde{A}(\nu) \qquad (28)$$

It is now obvious that the *apparent* spectral fringe broadening due to a pulse carrying a single carrier frequency, can be *thought of* as that due to the Fourier frequencies obtainable by applying the TF-FT on the pulse envelope. One can now appreciate why $I_{pls}(\nu_0,\tau)$ is generally considered to be the response due to a *transform limited pulse*; it is the spectrometer response due to a pulse with a single carrier frequency! However, this energy distribution in the broadened fringe $I_{pls}(\nu_0,\tau)$, or the width $\delta\nu$ of the *pulse response function,* must not be considered as due to the presence of many real physical frequencies; they correspond to the energy spread from the same carrier frequency ν_0. Based on the Fourier convolution relation of Eq.28, the bandwidth product $\delta\nu\delta t \approx 1$ (*time-frequency uncertainty* or the *bandwidth limit*) appears to be working! However, the extra broadening of the grating fringe is due to *partial overlap* of the train of finite pulses (see Fig.7). It has physically nothing to do with the time-frequency uncertainty relation; it is just a mathematical coincidence. The width of the fringe is a joint artifact of the instrument and the pulse envelope function $a(t)$.

This is one more example on how to appreciate the power of IPM-E over MDM-E in correctly identifying the real interaction processes that go on in nature to generate measurable data. While MDM-E makes us draw wrong physical conclusions, IPM-E guides us to extract out the correct interaction process behind spectrometers; they are just *linear pulse replicators*! Successes achieved through *curve-fitting* mathematical formulations makes us believe that TF-FT is a principle of nature; but it is not. We should consider the distribution of energies corresponding to the presence of real physical carrier frequencies as the physical spectrum, $b(\nu)$ of Eq, 26. A(v), the square modulus of the Fourier transform of the pulse envelope function a(t), does not represent any physical frequencies.

5. Summary

We have given six concrete cases of laser pulse parameters for which standard text books give ambiguous or incorrect physical interpretations of laser pulse characteristics based on the century old premise that light beams *interfere* (interact) and regroup their energy by themselves. We have presented a better methodology of thinking, the Interaction Process Mapping Epistemology (IPM-E). IPM-E helps us use our creative human logics, while leveraging our mathematical logics, to access a better glimpse at the unknown cosmic logics. But we need to refine that *glimpse* into a more and more accurate representation of the cosmic logics by relying upon refining the map of the various interaction process through repeated and iterative *visualization* of many related interaction processes through as many diverse observations as possible.

IPM-E helped us recognize the universal NIW-principle (Non-Interaction of Waves). Application of this NIW-principle resolves all the conceptual ambiguities identified in this chapter. In the process, we recognize that the key principle behind short pulse generation is essentially the insertion of an intracavity device whose transparency to the intracavity intensity can switch back and forth within the shortest possible time interval. This is why a properly arranged Kerr-medium, undergoing non-linear intensity dependent refractive index change as a *bulk medium,* makes the switch faster than saturable absorbers that goes through the quantum mechanical steps of excitation and de-excitation to create the oscillatory time gate. The concept of *mode locking* is only partially useful.

The fundamental science and the related technologies behind the field of ultra short pulse laser can advance much faster once we explicitly recognize that a reality based methodology of thinking, like IPM-E, is a much more powerful tool than the curve fitting methodology, like MDM-E. This is because IPM-E builds upon MDM-E while guiding us to search deeper into nature's reality. We must not forget that we can only discover cosmic logics and laws, not invent them. We can invent technologies by employing cosmic logics provided we have already discovered them. The best approach to get closer to cosmic logics is to keep on refining the map of interaction processes as they are more evident to our imaginations than our capacity to directly access the cosmic logics using only MDM-E.

6. Acknowledgement

The author is pleased to acknowledge support from Nippon Sheet Glass Corporation for some of the experiments presented here. Supports from Hemant Gupta and Scott Mather are also gratefully acknowledged.

7. References

Aczel, D. A. (1999). *God's Equation*, Four walls eight windows, New York.

Allen, L. B., Rice, R. R. & Mathews, R. F. (1969). "Two cavity mode-locking of a He-Ne laser", Appl. Phys. Lett.15, #12, pp.416-418.

Born, M. & Wolf, E. (1980). *Principle of Optics*, Cambridge U. Press, Cambridge, UK.

Dirac, P. A. M. (1974). *The Principles of Quantum Mechanics*, p.9, Oxford U. Press, Oxford.

Johnson, N. F.. (2009). *Simply Complexity: A Clear Guide to Complexity Theory*, Oneworld Publication, Oxford, England.

Lee, D. (2004). *A Comprehensive View of Temporal Domain Interference & Spectral Interpretation of Short Pulse.* PhD thesis. Physics Department, University of Connecticut.

Lee, D. and Roychoudhuri, C. (2003)."Measuring properties of superposed light beams carrying different frequencies"; *Optics Express* 11(8), 944-51; [http://www.opticsexpress.org/abstract.cfm?URI=OPEX-11-8-944].

Krausz, F. & Ivanov, M. (2009). "Attosecond Physics"; *Rev. Mod. Phys.* Vol.81, pp.163-234.

Kuhn, T. S. (1996). *The Structure of Scientific Revolutions*, Univ. of Chicago Press, Chicago.

Laughlin, R. (2006). *A Different Universe: Reinventing Physics from the Bottom Down*, Basic Books, New York.

Michelson, A. (1962). *Studies in Optics*, Univ. Chicago Press, Chicago.

Milonni, W. P. & Eberly, J. H. (2010). *Laser Physics*, Wiley, Hoboken, New Jersey.

OSA (2003). "The nature of light: What is a photon?" *Optics & Photonics News*, October Special Issue; OSA, Washington.

Penrose, R. (2005). *Road to Reality*, Alfred A. Knopf, New York.

Ronchi, V. (1970). *Nature of Light: An Historical Survey*, Harvard Univ. Press, Boston

Roychoudhuri, C. (2010a). "The consilient epistemology: structuring evolution of logical thinking," *Proc. 1st Interdisciplinary CHESS Interactions Conf.*, pp.273-295, World Scientific, London. More web info. at: evolutionandleadership.com

Roychoudhuri, C. (2010b). "Principle of non-interaction of waves", *J. Nanophoton.*, Vol. 4, 043512; doi:10.1117/1.3467504.

Roychoudhuri, C. (2009a). "Why we need to continue the 'What is a Photon?' conference: to re-vitalize classical and quantum optics." *Proc. SPIE* Vol. 7421-28; keynote presentation. More web info. at: 21stcenturyphysics.com

Roychoudhuri, C. (2009b). "Can photo sensors help us understand the intrinsic difference between quantum and classical statistical behavior?" *Proc. AIP* Vol.1101, pp. 167-177; Am. Inst. Phys., Melville, New York.

Roychoudhuri, C. & Prasad, N. (2009). "Light-matter interaction processes behind intracavity mode locking devices." *Proc. SPIE* Vol.7193-67, SPIE, Bellingham.

Roychoudhuri, C. (2008). *The nature of light: What is a photon?* See Chap. 6, "Inevitable Incompleteness of all theories: An epistemology to continuously refine human logics towards cosmic logics." CRC Press, New York.

Roychoudhuri, C. & Barootkoob, A. M. (2008). "Generalized quantitative approach to two-beam fringe visibility (coherence) with different polarizations and frequencies"; *Proc. SPIE* Vol. 7063-4.

Roychoudhuri, C. (2007a). "Shall we climb on the shoulders of the giants to extend the Reality horizon of physics?" *Proc. AIP* Vol.962 (2007), *Quantum Theory: Reconstruction of Foundations – 4;* Am. Inst. Phys., Melville, New York.

Roychoudhuri, C. (2007b). "Bi-centenary of successes of Fourier theorem! Its power and limitations in optical system designs"; *Proc. SPIE* Vol. 6667-18.

Roychoudhuri, C., Tirfessa, N., Kelley, C. & Crudo, R. (2007). "If EM fields do not operate on each other, why do we need many modes and large gain bandwidth to generate short pulses?" *Proc. SPIE* Vol.6468-53, SPIE, Bellingham.

Roychoudhuri, C. (2006a). "Locality of superposition principle is dictated by detection processes"; *Phys. Essays* 19 (3), September issue.

Roychoudhuri, C. (2006b). "Reality of superposition principle and autocorrelation function for short pulses," *Proc. SPIE* 6108, 61081E. [doi: 10.1117/12.670412].

Roychoudhuri, C. & Prasad, N. (2006). "Various ambiguities in re-constructing laser pulse parameters", *Proc. IEEE-LEOS,* Ann. Conf., Montreal.

Roychoudhuri, C. & Tayahi, M. (2006)."Spectral Super-Resolution by Understanding Superposition Principle & Detection Processes"; *Int. J. Microwave & Opt. Opt. Tech.;* Manuscript ID# IJMOT-2006-5-46;
http://www.ijmot.com/papers/membercheck.asp?id=IJMOT-2006-5-46

Roychoudhuri, C., Lee, D. & Poulos, P. (2006). "If EM fields do not operate on each other, how do we generate and manipulate laser pulses?" *Proc. SPIE Vol.* 6290-02.

Roychoudhuri, C. (2004). "Propagating Fourier frequencies vs. carrier frequency of a pulse through spectrometers and other media"; *Proc. SPIE* Vol.5531, pp.450-461.

Roychoudhuri, C., Lee, D., Jiang, Y., Kittaka, S., Nara, M., Serikov, V. & Oikawa, M. (2003). "Limits of DWDM with gratings and Fabry-Perots and alternate solutions"; *Proc. SPIE* Vol.5246, pp.333-344.

Roychoudhuri, C. (1976). "Is Fourier decomposition interpretation applicable to interference spectroscopy?" *Bol. Inst. Tonantzintla* 2, pp.101-107.
http://www.phys.uconn.edu/~chandra/

Siegman, A., E. (1986). *Lasers*, University Science Books, Sausalito, California.

Silverman, T. (2010). *Philosophical Solutions;* Int. Inst. For Advncd. Studies.

Smolin, L. (2006). *Trouble with Physics,* Houghton Mifflin, Boston.

SPIE (2005). "The nature of light: What are photons?" This is a biannual conference series under the auspices of SPIE Annual Confernce continuing since 2005. The fourth conference is set for August 2011. Past proceedings are: *Proc. SPIE* Vol.5866, 6664 and 7421; Bellingham.

Tommasini, D., Fernando, A., Michinel, H. & Seco, M. (2008). Detecting photon-photon scattering in vacuum at exawatt lasers", Phys. Rev. A 77, 042101(1-7).

Woit, P. (2007). *Not Even Wrong,* Basic Books, New York.

8

Infrared Thermo-Electric Photodetectors

W. Vandermeiren[1], J. Stiens[1], G. Shkerdin[2], V. Kotov[2],
C. De Tandt[1] and R. Vounckx[1]
[1]*Vrije Universiteit Brussel,*
[2]*Institute of Radio Engineering and Electronics of the Russian Academy of Science,*
[1]*Belgium*
[2]*Russia*

1. Introduction

During the last decade, pulsed laser radiation has gained interest by the material processing industry and the medical sector. For a growing set of applications (laser drilling, laser marking, laser surgery, semiconductor doping profiling, micro-structuring, layer deposition, etc.) it is advantageous to use pulsed laser radiation instead of continuous wave (CW) illumination as time limited exposure often results in reduced collateral damage and more precise processing (Phipps, 2007). In laser ablation for example, one aims to put an intense laser pulse on the surface of a target material in an as short time as possible. This short exposure time, which limits thermal diffusion inside the material, together with a carefully selected wavelength with a minimal absorption depth is required to ensure energy deposition in a small volume of the target material. Hence, for laser pulses which meet the ablation requirements, one can evaporate material in a very controlled fashion. Different methods exists as Q-switching, mode-locking and cavity dumping for achieving the required pulse characteristics for laser ablation.

Laser pulses span an enormously large parameter space in terms of wavelength, repetition rate, pulse duration and pulse energy, further referenced as pulse-parameters. Each of these pulse-parameters can be optimized for a given application and target material. Besides its dependence on the temporal characteristics of pulsed laser radiation, for some applications, the processing quality is also strongly dependent on the transverse laser beam profile. Consequently, there is a growing interest in detecting the spatio-temporal behavior of laser pulses.

In this chapter we briefly describe the most common infrared detector principles for measuring laser pulses and point out their respective advantages and disadvantages with respect to different pulse-parameters. Next, we show that Seebeck-effect based thermo-electric photodetectors can be designed to cover a relatively broad range of pulse-parameters (Stiens, 2006). Further, we discuss the working principle and operation regimes of the thermo-electric photo detector and explain the corresponding theoretical background in detail. Experimental results concerning short laser pulse induced thermo-voltages in n-GaAs are presented. This chapter is also concerned with the possibility of using the thermo-electric effect to measure the spatio-temporal behavior of laser pulses by means of linear focal plane arrays (LFPA). Certain related issues will be highlighted such as thermal cross-talk between pixels in case of pulse durations approaching CW illumination. A lock-in method is proposed to reduce the cross-talk level. The chapter will conclude with describing the future directions of research.

2. Infrared detector principles

This section deals with the most common infrared detector principles based on various physical mechanisms and materials for measuring laser pulses. Interest is mainly focused on wavelengths of the two atmospheric windows 3–5 μm (middle wavelength IR - MWIR) and 8–14 μm (low wavelength IR- LWIR) as in these bands the atmospheric transmission is highest, though in recent years, there has been increasing interest in longer wavelengths stimulated by space applications. Figure 1 (Rogalski, 2003)[1] shows the spectral detectivity curves for a number of commercially available IR detectors.

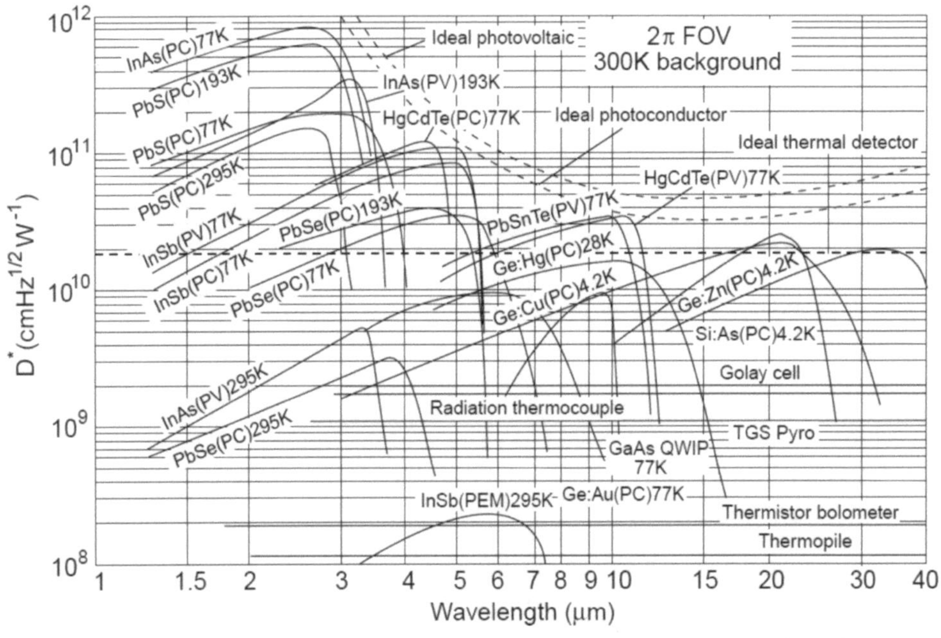

Fig. 1. Comparison of the detectivity D* of various commercially available infrared detectors when operated at the indicated temperature. Chopping frequency is 1000 Hz for all detectors except for the thermopile, thermocouple, thermistor bolometer, Golay cell and pyroelectric detector, where it is 10 Hz. Each detector is assumed to view a hemispherical surrounding at a temperature of 300 K. Theoretical curves for the background-limited D*(dashed lines) for ideal photovoltaic and photoconductive detectors and thermal detectors are also shown. PC—photoconductive detector, PV—photovoltaic detector, and PEM—photoelectromagnetic detector.

The main specifications that one has to consider while comparing different detector principles and materials are next to the spectral detectivity, the response speed, whether

[1] Reprinted from Progress in Quantum Electronics, Vol.27/2-3, Infrared detectors: status and trends, pp. 59-210, Copyright 2003, with permission from Elsevier.

cryogenic operation temperatures are required and the damage threshold. The relative importance of each of these detector parameters is typically strongly related to the application. In case of low power laser pulses, a high responsivity is of major importance while for measuring industrial high power laser pulses, a lower responsivity might be desirable as for most IR detectors, the power handling capability, which is limited by the maximum temperature increase on the detector surface, is inversely proportional to the absorption coefficient (see Eq. 23). Note that this is only true for short exposure times. At long exposure times, the damage threshold intensity becomes independent of the absorption coefficient.

Infrared detectors can be classified into thermal and photon detectors depending on their detection mechanism. A brief overview of IR detectors with their inherent advantages and disadvantages are classified and summarized in table 1 (Rogalski, 2003).

	Detector type	Advantages	Disadvantages
Thermal detectors	**Thermopile, bolometers, pyroelectric**	Light, rugged, reliable and low cost, room temperature operation	Low detectivity @ high frequency Slow response (ms order)
Photon detectors	**Intrinsic**		
	IV-VI (*e.g. PbS, PbSe, PbSnTe*)	Easy to prepare More stable materials	High thermal expansion coefficient Large permittivity
	II-VI (*e.g. HgCdTe*)	Easy bandgap tailoring High optical absorption Low thermal generation	Non-uniformity over large area High cost in growth and processing Surface instability
	III-V (*e.g. InGaAs, InAs, InSb, InAsSb*)	Good material and dopants Advanced technology Possible monolithic integration	Heteroepitaxy with large lattice mismatch Long wavelength cutoff limited to 7 μm
	Extrinsic (*e.g. Si:Ga, Si:As, Ge:Cu, Ge:Hg*)	Very long wavelength operation Simple technology	High thermal generation Extremely low temperature operation
	Free carriers (*e.g. PtSi, Pt$_2$Si, IrSi*)	Low cost, high yield Large & close packed 2D arrays	Low quantum efficiency Low temperature operation
	Quantum Wells		
	Type 1 (*e.g. GaAs/AlGAAs, InGaAs/AlGaAs*)	Matured material growth Good uniformity over large area	High temperature generation Complicated design & growth
	Type 2 (*e.g. InAs/InGaSb, InAs/InAsSb*)	Low Auger recombination rate Easy wavelength control	Complicated design & growth Sensitive to the interfaces
	Quantum dots (*e.g. InAs/GaAs, InGaAs/InGaP, Ge/Si*)	Normal incidence of light Low thermal generation	Complicated design & growth

Table 1. IR detector classification.

2.1 Thermal detectors

The output signal of thermal detectors is observed as a temperature dependent property change in the active detector material. This temperature dependent mechanism can result in a thermoelectric voltage, a resistance change or a pyroelectric voltage depending on the detector type. Table 2 shows different thermal detectors with the corresponding temperature dependency of the detector's output. Thermal detectors are often characterized by a relatively slow response time and low detectivity, which make them less suitable for short and low intensity optical pulses. Their photo sensitivity is typically independent on the wavelength as a broad band absorption layer is used as a window material to convert the optical intensity into heat. In contrast with most photon detectors, they can operate at room temperature. Forced cooling is only required for detectors which are designed for high average optical power levels.

Detector type	Signal (T)
Thermocouple	~ΔT
Bolometer	~T
Pyroelectric	~ dT/dt
Forward-bias diode	~T

Table 2. Different types of thermal detectors

Assuming a periodic optical radiation, the temperature change in a thermal based IR is given by:

$$\Delta T = \frac{\varepsilon\Phi_0}{(G_{th}^2 + \omega^2 C_{th}^2)^{1/2}} = \frac{\varepsilon\Phi_0 R_{th}}{(1+\omega^2\tau_{th}^2)^{1/2}} \quad \text{with} \quad \tau_{th} = \frac{C_{th}}{G_{th}} = R_{th}C_{th} \tag{1}$$

Where ε, Φ_0, R_{th}, ω, C_{th}, τ_{th} are the absoptivity, the incident radiative flux, the thermal resistivity, the angular frequency, the thermal capacitance and the thermal response time constant, respectively. In order to make ΔT as large as possible, one should use a detector material with a small thermal conductivity and a small thermal capacitance. The thermal response time is typically of the order of milliseconds. Hence, thermal detectors are less suitable for high speed measurements.
The voltage responsivity (R_v) is then defined as the ratio of the detector output signal to the input radiation power

$$R_v = \frac{\Delta V}{\Phi_0} = \frac{\Delta V}{\Phi_0 \Delta T}\frac{\varepsilon\Phi_0 R_{th}}{(1+\omega^2\tau_{th}^2)^{1/2}} = \frac{K\varepsilon R_{th}}{(1+\omega^2\tau_{th}^2)^{1/2}} \quad \text{with } K = \frac{\Delta V}{\Delta T} \tag{2}$$

The well-known detectivity criteria D* is widely used for comparing detectors, as D* is independent on the size and shape of the detector. The detectivity of thermal detectors can be calculated when the noise mechanisms are known. Hence, D* is a measure of the signal-to-noise (S/N) ratio of the detector. The main noise mechanism consists of Johnson noise. This noise in a Δf bandwidth for a resistor R is given by expression (3) where k_B and T are the Bolzmann constant and the temperature, respectively.

$$V_J^2 = 4k_B TR\Delta f \tag{3}$$

In addition to Johnson noise, other noise sources which contribute to the overall noise level are 1/f noise, thermal fluctuation noise and background noise. The thermal fluctuation noise and background noise can be shown to be (Rogalski, 2003)

$$V_{TF}^2 = \frac{4k_BT^2\Delta f}{1+\omega^2\tau_{th}^2}K^2R_{th} \qquad V_B^2 = \frac{8k_B\varepsilon\sigma A(T_d^2+T_b^2)}{1+\omega^2\tau_{th}^2}K^2R_{th} \tag{4}$$

Where σ is the Stefmann-Boltzmann constant. Hence, the detectivity D* of a thermal fluctuation noise limited and background fluctuation noise limited detector are given by expressions(5ab) (Rogalski, 2003), respectively, where A_d is the detector surface.

$$D_{TF}^* = \frac{(A_d\Delta f)^{1/2}}{V_{TF}}R_v = \left(\frac{A_d\varepsilon^2R_{th}}{4k_BT^2}\right)^{1/2} \qquad D_B^* = \left(\frac{\varepsilon}{8k_B\sigma(T_d^5+T_b^5)}\right)^{1/2} \tag{5}$$

Typical detectivities of thermal detectors (at 10Hz) are of the order of 10^8 - 10^9 $cmHz^{1/2}W^{-1}$.

2.2 Photon detectors

Photon detectors (e.g. photoconductive, photovoltaic, photo-electromagnetic detectors) absorb the radiation within the active material due to electron-hole creation. The observed electrical output signal is the consequence of a changed electronic energy distribution. In contrast with thermal detectors, photon detectors offer a higher detection performance and a faster response speed, although their photosensitivity is strongly wavelength dependent and most of them are cryogenically cooled for optimal performance in terms of sensitivity and signal-to-noise ratio.

The current responsivity R_i of photon detectors is proportional to the photoelectric gain (g) and the quantum efficiency (η). The quantum efficiency is a measure for the device's electrical sensitivity to light and is usually defined as the percentage of photons incident in the active area of the detector that produce an electron-hole pair. The photo-electric gain is the number of carriers collected at the contacts per generated pair. Hence the current responsivity equals

$$R_i = \frac{\lambda\eta}{hc}qg \tag{6}$$

Where λ is the wavelength, h is the Planck's constant, c is the light velocity and q is the elementary charge. The current noise associated to the generation and recombination can be defined as follows

$$I_n^2 = 2(G+R)A_et\Delta fq^2g^2 \tag{7}$$

Where G and R are the generation and recombination rates, Δf is the frequency band, A_e is the electrical area and t is the thickness of the detector. The detectivity can then be defined as

$$D^* = \frac{R_i(A_o\Delta f)^{1/2}}{I_n} = \frac{\lambda}{hc}\left(\frac{A_o}{A_e}\right)^{1/2}\frac{\eta}{(2(G+R)t)^{1/2}} \tag{8}$$

Where A_o is the optical surface. Assuming a single pass of the radiation and consider $A_o = A_e$, the expression for D* simplifies to

$$D^* = \frac{R_i(A_o\Delta f)^{1/2}}{I_n} = \frac{\lambda}{hc}\frac{\eta}{(2(G+R)t)^{1/2}} \quad \text{with } \eta = 1 - e^{-\alpha t} \tag{9}$$

Where α is the wavelength dependent absorption coefficient for a material under consideration. One can easily derive that a maximum detectivity is achieved for t = 1.26/α. In order to achieve an optimal detectivity, thermal generation should be reduced as much as possible. This is usually done with cryogenic cooling of the detector. Figure 1 shows that detectivities of photon detectors are of the order of 10^9 – 10^{11} $cmHz^{1/2}W^{-1}$, which is typically 1 to 2 orders of magnitude higher as compared to thermal detectors.

2.3 Comparison

Within the category of photon detectors, one sees that the performance of intrinsic IR detectors (e.g. HgCdTe), in terms of detectivity, is higher than other types of photon detectors. HgCdTe is characterized by a high optical absorption coefficient and quantum efficiency and relatively low thermal generation rate compared to extrinsic detectors and quantum well infrared photodetectors (QWIPs). The extrinsic photon detectors require more cooling than intrinsic photon detectors having the same long wavelength limit.

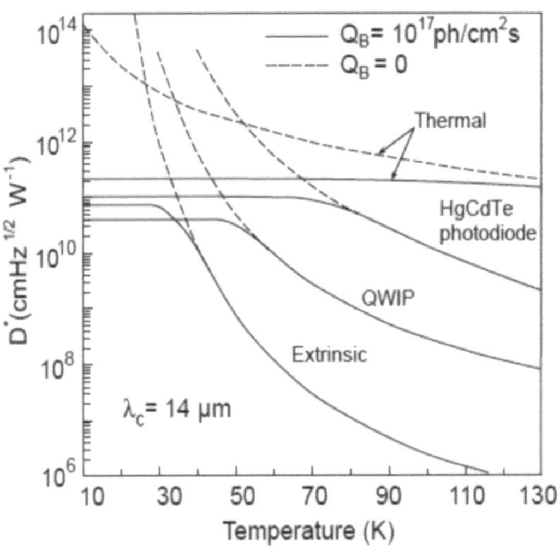

Fig. 2. Theoretical detectivity limits of LWIR photon and thermal detectors at a wavelength of 14 μm as a function of the detector temperature for zero background and a background of 10^{17} photons $cm^{-2}s^{-1}$.

One can deduce from figure 1 and 2 (Rogalski, 2003)[2] that the detectivity of thermal detectors is less temperature and wavelength dependent as compared to photon detectors. At temperatures below 50 K and zero background, LWIR thermal detectors are characterized by a theoretical performance limit (D*) lower than those of LWIR photon detectors. However, at temperatures above 60 K, the thermal detectors outperforms the LWIR photon detectors. At room temperature, the theoretical performance of thermal detectors is much better than LWIR photon detectors. The comparison of both types of detectors indicates that theoretical performance limits for thermal detectors are more favorable as wavelength of operation moves from the LWIR to the VLWIR.

---

[2] Reprinted from Progress in Quantum Electronics, Vol.27/2-3, Infrared detectors: status and trends, pp. 59-210, Copyright 2003, with permission from Elsevier.

Thermal detectors have a low time response of the order of milliseconds, limited by the thermal time constant of the detector. This property makes them less suitable for short pulse measurements. Intrinsic photon detectors on the contrary have a time response which is at least three orders of magnitude faster than for thermal detectors.

3. Seebeck infrared photodetector

A Seebeck infrared photodetector can be categorized as a special kind of thermal detector which can be operated at room temperature. The operation principle of this detector type is based on photon induced free carrier absorption in doped semiconductors. Detailed calculations have been performed for n-GaAs. The material choice is based on the performance and price level of n-GaAs with respect to other materials. The optical intensity decays exponentially along the axis of propagation inside the highly doped region of the semiconductor material (e.g. n-GaAs), as a result of the absorption process. Consequently, a temperature gradient is generated. This temperature gradient leads to free carrier diffusion in the conducting layers of the structure which provokes a change in the electron concentration and creates an electrical field inside the substrate. This thermo-electric effect is characterized by a Seebeck coefficient (S) which is dependent on temperature, the doping concentration and the crystal structure of the semiconductor material. The detector output is then given by:

$$V_{out} = S \cdot \Delta T_e \tag{10}$$

Where ΔT_e is the temperature difference between two electrodes on the substrate. Note that here, unlike most other thermal detectors, the absorbing window material and the active detector material are the same. Hence, the spectral window of semiconductor based Seebeck detectors is dependent on the spectral free carrier absorption process.

3.1 Seebeck coefficient

To calculate the Seebeck coefficient, one has to solve the kinetic equation for the electron distribution function in the presence of an electron temperature gradient. Taking into account the nonparabolicity of the Γ-valley and the anisotropy of the satellite L- and X-valleys of n-GaAs, the expression for S can be written as follows:

$$S = \frac{k_B(a_\Gamma \sigma_\Gamma + a_L \sigma_L + a_X \sigma_X)}{(\sigma_\Gamma \cdot \sigma_L \cdot \sigma_X)} \tag{11}$$

Where k_B and q are the Boltzmann constant and the elementary charge, respectively; $\sigma_{\Gamma,L,X}$ and $a_{\Gamma,L,X}$ are the electron conductivity and the dimensionless weighting coefficients for the Γ, L and X valley, accordingly. The contribution of the phonon-drag effect to the thermo-electric power is neglected as the active semiconductor layer is degenerate and the operation temperature is relatively high $T \geq 297$ K (George et al., 2001). The contributions of the satellite valleys are small near room temperature for doping concentrations $n_0 \leq 7\text{x}10^{18}$ cm^{-3}. In this case the expression for S simplifies to:

$$S \approx \frac{k_B}{q} a_\Gamma \tag{12}$$

Where a_Γ is a dimensionless weighting factor which can be calculated as follows:

$$a_\Gamma = \frac{1}{k_B T_e}\left(-\xi + \frac{\int_0^\infty d\varepsilon_\Gamma (\varepsilon_\Gamma{}^2 + \gamma\varepsilon_\Gamma{}^3 / E_g)^{3/2} (1+2\gamma\varepsilon_\Gamma / E_g)^{-1} \tau_\Gamma f'_{o,\Gamma}}{\int_0^\infty d\varepsilon_\Gamma (\varepsilon_\Gamma + \gamma\varepsilon_\Gamma{}^2 / E_g)^{3/2} (1+2\gamma\varepsilon_\Gamma / E_g)^{-1} \tau_\Gamma f'_{o,\Gamma}} \right) \tag{13}$$

where τ_Γ is the momentum relaxation time in the Γ-valley; $f'_{0,\Gamma} = df_0/d\varepsilon_\Gamma$ where f_0 is the 0th order Fermi-Dirac distribution function. T_e, E_g, γ, ε_Γ are the electron temperature, bandgap, band non-parabolicity factor and electron energy in the Γ-valley, respectively. The electron spectrum of the Γ-valley can be approximated by:

$$\varepsilon_\Gamma = \frac{E_g}{2\gamma}\left(\sqrt{1 + \frac{2\gamma}{E_g}\frac{\hbar^2 \vec{k}^2}{m_\Gamma}} - 1 \right) \tag{14}$$

Where ε_Γ is the electron energy counted from the bottom of the Γ-valley, k is the electron wave vector, m_Γ is the electron effective mass at the bottom of the Γ-valley.

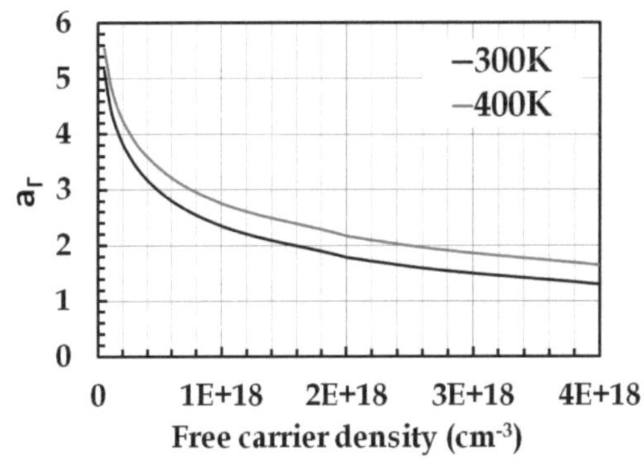

Fig. 3. Dependences of a_Γ on thefree electron density for two temperatures T_e = 300 K and 400 K (n-GaAs)

Dependences of a_Γ versus doping concentration for two temperatures T_e = 300 K and 400 K are shown in figure 3. Notice that a_Γ is positive and of the order of one which corresponds to a Seebeck coefficient of the order of 100 μV/K. Hence, the voltage responsivity of a n-GaAs based Seebeck detector, which can be calculated by means of Eq. 2, is of the order of a few mVW^{-1}. A logarithmic fit on these data at room temperature results in the following approximation for the Seebeck coefficient for $1x10^{17} < n_0 < 4x10^{18}$:

$$S \approx -72.9 \ln\left(\frac{n_0}{10^{18}}\right) + 207.4 \quad \left(\frac{\mu V}{K}\right) \tag{15}$$

One can also see that an increase of the electron gas degeneracy leads to a decrease of the Seebeck coefficient and an increase of the electrical conductivity. It is important to mention that the electrical conductivity should be high enough to probe the thermo-electromotive force (emf) induced potential. Therefore, a typical figure of merit for thermo-electric devices is given by:

$$Z = \frac{S^2 \sigma}{k_{th}} \tag{16}$$

Where, σ and k_{th} are the electrical and thermal conductivity, respectively. As the contributions of conduction carriers to the thermal conductivity are in general very small (Sze, 1981), Z (~$S^2\sigma$) will be maximal for a given doping concentration.

3.2 Thermo-voltage detector - detectivity

However, in order to maximize the detectivity of Seebeck based detectors, one should take into account the free electron absorption coefficient for the wavelength under consideration, which is strongly dependent on the doping concentration. Calculations were performed based on an multi-valley (Γ, X, L) model where non-parabolic and anisotropic effects are taken into account for the intravalley and (non)equivalent intervalley absorption mechanisms (Shkerdinet al., 1999). The absorption coefficient was calculated for wavelengths between 8 μm < λ < 19 μm and an electron density between 10^{17} cm^{-3} < n_0 < 10^{18} cm^{-3}. A compensation factor, given by Eq. 17,equal to one is assumed such that the contribution of impurity scattering on compensated donor and acceptor atoms to the absorption coefficient is neglected.

$$F_{comp} = \frac{N_D + N_A}{N_D - N_A} \tag{17}$$

Here, N_D and N_A are the donor and acceptor concentration, respectively. Figure 4 shows the absorption coefficient dependency on wavelength and electron density for n-GaAs. Note that the quantum mechanical model (Shkerdin et al., 1999), comprising a compensation factor dependent absorption coefficient, allows a good agreement with experimental data. However, no direct agreement can be found with experimental data from literature, e.g. (Blakemore, 1982), when no information is available about the compensation factor.

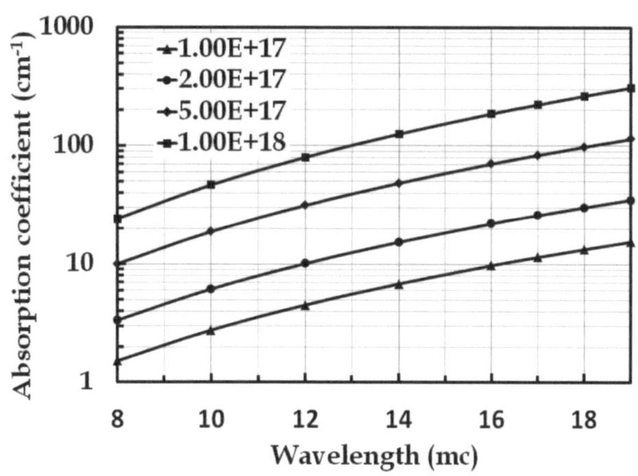

Fig. 4. Absorption coefficient dependence on wavelength and electron density for n-GaAs

Seebeck detectors can be categorized as thermal detectors. Assuming that only Johnson noise is present, we obtain:

$$D_J^* = \frac{(A_o \Delta f)^{1/2}}{V_J} R_v = S\varepsilon \mathrm{R_{th}} \left(\frac{A_d}{4k_B T R(1+\omega^2 \tau_{th}^2)} \right)^{1/2} \tag{18}$$

Where S is the Seebeck Coefficient. For a single pass, the absorptivity ε is given by

$$\varepsilon = 1 - e^{-\alpha_e t} \tag{19}$$

Where t and α_e are the highly doped layer thickness and the electron absorption coefficient.

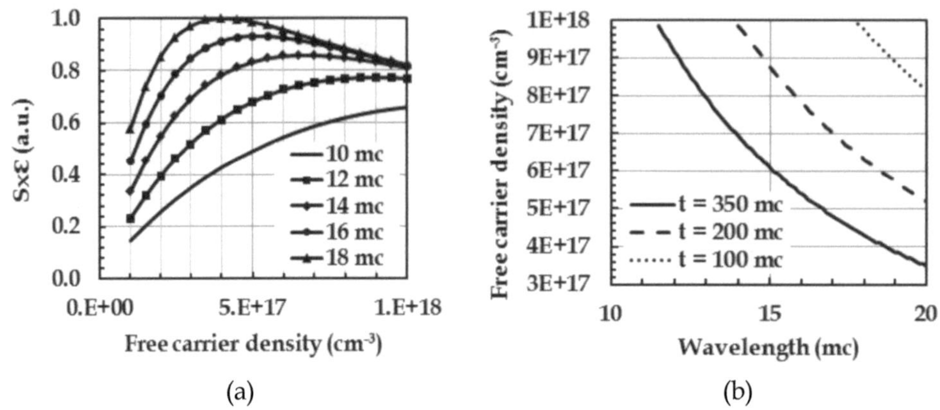

Fig. 5. (a) Detectivity optima in function of the free carrier density for different wavelengths for an absorption layer thickness of 350 μm; (b) Optimal free carrier density versus wavelength for different absorption layer thicknesses

Notice that, the Seebeck coefficient (S) and the absorptivity (ε) are the only parameters in Eq. 18 which are significantly dependent on the doping concentration. The latter is also strongly dependent on the wavelength and the absorption layer thickness. Hence, the detectivity D* of a n-GaAs based Seebeck detector can be optimized for a given wavelength and absorption layer thickness by tuning the doping concentration as illustrated in figure 5a and figure 5b. These detectivity maxima shift to lower doping levels as the $\alpha_e t$ product increases. The absorption coefficient was calculated in according with a multi-valley model (Shkerdin et al., 1999). As compared to measured absorption coefficients (Blakemore, 1982), which are slightly higher as explained above, the optimal detectivity will shift to lower carrier densities with Δn_o of the order of 0.5×10^{18} cm^{-3}. Notice that the doping level is generally not critical as these optima are relatively wide. Figure 5b shows the optimal free carrier concentration in n-GaAs versus wavelength for various absorption layer thicknesses. It indicates that for relatively small wavelengths (e.g. 10.6 μm) in combination with modest absorption layer thicknesses (t < 100 μm), high optimal doping levels are found. High doping levels, however (e.g. $N_{D,GaAs} > 4 \times 10^{18}$ cm^{-3}), are technologically difficult to realize in III-V semiconductors due to self-compensation. Although, it can be accomplished by the use of special growth techniques as "Delta-doping" (Schubert, 1996).

3.3 Operation regimes

The type of detector output, and consequently the information that can be obtained, is determined by the ratio between the laser pulse duration and various time constants of the detector. The operation regimes and most important time constants which define the operation regime transitions are schematically illustrated in figure 6.

The following time constants are important: the electron heating time, the electron energy relaxation time τ_e and the lattice thermal response time τ_{ia}. The relaxation time of optically heated electrons which is dependent on the doping concentration and the incident light

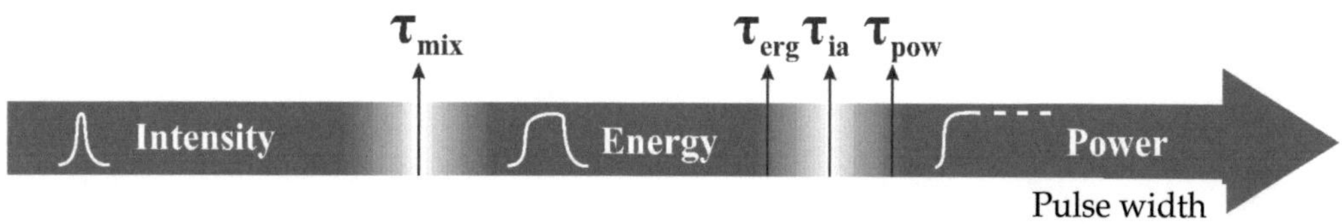

Fig. 6. Operation regimes of a semiconductor based thermo-eletric detector

intensity is of the order of a few picoseconds. For example, the relaxation time of longitudinal-optical phonon thermalization is about 3-7 picoseconds (Shah et al., 1970) (Vaissiere et al., 1992) Hence, for short incident laser pulses which are longer than a few electron energy relaxation time constants, the detector's voltage output will be proportional to the temporal evolution of the laser intensity. This "intensity"-regime is characterized by a high damage threshold of the order of 0.1-1 $GWcm^{-2}$ as the lattice temperature is almost unaffected for such short pulses. The damage threshold in terms of maximum intensity in function of the pulse full-width halve-maximum (FWHM) is illustrated in figure 10 for different detector dimensions. In this regime, the detector's output is independent on the thermal material properties as the heat capacity C_{th} and the thermal conductivity k_{th}. This regime is also characterized by a relatively low responsivity of the order of 4-8 $mV/MWcm^{-2}$ for highly doped n-GaAs. Consequently, this regime of operation is extremely suitable for direct measurements of short laser pulses having high optical intensities. The heating of free electron gas in doped n-GaAs is described in detail in (Shkerdin et al., 2007)
For systematically increased laser pulse widths, the electron gas will gradually transfer energy to the lattice of the doped semiconductor allowing the lattice temperature to increase. The response time of this lattice heating mechanism is much slower (10-100 μs) than electron heating. The pulse length for which the electron temperature equals the lattice temperature is called τ_{mix}. This is the regime of the mixed electron-lattice effects. τ_{mix} can be expressed as (20a) for low optical intensities, where τ_{en}, C_{th}, are the stationary energy relaxation time and the volumetric heat capacitance, respectively. For high optical intensities, this response time should be corrected for non-linear behavior, resulting in expression (20b)

$$\tau_{mix} = \frac{\tau_{en} C_{th}}{N k_B} \qquad \tau_{mix} = \frac{\delta T_e C_{th}}{\alpha_e W} \tag{20}$$

The dependence of τ_{mix} on electron concentration for small laser intensities was calculated using the model developed in (Shkerdin et al., 1999) (Shkerdin et al., 2002) and is illustrated in figure 7.
These numerical simulations show that within the doping concentration interval $0.7x10^{18}$ - $4x10^{18}$ cm^{-3}, the value τ_{mix} can drop from the range 325 - 40 ns down to 100 - 10 ns when the intensity varies from a low level to about 100 $MWcm^{-2}$. For a certain range of pulse widths longer than τ_{mix}, the lattice temperature will increase proportional to the pulse energy. However, the energy regime is limited by the thermal response time of the illuminated area (τ_{ia}). The illuminated area time constant τ_{ia} does not only depend on material parameters, but also on the pixel geometry. The thermal response time of the illuminated area is described by

$$\tau_{ia} = \frac{\beta C_{th}}{k_{th}\left(\frac{1}{L_{abs}^2} + \frac{4}{L_{illum}^2}\right)} \tag{21}$$

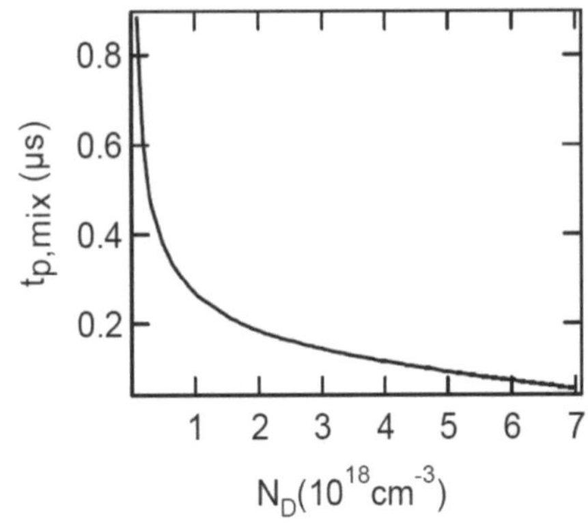

Fig. 7. Doping density dependence on τ_{mix}

Where L_{abs}, L_{illum}, are the absorption length inside the substrate and the illumination length which can be the characteristic size of a pixel or the laser beam width. β is a geometrical numerical factor of the order of one which depends on the pulse characteristics and the geometry. C_{th} and k_{th} are the volumetric heat capacitance and the thermal conductivity, respectively.

As the pulse length approaches τ_{ia}, the local lattice temperature will become less dependent on the pulse length due to thermal diffusion (lattice conductivity). Hence, for pulse lengths much larger than τ_{ia}, the lattice temperature is independent on the pulse width. This corresponds to a power measurement. The detector's maximum output can then be estimated using the following expression:

$$T_{surface,max} = \frac{W_0 L_{illum}^2}{k_{th}(L_{abs}+L_{illum})} \tag{22}$$

Where k_{th} and W_0 are the thermal conductivity and the optical intensity, respectively. In general, one can say that the power regime is characterized by a relatively high responsivity of the order of 0.01-0.1 $mVW^{-1}cm^{-2}$ for n-GaAs and a significantly lower damage threshold as compared to the intensity regime (see figure 10).

3.4 Study of a short pulse induced thermo-voltage in n-GaAs

By exposing an n-doped GaAs substrate to short optical CO_2 laser pulses with a pulse width smaller than τ_{mix}, one can distinguish the electron and lattice heating process. This section will describe the experimental setup and discuss the obtained results. The experimental setup is illustrated in figure 8.

A CO_2 laser produces short optical pulses with a pulse width of about 100 ns at a wavelength of 10.6 μm. These pulses propagate through an attenuating system consisting of two polarizers. About 50% of the power is reflected on a ZnSe beam splitter. This part is subsequently attenuated to a save intensity and measured by a reference photo electro magnetic (PEM) detector. The second part which propagates through the beam splitter is focused on a highly doped n-GaAs sample ($5x10^{17}$ cm^{-3}). This resulted in a beam width of about 170 μm on the top surface of the substrate. The incident pulse energy was limited to about 0.1 mJ to avoid non-linear thermal effects. The laser pulse induced thermo-voltage is then measured by means of a ground electrode at the backside of the substrate and a needle

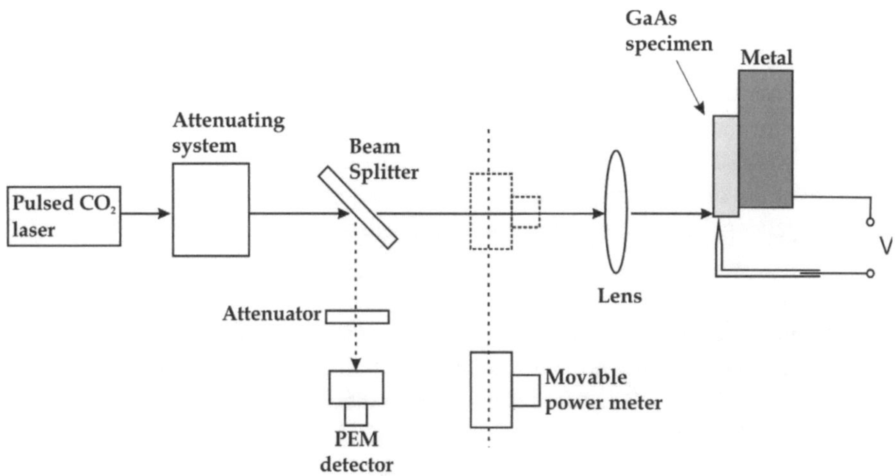

Fig. 8. Experimental setup for laser pulse induced thermo-voltage measurements

electrode positioned at about 50 μm from the top surface of the substrate. Note that the needle should be placed relatively close to the top surface of the substrate (typically smaller than the absorption length) to measure the electron heating process due to the absence of thermal diffusion in the intensity regime. The optical pulses expose the top surface of the substrate in the neighborhood of the needle. Special attention was paid to create an ohmic contact between the needle and the semiconductor material. The electrical signals of the PEM detector and the GaAs specimen are illustrated in figure 9. Here, the upper and lower curves corresponds to the laser pulse induced thermo-voltage in n-GaAs and the PEM detector output, respectively. The solid curves are experimental data, while the dashed curves corresponds to an analytical model for the laser pulse induced thermo-voltage behavior (Shkerdinet al., 2007). It was estimated that τ_{mix} is about 380 ns for a doping concentration of $5x10^{18}$ cm^{-3}. The laser pulse induced voltage consists of a first pulse which is a consequence of two mechanisms: the thermo-voltage effect due to electron gas heating and the photon drag effect. These mechanisms are fast and lead to the voltage linear temporal dependence that is proportional to the temporal intensity dependence of the incident optical pulse.

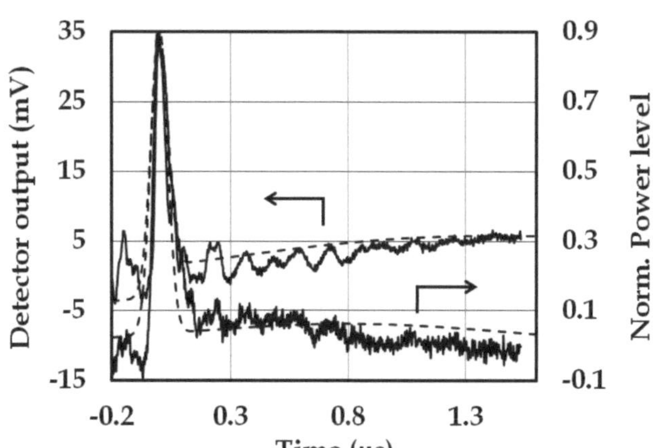

Fig. 9. Pulse induced voltage – n-GaAs versus PEM detector output

The measured combined sensitivity of the fast mechanisms is about 2.6 mVcm^2/MW, therefore these mechanisms can be useful to detect fast (sub-ns) pulses of large intensity (> 1MWcm^{-2}). The illustrated pulse is followed by a comparatively long tail which can be association to the electron-lattice interaction. The temporal voltage dependence in this region is proportional to the pulse energy temporal dependence for time duration smaller than a few microseconds when the lattice heat conduction effect is rather small. For the theoretical model describing this phenomena in detail, we refer to (Shkerdinet al., 2007)

3.5 Damage threshold

The damage threshold of a detector can be defined as the maximum intensity at which the detector may be exposed for a given exposure time and beam size. This intensity strongly depends on the wavelength dependent absorption coefficient, the beam size and the thermal parameters of the detector material. A simple analytical model can be used to estimate the energy density required to damage a particular material as reported by (Bartoli, 1977). This formula is valid for a semi-infinite solid irradiated by a Gaussian beam $P(r) = P_0 \exp(-r^2/a^2)$.

$$E_0 = \frac{\Delta T \rho c}{(1-R)\alpha_e}\left[1+\frac{\alpha_{th}\tau\alpha_e\pi^{1/2}}{a\tan^{-1}(4k\tau/a^2)^{1/2}}\right] \tag{23}$$

Here, ΔT is the minimum increase in surface temperature at the heating centre which is required for damage, τ is the pulse width, ρ is the density, c is the specific heat, α_{th} is the thermal diffusivity, α_e is the absorption coefficient, R is the reflection coefficient and a is the beam size. Figure 10a shows the resulting damage threshold intensity dependence on the pulse width for various beam sizes for n-GaAs with a doping concentration of $7\text{x}10^{17}$ cm^{-3}. Note that the first term in Eq. (23) will be dominant for short exposure times where the damage threshold is primarily determined by the absorption depth (α_e^{-1}). Hence, the second term is dominant at long exposure times where the heat diffusion distance $(k\tau)^{1/2}$ is much greater than the absorption depth (α_e^{-1}).

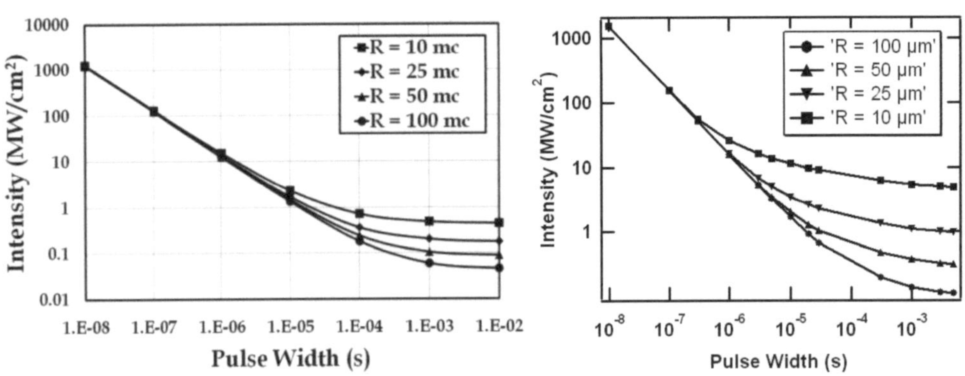

(a) (b)

Fig. 10. (a) Semi-infinite solid model based damage threshold intensity calculations for n-GaAs. (b) Finite solid model based damage threshold intensity calculations for n-GaAs (thickness 400 µm).

More accurate damage threshold intensities were calculated numerically by taken into account all relevant details of geometry and thermal configuration of the detector (n-GaAs substrate). Figure 10b illustrates the maximum allowable exposure time in function of the incident intensity for different pixel sizes. This figure was calculated for a 400 μm thick n-GaAs substrate with a doping concentration of 7×10^{17} cm^{-3}. The constraints on the curves are determined for a maximum lattice temperature increase of 500 K. Below 200 ns there is no more influence of the pixel size. This graph shows e.g. that for 10 μm radius pixels the maximum intensity of 10 μs pulses is 10 $MWcm^{-2}$.

Note that both models differs up to one order of magnitude at long exposure times. This can be attributed to the fact that the boundary conditions in terms of cooling efficiency play an important role at long exposure times for finite structures as the heat diffusion distance $(\kappa\tau)^{1/2}$ becomes considerable with respect to the structure dimensions. For example, a pulse width of 10^{-2} s incident on a n-GaAs substrate results in a heat diffusion distance $(\kappa\tau)^{1/2}$ of the order of 550 μm which is already larger than typical substrate thicknesses.

Bartoli et al. reported experimental damage threshold intensities of various IR detector materials. They show that intrinsic photodetectors (e.g. PbS, PbSe) have relatively low damage thresholds as compared to other IR detector materials due to their high absorption coefficient. Pyroelectric TGS damages at a much lower level than SBN or $LiTaO_3$ since the material is much more fragile and requires a smaller temperature increase for damage. HgCdTe, PbSnTe and InSb have comparable damage threshold intensities due to their similar material properties. Extrinsic photodetectors (e.g. Si:X, Ge:Y) are characterized by a high damage threshold as compared to other detector materials, primarily due to their low absorption coefficient. One can easily deduce for short exposure times, that the damage threshold intensity of a moderately doped (e.g. 7×10^{17} cm^{-3}) n-GaAs Seebeck detector is at least one order of magnitude higher than the threshold of HgCdTe based detectors.

The damage threshold of most thermal detectors can be fine-tuned by selecting the absorbing window material in function of the wavelength. Often a compromise will have to be made between detectivity and power handling capabilities.

3.6 Design possibilities

The main degrees of freedom that can be used in the design phase of a semiconductor based Seebeck detector are: the doping density, the doping layer thickness, the detector/pixel size, the type of electrode (finger contact or perimeter), and the implementation (single detector, four-quadrant position detectors, focal plane arrays).

As mentioned before, the type of detector output (intensity, energy or power) is determined by the ratio between the laser pulse FWHM and various time constants of the detector. Some of these time constants (e.g. τ_{ia}, τ_{mix}) are dependent on various design parameters, shown in expression 20 and 21, which can be tuned to extend a given regime of operation. For instance, the energy operation regime can be extended to longer pulse widths by increasing the absorption length and pixel size. Note that the absorption length can be increased in two ways: by increasing the highly doped absorption layer thickness for a thickness limited absorption length, or by decreasing the doping concentration (read absorption coefficient) for an absorption limited absorption length. This reasoning can be reversed for the extension of the power operation regime. Taken into account the doping level dependence of τ_{mix}, one can extend the energy regime towards shorter pulse widths by increasing the doping concentration.

The top electrode of the detector can be a finger contact (Figure 11a) which probes the local temperature in the neighborhood of the electrode or a perimeter design (Figure 11b) which measures the average temperature along the perimeter contact.

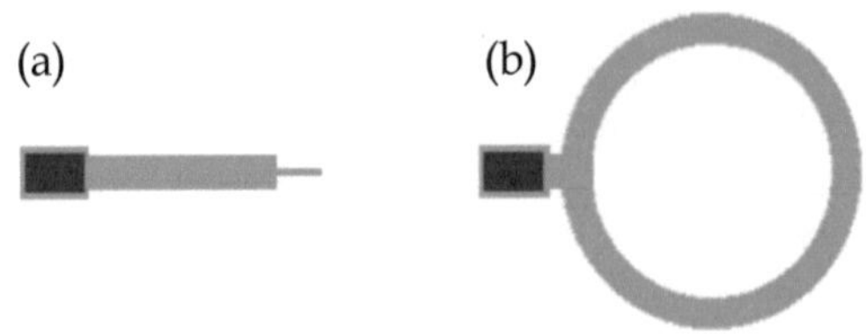

Fig. 11. (a) finger contact electrode; (b) perimeter contact electrode

Their design will determine the information one can get out of the detector. For pulse widths between τ_{mix} and τ_{ia}, the output of a finger contact electrode will be proportional to the local incident energy, whereas a perimeter contact, far from the heating center, allows to measure the total pulse energy incident within the perimeter of the detector. Note that energy measurements with a finger contact electrode allows, after post processing (differentiation), to obtain information about the temporal evolution of the local intensity. The total energy measurement obtained by use of the perimeter electrode on the contrary, allows, after differentiating the signal, to obtain information about the time evolution of the average intensity in the pixel. Special care should be taken in designing these electrodes in function of the pulse duration. Assuming a constant reference temperature, the detector's output will be proportional to the average potential on the contact area of the electrode. As it takes time to conduct heat through the substrate defined by the thermal diffusivity coefficient α_{th}, the electrode size should be carefully chosen in function of the pulse duration. The thermal diffusion length, which is dependent on the ratio between the pulse duration and the thermal time constant, can be taken as a measure (order of magnitude) for the maximum dimension of the contact electrode.

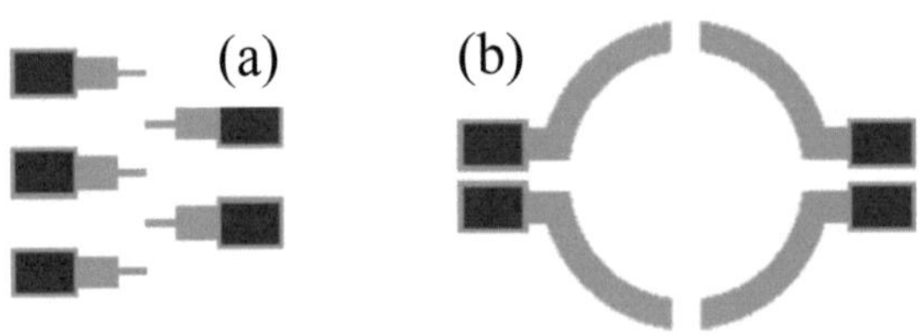

Fig. 12. (a) focal plane array detector; (b) four quadrant positioning detector

Seebeck based detectors can take many forms. In addition to the previously mentioned single pixel detectors, they can be implemented as a LFPA (e.g. for laser beam profilometry) or as a four quadrant positioning detectors as illustrated in figure 12a and 12b, respectively. Figure 13 shows a microscopic picture of a LFPA thermo-electric detector with squared perimeter contact pixels and a pixel pitch of 400 μm.

3.7 Thermal cross-talk

One of the major challenges regarding multi-pixel designs (e.g. laser beam profiler) are related to thermal cross-talk reduction. Thermal cross-talk can be defined as the unwanted signal measured by neighboring pixels as a consequence of thermal diffusion. Depending on the regime of operation, thermal cross-talk might be an important issue to cope with. This thermal cross-talk, which is mainly caused by heat transfer due to lattice vibrations can be

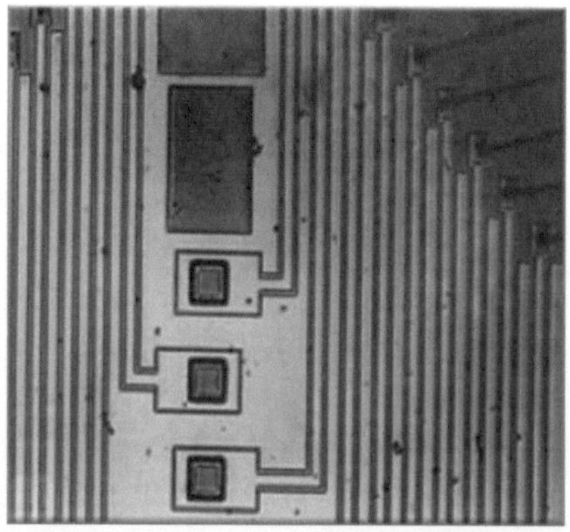

Fig. 13. LFPA thermo-electric detector with squared perimeter contact pixels

neglected in the intensity regime of operation, as there, the measured signal can be attributed to heated electrons. In the power regime on the other hand, the laser pulse width is long enough, with respect to the thermal time constant of the active material, to reach a thermal equilibrium between the input and output heat flux. Consequently, heat can diffuse inside the substrate resulting in a cross-talk signal in neighboring pixels. The penetration depth of heat inside the substrate can be approximated by (Ready, 1997)

$$D = \sqrt{4\alpha_{th} t_{FWHM}} \quad \text{with} \quad \alpha_{th} = \frac{k_{th}}{\rho C_p}$$

Where α_{th} is the thermal diffusivity, k_{th} is the thermal conductivity, ρ is the density, C_p is the volumetric heat capacity and t_{FWHM} is the pulse FWHM. For example, a Q-switched laser pulse of 100 ns incident on a GaAs based Seebeck detector results in a penetration depth of a few micrometers. The same detector reaches a thermal penetration depth of a few hundreds of micrometers for a one millisecond incident optical pulse. Hence, the thermal cross-talk level is strongly depend on the distance from the heating centre and the pulse duration. Hence, for linear focal plane arrays with a pixel pitch of a few hundred micrometers, this cross-talk level can be considerable for relatively long pulse widths (power operation regime).

It has been shown that the thermal cross-talk level can significantly be suppressed by applying a lock-in method (Vandermeiren et al., 2010). For lock-in periods significantly smaller than the thermal relaxation time, the thermal wave is strongly damped which limits it's propagation distance inside the substrate. Here, the optical pulse is amplitude-modulated (AM) as illustrated in figure 14 by means of a mechanical chopper or an external optical modulator (e.g. acousto-optic modulator).

The thermal diffusion length of the corresponding periodic heat source is then given by

$$\mu_{th} = \sqrt{\frac{\alpha_{th}}{\pi f_{Lock-in}}} \quad \text{with} \quad \alpha_{th} = \frac{k_{th}}{\rho C_p} \qquad (25)$$

Where α_{th}, k_{th}, ρ, C_p and $f_{Lock-in}$ are the thermal diffusivity, thermal conductivity, density, volumetric heat capacity and modulation frequency, respectively. The thermal diffusion length dependence on the lock-in frequency is illustrated in figure 15a. The periodic

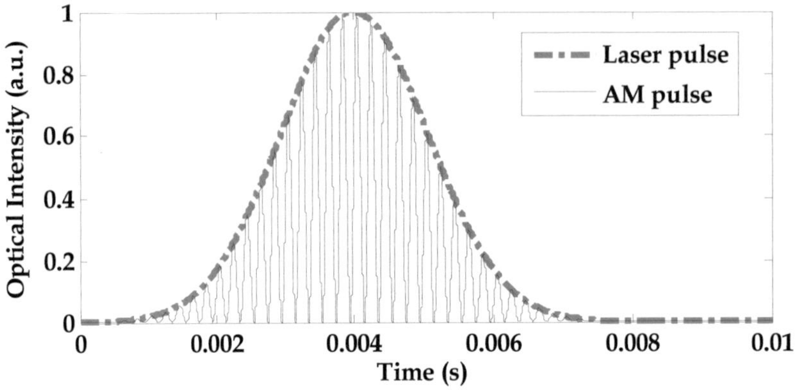

Fig. 14. Amplitude modulated laser pulse

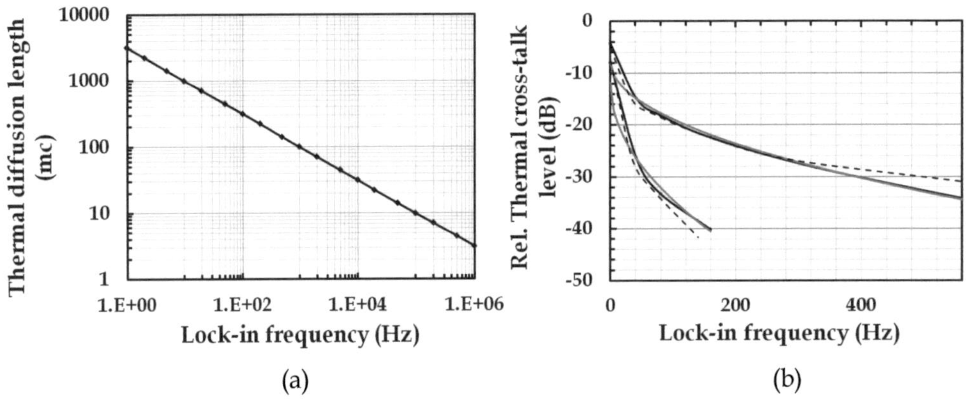

Fig. 15. (a) frequency dependent thermal diffusion length; (b)Lock-in frequency dependence on the thermal cross-talk level

temperature increase at a distance r from the point-like heat source is described by Eq. (26) (Breitenstein & Langenkamp, 2003). Note that this equation only holds when the incident laser beam is small compared to the thermal diffusion length.

$$\partial T(r,t) = \frac{Q}{4\pi k_{th} r} e^{\left(-(1+i)\frac{r}{\mu_{th}}\right)} e^{i(2\pi f_{Lock-in})t} \quad (26)$$

Where Q, k_{th}, μ_{th}, r and t are the power, thermal conductivity, thermal diffusion length, the distance from the heat source and time, respectively. Hence, the periodic temperature increase for a fixed distance r is proportional to $\exp(-\sqrt{f_{Lock-in}})$.

By means of relation (26) one can analytically extract the cross-talk reduction factor for GaAs as a function of the modulation frequency and the distance from the heat source. The cross-talk reduction factor was numerically and experimentally quantified in previous work (Vandermeiren et al., 2010). The modulation frequency dependence on the thermal cross-talk reduction factor at a distance of 400 μm and 800 μm is shown in figure 15b. The results of the analytical model (solid gray curves), a numerical finite element method (FEM) model (dashed gray curves) and experimental data (solid black curves) are in good agreement. For details about the numerical FEM model and experiments we refer to (Vandermeiren et al., 2010). Notice that the thermal cross-talk level can significantly be reduced by modulating the incident laser pulse. For example, in according with the analytical model, the cross-talk levels for the first and second adjacent pixels (r_{first} = 400 μm ; r_{second} = 800 μm), are attenuated by -10.5 dB and -19.3 dB respectively for a lock-in frequency of 140 Hz. The lock-in frequency should be chosen in function of the maximum allowable cross-talk level and the pixel pitch of a FPA-design. Expression 2 shows that the voltage responsivity of Seebeck detectors is inversely proportional to the lock-in frequency when $1 \ll \omega^2 r_{th}^2$. Figure 16 shows a measurement of the relative voltage responsivity dependence on the lock-in frequency for a n-GaAs based Seebeck detector. For lock-in frequencies exceeding about 400 Hz, the responsivity decreases with about 4 dB/kHz. The impact of this frequency dependent voltage responsivity reduction on the Johnson noise limited signal-to-noise ratio will be rather modest as the lock-in technique implies a narrow bandwidth as compared to a common broadband amplifier. This will significantly reduce the contribution of the Johnson noise which is approximately white.

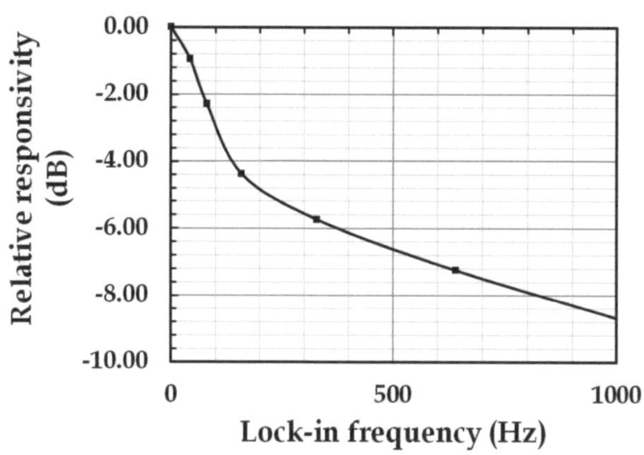

Fig. 16. Relative responsivity dependence on the lock-in frequency

3.7 Future directions of research

The ability of using thermo-electric LFPA for high power laser beam profilometry will be further investigated in the near future. A first aspect that will be considered when dealing with higher laser power levels is the impact of non-linear thermal effects on the thermo-voltage and thermal cross-talk level. These non-linearities include the temperature dependence of the absorption coefficient, the thermal conductivity, the Seebeck coefficient and the heat transfer coefficient. Furthermore, performance optimization in terms of electrical and thermal cross-talk, by improving the design and technology, will be tackled.
Another direction of research development will be dedicated to the implementation of high density 2D focal plane arrays. The main challenge here, in GaAs technology, is related to the implementation of multiplexing individual pixels to the electronics readout circuitry. Hence,

it might be worth to reconsider the material choice when dealing with 2D arrays. Silicon would be an obvious choice for implementing the readout electronics as one can benefit from standard CMOS technology. Hence, on the one hand we will investigate the ability of using silicon as the active detector material. However, on the other hand, we will consider the technological effort for using an arbitrary thermo-electric detector material stacked on a silicon die -comprising the readout electronics- by means of ball-bonding technology.

4. Conclusion

In this chapter we described, by means of introduction, most commonly used infrared detectors for measuring laser pulses used today with their respective advantages and disadvantages in terms of spectral detectivity, operation temperature and the damage threshold.

The theoretical background of semiconductor based Seebeck detectors was explained in detail. We showed that different operation regimes are applicable in function of the temporal characteristics of the optical input, due to various inherent time constants. These time constant permits to measure the intensity, energy or power in function of the incident pulse width. We also showed that the modelling and physical understanding of thermo-voltage detectors permits to optimize the detector design for a broad range of pulse parameters.

We showed that for most IR-detectors a trade-off exists between the responsivity and the power handling capabilities. We explained how one can control the responsivity and damage threshold of semiconductor based Seebeck detectors by means of the doping density and pixel geometry. It turns out that n-GaAs based Seebeck detectors with a moderate doping level ($7x10^{17}$ cm^{-3}) have a damage threshold which is about one order of magnitude higher than HgCdTe based detectors at short exposure times. This makes them suitable for short and intense pulse measurements.

We pointed out some design possibilities of Seebeck detectors in terms of functionality (single pixel, four quadrant or linear focal plane arrays) or the extension with respect to the pulse width interval of a particular regime of operation.

We showed quantitatively how one can improve the thermal cross-talk performance for LFPA's by means of a lock-in method. Hence, we demonstrated theoretically and experimentally that the thermal cross-talk amplitude can be reduced by -35 dB for the first neighboring pixel at a pixel pitch of 400 μm and at a lock-in frequency of 560 Hz. One can also conclude that the voltage responsivity reduction as a consequence of the lock-in frequency is small compared to the thermal cross-talk reduction. We expect the impact of the voltage responsivity reduction on the signal-to-noise ratio to be rather modest as the lock-in method implies that only a narrow bandwidth around the lock-in frequency can contribute to the noise level of the measured signal.

5. Acknowledgements

This work was funded by the Vrije Universiteit Brussel (VUB OZR1726, OZR1013BOF and GOA51/DEFIS42012-VUB) and the Brussels region in the context of the SOIB project (BRGEOZ128) on opto-electronic semiconductor components for CO_2 lasers.

6. References

Bartoli, F.; Esterowitz, L.; Kruer, M. & Allen, R. (1977). Irreversible laser damage in IR detector materials, *Applied Optics*, Vol. 16, 11, (November 1977), pp. 2934-2937, ISSN 0003-6935.

Breitenstein, O. & Langenkamp, M. (2003). *Lock-in thermography, basics and use for functional diagnostics of electronic components*, Springer-Verslag Berlin Heidelberg New York, ISBN 3 540 43439 9, Germany.

Blakemore, J. S. (1982). Semiconducting and other major properties of gallium arsenide. *J. Appl. Phys., Vol.* 53, 10, (October, 1982), pp. R123-R181, ISSN 0021-8979.

George, S.N.; Sharp, J. and Goldsmid, H.J. (2001) *Thermoelectrics: basic principles and new materials developments*, Springer-Verslag Berlin Heidelberg, ISBN 3-540-41245-x, Germany.

Phipps, C.R. (2007). *Laser Ablation and its Applications*, Springer Science, ISBN 0-387-30452-5, USA.

Razeghi, M. (1998). Current status and future trends of infrared detectors, *Opto-Electronics Review*, Vol. 6, 3, (September, 1998), pp. 155–194, ISSN 1230-3402.

Ready, J. F. (1997). *Industrial applications of lasers*, Academic Press, ISBN 0 12 583961 8, USA.

Rogalski, A. (2003). *Infrared Detectors: status and trends*, Progress in Quantum Electronics, Vol. 27, 2-3, pp. 59-210, (May 2003), ISSN 0079-6727.

Schubert, E. F. (1996). *Delta-doping of semiconductors*, Cambridge University Press, ISBN 0 521 48288 7, Great Britain.

Shah, J.; Leite, R.C.C. & Scott, J.F. (1970). Photoexcited hot LO phonons in GaAs, *Silid State Communications*, Vol. 8, 14, (May 1970), pp. 1089-1093, ISSN 0038-1098.

Shkerdin, G.; Stiens, J. & Vounckx, R. (1999). Comparative study of the intra- and intervalley contributions to the free-carrier induced optical nonlinearity in GaAs, *J. Appl. Phys*, Vol 87, 7, pp. 3807-3818, ISSN 0021-8979.

Shkerdin, G.; Stiens, J. & Vounckx, R. (2002) X-valley influence on hot free electron absorption and nonlinearities at 10.6 μm in highly doped GaAs, *European Phys. J. Appl. Phys*, Vol 19, 1, pp. 29-38, ISSN 1286-0042.

Shkerdin, G.; Stiens, J.; Kotov, V.; Vandermeiren, W. & Vounckx, R. (2007). Time dependence of CO_2 laser pulses recorded in the mixed detector regime of the photon drag and Seebeck effects in n-doped GaAs. *J. Appl. Phys.*, Vol. 102, 064509, (September, 2007), doi:10.1063/1.2779277, ISSN 0021-8979.

Stiens, J.; Shkerdin, G.; Vladimir, K.; Vandermeiren, W.; De Tandt, C.; Borghs, G. & Vounckx, R. (2006). Seebeck infrared photodetectors: an ultra wide dynamic range of design possibilities. *Proceedings of the SPIE*, pp. 61890, ISBN 0-8194-6245-4, Strasbourg, France, April 2006, SPIE-International Society for Optical Engineering, Bellingham, Washington.

Sze, S. M. (1981). *Physics of Semiconductor Devices 2nd edition*, John Wiley & sons, ISBN 0-471-05661-8, USA.

Vaissiere, J.C.; Nougier, J.P.; Fadel, M.; Hlou, L. & Kocevar, P. (1992) Numerical solution of coupled steady-state hot-phonon – hot-electron Boltzmann equation in InP,

Phys. Review B, Vol. 46, 20, (November 1992), pp. 13082-13099, ISSN 1098-0121.

Vandermeiren, W.; Stiens, J.; De Tandt, C.; Shkerdin, G.; Kotov, V.; Borghs, G.; Muys, P. & Vounckx, R. (2010). Thermal cross-talk reduction in IR thermo-electric photodetectors by lock-in method: 4D numerical simulations and experimental validation. *Proceedings of SPIE Photonics West – Physics and Simulation of Optoelectronic Devices XVIII*, pp. 75970Y-75970Y-9, ISBN 9780819479938, San Francisco USA, January 2010, SPIE-International Society for Optical Engineering, Bellingham, Washington.

9

Laser Pulses Characterization with Pyroelectric Sensors

Lorenzo Capineri[1] and Marina Mazzoni[2]
[1]*Dipartimento Elettronica e Telecomunicazioni,*
Università di Firenze
[2]*IFAC CNR*
Italy

1. Introduction

There are many industrial and medical applications of CO_2 (λ=10.6 μm) and Nd:YAG (λ=1.06 μm) infrared lasers for which the quality of the process are tightly connected to the characteristic of the laser pulse. These two types of lasers deliver pulses with duration, repetition frequency and power that can be controlled by means of a programmable electronic control unit. An open-loop control generally optimize the process performances by availing of a laser system model. However, this method cannot control that during the operation the laser source and the optical delivering system could deteriorate; moreover the laser beam characteristics and laser pulse temporal envelope could change by several factors like power supply variations, optical beam misalignments, dirty deposits on mirrors, changes in laser efficiency and many others .

For these reasons it is of crucial importance to provide array of sensors capable to measure both the laser pulse and beam characteristics on-line. These sensors generally require the design of front-end electronics for signal conditioning , processing and visualization at a rate equal to the laser pulse repetition frequency.

This chapter illustrates the capabilities offered by pyroelectric arrays of sensors for this aim. The main advantage of the pyroelectric thermal sensors is an extended spectral response beyond 100μm that includes the CO_2 gas infrared lasers.and the quantum cascade (QC) lasers. Chemical sensing at room temperature with pulsed QC-DFB lasers operating at 15.6 μm has been developed (Kosterev et al., 2002); in this paper also a spatial distribution of the laser power was made with a pyroelectric sensor, as in other spectroscopic applications which employed infrared lasers (Toci G. et al., 2000). This type of sensors can be fabricated with low-cost pyroelectric materials. The simulated time and frequency responses are particularly useful for their design which in turn must be tailored for the reproducibility characterization of the laser pulses. The sensor signals depend on the adopted fabrication, packaging technology, and on and the gain-to-frequency factor of the front-end electronics beside the pyroelectric material properties. A limited bandwidth, sensitivity and linearity *vs* power density generate distortions of the output signal. Consequently, the designer has to consider a certain amount of distortion of the laser pulse that must evaluated for a proper monitoring operation and eventually minimized. The analysis of this type of distortions is fundamental to optimize the design of pyroelectric sensors. Moreover, the front-end electronics can include the filtering

(analog or digital) of the signals to compensate distortions. Among the several ferroelectric materials, in this chapter we focus on PVDF (polynivilidene - fluoride) array of sensors.
The chapter illustrates examples of the applications of these sensors, which were derived from academic research projects in collaboration with companies. It will be shown how to use, design and built monitoring systems for mid infrared (IR) laser pulses. Moreover, we provide the basic assemblying principles of pyroelectric transducers and front-end electronics for a quick and easy reproduction by professionals involved in laser pulse monitoring and laser processes optimization.

2. Pyroelectric sensors modelling

The use of CO_2 power lasers systems for industrial (welding, marking, etc.) and medical (surgery and therapy, etc.) applications are now numerous and of increasing importance. These systems necessitate of low-cost detectors able to measure the laser beam characteristics for process optimisation. Pyroelectric sensors are the most suitable devices for the measurement of industrial lasers characteristics (power and energy) because of their broad spectral response and ruggedness. The pyroelectric detectors can be conveniently utilized for the radiation detection by high power (> 1 kW) CO_2 continuous wave (CW) lasers, by means of mechanical chopper for the detected radiation modulation necessary for their operation. With the recent generation of CO_2 pulsed lasers, which can achieve pulse repetition frequency higher than 100 Hz, a temperature gradient on the sensor can be used in place of the mechanical modulation. This section is aimed to describe the design a pyroelectric sensor with response optimised for high-power CO_2 laser applications for which the requirement of a short settling time is much more stringent than voltage responsivity.
Pyroelectric sensors based on PVDF polymer foils are suitable for designing detectors, in the mid mid IR region, with response time sufficiently fast for monitoring and controlling CO_2 lasers at high pulse repetition frequency (PRF >100Hz). In the following we describe how to design and built single and multi-element detectors, capable of following the laser pulses evolution with rising times of about 0.05 ms and about 2 ms (FWHM) duration. For monitoring these lasers pulses it is necessary to provide sensors with sufficient short time response to follow these rapid rising edges.
The choice of PVDF pyroelectric materials offers the possibility to produce low cost, large area detectors which can be installed inside the laser system or close to the target under laser processing. These detectors, assembled in a matrix array configuration can be also used for real-time control of the laser beam profile.

2.1 Materials and methods

Pyroelectric detectors have been designed by using a theoretical sensor model with multilayer structure (Setiadi & Regtien, 1995). The computational results reported are relative to the configuration shown in Figure 1, where the multilayer structure is composed by a pyroelectric film, a bonding layer and a substrate. The heat sink at the top at the bottom of the structure is represented by the air surrounding the sensor.
An equivalent circuit of a multylayer sensor can be obatined, see Figure 2, where the thermal and electrical characteristics can be represented by resistive-capacitive impedances.
The pyroelectric current $I(t)$ results proportional to the time derivative of the excess mean temperature of the pyroelectric layer. For a sinusoidal incident power density $P_s(t) = P_i e^{(j\omega t)}$ at angular frequency ω, the current responsivity in the frequency domain results in the following expression:

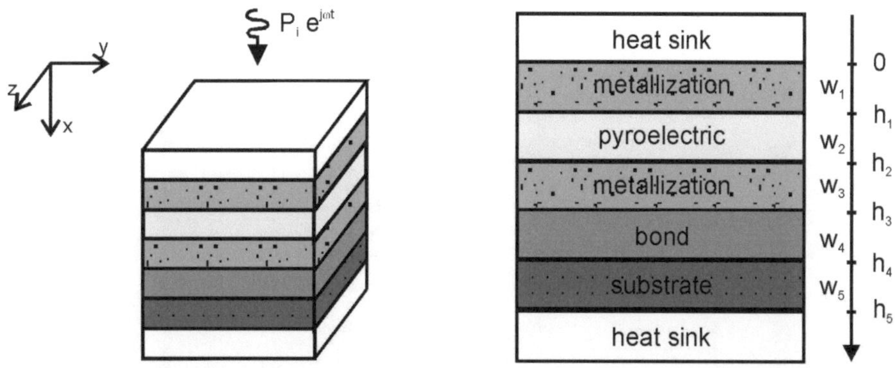

Fig. 1. Multilayer model of a pyroelectric detector; w_i is the thickness of layer i , A is the active area of the sensor exposed to a modulated incident radiation $P_s(t) = P_i e^{(j\omega t)}$ along the x direction

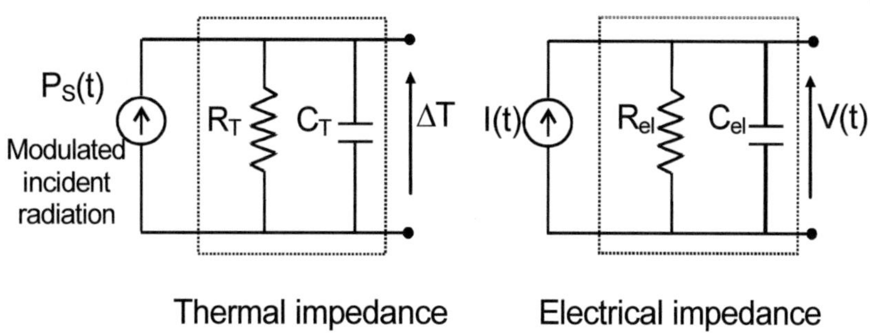

Fig. 2. Pyroelectric sensor equivalent circuit.

$$I(\omega) = \frac{j\omega Ap\eta P_i}{w_p \alpha_p} \cdot \left[A_p(\exp(w_p \alpha_p) - 1)) + B_p(1 - \exp(-w_p \alpha_p)) \right] \quad (1)$$

where η is the absorbance, p is the pyroelectric coefficient, *A* the sensor active surface; w_p and α_p are the thickness and the reciprocal of the thermal diffusion distance of the pyroelectric layer, respectively. For a structure of the sensor with N layers, the coefficients A_p and B_p are derived from the steady-state solutions of a linear system of N equations (Setiadi & Regtien, 1995).

The current responsivity optimization can be analyzed as a function of the different physical parameters of the multilayer structure. Figure 3 and Figure 4 show the frequency responses of a typical detector as a function of the thicknesses of the PVDF film and of the metallization, respectively. These last two parameters revealed the most significant in the design optimization. The results which we show refer to an active area of the sensor of A=1cm^2, and a pyroelectric coefficient of λp=20μC/ m^{2}K. The measurements were performed at a power density of 1 mW/mm^2. The values of the materials properties deduced from literature (Giacoletto and Landee et. al., 1977)(Regtien et. al., 1992) are reported in Table 1.

The current amplitude generated by a given incident power density can be estimated by means of simulations involving Equation (1). For the values reported previously, the current amplitude varies from nA to tens μA in the frequency range 1Hz-100kHz. These pyroelectric current values require a front-end electronic design with precision, low bias-current operational amplifiers.

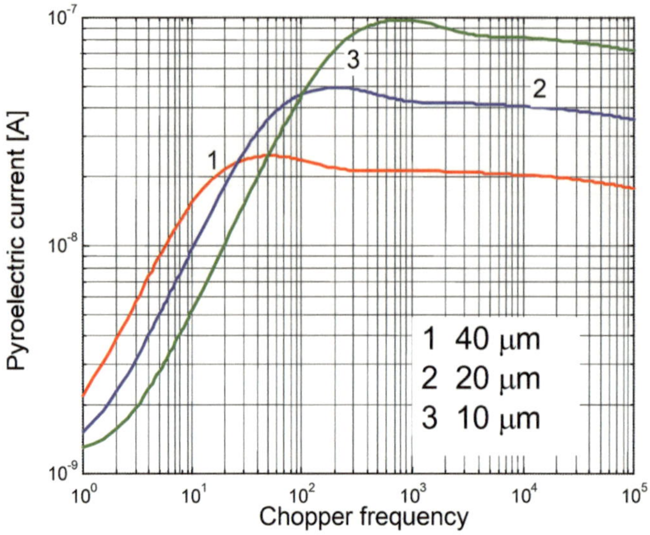

Fig. 3. Pyroelectric current responsivities for different thicknesses of the PVDF layer.

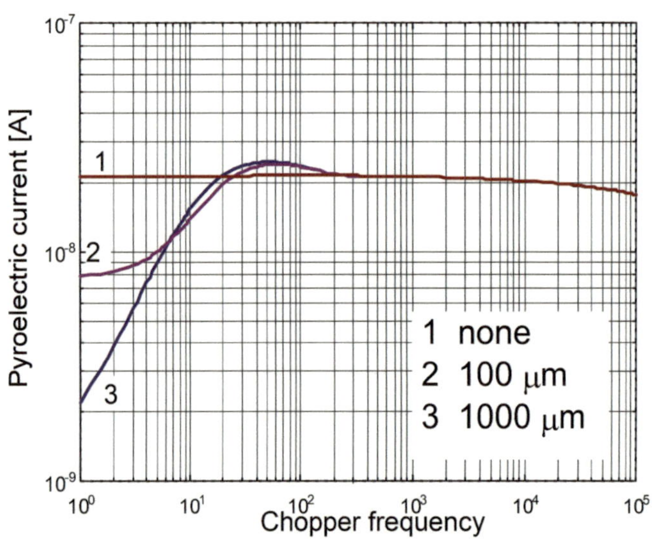

Fig. 4. Pyroelectric current responsivities for different thicknesses of the copper substrate.

When sensors are built in a matrix array, the thermal cross-talk between adjacent elements must also be evaluated. and the lateral heat conduction (Schopf et al.,1989), which was omitted in (1), must be taken into account in the simulations. To account for the lateral heat conduction along the direction y (and similarly for direction z), a basic relationship is reported:

$$T(y,t) = T_b \cdot e^{-\frac{y}{\lambda_D}} \cdot e^{j\left(\omega \cdot t - \frac{y}{\lambda_D}\right)} \tag{2}$$

where T_b is the temperature at the boundary of a pixel and $\omega=2\pi f_{chop}$. The thermal cross-talk results reported in Figure 5 for different values of chopper frequency (f_{chop}), were obtained with the following parameters: pixel area=200μm x 200 μm, $\lambda_D=(2D/2\pi f_{chop})^{0.5}$ penetration depth, D=0.0056 cm^2/s temperature diffusivity.The results shown in Figure 5 point out the advantage of working with a high pulse repetition rate (>100 Hz) in the industrial and medical laser applications that require arrays of sensors.

Layer	Material	Thermal conductivity [W/cm K]	Thermal capacitance [J/cm³ K]	Thickness [μm]
Metallization	Gold	3.10	2.83	0.07
Pyroelectric	PVDF	0.0013	2.30	40
				20
				10
Metallization	Gold	3.10	2.83	0.07
Bonding		1	1.65	1
Substrate	Silicon	1.49	1.64	1000
				100
				0

Table 1. Physical parameters used in the simulations shown in Figures 3 and 4.

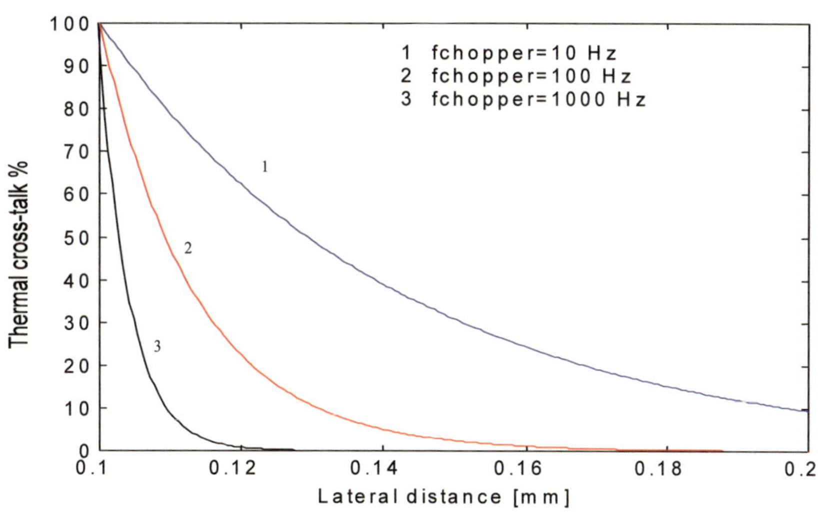

Fig. 5. Thermal cross-talk between two elements (area 0.2x0.2 mm^2 each) as function of the lateral distance from boundary (y=0).

The designed front-end electronics is based on a low-noise transimpedance amplifier in order to convert the current generated by the detector in a voltage signal with signal-to-noise ratio suitable for a successive analog or digital signal processing. In Table 2, Z_{el} is the feedback impedance formed by the parallel circuit R_a and C_c of the basic configuration of the transimpedance amplifier shown in Figure 6. It can be easily demonstrated that, for an ideal operational amplifier, the voltage response can be written as Vo=Ip * Z_{el} , and the -3dB cut-off frequency as $f_{cut\text{-}off} = 1 / 2\pi R_a C_c$. According to the values shown in Table 2, the cut-off frequency resulted 10.6 kHz.

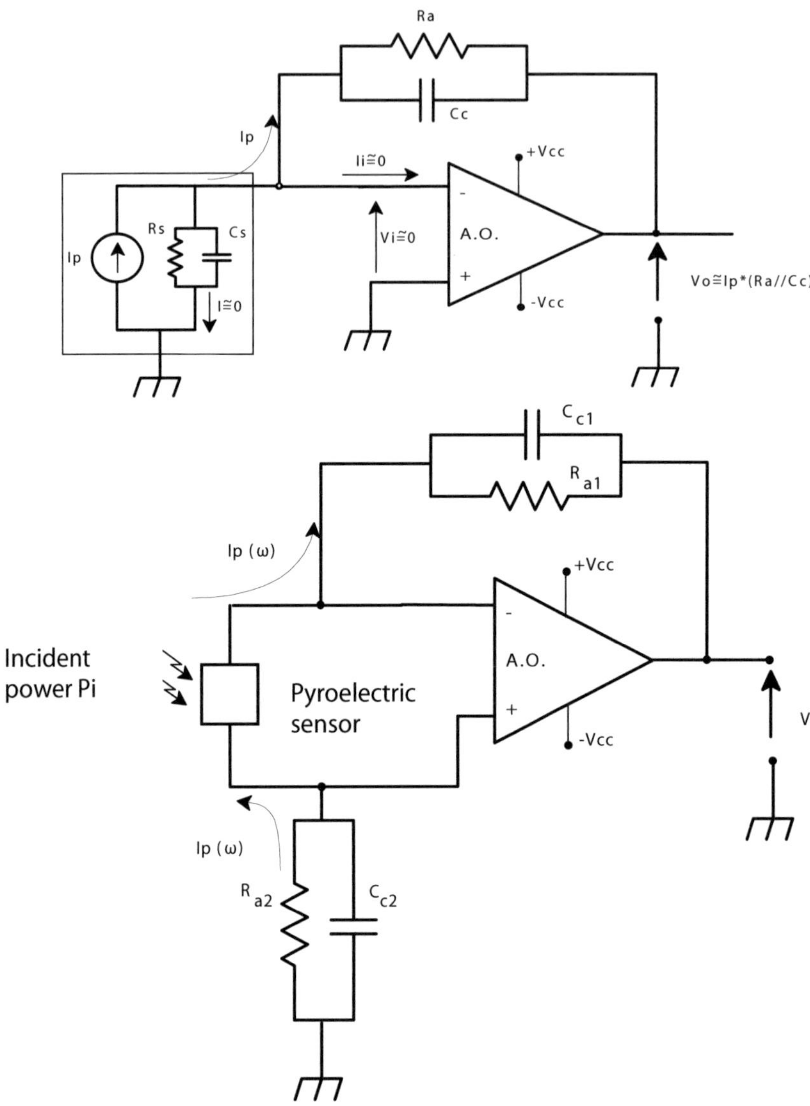

Fig. 6. Top) Basic transimpedance amplifier for a pyrolectric transducer with single ended (unbalanced) connetion; Bottom) Differential (balanced) configuration of transimpedance amplifier with $R_{a1}= R_{a2}$ and $C_{a1}= C_{a2}$. The differential configuration has a gain twice the single ended configuration and improved common mode noise rejection.

The usefulness of sensor and preamplifier modeling with electronic CAD programs is the possibility of calculating the current or voltage responsivity function for a given shape of the laser pulse. According to this approach (Capineri et al. , 2000), a temporal and frequency analysis of arbitrary modulated laser intensity have been demonstrated possible. The experimental detector pulse time response of 24 μs to a simulated 17 μs rising edge of a Nd-YAG laser pulse is shown in Figure 7. By a fitting process based on the root mean square error the model parameters can be retrieved with good accuracy .

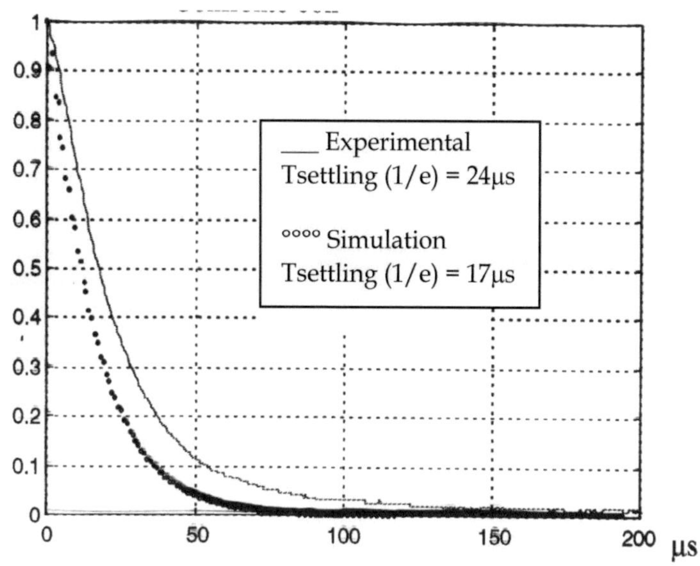

Fig. 7. Comparison of pyroelectric sensor normalized voltage response between simulated model and experimental sensor.

Several single-element detectors were built, which were able to follow laser pulses with rise time up to 0.003 ms. Figure 8 shows an example of the time response to a CO_2 laser pulse for the values reported in Table 2.

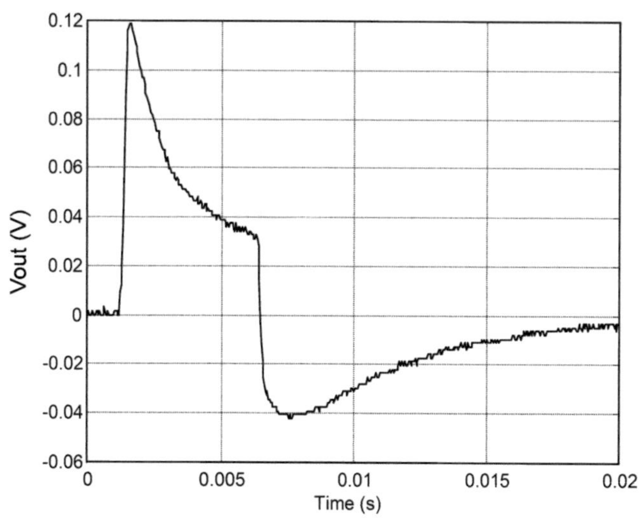

Fig. 8. Pyroelectric signal in response to a pulsed CO_2 laser

Active area of detector	9 mm^2
Thickness(PVDF)	40 μm
Gold metallisation	0.1 μm
Z_{el}	Ra= 1 MΩ parallel with Cc= 15 pF
Pulsed laser characteristics	
PRF	50 Hz
Duty cycle	22%

Table 2. Parameter values characterising detector and laser source for the measurement shown in Figure 8

The time response is characterized by a rising time 0.2 ms like that of the laser pulse and the undershot is characteristic of a rapid cut-off. The recovery time (7 ms at 1/e of Vout max) is governed by the detector thermal time constant R_T C_T.
The settling time to zero value is mainly determined by the undershoot and it is approximately 15 ms.
This basic example demonstrates the feasibility of the pyrolectric PVDF film sensor technology for monitoring IR laser pulses (Capineri et. al., 1992). Another technology that has been demonstrated useful for sensor fabrication availed of a screen printed pyroelectric paste (Capineri[b] et al., 2004). Both pyroelectric materials have been employed to design and build array of sensors with different configuration and size, depending on the application (Capineri et al. 1998)(Capineri et al., 2005)(Mazzoni et al.,2007). Some example of pyroelectric arrays used to design monitoring devices for CO_2 power laser systems are described in the following section.

3. Technologies for PVDF pyroelectric sensor arrays

Commercially available pyroelectric arrays mostly employ ferroelectric materials as BST, PbSe, $LiNbO_3$ and $LiTaO_3$. These sensors are fabricated with technologies which are compatible with integrated electronics. Their spatial resolution is determined by the pitch between elements, typically 50 μm wide for arrays in the order of 320x240 pixels. Their performances in terms of sensitivity and NEP are suitable for thermal imaging applications and for remote temperature measurements (Muralt, 1996)(Capineri[b], 2004). The aim of this section is to describe enabling technologies for the development of low-cost pyroelectric sensor arrays for the beam characterization of CO_2 power lasers (λ=10.6 μm). A low-cost pyroelectric material PVDF is commercially available in the form of thin foils that can be metalized by means of evaporation or sputtering. The polymer foils are mechanically flexible and necessitates of fabrication technologies suitable for realizing the electrical contacts; rigid carrier substrates and low temperature conductive epoxy are usually employed for this aim. In this section, we describe some solutions that exploit printed circuits boards technology. The array of sensors should sustain relatively high power densities even if a beam power partitioning system is considered. Experimental characterization of sensors with PVDF foils with gold metallization in different conditions of laser pulses (peak power, duty cycle and pulse repetition frequency), showed that an average power density of 1 W/cm^2 should not be exceeded. An array element pitch of 1 mm was estimated sufficient to detect most of the significant anomalies of the laser beam intensity spatial distribution of a CO_2, 40 W continuos power laser.

A fabrication technology that can be adopted for a fast production of small series of sensors is the laser ablation (Capineri[a] et al. , 2004). In the following we describe the main features of the laser microfabrication for patterning electrodes on the film, and the line connections routing strategy. Two examples are shown: a matrix array (8x8 elements, pitch 1.45 mm) and a linear array (10x1 elements, pitch 1 mm). Preliminary experimental results on laser microdrilling of the PVDF material will be presented for microvias fabrication aimed to make individual contacts of each front electrode element. For the packaging we adopted the bonding of the sensor array to printed circuit boards and standard connectors for the external contacts to the front-end electronics board.

3.1 Laser microfabrication for ferroelectric polymer (PVDF) sensors

Polymer ferroelectric materials like PVDF are now commercially available from several manufacturers and are used for fabricating pyroelectric and ultrasonic piezoelectric sensors (Binnie et al., 2000)(Ritter et al. 2001). The relative merit of polymers respect to ceramics is their low weigh, mechanical flexibility, non reactivity to chemical agents and relative low cost with respect to piezoelectric ceramics. On the contrary, they have a limited operating range (T_{MAX}=80°C) and generally a lower figure of merit with respect to other piezoelectric or pyroelectric materials (De Cicco et al., 1999). In our application the choice of PVDF was mandatory for the large area required to monitor the position and intensity spatial distribution of a laser spot of about 1 cm^2.

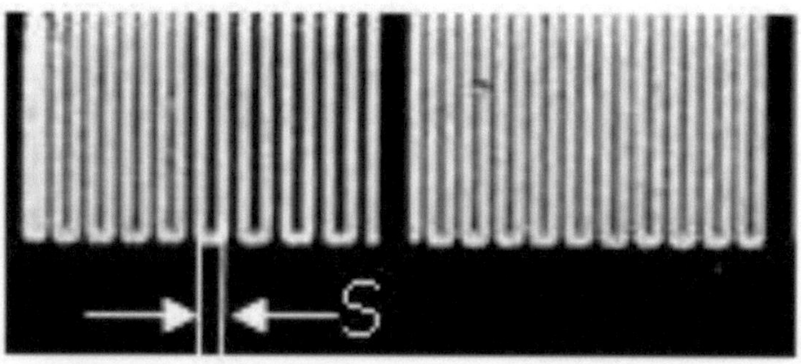

Fig. 9. Example of laser ablation of a set of parallel lines at two different separation distances S on a 40μm thick gold metallized PVDF film: (Left) S_1=150 μm , (Right) S_2=100 μm.

Considering the high incident power available, the sensor current responsivity requirements are not stringent and the transimpedance amplifiers can be designed with feedback impedances in the range 10MΩ-1GΩ; these values are not so large to be influenced by parasitic capacitances due to circuit layout or connection lines through the packaging. For the temporal diagnostics of the CO_2 laser pulses a response time better than 10 μs is needed. The use of a plastic film as active pyroelectric material requires a suitable technology to transfer the design of the electrodes pattern on one or both sides of the film. The routing of electrical lines from the central elements of the matrix array to the external connector pins asked also for solutions adequate to the element miniaturization which needs line width negligible respect to the element size. In our approach we used a Nd:YAG laser (λ=1.064μm) marking tool (mod. Lasit, EL.EN. s.p.a., Italy) to ablate the metallizations of the PVDF film which are typically made with gold, aluminum, or even conductive silver ink, according to the optical and electrical requirements. The process has been developed for metallization with thickness ranging from 0.1μm to 10μm which are typical of evaporation and screen-printing respectively. The laser ablation process needs to be optimized by successive refinements of the laser marking parameters such as the pulse repetition frequency, laser pumping current, pulse duration and focal distance. The laser setting was tuned according to the trace width (microfabrication features), the minimum induced mechanical film damage, the process repeatability and the electrodes design flexibility.

An interesting characteristic of the laser microfabrication is the contemporary ablation of the metallization on both sides of the film (Capineri[a] et al., 2004). After the ablation of the front electrodes metallization, the laser beam reaches the bottom side of the PVDF film without being absorbed by the bulk. This is possible due to the low absorption of the thin PVDF film at the Nd:YAG emission wavelength. In this way the patterning of the electrodes on both sides is attained with only one laser ablation run. The replica of the same pattern on both sides of the PVDF film is useful when differential connections to individual elements of the array are needed; differential transimpedance amplifiers can be employed for improving the common mode noise rejection as shown in Figure 6. The laser microfabrication method has been successfully demonstrated for different PVDF film thickness ranging from 9 μm to 110 μm.

In Figure 9 the results of a spatial resolution test is shown. The minimum distance S between two lines or two array elements should result higher than about S = 140 μm. In Figure 10, the zoom over a portion of the linear array reported in Figure 11(A) shows a detail of the gold metallized areas with rounded ablated corners.

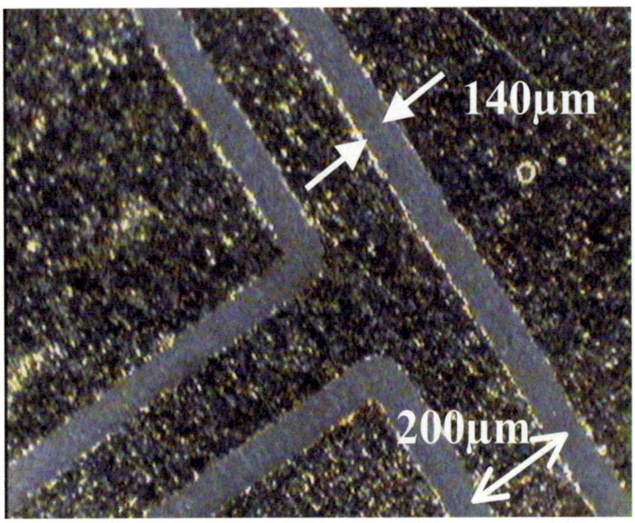

Fig. 10. Example of electrodes patterning by laser ablation.

Because of the low capability of this type of film to sustain overheating beyond 80°C, a study was performed to verify the presence of an eventual damage to the PVDF material. In particular, we compared the pyroelectric responses of single elements obtained by two different techniques, i.e. laser ablation and gold metal evaporation. No significant difference was observed. Some examples of fabricated pyroelectric arrays on 40 μm thick gold metallized PVDF film are reported in Figure 11 (A) and (B).

In Figure 11(A) the box indicates the active area of a linear array with 10x1 elements of dimensions 0.9x2mm^2 each, pitch 1 mm and connection lines width 0.2 mm. Four such arrays were mounted at 90° angle on an electronic board in order to monitor the position and dimensions of a CO_2 laser beam in real-time. In Figure 11 (B) a fine pitch matrix array for beam spatial intensity distribution measurements is shown; it is provided with 8x8 elements, of area 1.25x1.25mm^2 and pitch 1.45 mm.

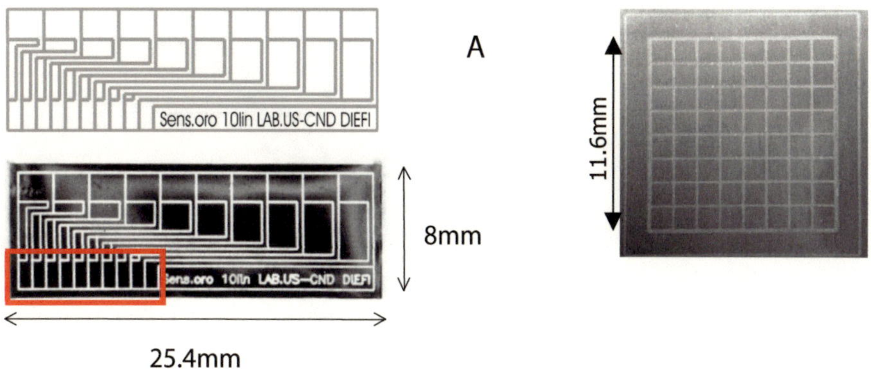

Fig. 11. A) AUTOCAD drawing for design the electrodes geometry of a linear array (top) and resulting sensor microfabrication (bottom). The rectangle in red color indicates the array of 10 active elements. B) matrix array: 8x8 square elements, side 1.25 mm, pitch 1.45mm.

The solution adopted for bonding the PVDF pyroelectric arrays to a rigid substrate utilizes two PCBs, called here top and bottom Printed Circuit Boards (PCB), called here top and bottom PCBs. The electrical connections between the film and PCBs are obtained by conductive epoxy (type EP21TDC/N, MasterBond, USA) and curing at room temperature. The PCBs have copper pads which overlap the gold pads on the PVDF film. This bonding technique proven to be reliable having used the sensors over a period of at least two years with no change in characteristics and performances. The routing from the external pads towards the active elements is not a problem for the linear array geometry.

On the contrary, the routing of the connection lines of the two-dimensional array poses the problem of individually contacting the front electrodes exposed to the radiation. Moreover, the connection line surface acts as a spurious sensor that creates cross-talk effects and ghost signals at the outputs of the sensor array. At present, our laser microfabrication technology with a Nd:Yag laser (not specifically devoted to this application) provides an ablated line width of 140 μm, which is the minimum pitch between matrix elements or conducting lines. Looking for novel solutions to this problem, we investigated a new structure for assembling matrix arrays that retains the advantages of the laser microfabrication and the packaging techniques previously described. We also developed a fabrication process for electrodes patterning on a PVDF film metallized only on one side. The opposite side was metallized in a second step by evaporating a single continuos semitransparent gold electrode of thickness less than 100 nm. This process provides a common front electrode for all elements which is connected to a top PCB and then to ground. The exposure of this front electrode to the incident beam occurs through a protection window (ZnSe or Ge) in the top PCB (see Figure 12). The front common electrode is grounded and the 64 single ended transimpedance amplifiers are connected by a standard PGA 84 pin connector.

The PVDF sensor was then bonded on the 64 central pads of the bottom PCB by using a programmable robot provided with a dispenser. This step of the fabrication is critical because the uniformity and reliability of the bonding process can be easily affected by the conductive epoxy viscosity variability during the dispensing and curing phases. The sandwich of the two PCBs and sensor in between is then soldered to the PGA 84 pins connector. The photo in Figure 13 shows one prototype of the matrix pyroelectric array.

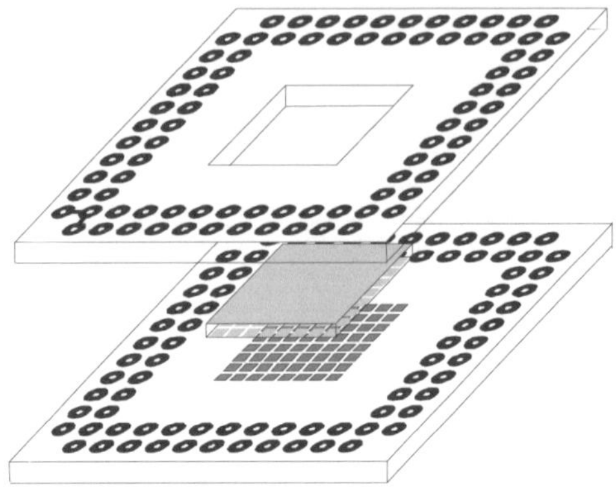

Fig. 12. Assembly for the pyroelectric matrix array.

Fig. 13. Packaging for the pyroelectric matrix array.

The 64 elements matrix array have been characterized in terms of voltage responsivity and response uniformity. A thermal cross-talk ranging from -33dB to -41dB was found in the frequency range 10Hz-200Hz. The diagram in Figure 14 is an example of measured cross-talk on one element with side L=2.25 mm. It was obtained with a modulated laser diode at repetition frequency of 185 Hz and a laser spot diameter 500 μm. The results indicate that the lateral heat conduction of the front semi-transparent electrode is modest. We also found that it is slightly dependent on the beam modulation frequency. However, in the perspective of increasing the number of elements, the modification of the original design of the matrix array will consist of square elements in the front electrode contacted to a bottom PCB through microvias. A reasonable value for the microvias diameter is in the range 10-50 μm, according to the minimized pitch of the array. Preliminary results of microdrilling with a duplicated Nd:YAG source have produced a line of through holes with diameters ranging from 20μm to 40μm (see Figure 15). The variation of the holes diameter is due to different settings during the laser process. Similar processing methods have been also explored more recently from other authors (Rabindra et al., 2008) (Lee et al., 2008).

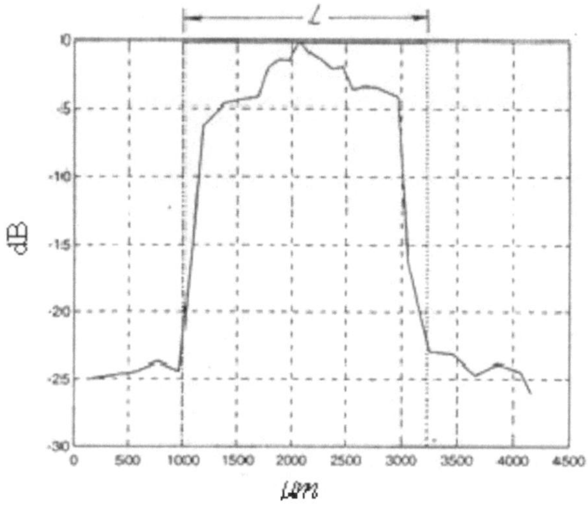

Fig. 14. Cross-talk measured on a single element at laser beam modulation frequency of 185 Hz.

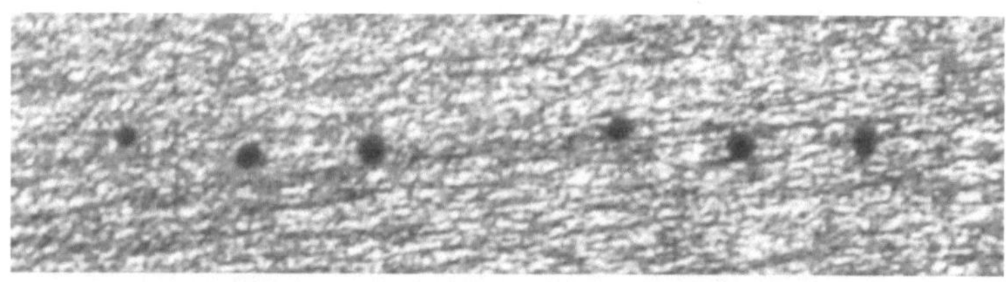

Fig. 15. Laser microdrilling through a 40 μm thick gold metallized PVDF film. The holes diameter varies from 20μm to 40μm.

4. Applications of PVDF pyroelectric array of sensors for CO_2 laser monitoring

In this section we explore the main applications of pyroelctric arrays in a linear and matrix configuration.

Pyroelectric sensor linear arrays of PVDF were found particularly suitable for the control of the spot dimensions of high power infrared laser beams. The sensors were tested for maximum power density in temporal cycles of tenths hours each.

We designed an optoelectronic instrument for the *on line* measure of the dimensions of the laser spot emitted by a multikilowatt CO_2 industrial laser. Due to the high power and long service time the optical components are subjected to thermal stresses which cause variation of the laser beam characteristics (shape and position).

In Figure 16 we show the schematic diagram of the experimental apparatus which consists of the laser source, a beam expander, a beam deflector and a focussing lens. The main beam of continuos power Pi is sampled after the beam expander by using a diffractive optics which splits the beam into a reflected beam, of power Pr=98.8% Pi, and a sampled beam of lower power and equal to 0.5% Pi. This low power beam of about 15 W (for a Pi=3 kW) has a typical diameter of 25 mm and follows the variations of the main one. We could measure its dimensions along two perpendicular directions with the linear array configuration shown in figure 17. The minimum required spatial resolution was 1mm and the variation of the dimensions were in the range of 20 mm - 30 mm.

We verified the damage threshold of the sensors made of gold metallized PVDF film with an experimental set-up in which the power density on the sensor was varied by changing the repetition frequency and duty cycle of an average power equal to 30 W which was delivered by the CO_2 laser source. A single sensor was irradiated through a metal diaphragm in cycles lasting tenths of hours each at increasing power density ranging from 0.15 W/cm^2 to 3 W/cm^2. The voltage response of the sensor was tested during each phase and the results are shown in Figure 18. The sensor response remained constant for a fixed value of the power density and it decreased for higher power density values owing to the increase of the sensor average temperature. At a value of 3.6 W/cm^2 we observed the destruction of the sensor, hence we safely reduced the power threshold value to 3 W/cm^2.

In Figure 19, we show an assembled linear array prototype; each of the four arrays is composed of ten elements with pitch 1 mm. Other measured characteristics of the fabricated linear array sensors are:

- Thermal cross-talk better than -40 dB at 200 Hz
- Bandwidth (-3dB): 257 Hz
- Current responsivity max: 190nA/W

The linear arrays in cross configuration have been experimented for real-time beam diameter monitoring but their use was extended also to laser power monitoring according to their useful bandwidth. It has been demonstrated that at fixed pulsed repetition frequency these sensors provide a reliable estimation of the incident laser power. Moreover, the fabrication technology explained in the previuos section, allowed the realization of pitches between elements of about 150 μm. This value is adequate also for real-time imaging of power laser beams by devising a rotating reflector that scanned the beam section at an angular velocity adequate for granting an accurate imaging of the laser pulse (Coutouly et al., 1999)(Akitt et al., 1992)(Mann et al., 2002), (Mazzoni et al., 2007).

4.1 Dual use of pyroelectric arrays for CO_2 and Nd:YAG laser pulses: laser pulse characterization and beam positioning

Industrial and medical CO_2 laser equipment are controlled for the optimization of the power emission according to the process. This normally implies two operation modes: continuous (CW) and pulsed (PW). In both cases it is important to monitor some beam parameters in real-time for maintaining the quality of the process or for diagnostic purposes (to check the functional anomalies). For both modes sensors are necessary that can operate at the laser wavelength (mid-IR) with an electronic instrument suitable for acquiring, processing and visualizing the beam parameters. The considered parameters were: the beam point stability, the beam spatial intensity distribution and the laser pulse shape related to the instantaneous emitted power. The measurements of these parameters are standardized (ISOFDIS

11146,11670,11554) and each one requires specific characteristic of the sensor and processing electronics. The pyroelectric array of sensors described in the previuos sections are suitable for these applications and represent a good compromise between cost and performances.

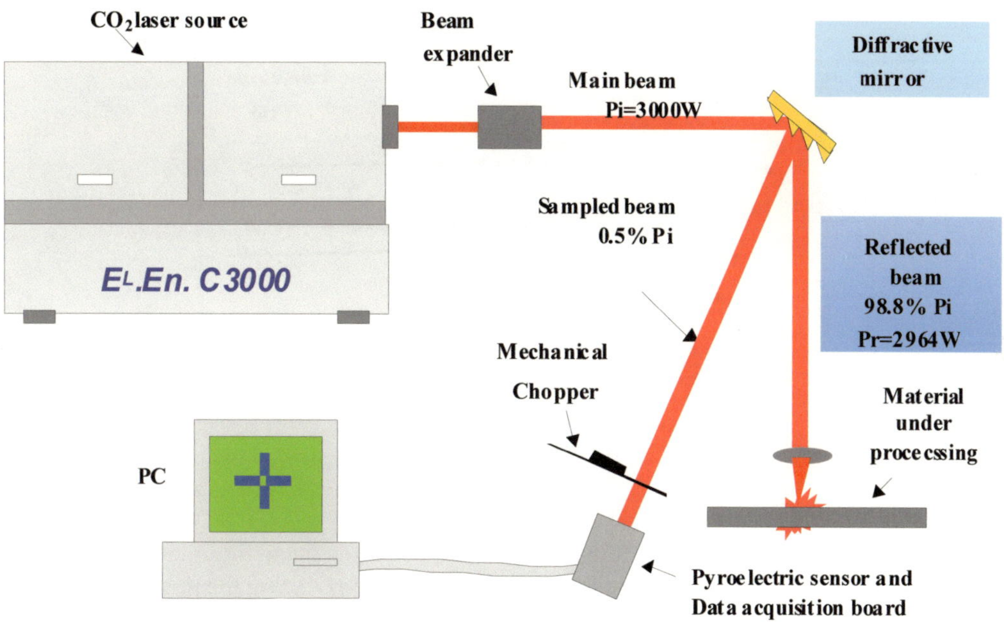

Fig. 16. Schematic diagram of diagnostic system of laser beam dimensions.

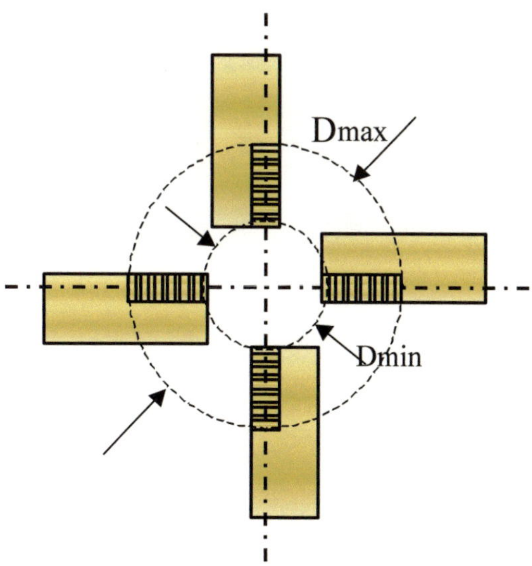

Fig. 17. Configuration of linear arrays for measuring the beam dimensions in the range Dmin-Dmax.

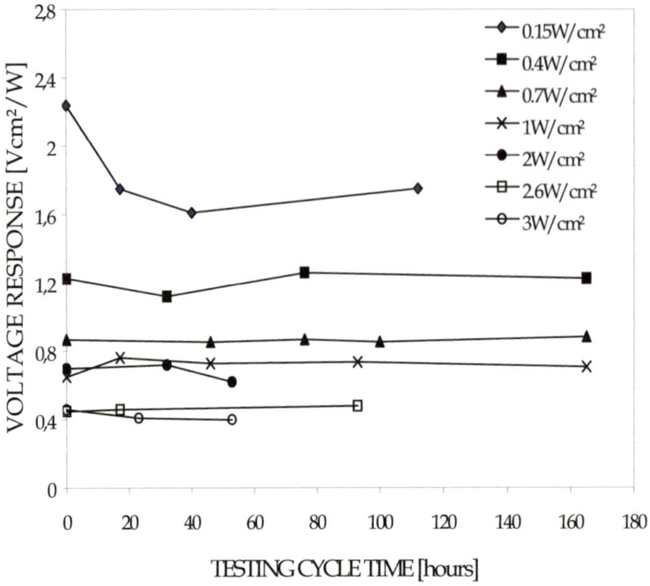

Fig. 18. Voltage response for different incident power densities during life tests.

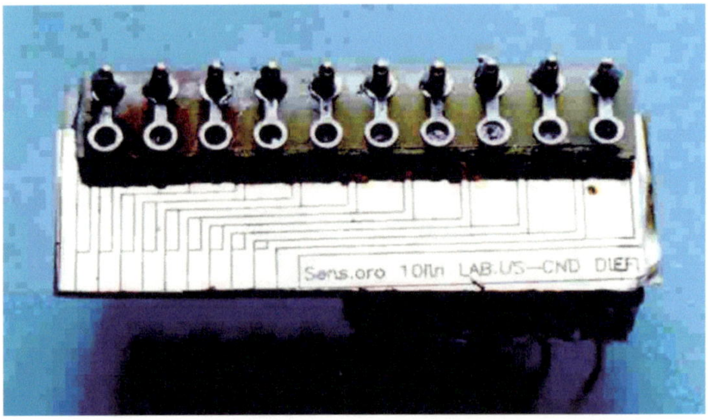

Fig. 19. Assembled linear array of 10x1 elements.

In this section it will be shown that a versatile instrument can be interfaced to different measuring modules provided with linear or matrix arrays of pyroelectric sensors. The two measuring modules were: Module "BeamScan64" for the laser spatial intensity characterization, and Module "PosIRix" for the laser beam point stability and pulse shape characterization (Capineri et al. 1999)(Capineri et al., 2005). The architecture can be replicated with other choices of the analog electronic components and with a microcontroller with upgraded performances.

4.2 Portable electronic instrument architecture

The instrument operates in a stand-alone mode and automatically switches the running program depending on the connected external module. The analog signals from up to 64 channels are digitally converted by two parallel ADCs on chip of a microcontroller Hitachi SH7044 and presented on a QVGA LCD with 256 colors. The instrument was tested with two sensor modules: an 8x8 matrix array for laser beam mapping with 64 high gain (1GΩ) transimpedance amplifiers, and a large area four-quadrant sensor for the beam point stability (Capineri et al. ,1999) control and laser pulse monitoring. The complete architecture of the analog-to-digital mother board is shown in Figure 20 and a photo of the prototype system is shown in Figure 22. The instrument is interfaced to external modules by a versatile bus (V-Bus) that includes several I/O digital lines, 64 analog lines, and several auxiliary lines for power supplies and remote sensing/controls. Inputs for an automatic identification of the plugged-in modules were also provided.

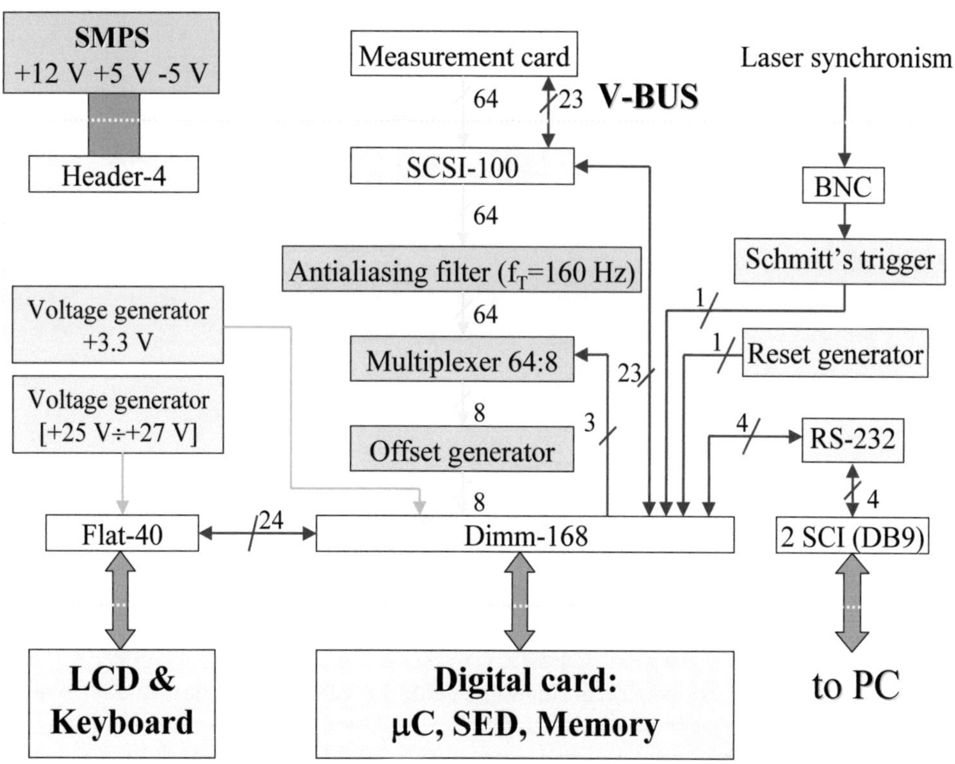

Fig. 20. Block scheme of the mother board of the electronic instrument

Fig. 21. Photograph of the components of the electronic instrument for IR laser beam characterisation

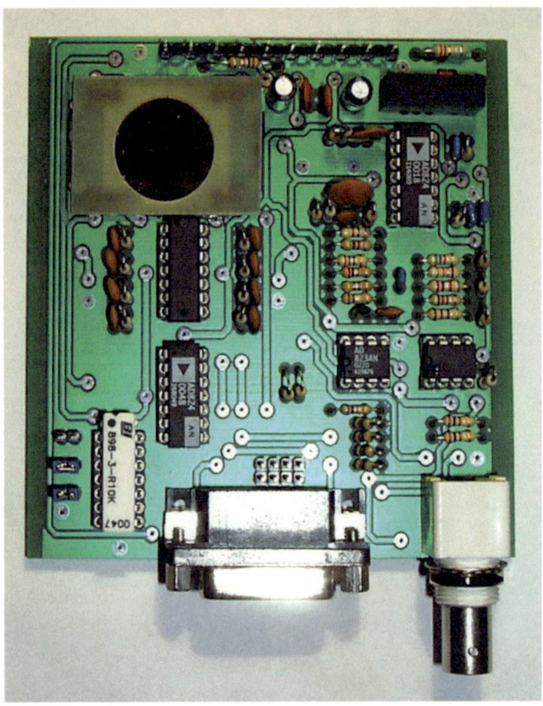

Fig. 22. Electronic board of the module PosIRix. In the bottom side, the DB5 type connector for the instrument V-Bus connection is sided in a central position; on the left side is shown the BNC connector for laser pulse monitoring by an oscilloscope. On the top left a 28mm x 28 mm PVDF four quadrant sensor with a circular ZnSe window for spectral filtering the CO_2 wavelength.

5. Module "PosIRix" for laser beam point stability and pulse shape characterization

This module consists of a 28 mm x 28 mm sensor divided in four quadrants by laser ablation of a gold metallized PVDF ferroelectric film of thickness equal to 40 μm. The pyroelectric material was bonded to a FR-4 epoxy rigid substrate with thermal conductive glue. The substrate was also used to make electrical contacts with bottom electrodes. The sensor fabrication was optimized in order to achieve the maximum sustainable power density D_{PMAX} and the maximum bandwidth of the voltage responsivity BW_{MAX}. This choice of the sensor design parameters (dimensions, substrate, bonding) is an example of good compromise among cost, bandwidth, sustainable power density and mechanical robustness. The sensor and front-end electronics were characterised with different powers, duty cycles and pulse repetition frequencies of a CO_2 laser source. Values of D_{PMAX} =2W/cm^2 and BW_{MAX} (-3dB) =18 kHz were found with a 4.4 MΩ transimpedance amplifier. We also demonstrated the adaptability of this sensor to a specific medical application of the laser by designing an electronic equalization filter of the amplitude of the frequency response in order to achieve a flat bandwidth (±1dB) between 10Hz and 18 kHz. In this way the laser pulse shape was reproduced with high fidelity, even for PRF as low as 10 Hz, in a range where the responsivity of the sensor is not flat. Two examples are reported in Figure 23 and 24. They show the reconstruction of the pulse shape of a CO_2 laser modulated at a PRF equal to 30 Hz and 100 Hz, respectively. In the same figures, we showed the response measured with a large bandwidth (20 MHz), small-size (i.e. 1mm^2), commercial HgCdTe photovoltaic sensor for comparison.

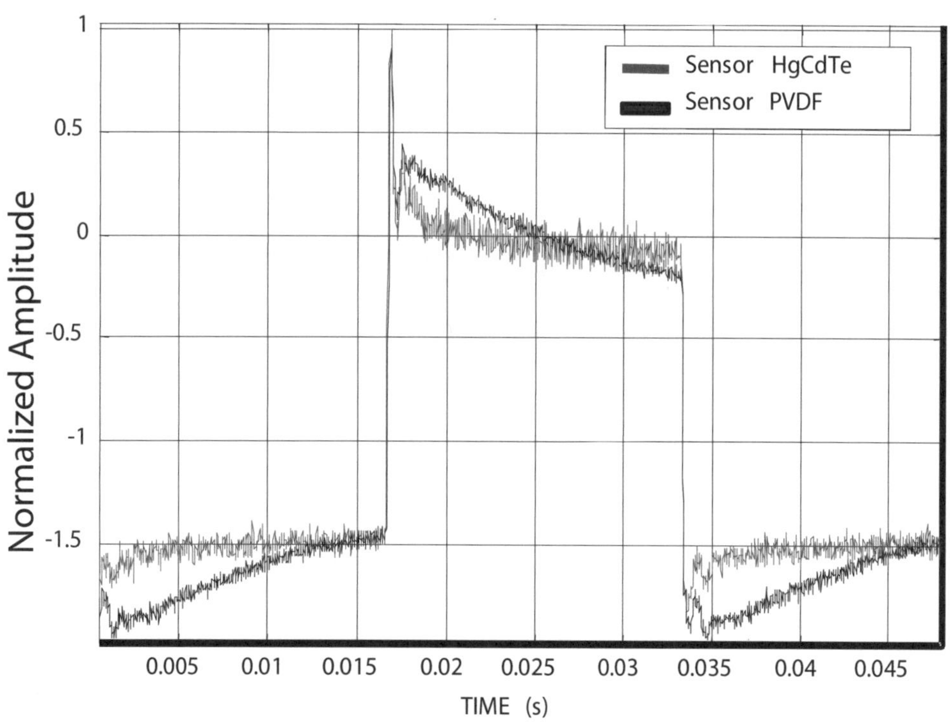

Fig. 23. CO_2 Laser pulse shape at PRF 30 Hz.

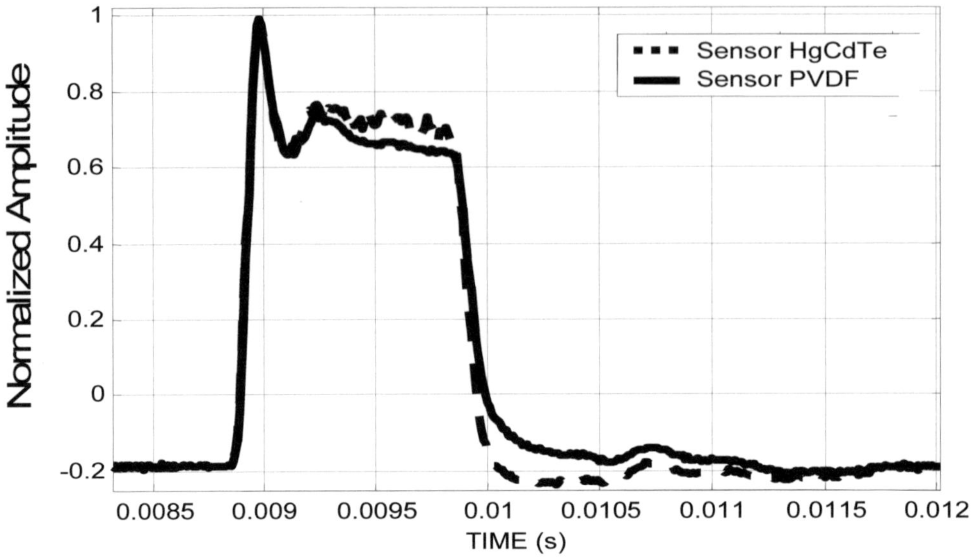

Fig. 24. CO_2 Laser pulse shape at PRF 100 Hz.

The same module was also used for monitoring the laser beam point stability by designing a programmable narrow band filter centered at the PRF of the laser source; this narrowband signal was digitized and fed to an algorithm that estimates the centroid of the intensity spatial distribution on the sensor plane (Capineri et al. 1999) with four quadrant signals. The algorithm is implemented on the microcontroller used in the portable instrument described in Section 4.2. The complete block scheme of this module is reported in Figure 25.

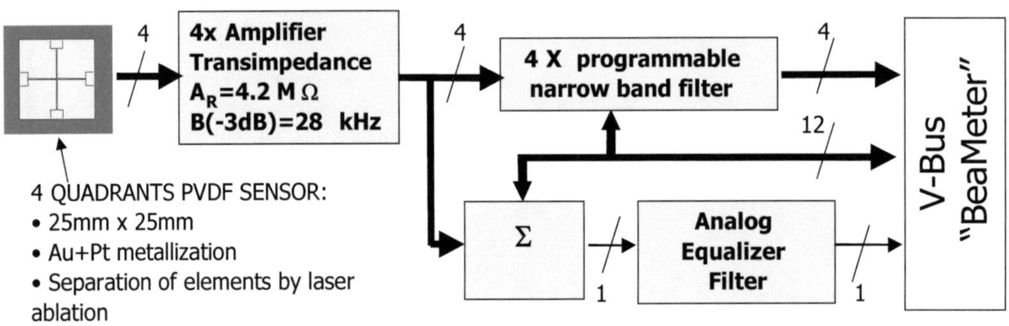

Fig. 25. Electronic analog signal processing carried out by module "PosIRix".

5.1 Signal filtering for limited bandwidth sensors

Two new implementations were developed for the processing and visualization of signals generated by PVDF pyroelectric sensor arrays with compensation filtering (Capineri et al. 2005) aimed to improve the reconstruction accuracy of CO_2 laser pulses. These implementations were especially devoted to biomedical applications for which there is a stringent demand for an accurate reproduction of both the fast and slow components of the laser pulse for the evaluation of the intensity in these two temporal regimes. The implementations were realised for the module "Posirix" which was described in the previous section. It was primarily designed for laser beam positioning and allows the visulization of the laser pulse by an oscilloscope or by a dedicated instrument with real-time display. For the laser pulse envelope evaluation we used the sum of the signals from the four pixels to make the first temporal information independent of the beam centroid position within the sensor matrix array. For this solution, a requirement for achieving an accurate pulse reconstruction are four elements with the same frequency response.

5.2 Design of the analog filter

For the filter project of the bandwidth limited sensor we used the ideal compensation filter consisting in the classical inverse filter $H_c(f)$ defined as:

$$H_C(f) = \frac{K}{H(f)} \tag{3}$$

where K is a gain factor for a flat frequency response of the summing amplifier, and $H(f)$ is the sensor voltage frequency response. For the fitting function H_{fit} we used a bi-quadratic form in order to keep its realization simple by means of an analog filter. The fitting program was developed in Matlab (Mathworks, USA) and calculated the vector of coefficients a_i of the biquadratic function resulting from the minimization of the mean square error (*err*). This program required the following input parameters:

- a *vector* with the initial values of a_i;
- the frequency values *fmin* and *fmax* delimiting the range for the fitting of $H_{fit}(f)$ with $H_C(f)$;
- the vector with input data $H_c(f)$ interpolated in the range 1Hz - 50 kHz at 10 Hz steps.
- the tolerance on the functional value *(err)* and on the coefficient values a_i, the maximum number of iterations and of elaboration on *err*.

The *vector* with the initial a_i values was found with a trial procedure of few iterations using the minimization function *"fminsearch"* which starts from an initial guess of the coefficients and a rather high tolerance value to grant an uniform error density also in the frequency region with less data The program progressively decreases the tolerance value to increase the precision in the determination of the optimal vector of coefficients a_i.

The values *fmin* and *fmax* have been chosen to get a small ripple in the sensor bandwidth, particularly sensitive to the pole positions. After some trials they were set to 100 Hz and 7 kHz, respectively, so as it was impossible to cover the full range with only one biquadratic function. We had to use another filter function to complete the filter project.

For obtaining the complete transfer function of the compensation filter, we found the biquaquadratic coefficients for the following function,

$$H_{fit2}(f) = \frac{1}{H(f) \cdot H_{fit}(f)} \tag{4}$$

that we multiplied by the "high frequency" filter function $H_{fit}(f)$ to find the final filter function. By using the procedure described above to cover the remaining "low frequency" regions, the *fmin* and *fmax* values were set to 15 Hz and 250 Hz this time, with a superposition of the minimizing frequencies ranges of the two fitting functions of about 150 Hz. The final fitting function $H_{filt}(S=j2\pi f)$ resulted:

$$H_{filt}(S) = \frac{1,89S^4 + 48957S^3 + 2,16\cdot 10^7 S^2 + 3,06\cdot 10^9 S + 5,04}{S^4 + 3,82\cdot 10^4 S^3 + 1,79\cdot 10^7 S^2 + 1,89\cdot 10^9 S + 3,25\cdot 10^9}. \quad (5)$$

With H_{filt} the compensated filter bandwidth at -3dB extended from 4.4 Hz to 17.8 kHz with a ripple in band of 0.43 dB.
The function can be factorized into four terms which have a direct correspondence with the four building blocks A, B, C, D shown in Figure 26.
The analog design considered components values and tolerances commercially available, and it was started from a six order function H_{cf} with two nearly equal poles and zeroes that allowed more flexibility and no substantial filtering performance variation as shown in Figure 27.

5.3 Design of the digital filter

The digital filtering has the advantage of a circuit reduced dimension and uses the same analytical transfer function found for the analog implementation. Its capability is limited by the Hitachi SH2 microprocessor implementation on board of the same instrument. With a 128kByte RAM it is possible to use only numerical filter of the type IIR for their reduced computational request with respect to FIR ones. Furthermore, owing to the precision limitation to 32 bit of the microprocessor, the implementation of the transfer function resulting from the bilinear transformation of the sampled $H_{filt}(f)$ function at f_{sampl}=115.2 kHz requires an accurate analysis of the zeroes and poles position for the filter stability determination. We found that this implementation made the low frequency filtering worse and required the elimination of a zero-pole couple on the unitary circle corresponding to a frequency of about 10 Hz. We also evaluated the artefacts introduced in the transformation from the analog to the digital masks consisting in modulus and phase differences between the implemented and bilinearly transformed functions above 20 kHz as shown in Figure 28. With a cascade of two filter cells of the second order, the execution time to perform the complete filtering of one laser pulse was about 7.59 μs, slightly less than the time between two samples ($1/f_{sampl}$ = 8.68 μs). Hence it was possible to perform the filtering in real time, and successively give a representation of the pulse envelope on a LCD display. Owing to the reduced dynamic of this monitor, the comparisons with the analog and digital filtering where performed on a PC, after acquisition of the signals from the sensor with an oscilloscope. The digital filter was realized with Matlab functions (Filter, qfilt), in this case.
Experimental results obtained with modulated CO_2 laser beams, at pulse repetition rates from 10 Hz to 1000 Hz and variable duty cycle, proved an accuracy in the laser pulses reconstruction that is not available in the commercial IR beam positioning sensors.
The analog implementation results much more noisy, but the digital implementation suffer for the imposed limitations that make the low frequency components reproduction worse.

6. Conclusions

In this chapter we described the capabilities of pyroelectric sensors built by means of low-cost hybrid technologies based on PVDF films for monitoring pulses of IR lasers. The technologies presented here can be used to design large area sensors for measuring the beam characteristics of pulsed CO_2 power lasers. Details and useful references are provided to build measuring modules both for the beam centroid positioning and the temporal monitoring of the laser pulses. Criterions for designing analog or digital compensation

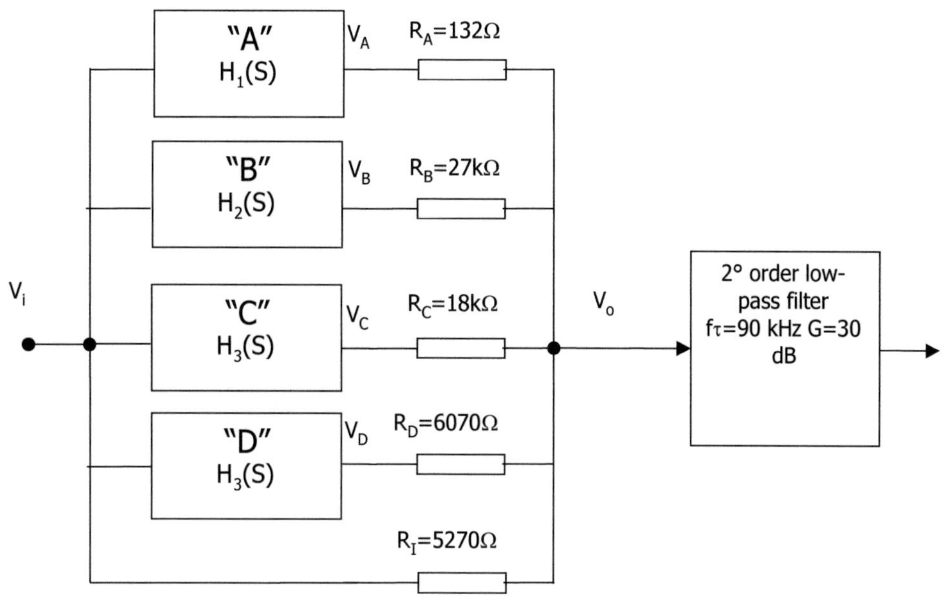

Fig. 26. (Top) Analog implementation of the filter function H_{filt} with four building blocks A, B, C, D. (Bottom) The Sallen-Key low-pass filter reduces the high frequency noise.

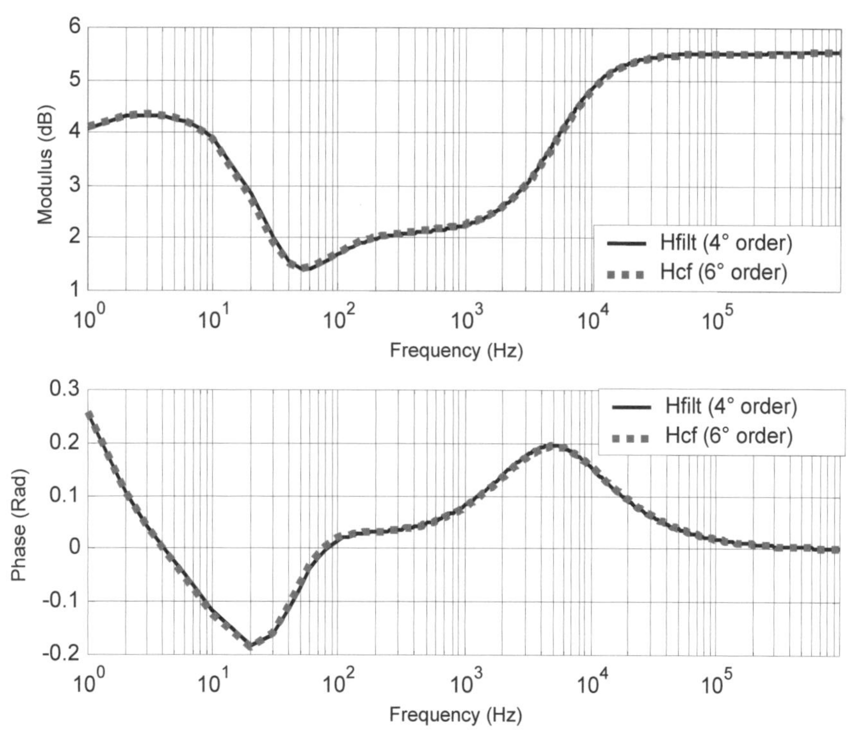

Fig. 27. Comparisons between computed fourth order H_{filt} and sixth order $H_c(f)$ implementations of the compensation filter function $H_c(f)$.

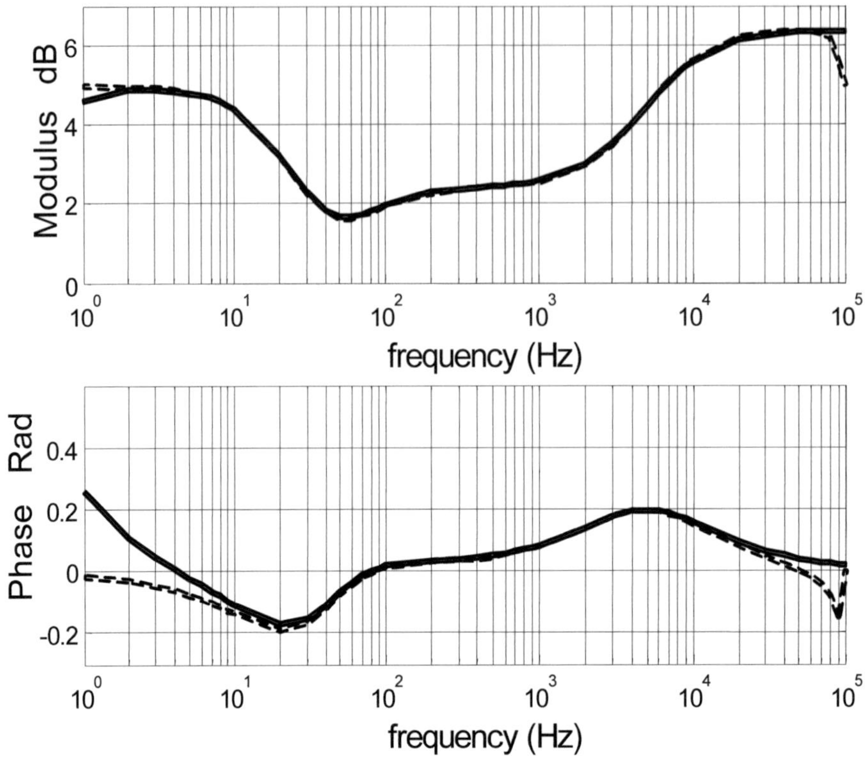

Fig. 28. Computed H_{filt} (continuos line) and numerical H_d (dashed line) implementations.

filters were provided in order to minimize the effect of the typical bandwidth of the pyroelectric thermal PVDF sensors. In this perspective the designed sensors can be seen as an external active probe of an oscilloscope and become an useful instrument for laboratories and companies where the IR laser sources are employed. The fabrication technology of PVDF pyroelectric arrays was reported and low-cost assembling and packaging solutions were presented. Future research for this type of sensors will deal with the analysis of a closed-loop control in real time of the laser system made now possible thanks to the computational power and versatility of commercially available microcontrollers.

7. Acknowledgments

The authors wish to acknowledge the support of CNR project MADESS II and Tuscany Region for having supported this project and the precious scientific and technical collaboration of Prof Leonardo Masotti (Università di Firenze, Italy), Dr Ing Giovanni Masotti (El.En. s.p.a.) and of all the master thesis students that made possible the realization of the projects.

8. References

Akitt D.R et al., (1992), Highperformance automatic alignment and power stabilization system for a multikilowatt CO_2 laser, *Rev. Sci. Instrum.*, vol. 63, pp.1859–1866, 1992.

Binnie T.D. et al., (2000) An integrated 16x16 PVDF pyroelectric sensor array, IEEE Transactions on UFFC, no. 47, pp 1413 -1420

Capineri L. et al. (1998), A 3x3 matrix of thick-film pyroelectric transducers, *Electronics Letters*, Vol. 34, pp 1486-1487

Capineri L. et al. (1999), A beam position sensor for low power infrared laser diodes, *Review of Scientific Instruments*, Vol. 70, pp. 1-8

Capineri L. et al., (2000), Pyroelectric PVDF sensor modeling of the temporal voltage response to arbitrarily modulated radiation, *IEEE Transactions on Ultrasonic and Frequency Control*, Vol. 47, pp. 1406-1412

Capineri[a] L. et al., (2004), European patent EP 1380821 "Matrix-type pyroelectric sensor, method for its fabrication and device for characterizing laser beams comprising said sensor "

Capineri[b] L. et al. (2004), Comparison between PZT and PVDF thick films technologies in the design of low-cost pyroelectric sensors, *Review of Scientific Instruments*, Vol. 75, , pp 4906-4910

Capineri L. et al. (2005), CO_2 laser pulse monitoring instrument based on PVDF pyroelectric array. *IEEE Sensors Journal*, Vol. 5, pp 520-529

Coutouly J.F. et al (1999), Simple is best for real-time beam analysis, *Opto Laser Europe*, n. 58, pp.34–37

De Cicco G. et al. (1999), Pyroelectricity of PZT-based thick-films, *Sensors and Actuators*, Vol. 76, pp. 409–415

Giacoletto L.J. & Landee R. W., (1977), *Electronics Designers Handbook* ed. McGraw-Hill, 0070231494, New York

Hammes P.C.A. & Regtien P.P.L., (1992), An integrated infrared sensor using the pyroelectric polymer PVDF, *Sensors and Actuators A*, Vol. 32, pp. 396-402

Kosterev A.A. et al. (2002), Chemical sensing with pulsed QC-DFB lasers operating at 15.6 μm, *Appl.Phys. B*, Vol. 75, pp.351-357

Lee S. et al. (2008), Femtosecond laser micromachining of polyvinylidene fluoride (PVDF) based piezo films, *Journal of Micromechanics and Microengineering*, Vol. 18, doi: 10.1088/0960-1317/18/4/045011

Mann S. et al., (2002), Automated beam monitoring and diagnosis for CO_2 lasers, Proceedings of SPIE 4629, Laser Resonators and Beam Control V, June 2002, pp. 112–121

Mazzoni M. et al., (2007), A large area PVDF pyroelectric sensor for CO_2 laser beam alignment, *IEEE Sensor Journal*, Vol. 7, pp. 1159-1164

Muralt P., (1996), Piezoelectric and pyroelectric microsystems based on ferroelectric thin films, *Proceedings of the Tenth IEEE International Symposium on Applications of Ferroelectrics*, Aug. 1996, pp. 145–151

Rabindra N. D. et al. (2008), Laser processing of materials: a new strategy toward materials design and fabrication for electronic packaging, *Circuit World*, Vol. 36 , ISSN: 0305-6120

Ritter T.A. et al. (2001), Development of high frequency medical ultrasound arrays, *2001 IEEE Ultrasonics Symposium*, August 2001, pp 1127 -1133

Rocchi S. et. al., (1992), A transducer modelling technique for the identification of the transfer function and driving-point impedance, *Sensors and Actuators A*, Vol. 32, pp. 361-365

Schopf H. et al., (1989), A 16-element linear pyroelectric array with NaNO2 thin films, *Infrared Physics*, Vol. 29, pp. 101-106

Setiadi D. & Regtien P.P.L., (1995), *Sensors and Actuators A*, Vol 46-47, pp. 408-412

Toci G. et al. (2000), Use of a PVDF pyroelectric sensor for beam mapping and profiling of a mid-infrared diode laser , *Rev. Sci. Instr.*, Vol. 71, pp. 1635 - 1637

10

Time-gated Single Photon Counting Lock-in Detection at 1550 nm Wavelength

Liantuan Xiao, Xiaobo Wang, Guofeng Zhang and Suotang Jia
State Key Laboratory of Quantum Optics and Quantum Optics Devices, College of Physics and Electronics Engineering, Shanxi University, Taiyuan 030006, China

1. Introduction

Time-gated single photon counting (TGSPC), which employs a single photon detector as the detection apparatus (Poultney, 1972; 1977), has received increasing attention because of its superior spatial resolution and the absence of the so-called classical dead zones (Forrester & Hulme, 1981). TGSPC is a repetitively pulsed statistical sampling technique that records the time of arrival of photons and logs this against the time of emission of a laser pulse. TGSPC have become increasingly important in a number of applications such as time-resolved photoluminescence (Dixon, 1997; Leskovar & Lo, 1976), optical time-domain reflectometry (Lacaita et al., 1993; Benaron & Stevenson, 1993; Wegmüller et al., 2004), time-of-flight laser ranging (Pellegrini et al., 2000; Carmer & Peterson, 1996) and 3D imaging (Moring et al., 1989; Mäkynen et al., 1994).

Ultrasensitive detection with single photon detection capability requires detectors high quantum efficiency and low dark noise. Operation in the 1550 nm spectral region enables it to be worked in fiber, and the eye-safe ranging brings it to be carried out in daylight conditions. In the 1550 nm wavelength implementations, InGaAs/InP avalanche photodiode detectors (APDs) are commonly used (Pellegrini, et al., 2006; Hiskett, et al., 2000; Lacaita, et al., 1996). However, these APDs have low quantum efficiency because the photons may pass through the very thin depletion layer without being absorbed. In addition, these single-photon detectors exhibit high afterpulse probability, which can cause significant distortion for the measurements. In order to reduce this effect these detectors have to be operated in a time-gated mode. As each photon's arrival time is an independent measure of range, and accuracy can be improved by increasing the number of samples. Unfortunately, direct photon counting will induce the quantum fluctuation (i.e. shot noise).

Time-correlated single-photon counting (TCSPC) is a repetitively pulsed statistical sampling technique that records the time of arrival of photons reflected from a target and logs this against the time of emission of a laser pulse (Becker, 2005). Each photon's arrival time is an independent measure of flight, and accuracy can be improved by increasing the number of samples. However, the technique's main disadvantage is an extended data-acquisition time being required where the illumination noise is a serious problem.

Weak light detection can be improved by use of the lock-in principle (Stanford Research Systems, 1999). A lock-in detects a signal at a known modulation frequency in amplitude and phase and suppresses noise at other frequencies. The lock-in detection principle can enhance the signal-to-noise ratio (SNR) by orders of magnitude. The lock-in principle was first applied to photon-counting detection by Arecchi *et al.* (Arecchi et al. 1966) and was used subsequently in many low-light measurements (Murphy et al., 1973; Alfonso & Ockman, 1968). A dual-phase implementation of the gated photon counting is hampered by signal pick up from harmonics under nonsinusoidal modulation (Stanford Research Systems, 1995). To obtain a precise phase signal, photon counts were reconverted to analog signals that feed into a lock-in amplifier (Braun & Libchaber, 2002). In a previous publication we have demonstrated that the wavelength modulation lock-in can improve the SNR of photon counting for weak fluorescence effectively and eliminate the quantum fluctuation (Huang et al., 2006).

In this chapter, we present an overview of the principle of single-photon detection at 1550nm. And then we focus on the question of illumination noise, detector dark count noise and the detection efficiency of single-photon detector, and we show that the novel method of photon-counting lock-in for TGSPC detection can suppress background noise, and importantly, enhance the detection efficiency of single photon detector.

2. Single photon detection at 1550nm

2.1 Single photon avalanche diodes

An avalanche photodiode reverse-biased above its breakdown voltage, V_{bd}, allows single photon detection (Ribordy et al., 1998). When such a diode is biased above V_{bd}, it remains in a zero current state for a relatively long period of time, usually in the millisecond range. During this time, a very high electric field exists within the p-n junction forming the avalanche multiplication region.

Under these conditions, if a primary carrier enters the multiplication region and triggers an avalanche process, several hundreds of thousands of secondary electron-hole pairs are generated by impact ionization, thus causing the diode's depletion capacitance to be rapidly discharged (Stucki, 2001). As a result, a sharp current pulse is generated and can be easily measured. This mode of operation is commonly known as Geiger mode (Ribordy et al., 2004). Unfortunately, typical photodiodes, as those used in conventional imagers, are not compatible with this mode of operation since they suffer from a premature breakdown when the bias voltage approaches V_{bd}. Premature breakdown occurs since the peak electric field is located only in the diode's periphery rather than in the planar region. A single photon avalanche diode (SPAD), on the other hand, is a specifically designed photodiode in which premature breakdown is avoided and a planar multiplication region is formed within the whole junction area (Hadfield, 2009).

Linear mode avalanche photodiodes, which are biased just below V_{bd}, have a finite multiplication gain. Statistical variations of this finite gain produce an additional noise contribution known as excess noise (Tilleman & Krishnaswami, 1996; Yano et al., 1990). SPADs, on the other hand, are not concerned with these gain fluctuations since the optical gain is virtually infinite (Takesue et al., 2006). Nevertheless, the statistical nature of the avalanche buildup is translated onto a detection probability. Indeed, the probability of detecting a photon hitting the SPAD's surface depends on the diode's quantum efficiency and the probability for an electron or for a hole to trigger an avalanche (Legre et al., 2007). Intensity information is obtained by counting the pulses during a certain period of time or by measuring the mean time interval between successive pulses. The same mechanism may be used to evaluate noise. Thermally or tunneling generated carriers within the p-n junction, which produce dark current in linear mode photodiodes, can trigger avalanche pulses. In Geiger mode, they are indistinguishable from regular photon-triggered pulses and they produce spurious pulses at a frequency known as *dark count rate* (DCR). DCR strongly depends on temperature and it is an important parameter for a TGSPC since it generates false measurements (Thew et al., 2007).

The practical detection efficiency, η, is defined as the overall probability of registering a count if a photon arrives at the detector. In most photon-counting applications a high value of η is certainly desirable. The higher the value of η, the smaller the signal loss, thus results more efficient and accurate measurements. DCR and detection efficiency determine the lowest power that is detectable by the device through the noise equivalent power (NEP) which is defined as $NEP = h\nu\sqrt{2D} / \eta$, here $h\nu$ is the energy of the signal photon, and D is the DCR (Hiskett, 2001; Gisin et al., 2002).

Another source of spurious counts is represented by *after-pulses* (Roussev et al., 2004). They are due to carriers temporarily trapped after a Geiger pulse in the multiplication region that are released after a short time interval, thus re-triggering a Geiger event. After-pulses depend on the trap concentration as well as on the number of carriers generated during a Geiger pulse. The number of carriers depends in turn on the diode's parasitic capacitance and on the external circuit, which is usually the circuit used to quench the avalanche. Typically, the quenching process is achieved by temporarily lowering the bias voltage below V_{bd}. Once the avalanche has been quenched, the SPAD needs to be recharged again above V_{bd} so that it can detect subsequent photons. The time required to quench the avalanche and recharge the diode up to 90% of its nominal excess bias is defined as the *dead time*. This parameter limits the maximal rate of detected photons, thus producing a saturation effect (Dixon, et al., 2008).

The commercially available InGaAs/InP avalanche photodiode has been the most practical device for SPADs at 1550nm telecommunication wavelength (Warburton et al., 2009). Since a photo-excited carrier grows into a macroscopic current output via the carrier avalanche multiplication in an APD operated in the Geiger mode, a single-photon can be detected efficiently. However, fractions of the many carriers trapped in the APD are subsequently emitted, and trigger additional avalanches that cause erroneous events. The InGaAs/InP SAPD in Geiger mode has a particularly high probability that afterpulses occur. Therefore, the InGaAs/InP SAPD is usually operated in the gated mode in which the gate duration (gate-on time) is generally set to a few nanoseconds (Namekata et al., 2006; Yoshizawa et al., 2004). Then the interval between two consecutive gates is set to more than the lifetime (in orders of microseconds) of the trapped carriers so that the afterpulse is suppressed. As a result, the repetition frequency of the gate has been limited to several megahertz, which is unsuitable for applications such as the high-speed detection (Hadfield et al., 2006).

2.2 The block diagram for single photon detector at 1550nm

Fig. 1 shows a typical block diagram scheme for a commercially single photon detector, Photon Counting Receiver PGA 600 manufactured by Princeton Lightwave Inc (Princeton Light Wave, 2006). The receiver has four major functional elements. These are the InGaAs SPAD, analog signal processing circuitry, a discriminator circuitry, and triggering, biasing and blanking circuitry.

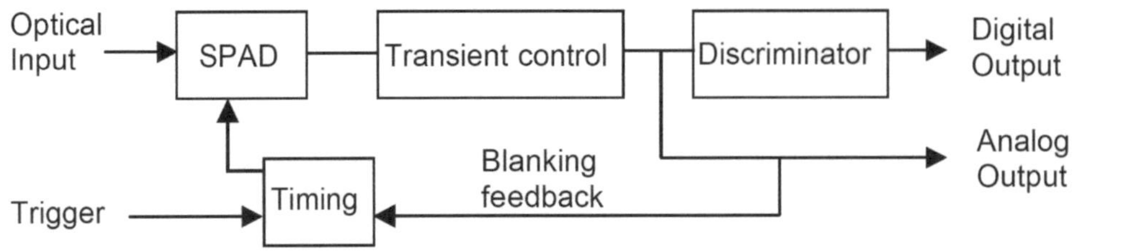

Fig. 1. The typical block diagram scheme for 1550nm single photon detector.

The SAPD is operated at ~ 218 K to reduce the probability of DCR. When the detector is triggered, the APD bias voltage is raised above its reverse V_{bd} to operate in Geiger mode. A short time later the bias is reduced below V_{bd} again to prevent false events.

The analog signal processing circuitry eliminates the transient noise created when a short bias pulse is applied to the SPAD, and isolates the charge pulse that results when a photon trigger an avalanche event.

The discriminator circuitry generates a digital logic pulse when the pulse-height of an analog charge signal exceeds a threshold level set to reject electronic noise. In a typical photon counting system, the SPAD output exhibits fluctuations in the pulse height and these pulses are amplified and directed into the discriminator. The discriminator compares the input pulses with the preset reference threshold voltage, where the lower pulses are eliminated. The higher pulses output at a constant level, usually as transistor-transistor logic (TTL) level from 0V to 5 V, allowing counting the discriminated pulses. To increase the detection efficiency, it is advantageous to set the level discrimination at a lower position, but this is also accompanied by a noise increase thus increasing dark count and the NEP.

The triggering circuitry initiates bias pulse generation when a trigger pulse reaches a set threshold level.

The delay between triggering and bias pulse generation can be adjusted so that the bias pulses accurately coincide with the expected arrival times of the photons. By using short bias pulses, the probability of dark counts can be significantly reduced, improving the detector's SNR performance.

When the detector is triggered, the SPAD bias voltage is raised above its reverse breakdown voltage to operate in Geiger mode. This feature is useful to suppress afterpulsing of the SPAD. The detector has both of digital and analog output. The discriminator circuitry generates a digital logic pulse when the pulse-height of an analog charge signal exceeds a threshold level set to reject electronic noise. The threshold is set as the cross-over voltage at which background noise and the single photon make equal contributions to the pulse height distribution. With the certain threshold, the receiver provides 20% detection efficiency and 10^{-5} dark count probability per 1 ns gating pulse.

2.3 Quantum fluctuations and SNR of photon counting

The SPAD records the incident photons in the sampling time τ. Suppose the average photon count is a, the quantum fluctuations of the photon counting distribution can be expressed as (Lee et al., 2006)

$$I_{sn} = \sqrt{\sum_n P(n)(n-\alpha)^2}\,, \tag{1}$$

where n is the actual photon numbers measured during the experiment.
For the coherent light field, the photon counting distribution obeys Poisson distribution

$$P_c(n) = \alpha^n e^{-\alpha} / n! . \quad (2)$$

The quantum fluctuations of coherent light field should be

$$I_{sn}^c = \sqrt{\alpha} . \quad (3)$$

The SNR of photon counting can be expressed as

$$SNR = \frac{\alpha\tau}{\sqrt{(\alpha + B + D)}} = \frac{\alpha}{\sqrt{(\alpha + B + D)}}\sqrt{\tau} . \quad (4)$$

Where B is the photon count rate caused by illumination noise light. When the background stray light and the dark counts of detector, working in low-temperature environment, can be ignored compared to signal counts, the maximum value $(\alpha\tau)^{1/2}$ of spectral SNR can be obtained. Increasing τ could get a higher SNR, but the temporal resolution should be decreased in this way.

2.4 The principle of photon counting lock-in

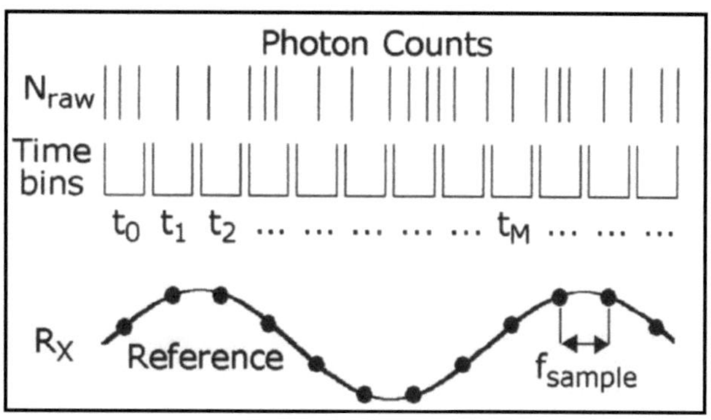

Fig. 2. Implementation of the photon-counting lock-in.

Lock-in amplifier is a synchronous coherent detector using principle of cross-correlation, extracting useful signals from noise because the reference signal frequency related to the input signal frequency but not related to noise frequency. It is equivalent to a very narrow bandwidth band-pass filter, and it is necessary to compress the filter bandwidth as much as possible in order to suppress noise.

When the incident photons were intensity modulated by the sine-wave of frequency f_s, the instantaneous photon counts at time t is expressed as $r_0 + m\cos(2\pi f_s t)$, where r_0 is average photon counts, m is depth of modulation. Then within the sampling time τ the effective photon counts can be expressed as (Huang et al., 2006)

$$r_t(\tau) = r_0 + m\left[\frac{\sin(\pi f_s \tau)}{\pi f_s \tau}\right]\cos(2\pi f_s t + \pi f_s \tau) . \quad (5)$$

The probability of n photon being detected in the sampling time τ is

$$p(n) = \frac{e^{-r_t(\tau)}(r_t(\tau))^n}{n!} . \quad (6)$$

As shown in Fig. 2, projections of time-binned photon counts N_{raw} to sinusoidal reference R_X are time averaged to yield the small I_X signal of the lock-in.

The photon counts lock-in needs the signal to be converted to analog signal to fulfill the input demand of lock-in amplifier, where only the frequency components corresponding with the demodulation filter bandwidth will be retained. Reduction of noise imposed upon a useful signal with frequency f_s, is proportional to the square root of the bandwidth of a bandpass filter Δf, centre frequency f_s.

The *SNR* of demodulated signal is

$$SNR = \frac{S_N}{(I_{sn} + I_{ex})(\Delta f / f_n)} . \quad (7)$$

Where S_N is the analog signal of the photon counts, I_{ex} is excess noise. f_n is the noise distribution bandwidth of need to be measured signal. Compressing the filter bandwidth Δf makes the corresponding noise signal decreases, as a result enhance the *SNR* of TGSPC measurement.

3. Photon-counting lock-in for TGSPC detection

3.1 Experiment setup

The diagram of the photon counting lock-in for TGSPC measurement is as shown in Fig. 3. A 1550nm wavelength, 300-ps pulse length laser (id300, ID Quantique), external triggered

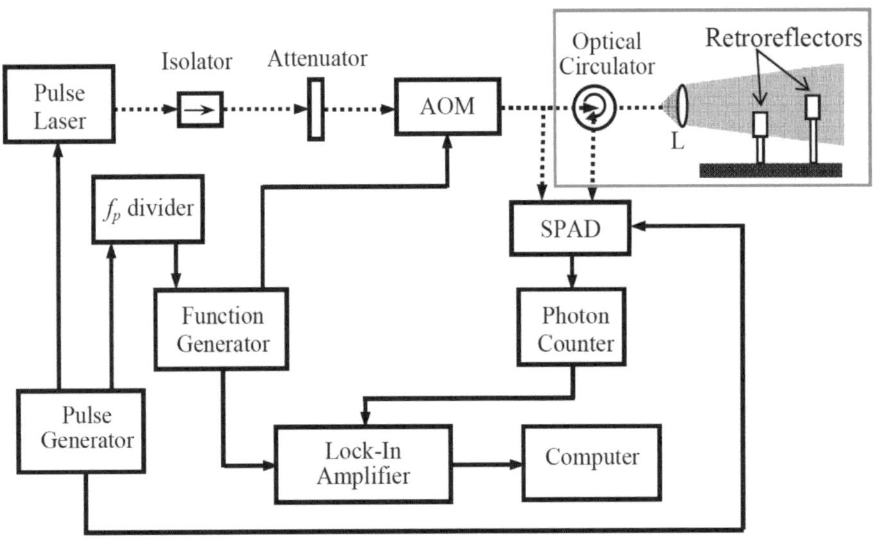

Fig. 3. Schematic diagram of photon counting lock-in experiment. AOM is acousto-optic modulator. SPAD is Photon Counting Benchtop Receiver with an InGaAs single photon detector. The dashed line range is the setup for TGSPC measurement.

by a pulse generator (SRS DG645) at f_p=4MHz repetition rate, is attenuated to produce a suitable (~5-100kcps) counting rate in the detector. The weak laser pulses were intensity modulated with a frequency-downshifted (200MHz) acousto-optic modulator (AOM, M200-2J-F2S Gooch & Housego). A sine-wave function generator is triggered by the variable divider from laser pulses frequency, worked as the modulation signal and added onto the AOM's driver. The weak pulses are launched into the PCBR and incidence on the SPAD. Meanwhile, the synchronizing TTL pulses from DG535 were used to trigger the SPD. The analog output of the SPD is connected to the photon counter (SRS SR400); and the photon counting signal is demodulated by the lock-in amplifier (SRS SR830). The experiment control and the data collection were completed by Labview software.
Two identical air-filled retroreflecting corner cubes were attached to translation stages and mounted on an optomechanical rail that was positioned at a range of approximately 10 m from the fiber coupler lens. The optics of the laser diode was configured so that both corner cubes were illuminated, and the photon returns were collected by the same lens.

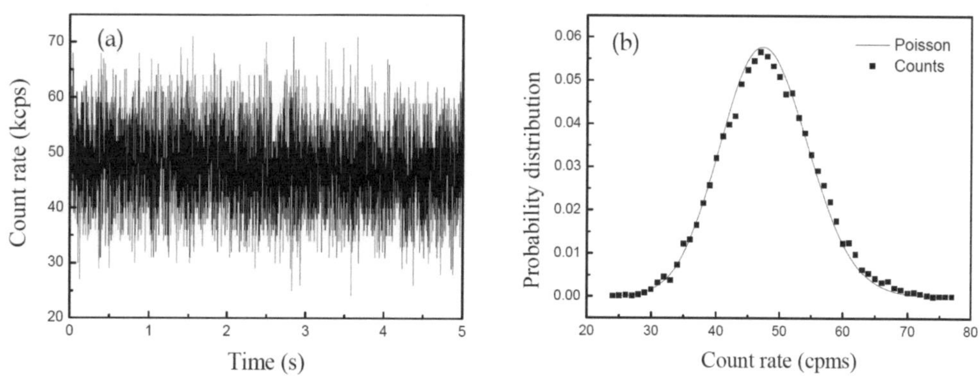

Fig. 4. Quantum fluctuation of photons counting (a) and its statistics characteristics (b).

3.2 Photon counts lock-in results

The photon counts were converted to analog signal by using SR400 with the digital-analog conversion factor g=0.1mV/count. A sine-wave signal, with repetition frequency f_s=10Hz, peak-to-peak voltage V_{pp}=300mV and offset voltage V_{off}=370mV, worked as the modulation signal and was added onto the AOM's driver.
The directly photon counting TGSPC measurement is shown in Fig. 4 (a), the average photon number is a=50kcps, sampling time τ=1ms. It is found of the quantum fluctuations of photon counting obviously. The data points in Fig.4 (b) are statistical characteristics of photon counting in Fig.4 (a). The solid line is fitted curve of Poisson distribution function. As can be seen from the figure, the photon counting of coherent light field obeys Poisson distribution.
Fig. 5 (a) is an analog signal of the photon counts after the digital-analog converted in photon counting TGSPC of the retroreflector. Fig. 5 (b) is the Fourier transforms (FFT) results of analog signal in Fig. 5 (a). We found that the noise amplitude distribution of the photon counting quantum fluctuations is a uniform distribution in the frequency domain. Lock-in method here will not be affected by the low frequency $1/f$ noise. The magnitude of carrier signal is about -8dB at the location of 10Hz modulation frequency, two orders of

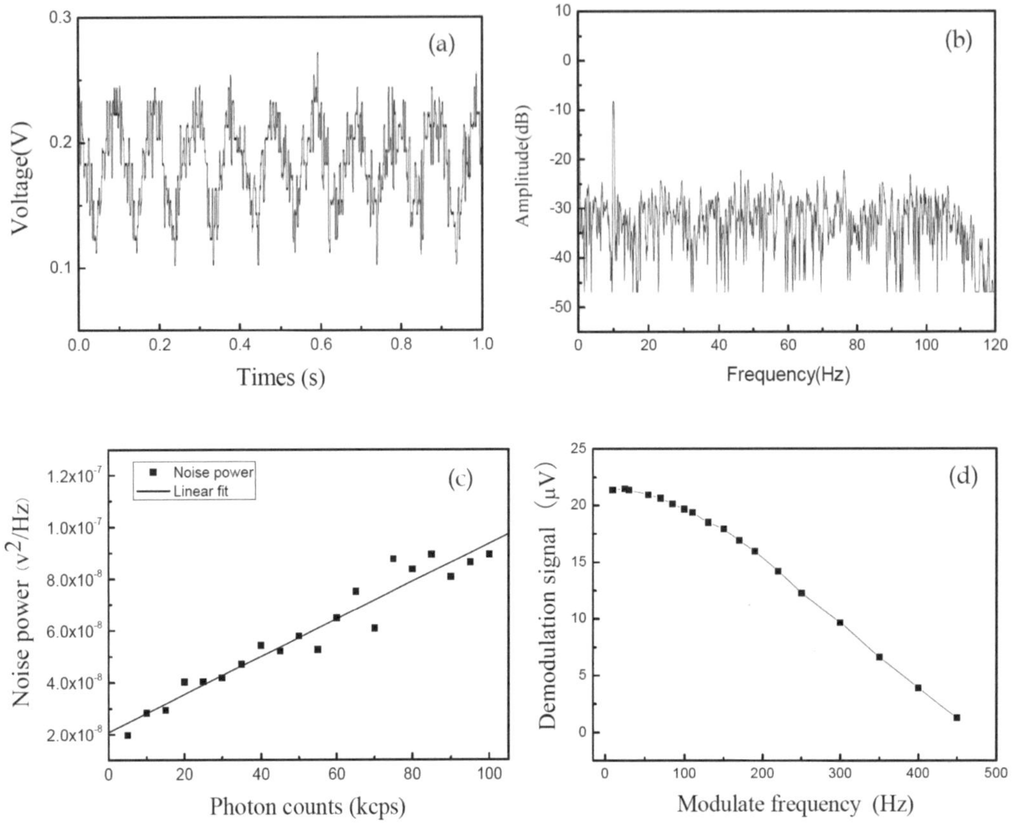

Fig. 5. The digital-to-analog signal for photon counts (a) and its noise amplitude spectral characteristics (b). (c) shows the linear relationship between noise power and the mean photon number. (d) is the demodulation output from the lock-in amplifier vs. modulation frequency.

magnitude higher than the corresponding noise amplitude. There is a sudden decrease at 110 kHz only because of the 110 kHz wideband of the lock-in amplifier. Fig. 5 (c) shows the linear relationship between noise power and the mean photon number. The slope is 7.2^{-8} $V^2/Hz^{1/2}$/photon.

Nevertheless, the choice of modulation frequency also affected the measurement results. As shown in Fig. 5 (d), the output signal value of lock-in amplifier will gradually decrease as the modulation frequency increasing. If the modulation frequency is lower, the data processing will need longer time, which limits the measurement speed. Increasing the modulation frequency would increase the speed of data processing, but reduce the average photon number and then reduce the measurement sensitivity. In practical applications, we need to select appropriate modulation frequency according to the average photon number.

Fig. 6 is the measurement results of the TGSPC at one of retroreflectors. The dashed line is the time distribution characteristics of TGSPC using photon counting methods directly. The counting time τ=10ms and the step of time delay is 5ps. We found the bigger counts induce the greater fluctuation. This is determined by the quantum fluctuations $(\alpha\tau)^{1/2}$ in photon counting. The solid line is the photon counts modulation TGSPC results measured by the lock-in amplifier where the quantum fluctuations are eliminated effectively.

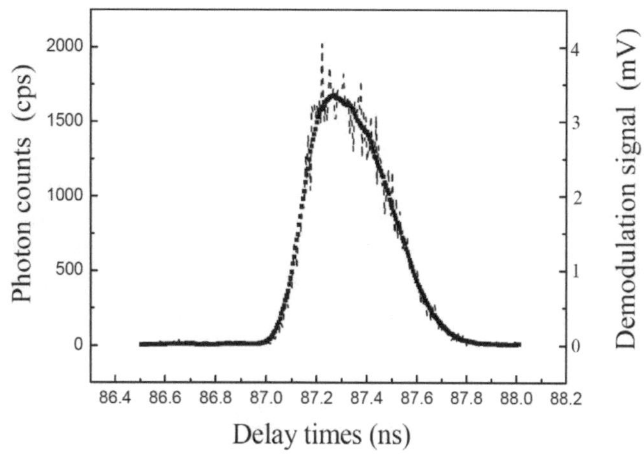

Fig. 6. TGSPC results from direct measurement of photon counting (dashed line) and photon counts modulation (dots).

Table 1 is the relationship between the slope of the low-pass filter and equivalent noise bandwidth (ENBW). Where T is integration time constant of lock-in amplifier. Filter slope determines the extent of the noise filter, the greater slope namely the smaller the noise equivalent bandwidth, the ability to filter out the noise being stronger. Improving SNR could be achieved by selecting the integration time of the lock-in amplifier and then changing the filter bandwidth.

Slope	*ENBW*
6dB/oct	$1/(4T)$
12dB/oct	$1/(8T)$
18dB/oct	$3/(32T)$
24dB/oct	$5/(64T)$

Table 1. The relationship between the slope of the low-pass filter and equivalent noise bandwidth *ENBW*.

The slope of the filter used in our experiments is 18 dB, integral time is T=100ms. The filter bandwidth of lock-in amplifier that of corresponding the noise equivalent bandwidth can be calculated *ENBW*=0.94Hz. In the place of 10 Hz modulation frequency of Fig. 5 (b), the corresponding voltage noise spectral density is $10^{-3}\,V/\sqrt{Hz}$. Then the SNR of photon counts modulated TGSPC is

$$SNR_M = 0.15V/(10^{-3}\,V/\sqrt{Hz} \times \sqrt{0.94Hz}) = 159\,. \tag{8}$$

With the average photon number at the peak $\alpha = 1700cps$, maximum SNR of photon counting can be obtained from equation (3)

$$SNR_{PC} = \sqrt{\alpha\tau} = 4.12\,. \tag{9}$$

The signal-to-noise improvement ratio corresponds to the photon counting modulation is

$$SNIR = 20\log(SNR_M / SNR_{PC}) = 20\log(159/4.12) = 31.7dB. \quad (10)$$

3.3 Single photon lock-in and optimal threshold for the discriminator

We have shown that the photon counts lock-in can improve the SNR of TGSPC. However, the analog conversion also makes it difficult for one to measure relative small-signal amplitude by normalizing it against the DC background. We will demonstrate experimentally the single photon lock-in and the optimal discriminate determination.

The experimental setup is shown in Fig. 3. A sine-wave function generator is triggered by the 1/40 divider from laser pulses frequency f_P, with repetition frequency f_s=100 kHz, as the modulation signal and added onto the AOM's driver. The analog output of the single photon detector is connected to the photon counter (SRS SR400). Discriminators of SR400 are provided with a selectable threshold in 0.2 mV steps. And the outputs from SR400 are 100 ns pulses. It is found that SR400 output with the discriminate voltage at 184 mV has the same photon counting value as that from the detector digital output. And the photon counting will be carried out at the 184 mV threshold.

For a sine-wave modulation, the probability for single photon being detected is

$$P = \frac{\eta}{1+d}\left(1 + d\cos(2\pi f_s + \phi)\right), \quad (11)$$

where d represents the depth of modulation, ϕ is the signal phase. Here we assumed η is the maximum detection efficiency during the modulation. With the experimental parameter above, we have $d \approx 1$.

The single photon lock-in method we used here means that the pulses from the SR400 output are attenuated and then directly demodulated by the lock-in amplifier. The synchronous 100 kHz sine-wave was added onto the lock-in amplifier as the reference signal.

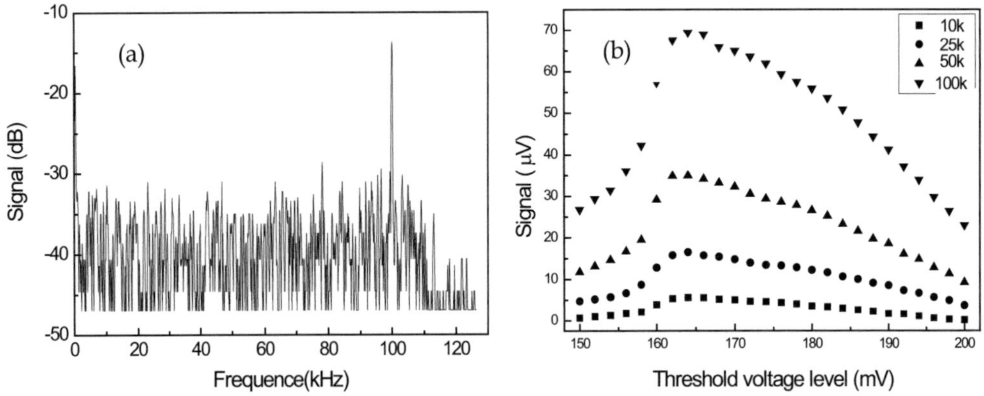

Fig. 7. (a) The frequency spectrum of single photons lock-in. (b) The single photon lock-in output corresponding to different mean photon counts, 10 kcps, 25 kcps, 50 kcps and 100 kcps, respectively.

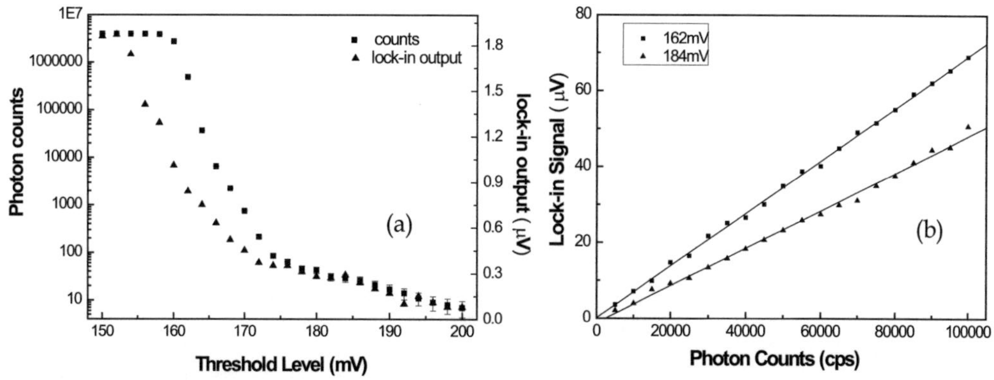

Fig. 8. (a) Dark count and its lock-in measurements vs discriminate threshold. (b) Single photon lock-in outputs vs threshold with different photon counts. The single photon lock-in outputs vs photon counts for threshold being 162 mV and 184 mV, respectively.

The frequency spectrum of the monitor out of lock-in amplifier is shown in Fig. 7 (a). Here, the mean photon number is 100 kcps and the SR400 threshold is 184mV. Note that the single photon modulation signal at the place of frequency 100 kHz. As the dark counts of the SPAD follow Poisson statistics, i.e., dominating shot noise with white noise spectral density, we found the uniform distribution of the background noise. The effect of Flicker noise ($1/f$) noise on the accuracy of measurements can be ignored. At lower threshold, the ($1/f$) noise may become dominant, so we choose the 100 kHz for single photon modulation, due to the higher noise in the low-frequency region.

As shown in Fig. 7 (b), when we change the level discrimination from 184 mV to 162 mV, it is found that the dark counts increase quickly which cover 4 orders of magnitude where the weak photon signals will be immerged in the case at lower threshold. The limit to detection efficiency is primarily device saturation from dark counts.

In Fig. 7 (b), we show the single photon lock-in output corresponding to different mean photon counts, 10 kcps, 25 kcps, 50 kcps and 100 kcps, respectively. The data are obtained by first setting the discriminate voltage, and measuring the mean photon counts and lock-in output respectively. The traces show the discriminate threshold can be optimized at 162 mV where the lock-in has the maximum output.

Accordingly, we have measured the lock-in output with the lock-in integrated time 100 ms, and the equivalent noise bandwidth for bandpass filter Δf =1 Hz. It is interesting to note that the lock-in output increase only 4 times from 184 mV to 162 mV in Fig. 8 (a).

The demodulated signals versus photon counts for discriminate threshold being 162 mV and 184 mV are shown in Fig. 8 (b). The two curves show that the intensity of single photon lock-in signals are increasing linearly as the photon counts increased. The slope for the fitted line is 1.24 μV/kcps at 184 mV threshold, and 2.32 μV/kcps at 162 mV, respectively. It is shown that the detected efficiency with single photon lock-in at 162 mV is 1.87 times bigger than that of the photon counting method at 184 mV.

We have demonstrated our measurement system in TGSPC experiment for a 3m-length displacement between the two retroreflectors. The backscattered photons reach to the InGaAs single photon detectors through a fiber optical circulator, as shown in Fig. 9. With the 162 mV optimal threshold, the single photon lock-in for TGSPC experiment is shown as Fig. 9, where the backscattered signal is presented as a function of length. Here it is found that the dark count and the photon shot noise are restrained, and clearly the conventional photon counting is dogged by a high dark count rate at this low threshold.

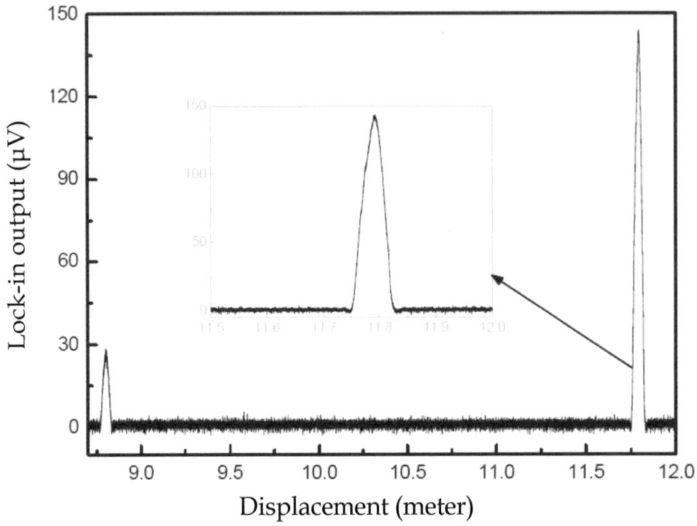

Fig. 9. The TGSPC measurement by using single photons lock-in with the optimal threshold 162 mV.

4. Conclusion and outlook

The single photon detection for TGSPC which has some features of broad dynamic range, fast response time and high spatial resolution, remove the effect of the response relaxation properties of other photoelectric device. We present a photon counting lock-in method to improve the SNR of TGSPC. It is shown that photon counts lock-in technology can eliminate the effect of quantum fluctuation and improve the SNR. In addition, we demonstrate experimentally to provide high detection efficiency for the SPAD by using the single photon lock-in and the optimal discriminate determination. It is shown that the background noise could be obviously depressed compared to that of the conventional single photon counting. The novel method of photon-counting lock-in reduces illumination noise, detector dark count noise, can suppress background, and importantly, enhance the detection efficiency of single-photon detector.

The conclusions drawn give further encouragement to the possibility of using such ultra sensitive detection system in very weak light measurement occasions (Alfonso & Ockman, 1968; Carlsson & Liljeborg, 1998). This high SNR measurement for TGSPC could improve the dynamic range and time resolution effectively, and have the possibility of being applied to single-photon sensing, quantum imaging and time of flight.

5. Acknowledgments

The project sponsored by the 863 Program (2009AA01Z319), 973 Program (Nos.2006CB921603, 2006CB921102 and 2010CB923103), Natural Science Foundation of

China (Nos. 10674086 and 10934004), NSFC Project for Excellent Research Team (Grant No. 60821004), TSTIT and TYMIT of Shanxi province, and Shanxi Province Foundation for Returned Scholars.

6. References

Alfonso, R. R. & Ockman, N. (1968). Methods for Detecting Weak Light Signals. *J. Opt. Soc. Am.*, Vol. 58, No. 1, (January 1968) page numbers (90-95), ISSN: 0030-3941

Arecchi, F. T.; Gatti, E. & Sona, A. (1966). Measurement of Low Light Intensities by Synchronous Single Photon Counting. *Rev. Sci. Instrum.*, Vol. 37, No. 7, (July 1966) page numbers (942-948), ISSN: 0034-6748

Becker, W. (2005). *Advanced Time-Correlated Single Photon Counting Techniques*, Springer, ISBN-10 3-540-26047-1, Germany

Benaron, D. A. & Stevenson, D. K. (1993). Optical Time-of-flight and Absorbance Imaging of Biologic Media. *Science*, Vol. 259, No. 5100, (March 1993) page numbers (1463-1466), ISSN: 0036-8075

Braun, D. & Libchaber, A. (2002). Computer-based Photon-counting Lock-in for Phase Detection at the Shot-noise Limit. *Opt. Lett.*, Vol. 27, No. 16, (August 2002) page numbers (1418-1420), ISSN: 0146-9592

Carmer, D. C. & Peterson, L. M. (1996). Laser Radar in Robotics. *Proceedings of the IEEE*, Vol. 84, No. 2, (February 1996) page numbers (299-320), ISSN: 0018-9219

Dixon, A. R.; Yuan, Z. L.; Dynes, J. F.; Sharpe, A. W. & Shields, A. J. (2008). Gigahertz Decoy Quantum Key Distribution with 1 Mbit/s Secure Key Rate. *Opt. Express*, Vol. 16, No. 23, (October 2008) page numbers (18790-18970), ISSN: 1094-4087

Dixon, G. J. (1997). Time-resolved Spectroscopy Defines Biological Molecules. *Laser Focus World*, Vol. 33, No. 10, (October 1997) page numbers (115-122), ISSN: 1043-8092

Forrester, P. A. & Hulme J. A. (1981). Laser Rangefiders. *Opt. Quant. Electron.*, Vol. 13, No. 4, (April 1981) page numbers (259-293), ISSN: 0306-8919

Gisin, N.; Ribordy, G.; Tittel, W. & Zbinden, H. (2002). Quantum Cryptography. *Rev. Mod. Phys.*, Vol. 74, No. 1, (January 2002) page numbers (145–195), ISSN: 0034-6861

Hadfield, R. H. (2009). Single-photon Detectors for Optical Quantum Information Applications. *Nat. Photonics*, Vol. 3, No. 12, (December 2009) page numbers (696-705), ISSN: 1749-4885

Hadfield, R. H.; Habif, J. L.; Schlafer, J.; Schwall, R. E. & Nam, S. W. (2006). Quantum Key Distribution at 1550 nm with Twin Superconducting Single-photon Detector. *Appl. Phys. Lett.*, Vol. 89, No. 24, (December 2006) page numbers (241129-1~3), ISSN: 0003-6951

Hiskett, P. A.; Buller, G. S.; Loudon, A. Y.; Smith, J. M.; Gontijo, I.; Walker, A. C.; Townsend, P. D. & Robertson, M. J. (2000). Performance and Design of InGaAs-InP Photodiodes for Single-photon Counting at 1.55μm. *Appl. Opt.*, Vol. 39, No. 36, (December 2000) page numbers (6818–6829), ISSN: 0003-6935

Hiskett, P. A.; Smith, J. M.; Buller, G. S. & Townsend, P. D. (2001). Low-noise Single-photon Detection at a Wavelength of 1.55 μm. *Electron. Lett.*, Vol. 37, No.17, (August 2001) page numbers (1081–1083), ISSN: 0013-5194

Huang, T.; Dong, S. L.; Guo, X. J.; Xiao, L. T. & Jia, S. T. (2006). Signal-to-noise Ratio Improvement of Photon Counting using Wavelength Modulation Spectroscopy. *Appl. Phys. Lett.*, Vol. 89, No. 6, (August 2006) page numbers (061102-1~3), ISSN: 0003-6951

Lacaita, A. L.; Francese, P. A. & Cova, S. (1993). Single-photon Optical-time Domain Reflectometer at 1.3 μm with 5 cm Resolution and High Sensitivity. *Opt. Lett.*, Vol. 18, No. 13, (July 1993) page numbers (1110–1112), ISSN: 0146-9592

Lacaita, A.; Zappa, F.; Cova, S. & Lovati, P. (1996). Single-photon Detection Beyond 1 μm: Performance of Commercially Available InGaAsInP Detectors. *Appl. Opt.*, Vol. 35, No. 16, (June 1996) page numbers (2986–2996), ISSN: 0003-6935

Lee D.; Yoon, H. & Park, N. (2006). Optimization of SNR Improvement in the Noncoherent OTDR Based on Simplex Codes. *J. Lightwave Technol.*, Vol. 24, No. 1, (January 2006) page numbers (322-328), ISSN: 0733-8724

Legre, M.; Thew, R.; Zbinden, H. & Gisin, N. (2007). High Resolution Optical Time Domain Reflectometer Based on 1.55μm Up-conversion Photon-counting Module. *Opt.Express,* Vol. 15, No. 13, (June 2007) page numbers (8237 -8242), ISSN: 1094-4087

Leskovar, B. & Lo, C.C. (1976). Photon Counting System for Subnanosecond Fluorescence Lifetime Measurements, *Rev. Sci. Instrum.,* Vol. 47, No. 9, (September 1976) page numbers (1113 - 1121), ISSN: 0034-6748

Mäkynen, A. J.; Kostamovaara, J. T. & Myllylä, R. A. (1994). Tracking Laser Radar for 3-D Shape Measurements of Large Industrial Objects Based on Time-of-Flight Laser Rangefinding and Position-Sensitive Detection Techniques. *IEEE T. Instrum. Meas.*, Vol. 43, No. 1, (February 1994) page numbers (40-49), ISSN: 0018-9456

Moring, I.; Heikkinen, T.; Myllylä, R. & Kilpelä, A. (1989). Acquisition of Three Dimensional Image Data by a Scanning Laser Range Finder. *Opt. Eng.*, Vol. 28, No. 8, (August 1989) page numbers (897 - 902), ISSN: 0091-3286

Murphy, M. K.; Clyburn, S. A. & Veillon, C. (1973). Comparison of Lock-in Amplification and Photon Counting with Low Background Flames and Graphite Atomizers in Atomic Fluorescence Spectrometry. *Anal.Chem.,* Vol. 45, No. 8, (July 1973) page numbers (1468-1473), ISSN: 0003-2700

Namekata, N.; Sasamori, S. & Inoue, S. (2006). 800 MHz Single-photon Detection at 1550-nm using an InGaAs/InP Avalanche Photodiode Operated with a Sine Wave Gating. *Opt. Express,* Vol. 14, No. 21, (September 2006) page numbers (10043–10049), ISSN: 1094-4087

Pellegrini, S.; Buller, G. S.; Smith, J. M.; Wallace, A. M. & Cova, S. (2000). Laser-based Distance Measurement using Picosecond Resolution Time-correlated Single-photon Counting. *Meas. Sci. Technol.*, Vol. 11, No. 6, (June 2000) page numbers (712–716), ISSN: 0957-0233

Pellegrini, S.; Warburton, R. E.; Tan, L. J. J.; Jo Shien Ng; Krysa, A. B.; Groom, K.; David, J. P. R.; Cova, S.; Robertson, M. J. & Buller, G. S. (2006). Design and Performance of an InGaAs-InP Single-photon Avalanche Diode Detector. *IEEE J. Quant. Elect.*, Vol. 42, No. 4, (April 2006) page numbers (397–403), ISSN: 0018-9197

Poultney, S. K. (1972). Single-photon Detection and Timing: Experiments and Techniques, In: *Adv. Electron. El. Phys.*, L. Marton and Claire Marton (Ed.), Vol. 31, page numbers (39–117), Elsevier, ISBN: 9780120145317, Netherlands

Poultney, S. K. (1977). Single Photon Detection and Timing in the Lunar Laser Rranging Experiment. *IEEE T. Nucl. Sci.*, Vol. 19, No. 3, (June 1972) page numbers (12-17), ISSN: 0018-9499

Princeton Light Wave, (2006). http://princetonlightwave.com/content/PGA-600%20V1.01.pdf.

Ribordy, G.; Gautier, J. D.; Zbinden, H. & Gisin, N. (1998). Performance of InGaAs/InP Avalanche Photodiodes as Gated-mode Photon Counters. *Appl. Opt.* Vol. 37, No. 12, (January 1998) page numbers (2272–2277), ISSN: 0003-6935

Ribordy G.; Gisin, N.; Guinnard, O.; Stucki, D.; Wegmuller, M. & Zbinden, H. (2004). Photon Counting at Telecom Wavelengths with Commercial In-GaAs/InP Avalanche Photodiodes: Current Performance. *J. Mod. Opt.*, Vol. 51, No. 9, (May 2004) page numbers (1381–1398), ISSN: 0950-0340

Roussev, R. V.; Langrock, C.; Kurz, J. R. & Fejer, M. M. (2004). Periodically Poled Lithium Niobate Waveguide Sum-frequency Generator for Efficient Single-photon Detection at Communication Wavelengths. *Opt. Lett.* Vol. 29, No. 13, (July 2004) page numbers (1518-1520), ISSN: 0146-9592

Stanford Research Systems, (1995). "Signal recovery with photomultiplier tubes," Application Note 4, http://www.srsys.com.

Stanford Research Systems, (1999). "DSP Lock-in Amplifier SR850," Chap. 3, http://www.srsys.com.

Stucki, D.; Ribordy, G.; Stefanov, A.; Zbinden, H.; Rarity, J. G. & Wall, T. (2001). Photon Counting for Quantum Key Distribution with Peltier Cooled InGaAs–InP APDs. *J. Mod. Opt.*, Vol. 48, No. 13, (November 2001) page numbers (1967–1981), ISSN: 0950-0340

Takesue, H.; Diamanti, E.; Langrock, C.; Fejer, M. M. & Yamamoto, Y. (2006). 1.5 μm Photon-counting Optical Time Domain Reflectometry with a Single-photon Detector based on Upconversion in a Periodically Poled Lithium Niobate Waveguide. *Opt. Lett.*, Vol. 31, No. 6, (March 2006) page numbers (727–729), ISSN: 0146-9592

Thew, R. T.; Stucki, D.; Gautier, J. D.; Zbinden, H. & Rochas, A. (2007). Free-running InGaAs/InP Avalanche Photodiode with Active Quenching for Single Photon Counting at Telecom Wavelengths. *Appl. Phys. Lett.*, Vol. 91, No. 20, (November 2007) page numbers (201114-1~3) ISSN: 0003-6951

Tilleman, M. M. & Krishnaswami, K. K. (1996). Design of Fibre Optic Relayed Laser Radar. *Opt. Eng.*, Vol. 35, No. 11, (May 1996) page numbers (3279-3284), ISSN: 0091 3286

Warburton, R. E.; Itzler, M. & Buller, G. S. (2009). Free-running Room Temperature Operation of an InGaAs/InP Single-photon Avalanche Diode. *Appl. Phys. Lett.*, Vol. 94, No. 7, (February 2009) page numbers (071116-1~3), ISSN: 0003-6951

Wegmüller, M.; Scholder, F. & Gisin, N. (2004). Photon-counting OTDR for Local Birefringence and Fault Analysis in the Metro Environment. *J. Lightwave Technol.*, Vol. 22, No. 2, (February 2004) page numbers (390-400), ISSN: 0733-8724

Yano, H.; Aga, K.; Kamei, H.; Sasaki, G. & Hayashi, H. (1990). Low-Noise Current Optoelectronic Integrated Receiver with Internal Equalizer for Gigabit-per-Second Long-Wavelength Optical Communications. *J. Lightwave Technol.,* Vol. 8, No.9, (September 1990) page numbers (1328-1333), ISSN: 0733-8724

Yoshizawa, A.; Kaji, R. & Tsuchida, H. (2004). Gated-mode Single-photon Detection at 1550 nm by Discharge Pulse Counting. *Appl. Phys. Lett.,* Vol. 84, No. 18, (April 2004) page numbers (3606–3608), ISSN: 0003-6951

11

Laser Beam Diagnostics in a Spatial Domain

Tae Moon Jeong and Jongmin Lee
Advanced Photonics Research Institute,
Gwangju Institute of Science and Technology
Korea

1. Introduction

The intensity distribution of laser beams in the focal plane of a focusing optic is important because it determines the laser-matter interaction process. The intensity distribution in the focal plane is determined by the incoming laser beam intensity and its wavefront profile. In addition to the intensity distribution in the focal plane, the intensity distribution near the focal plane is also important. For a simple laser beam having a Gaussian or flat-top intensity profile, the intensity distribution near the focal plane can be analytically described. In many cases, however, the laser beam profile cannot be simply described as either Gaussian or flat-top. To date, many researchers have attempted to characterize laser beam propagation using a simple metric for laser beams having an arbitrary beam profile. With this trial, researchers have devised a beam quality (or propagation) factor capable of describing the propagation property of a laser beam, especially near the focal plane. Although the beam quality factor is not a magic number for characterizing the beam propagation, it can be widely applied to characterizing the propagation of a laser beam and is also able to quickly estimate how small the size of the focal spot can reach. In this chapter, we start by describing the spatial profile of laser beams. In Section 2, the derivation of the spatial profile of laser beams will be reviewed for Hermite-Gaussian, Laguerre-Gaussian, super-Gaussian, and Bessel-Gaussian beam profiles. Then, in Section 3, the intensity distribution near the focal plane will be discussed with and without a wavefront aberration, which is another important parameter for characterizing laser beams. Although the Shack-Hartmann wavefront sensor is widely used for measuring the wavefront aberration of a laser beam, several other techniques to measure a wavefront aberration will be introduced. Knowing the intensity distributions near the focal plane enables us to calculate the beam quality (propagation) factor. In Section 4, we will review how to determine the beam quality factor. In this case, the definition of the beam quality factor is strongly related to the definition of the radius of the intensity distribution. For a Gaussian beam profile, defining the radius is trivial; however, for an arbitrary beam profile, defining the beam radius is not intuitively simple. Here, several methods for defining the beam radius are introduced and discussed. The experimental procedure for measuring the beam radius will be introduced and finally determining the beam quality factor will be discussed in terms of experimental and theoretical methods.

2. Spatial beam profile of the laser beam

In this section, we will derive the governing equation for the electric field of a laser beam. The derived electric field has a special distribution, referred to as beam mode, determined by the boundary conditions. Two typical laser beam modes are Hermite-Gaussian and Laguerre-Gaussian modes. In this chapter, we also introduce two other beam modes: top-hat (or flat-top) and Bessel-Gaussian beam modes. These two beam modes become important when considering high-power laser systems and diffraction-free laser beams. These laser beam modes can be derived from Maxwell's equations.

2.1 Derivation of the beam profile

When the laser beam propagates in a source-free (means charge- and current-free) medium, Maxwell's equations in Gaussian units are:

$$\nabla \times \vec{E} + \frac{1}{c}\frac{\partial \vec{B}}{\partial t} = 0, \tag{2.1}$$

$$\nabla \times \vec{H} - \frac{1}{c}\frac{\partial \vec{D}}{\partial t} = 0, \tag{2.2}$$

$$\nabla \cdot \vec{D} = 0, \tag{2.3}$$

$$\text{and } \nabla \cdot \vec{B} = 0 \tag{2.4}$$

where $\vec{E}$ and $\vec{H}$ are electric and magnetic fields. In addition, $\vec{D}$ and $\vec{B}$ are electric and magnetic flux densities defined as

$$\vec{D} = \vec{E} + 4\pi\vec{P} \text{ and } \vec{B} = \vec{H} + 4\pi\vec{M}. \tag{2.5}$$

Polarization and magnetization densities ($\vec{P}$ and $\vec{M}$) are then introduced to define the electric and magnetic flux densities as follows:

$$\vec{P} = \chi\vec{E} \text{ and } \vec{M} = \eta\vec{H}. \tag{2.6}$$

As such, the electric and magnetic flux densities can be simply expressed as

$$\vec{D} = \varepsilon\vec{E}, \text{ and } \vec{B} = \mu\vec{H}. \tag{2.7}$$

where ε and μ are the electric permittivity and magnetic permeability, respectively. Note that if there is an interface between two media, $\vec{E}$, $\vec{H}$, $\vec{D}$, and $\vec{B}$ should be continuous at the interface. This continuity is known as the continuity condition at the media interface. To be continuous, $\vec{E}$, $\vec{H}$, $\vec{D}$, and $\vec{B}$ should follow equation (2.8).

$$\hat{n} \times (\vec{E}_2 - \vec{E}_1) = 0, \ \hat{n} \times (\vec{H}_2 - \vec{H}_1) = 0, \ \hat{n} \cdot (\vec{D}_2 - \vec{D}_1) = 0, \text{ and } \hat{n} \cdot (\vec{B}_2 - \vec{B}_1) = 0 \tag{2.8}$$

Next, using equation (2.5), and taking $\nabla\times$ in equations (2.1) and (2.2), equations for the electric and magnetic fields become

$$\nabla \times \nabla \times \vec{E} + \frac{1}{c^2}\frac{\partial^2 \vec{E}}{\partial t^2} = -\frac{4\pi}{c}\frac{\partial}{\partial t}\left[\frac{1}{c}\frac{\partial \vec{P}}{\partial t} + \nabla \times \vec{M}\right], \tag{2.9}$$

$$\text{and } \nabla\times\nabla\times\vec{H}+\frac{1}{c^2}\frac{\partial^2\vec{H}}{\partial t^2}=\frac{4\pi}{c}\left[\nabla\times\frac{\partial\vec{P}}{\partial t}-\frac{1}{c}\frac{\partial^2\vec{M}}{\partial t^2}\right]. \tag{2.10}$$

Because the electric and magnetic fields behave like harmonic oscillators having a frequency ω in the temporal domain, $\frac{\partial}{\partial t}$ can be replaced with $-i\omega$. Then, using the relation $k=\frac{\omega}{c}$ (c is the speed of light), equations (2.9) and (2.10) become

$$\nabla\times\nabla\times\vec{E}(\vec{r})-k^2\vec{E}(\vec{r})=4\pi\left[k^2\vec{P}(\vec{r})+ik\nabla\times\vec{M}(\vec{r})\right], \tag{2.11}$$

$$\text{and } \nabla\times\nabla\times\vec{H}(\vec{r})-k^2\vec{H}(\vec{r})=4\pi\left[-ik\nabla\times\vec{P}(\vec{r})+k^2\vec{M}(\vec{r})\right]. \tag{2.12}$$

If we assume that the electromagnetic field propagates in free space (vacuum), then polarization and magnetization densities ($\vec{P}$ and $\vec{M}$) are zero. Thus, the right sides of equations (2.11) and (2.12) become zero, and finally,

$$\nabla\times\nabla\times\vec{E}(\vec{r})-k^2\vec{E}(\vec{r})=0, \tag{2.13}$$

$$\text{and } \nabla\times\nabla\times\vec{H}(\vec{r})-k^2\vec{H}(\vec{r})=0. \tag{2.14}$$

By using a BAC-CAB rule in the vector identity, equation (2.13) for the electric field becomes

$$\nabla\nabla\cdot\vec{E}(\vec{r})-\nabla\cdot\nabla\vec{E}-k^2\vec{E}(\vec{r})=0. \tag{2.15}$$

We will only consider the electric field because all characteristics for the magnetic field are the same as those for the electric field, except for the magnitude of the field. Because the source-free region is considered, the divergence of the electric field is zero ($\nabla\cdot\vec{E}(\vec{r})=0$). Finally, the expression for the electric field is given by

$$\nabla\cdot\nabla\vec{E}+k^2\vec{E}(\vec{r})=0. \tag{2.16}$$

This is the general wave equation for the electric field that governs the propagation of the electric field in free space. In many cases, the propagating electric field (in the z-direction in rectangular coordinates) is linearly polarized in one direction (such as the x- or y-direction in rectangular coordinates). As for a linearly x-polarized propagating electric field, the electric field propagating in the z-direction can be expressed in rectangular coordinates as

$$\vec{E}(\vec{r})=\hat{i}E_0(x,y,z)\exp(ikz). \tag{2.17}$$

By substituting equation (2.17) into equation (2.18), the equation becomes

$$\left(\frac{\partial^2}{\partial x^2}+\frac{\partial^2}{\partial y^2}+\frac{\partial^2}{\partial z^2}\right)\hat{i}E_0(x,y,z)\exp(ikz)+\hat{i}k^2E_0(x,y,z)\exp(ikz)=0. \tag{2.18}$$

Equation (2.18) is referred to as a homogeneous Helmholtz equation, which describes the wave propagation in a source-free space. By differentiating the wave in the z-coordinate, we obtain

$$\frac{\partial}{\partial z}E_0(x,y,z)\exp(ikz) = ikE_0(x,y,z)\exp(ikz) + \frac{\partial E_0(x,y,z)}{\partial z}\exp(ikz), \tag{2.19}$$

and

$$\begin{aligned}\frac{\partial^2}{\partial z^2}E_0(x,y,z)\exp(ikz) = -k^2E_0(x,y,z)\exp(ikz) + 2ik\frac{\partial E_0(x,y,z)}{\partial z}\exp(ikz) \\ + \frac{\partial^2 E_0(x,y,z)}{\partial z^2}\exp(ikz)\end{aligned}. \tag{2.20}$$

In many cases, the electric field slowly varies in the propagation direction (z-direction). The slow variation of the electric field in z-direction can make possible the following approximation (slowly varying approximation):

$$\left|\frac{\partial^2 E_0(x,y,z)}{\partial z^2}\right| \ll 2\left|k\frac{\partial E_0(x,y,z)}{\partial z}\right|. \tag{2.21}$$

By inserting equation (2.20) into equation (2.18) and using the assumption of equation (2.21), equation (2.18) becomes

$$\frac{\partial^2 E_0(x,y,z)}{\partial x^2} + \frac{\partial^2 E_0(x,y,z)}{\partial y^2} + 2ik\frac{\partial E_0(x,y,z)}{\partial z} = 0. \tag{2.22}$$

Equation (2.22) describes how the linearly polarized electric field propagates in the z-direction in the Cartesian coordinate.

2.2 Hermite-Gaussian beam mode in rectangular coordinate

In the previous subsection, we derived the equation for describing the propagation of a linearly polarized electric field. Now, the question is how to solve the wave equation and what are the possible electric field distributions. In this subsection, the electric field distribution will be derived as a solution of the wave equation (2.22) with a rectangular boundary condition. Consequently, the solution of the wave equation in the rectangular coordinate has the form of a Hermite-Gaussian function. Thus, the laser beam mode is referred to as Hermite-Gaussian mode in the rectangular coordinate; the lowest Hermite-Gaussian mode is Gaussian, which commonly appears in many small laser systems.

Now, let us derive the Hermite-Gaussian beam mode in the rectangular coordinate. The solution of equation (2.22) in rectangular coordinates was found by Fox and Li in 1961. In that literature, they assume that a trial solution to the paraxial equation has the form

$$E_0(x,y,z) = A(z)\times\exp\left[-ik\frac{x^2+y^2}{2q(z)}\right] \tag{2.23}$$

where A(z) is the electric field distribution in z-coordinate and $q(z)$ is the general expression for the radius of the wavefront of the electric field to be determined. For the time being, let us assume that the electric field distribution in x- and y-coordinates is constant. Then, if $q(z)$ is complex-valued, $q(z)$ can be expressed with real and imaginary parts as follows:

$$\frac{1}{q(z)} = \frac{1}{q_r(z)} - i\frac{1}{q_i(z)}. \tag{2.24}$$

By inserting equation (2.24) into equation (2.23), the resulting equation will be

$$E_0(x,y,z) = A(z) \times \exp\left[-ik\frac{x^2+y^2}{2q_r(z)}\right] \times \exp\left[-k\frac{x^2+y^2}{2q_i(z)}\right]. \tag{2.25}$$

The real part of equation (2.25) determines the magnitude distribution of the electric field and the imaginary part gives the spatial phase or wavefront profile. In a specific case such as the Gaussian beam profile, $q_i(z)$ determines the radius of the Gaussian beam, defined as

$$q_i(z) = \frac{\pi w^2(z)}{\lambda} \tag{2.26}$$

where $w(z)$ is the radius of the Gaussian beam profile. By calculating $\frac{\partial}{\partial x}$, $\frac{\partial}{\partial y}$, $\frac{\partial^2}{\partial x^2}$, $\frac{\partial^2}{\partial y^2}$, and $\frac{\partial}{\partial z}$ using equation (2.23), we can obtain

$$\frac{\partial E_0}{\partial x} = -ik\frac{x}{q(z)} \times A(z) \times \exp\left[-ik\frac{x^2+y^2}{2q(z)}\right], \tag{2.27}$$

$$\frac{\partial E_0}{\partial y} = -ik\frac{y}{q(z)} \times A(z) \times \exp\left[-ik\frac{x^2+y^2}{2q(z)}\right], \tag{2.28}$$

$$\frac{\partial^2 E_0}{\partial x^2} = -ik\frac{A(z)}{q(z)} \times \exp\left[-ik\frac{x^2+y^2}{2q(z)}\right] - k^2\frac{x^2}{q^2(z)}A(z) \times \exp\left[-ik\frac{x^2+y^2}{2q(z)}\right], \tag{2.29}$$

$$\frac{\partial^2 E_0}{\partial y^2} = -ik\frac{A(z)}{q(z)} \times \exp\left[-ik\frac{x^2+y^2}{2q(z)}\right] - k^2\frac{y^2}{q^2(z)}A(z) \times \exp\left[-ik\frac{x^2+y^2}{2q(z)}\right], \tag{2.30}$$

and

$$\frac{\partial E_0}{\partial z} = \frac{dA(z)}{dz} \times \exp\left[-ik\frac{x^2+y^2}{2q(z)}\right] + ikA(z)\frac{x^2+y^2}{2q^2(z)}\frac{dq(z)}{dz} \times \exp\left[-ik\frac{x^2+y^2}{2q(z)}\right]. \tag{2.31}$$

And, by inserting equations (2.27)-(2.31) into equation (2.22), equation (2.22) becomes

$$\left[k^2\left(\frac{dq(z)}{dz}-1\right)\frac{x^2+y^2}{q^2(z)}-\frac{2ik}{q(z)}\left(\frac{q(z)}{A(z)}\frac{dA(z)}{dz}+1\right)\right]A(z)=0 \cdot \tag{2.32}$$

All relations in the parentheses on the left side of equation (2.32) should be zero in order to satisfy the above equation for any condition, i.e.

$$\frac{dq(z)}{dz}=1 \text{ and } \frac{q(z)}{A(z)}\frac{dA(z)}{dz}=-1 \text{ or } \frac{dA(z)}{A(z)}=-\frac{dz}{q(z)}=-\frac{dq(z)}{q(z)}\frac{dz}{dq(z)}=-\frac{dq(z)}{q(z)}. \tag{2.33}$$

By integrating equation (2.33), the following relationship is obtained:

$$q(z)=q(z_0)+z-z_0 \text{ and } \frac{A(z)}{A(z_0)}=\frac{q(z_0)}{q(z)}. \tag{2.34}$$

Now, let us consider the case that the electric field has a distribution in the x- and y-directions. In this case, it is convenient to separate variables and the electric field can be rewritten as

$$E_0(x,y,z)=E_{mn}(x,y,z)=A(z)E_m(x)E_n(y)=A[q(z)]E_m(x)E_n(y). \tag{2.35}$$

Here, if we only consider the electric field in x-z plane, then

$$E(x,z)=A[q(z)]\times E_m(x)\times\exp\left[-ik\frac{x^2}{2q(z)}\right]. \tag{2.36}$$

And by differentiating the electric field, we obtain

$$\begin{aligned}\frac{\partial^2}{\partial x^2}E(x)=A[q(z)]\times\frac{\partial^2}{\partial x^2}E_m(x)\times\exp\left[-ik\frac{x^2}{2q(z)}\right]\\ +2A[q(z)]\times\frac{\partial}{\partial x}E_m(x)\times\left[-ik\frac{x}{q(z)}\right]\times\exp\left[-ik\frac{x^2}{2q(z)}\right]\\ +A[q(z)]\times E_m(x)\times\left[-ik\frac{1}{q(z)}\right]\times\exp\left[-ik\frac{x^2}{2q(z)}\right]\\ +A[q(z)]\times E_m(x)\times\left[-k^2\frac{x^2}{q^2(z)}\right]\times\exp\left[-ik\frac{x^2}{2q(z)}\right]\end{aligned}, \tag{2.37}$$

and

$$\frac{\partial}{\partial z}E(x,z)=\frac{d}{dq}A[q(z)]\times E_m(x)\times\exp\left[-ik\frac{x^2}{2q(z)}\right]+A[q(z)]\times E_m(x)\times\left[ik\frac{x^2}{2q^2(z)}\right]\times\exp\left[-ik\frac{x^2}{2q(z)}\right]. \tag{2.38}$$

By inserting equations (2.37) and (2.38) into equation (2.22), we obtain

$$\frac{\partial^2}{\partial x^2}E_m(x)-2ik\frac{x}{q(z)}\frac{\partial}{\partial x}E_m(x)+ik\left[\frac{2}{A[q(z)]}\frac{d}{dq}A[q(z)]-\frac{1}{q(z)}\right]E_m(x)=0\,. \tag{2.39}$$

Next, by only considering the imaginary part in the beam parameter (we can assume the electric field is plane parallel in this case), the beam parameter becomes

$$\frac{1}{q(z)} = -i\frac{\lambda}{\pi w^2(z)}. \tag{2.40}$$

And, by inserting equation (2.40) into equation (2.39), we have

$$\frac{w^2(z)}{2}\frac{\partial^2}{\partial x^2}E_m(x) - 2x\frac{\partial}{\partial x}E_m(x) + ik\frac{w^2(z)}{2}\left[\frac{2}{A[q(z)]}\frac{d}{dq}A[q(z)] - \frac{1}{q(z)}\right]E_m(x) = 0. \tag{2.41}$$

Then, by substituting the variable with the relation $\sqrt{2}x/w(z) = u$, we finally obtain

$$\frac{\partial^2}{\partial u^2}E_m(u) - 2u\frac{\partial}{\partial u}E_m(u) + ik\frac{w^2(z)}{2}\left[\frac{2}{A[q(z)]}\frac{d}{dq}A[q(z)] - \frac{1}{q(z)}\right]E_m(u) = 0. \tag{2.42}$$

Note that equation (2.42) is similar to the differential equation for Hermite polynomials, $H_m(x)$.

$$\frac{d^2H_m(x)}{dx^2} - 2x\frac{dH_m(x)}{dx} + 2mH_m(x) = 0 \tag{2.43}$$

Thus, the electric field distribution has the form of a Hermite polynomial, i.e.,

$$E_m(x,z) = A[q(z)] \times H_m\left(\frac{\sqrt{2}x}{w(z)}\right) \times \exp\left[-ik\frac{x^2}{2q(z)}\right]. \tag{2.44}$$

In the same way, we can calculate the electric field distribution in the y-direction, and obtain the electric field distribution in the y-direction as

$$E_n(y,z) = A[q(z)] \times H_n\left(\frac{\sqrt{2}y}{w(z)}\right) \times \exp\left[-ik\frac{y^2}{2q(z)}\right]. \tag{2.45}$$

Thus, generally, the electric field distribution in the x- and y-directions is

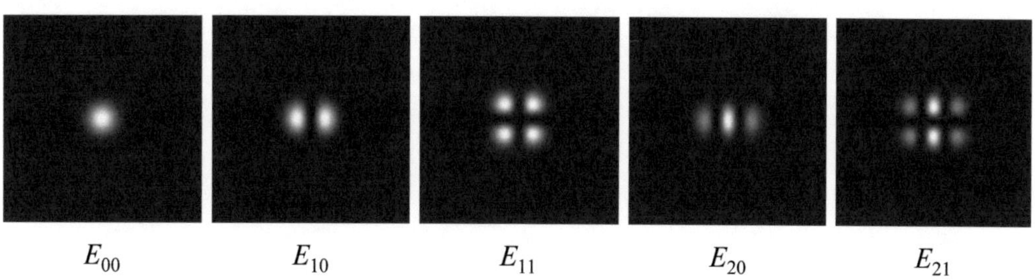

Fig. 1. Intensity distributions for several Hermite-Gaussian laser beam modes.

$$E_{mn}(x,y,z)=A\left[q(z)\right]\times H_m\left(\frac{\sqrt{2}x}{w(z)}\right)\times H_n\left(\frac{\sqrt{2}y}{w(z)}\right)\times \exp\left[-ik\frac{x^2+y^2}{2q(z)}\right], \quad (2.46)$$

though some Hermite polynomials of low order are given by

$$H_0(x)=1,\ H_1(x)=x,\ H_2(x)=4x^2-2, \text{ and } H_3(x)=8x^3-12x. \quad (2.47)$$

Figure 1 shows some low order Hermite-Gaussian beam modes in the rectangular coordinate. The intensity distribution of the lowest beam mode ($m=n=0$) is Gaussian and the Gaussian intensity profile is called either the TEM_{00} mode or the fundamental mode.

2.3 Laguerre-Gaussian beam mode in cylindrical coordinate

We can also solve the differential equation (2.16) in the cylindrical coordinate with a radially symmetric boundary condition. The solution of the wave equation in the cylindrical coordinates has the form of a Laguerre function; thus, the solution is called the Laguerre-Gaussian beam mode. In the cylindrical coordinates, the electric field propagating in the z-direction is given by

$$\frac{\partial^2 E(r,\phi,z)}{\partial r^2}+\frac{1}{r}\frac{\partial E(r,\phi,z)}{\partial r}+\frac{1}{r^2}\frac{\partial^2 E(r,\phi,z)}{\partial \phi^2}+\frac{\partial^2 E(r,\phi,z)}{\partial z^2}+k^2E(r,\phi,z)=0. \quad (2.48)$$

The solution for the differential equation (2.48) has the form of Laguerre polynomials. As such, the solution of the differential equation is given by

$$E_{mn}(r,\phi,z)=E_0\left(\frac{\sqrt{2}r(z)}{w(z)}\right)^n\times L_m^n\left(\frac{2r^2(z)}{w^2(z)}\right)\times\begin{Bmatrix}\cos(m\phi)\\ \sin(m\phi)\end{Bmatrix}\times\exp\left(-\frac{r^2(z)}{w^2(z)}\right). \quad (2.49)$$

Note that some low order Laguerre polynomials are given by

$$L_0^l(x)=1,\ L_1^l(x)=l+1-x,\ L_2^l(x)=(l+1)(l+2)/2-(l+2)x+x^2/2,$$

$$\text{and } L_3^l(x)=(l+1)(l+2)(l+3)/6-(l+2)(l+3)x/2+(l+3)x^2/2-x^3/6. \quad (2.50)$$

Figure 2 shows some low order Laguerre-Gaussian beam modes in the cylindrical coordinate. Note that as for the Hermite-Gaussian beam, the lowest beam mode is Gaussian and is also called the fundamental mode.

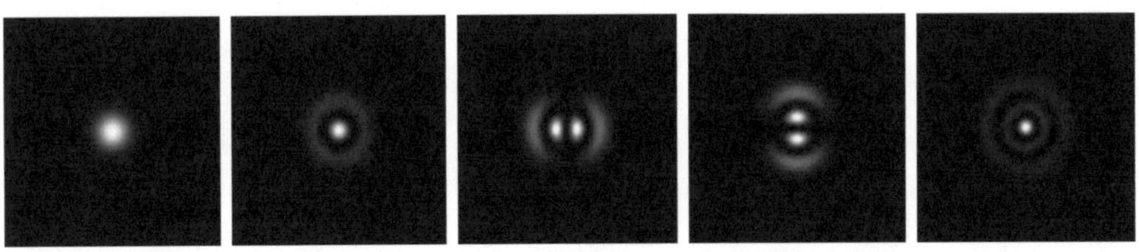

E_{00} E_{10} E_{11} (cos) E_{11} (sin) E_{20}

Fig. 2. Intensity distributions for several Laguerre-Gaussian laser beam modes.

2.4 Other beam modes

2.4.1. Flat-top beam profile and super Gaussian beam profile

In high-power laser systems, a uniform beam profile is required in order to efficiently extract energy from an amplifier. The uniform beam profile is sometimes called a flat-top (or top-hat) beam profile. However, the ideal flat-top beam profile is not possible because of diffraction; in many cases, a super-Gaussian beam profile is more realistic. The definition of the super-Gaussian beam profile is given by

$$\exp\left[-2r^{2n}/w_0^2\right]. \tag{2.51}$$

Here, n is called the order of the super-Gaussian beam mode and w_0 is the Gaussian beam radius when n is 1. Figure 3 shows the intensity profiles for several super-Gaussian beam profiles having different orders. As shown in the figure, the intensity profile becomes flat in the central region as the order of the super-Gaussian beam profile increases. Note that the flat-top beam profile is a specific case of the super-Gaussian beam profile having an order of infinity.

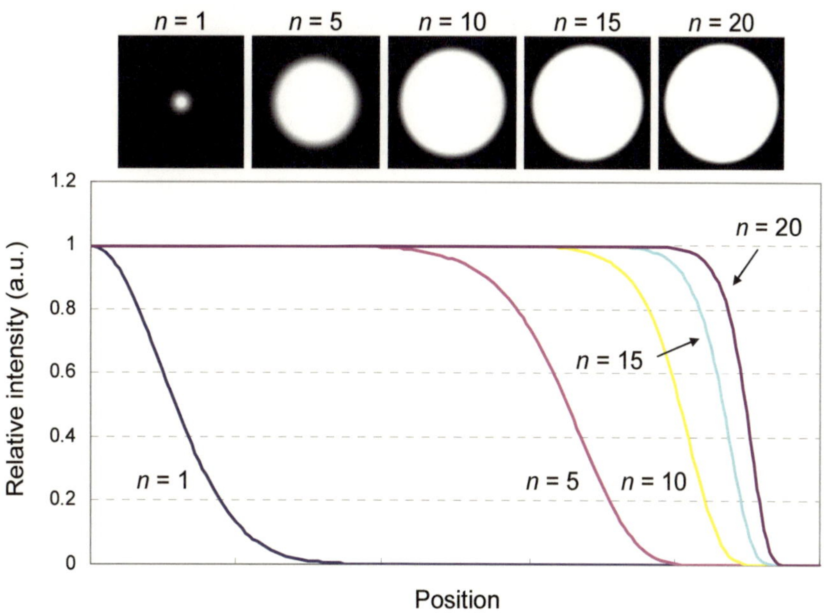

Fig. 3. Intensity distributions and their line profiles for several super-Gaussian laser beam modes having a different super-Gaussian order n.

2.4.2 Bessel-Gaussian beam profile

In this subsection, we will introduce a special laser beam mode called a Bessel beam. The Bessel function is a solution of the wave equation (2.16) in the cylindrical coordinate. Until 1987, the existence of the Bessel laser beam was not experimentally demonstrated. Theoretically, the Bessel laser beam has a special property that preserves its electric field distribution over a long distance. This is why the Bessel laser beam is referred to as a diffraction-free laser beam mode. However, in real situations, the Bessel laser beam mode preserves its electric field distribution for a certain distance because of the infinite power problem. Now, let us derive the Bessel laser beam mode from the wave equation. The wave equation in the cylindrical coordinate can be rewritten as

$$\frac{\partial^2 E(r,\phi,z)}{\partial r^2}+\frac{1}{r}\frac{\partial E(r,\phi,z)}{\partial r}+\frac{1}{r^2}\frac{\partial^2 E(r,\phi,z)}{\partial \phi^2}+\frac{\partial^2 E(r,\phi,z)}{\partial z^2}+k^2 E(r,\phi,z)=0\,. \qquad (2.48)$$

Then, using the separation of variables, the solution for equation (2.48) is

$$E(r,\phi,z)=E(r,\phi)\times\exp(-i\beta z)\,, \qquad (2.52)$$

and by inserting equation (2.52) into equation (2.48), equation (2.48) becomes

$$\frac{\partial^2 E(r,\phi,z)}{\partial r^2}+\frac{1}{r}\frac{\partial E(r,\phi,z)}{\partial r}+\frac{1}{r^2}\frac{\partial^2 E(r,\phi,z)}{\partial \phi^2}+\left(k^2-\beta^2\right)E(r,\phi,z)=0\,. \qquad (2.53)$$

If the electric field is radially symmetric, then the electric field $E(r,\phi,z)$ becomes $E(r,z)$ and the derivative with respect to the angular direction vanishes, i.e.,

$$r^2\frac{\partial^2 E(r,z)}{\partial r^2}+r\frac{\partial E(r,z)}{\partial r}+\left(k^2-\beta^2\right)r^2 E(r,z)=0\,. \qquad (2.54)$$

Note that equation (2.54) is similar to Bessel's differential equation with an order of 0. The Bessel's differential equation is expressed as

$$\rho^2\frac{d^2}{d\rho^2}Z_v(k\rho)+\rho\frac{d}{d\rho}Z_v(k\rho)+\left(k^2\rho^2-v^2\right)Z_v(k\rho)=0\,. \qquad (2.55)$$

And, the solution of equation (2.54) is given by

$$E(r,z)=E_0\times J_0\left(\sqrt{k^2-\beta^2}\,r\right)\times\exp(-i\beta z)\,. \qquad (2.56)$$

Thus, the solution in the cylindrical coordinate for the differential equation for the electric field is shown to be the Bessel function. Figure 4 presents the intensity distribution and profile for the Bessel laser beam mode.

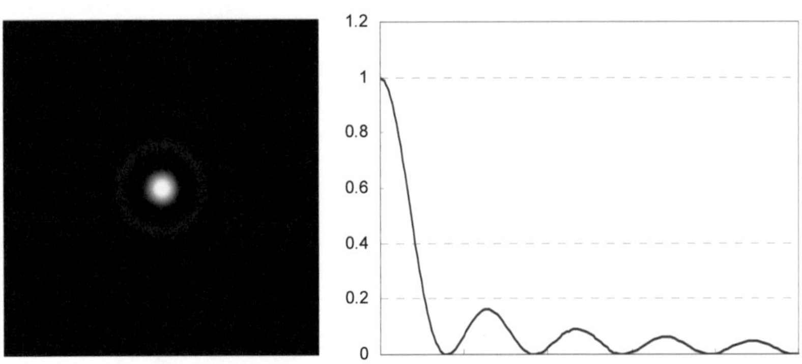

Fig. 4. Intensity distribution and its line profile for Bessel laser beam mode.

In 1987, Gori *et al.* introduced the Bessel-Gaussian laser beam mode to avoid the infinite power problem. In the Bessel-Gaussian laser beam mode, the electric field is given by

$$E(r,z_0) = E_0 \times J_0(\beta r) \times \exp\left(-\frac{r^2(z_0)}{w^2(z_0)}\right). \tag{2.57}$$

3. Intensity distribution of the focused laser beam

In the previous section, we derived the electric field distribution referred to as the laser beam mode. In order to determine the beam quality (or propagation) factor for the laser beam mode, we need to know the focusing property of the laser beam. The electric field distribution of a focused laser beam can be theoretically calculated from an incident beam profile. In this section, the calculation of the electric field distribution of a focused laser beam is introduced. For this task, three different approaches are used: a geometrical approach using a ray transfer matrix, a wave optics approach using diffraction theory, and a Fourier transform approach. From the geometrical approach, the physical insight and useful relationships for Gaussian beam parameters before and after a focusing optic can be easily obtained. However, if the incident beam profile is not Gaussian and has a wavefront aberration, it becomes more difficult to use the geometrical approach to explain the focusing property of the laser beam. In this case, the wave optics approach using diffraction theory gives more accurate calculation results. The wave optics approach offers analytic solutions for a Gaussian and uniform laser beams. However, the wave optics approach does not provide an analytical solution for an arbitrary incident laser beam mode. In this case, the Fourier transform approach becomes very useful. By using the Fourier transform method, the electric field distribution of the focused laser beam can be easily obtained and, together with the focus shift method, the electric field distribution near the focal plane can be quickly obtained. In particular, the Fourier transform approach is more useful for a laser beam having a wavefront aberration.

3.1 Geometrical approach

Propagation of the electric field can be described by the ray transfer matrix of an optical element. The ray transfer matrix determines the deviation angle at a location of the optical element; this matrix is also called the ABCD matrix and is expressed by

$$\begin{bmatrix} A & B \\ C & D \end{bmatrix}. \tag{3.1}$$

Let us now consider the case in which a Gaussian laser beam passes through an optical element having ABCD elements. The Gaussian beam mode [$E_1(x_1, y_1, z_1)$] before the optical element is again

$$E_1(x_1,y_1,z_1) = A(z_1) \times \exp\left[-ik\frac{x_1^2 + y_1^2}{2q(z_1)}\right] \text{ and } \frac{1}{q(z_1)} = \frac{1}{R(z_1)} - i\frac{\lambda}{\pi w_1^2(z_1)}. \tag{3.2}$$

Then, when the Gaussian laser beam passes through the optical element, the electric field right after the optical element is determined by

$$\begin{aligned} E_2(x_2,z_1) &\sim A(z_1)\int \exp\left[-\frac{ik}{2q(z_1)}x_1^2\right]\exp\left[-i\frac{\pi}{\lambda B}\left(Ax_1^2+Dx_2^2-2x_1x_2\right)\right]dx_1 \\ &= A(z_1)\int \exp\left[-\left(\frac{ik}{2q(z_1)}+\frac{i\pi}{\lambda B}A\right)x_1^2-i\frac{\pi}{\lambda B}Dx_2^2+2i\frac{\pi}{\lambda B}x_1x_2\right]dx_1 \end{aligned} \quad (3.3)$$

Here, only the electric field in the x-direction is considered, though the electric field in the y-direction can be calculated in the same manner. After some calculation, the electric field after the optical element is given by

$$E_2(x_2)=\sqrt{\frac{\pi}{\frac{ik}{2q_1}+\frac{i\pi}{\lambda B}A}}\exp\left[\frac{ik}{2B}x_2^2\left(-D+\frac{1}{B/q_1+A}\right)\right]. \quad (3.4)$$

The following definite integral formula (3.5) is used to derive equation (3.4).

$$\int_{-\infty}^{\infty}\exp\left(-ax^2-2bx\right)dx=\sqrt{\frac{\pi}{a}}\exp\left(\frac{b^2}{a}\right). \quad (3.5)$$

Because the determinant of the matrix is 1 (i.e., $AD-BC=1$), the electric field distribution after the optical element can be rewritten as follows:

$$E_2(x_2)=\sqrt{\frac{\pi}{\frac{ik}{2q_1}+\frac{i\pi}{\lambda B}A}}\exp\left[-\frac{ik}{2}x_2^2\left(\frac{q_1C+D}{q_1A+B}\right)\right]. \quad (3.6)$$

By defining $\frac{1}{q_2}$ as $\frac{q_1C+D}{q_1A+B}$, the resultant electric field distribution again has the same expression as the incident electric field except for the laser beam parameter q_2, such that

$$E_2(x_2)\sim\exp\left[-\frac{ik}{2q_2}x_2^2\right]. \quad (3.7)$$

Again, let us assume the Gaussian laser beam is focused by a focusing optic having a focal length of f. Then, we need to determine the electric field distribution at the focal plane of the focusing optic; the ray transfer matrix for a focusing optic and a free distance is given by

$$\begin{bmatrix} A & B \\ C & D \end{bmatrix}=\begin{bmatrix} 1 & f \\ 0 & 1 \end{bmatrix}\begin{bmatrix} 1 & 0 \\ -1/f & 1 \end{bmatrix}=\begin{bmatrix} 0 & f \\ -1/f & 1 \end{bmatrix}. \quad (3.8)$$

If the incident Gaussian laser beam is an ideal plane wave (i.e. $R(z_1)=\infty$), $q(z_1)$ is simply defined as $i\frac{\pi w_1^2(z_1)}{\lambda}$. Then, $\frac{1}{q_2}$ can be quickly calculated as

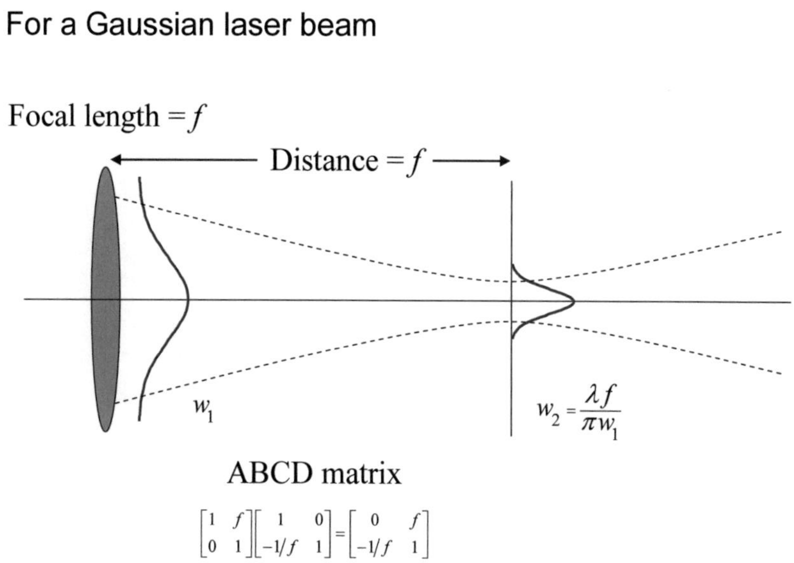

Fig. 5. Focusing Gaussian laser beam mode having a focusing optic with a focal length of f.

$$\frac{1}{q_2}=\frac{1}{R_2}-i\frac{\lambda}{\pi w_2^2}=-i\frac{\pi w_1^2}{\lambda}\frac{1}{f^2}+\frac{1}{f}. \tag{3.9}$$

Thus, the intensity profile of a focused Gaussian beam is again Gaussian with a new Gaussian width w_2, which is given by

$$w_2=\frac{\lambda f}{\pi w_1}=\frac{w_1 f}{z_R^2}, \tag{3.10}$$

where z_R is defined as $w_2=\frac{\lambda f}{\pi w_1}=\frac{w_1 f}{z_R^2}$ and called the Rayleigh range, at which point the area of the laser beam increases by a factor of 2. The electric field distribution near the focal plane can be calculated by replacing the focal length f with the distance d in equation (3.8). However, even if an arbitrary optical element having an arbitrary wavefront aberration can be represented by a ray transfer matrix, the general description of the electric field distribution for an arbitrary electric field cannot be simply expressed by the geometrical approach.

3.2 Wave optics approach using diffraction theory

Now, in this section, we will directly calculate the electric field distribution based on the diffraction integral. Again, consider that an electric field converges from a focusing optic having a focal length of f to the axial focal point. Then, the electric field distribution at a point (x_2, y_2) in the focal plane is given by

$$E(x_2,y_2)=-\frac{i}{\lambda}\frac{e^{-ikf}}{f}\int E_1(x_1,y_1)\frac{e^{iks}}{s}dx_1dy_1. \tag{3.11}$$

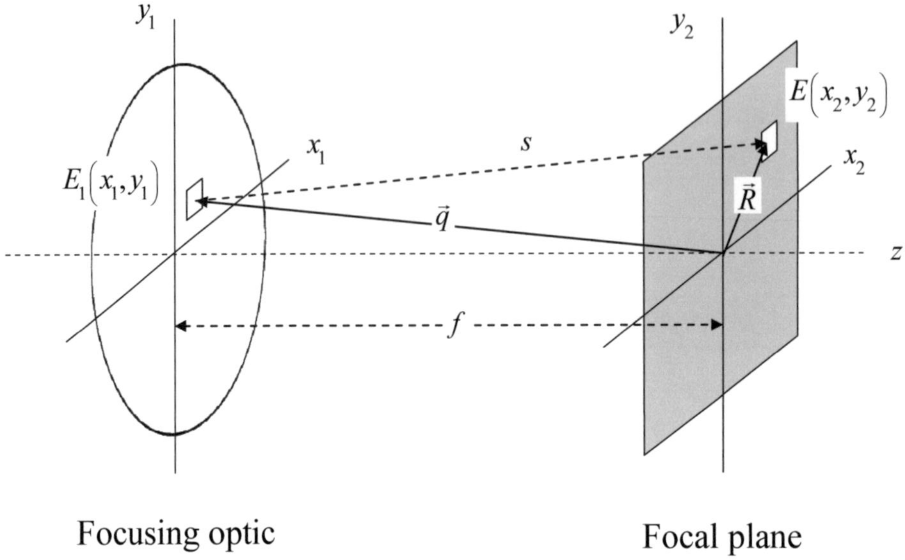

Fig. 6. Diffraction of electric field at a focusing optic having a focal length f .

To evaluate equation (3.11), let us assume that the focusing optic is circular and that the radius of the focusing optic is a. Then, it is convenient to express (x_1, y_1, z_1) and (x_2, y_2, z_2) in the cylindrical coordinate as follows:

$$x_1 = a\rho\sin\theta\,,\ y_1 = a\rho\cos\theta\,, \text{and}\ x_2 = r\sin\varphi\,,\ y_2 = r\cos\varphi \tag{3.12}$$

where ρ extends from zero to 1. In this expression, from Fig. 6, the difference $s-f$ and the small area of dx_1dy_1 can be, via an approximation, expressed as

$$s - f = -\vec{q}\cdot\vec{R}\ \text{and}\ dx_1dy_1 = f^2 d\Omega \tag{3.13}$$

where $d\Omega$ is the infinitesimal solid angle. Then, using the approximation $s \approx f$, the electric field distribution at a point (x_2, y_2) becomes

$$E(x_2, y_2) = -\frac{i}{\lambda}\int E_1(x_1, y_1)e^{-k\vec{q}\cdot\vec{R}}d\Omega\,. \tag{3.14}$$

Equation (3.14) is known as the Debye integral and expresses the electric field as a superposition of plane wave components having different directions of propagation, known as angular spectrums. The phase component in the integral is

$$\vec{q}\cdot\vec{R} = \frac{x_1x_2 + y_1y_2 + z_1z_2}{f}\,, \tag{3.15}$$

and the axial position z_1 of element dx_1dy_1 from the origin of (x_2, y_2, z_2) is

$$z_1 = -\sqrt{f^2 - a^2\rho^2} = -f\left[1 - \frac{1}{2}\frac{a^2\rho^2}{f^2} + \frac{3}{8}\frac{a^4\rho^4}{f^4} - \cdots\right]. \tag{3.16}$$

By inserting equation (3.16) into equation (3.15) and using $f \gg a$, we obtain the following expression for the phase component in the Debye integral:

$$k\vec{q}\cdot\vec{R} = k\frac{x_1x_2 + y_1y_2 + z_1z_2}{f} = \frac{2\pi}{\lambda}\frac{a\rho r\cos(\theta-\varphi)}{f} - z\frac{2\pi}{\lambda}\left[1 - \frac{1}{2}\frac{a^2\rho^2}{f^2}\right]. \quad (3.17)$$

Then, by introducing dimensionless variables u and v in the focal plane, defined as

$$u = \frac{2\pi}{\lambda}\left(\frac{a}{f}\right)^2 z \text{ and } v = \frac{2\pi}{\lambda}\left(\frac{a}{f}\right)r, \quad (3.18)$$

the phase component in the Debye integral is

$$k\vec{q}\cdot\vec{R} = v\rho\cos(\theta-\varphi) - \left(\frac{f}{a}\right)^2 u + \frac{1}{2}u\rho^2. \quad (3.19)$$

Thus, equation (3.14), which expresses the electric field distribution in the focal plane, becomes

$$E(u,v) = -\frac{i}{\lambda}\frac{a^2}{f^2}\exp\left[i\left(\frac{f}{a}\right)^2 u\right]\int_0^1\int_0^{2\pi} E(\rho,\theta)\exp\left\{-i\left[v\rho\cos(\theta-\varphi) + \frac{1}{2}u\rho^2\right]\right\}\rho d\rho d\theta . \quad (3.20)$$

And, if the incoming electric field is radially symmetric, equation (3.20) can be simply expressed as

$$E(u,v) = -\frac{2\pi i}{\lambda}\frac{a^2}{f^2}\exp\left[i\left(\frac{f}{a}\right)^2 u\right]\int_0^1 E(\rho)J_0(v\rho)\exp\left[-i\frac{1}{2}u\rho^2\right]\rho d\rho, \quad (3.21)$$

using the definition of Bessel function.

$$J_0(x) = \frac{1}{2\pi}\int_0^{2\pi}\exp(ix\cos\theta)d\theta. \quad (3.22)$$

If we consider the uniform intensity profile, then the incoming electric field is constant with respect to the position (i.e. $E(\rho,\theta) = C$). In this case, the electric field distribution in the focal plane has the form

$$E(u,v) = -i\frac{2\pi}{\lambda}\frac{a^2C}{f^2}\exp\left[i\left(\frac{f}{a}\right)^2 u\right]\int_0^1 J_0(v\rho)\exp\left(-\frac{1}{2}iu\rho^2\right)\rho d\rho. \quad (3.23)$$

To calculate equation (3.35) further, we separate the integral into the real and imaginary parts.

$$E(u,v) = -i\frac{2\pi}{\lambda}\frac{a^2C}{f^2}\exp\left[i\left(\frac{f}{a}\right)^2 u\right]\left[\int_0^1 J_0(v\rho)\cos\left(\frac{1}{2}u\rho^2\right)\rho d\rho - i\int_0^1 J_0(v\rho)\sin\left(\frac{1}{2}u\rho^2\right)\rho d\rho\right]. \quad (3.24)$$

There are two different cases in evaluating the integrals in equation (3.24). In the first case, when $|u/v|<1$ (i.e. inside the geometrical shadow), we use the relation for the Bessel function to obtain

$$2\int_0^1 J_0(v\rho)\cos\left(\frac{1}{2}u\rho^2\right)\rho d\rho = \frac{2}{v}\int_0^1 \frac{d}{d\rho}\left[\rho J_1(v\rho)\right]\cos\left(\frac{1}{2}u\rho^2\right)\rho d\rho$$

$$= \frac{2}{v}\left[J_1(v)\cos\left(\frac{1}{2}u\right) + u\int_0^1 \rho^2 J_1(v\rho)\sin\left(\frac{1}{2}u\rho^2\right)\rho d\rho\right]$$

$$= \frac{\cos(u/2)}{u/2}\left[\left(\frac{u}{v}\right)J_1(v) - \left(\frac{u}{v}\right)^3 J_3(v) + \cdots\right] + \frac{\sin(u/2)}{u/2}\left[\left(\frac{u}{v}\right)^2 J_2(v) - \left(\frac{u}{v}\right)^4 J_4(v) + \cdots\right]. \quad (3.25)$$

$$= \frac{\cos(u/2)}{u/2}U_1(u,v) + \frac{\sin(u/2)}{u/2}U_2(u,v)$$

where the following definition of the Lommel function $U_n(u,v)$ and the relation for the Bessel function are used to obtain equation (3.25):

$$U_n(u,v) = \sum_{s=0}^{\infty}(-1)^s\left(\frac{u}{v}\right)^{n+2s} J_{n+2s}(v), \quad (3.26)$$

$$\text{and } \frac{d}{dx}\left[x^{n+1}J_{n+1}(x)\right] = x^{n+1}J_n(x). \quad (3.27)$$

In a similar way, we obtain the expression for the imaginary part:

$$2\int_0^1 J_0(v\rho)\sin\left(\frac{1}{2}u\rho^2\right)\rho d\rho = \frac{\sin(u/2)}{u/2}U_1(u,v) - \frac{\cos(u/2)}{u/2}U_2(u,v). \quad (3.28)$$

In the second case, when $|u/v|>1$ (i.e. outside the geometrical shadow), we evaluate equation (3.24) by integrating by parts with respect to the trigonometric function in order to finally obtain the expressions for the real and imaginary parts as follows:

$$2\int_0^1 J_0(v\rho)\cos\left(\frac{1}{2}u\rho^2\right)\rho d\rho = \frac{\sin\left(v^2/2u\right)}{u/2} + \frac{\sin(u/2)}{u/2}V_0(u,v) - \frac{\cos(u/2)}{u/2}V_1(u,v), \quad (3.29)$$

$$\text{and } 2\int_0^1 J_0(v\rho)\sin\left(\frac{1}{2}u\rho^2\right)\rho d\rho = \frac{\cos\left(v^2/2u\right)}{u/2} - \frac{\cos(u/2)}{u/2}V_0(u,v) - \frac{\sin(u/2)}{u/2}V_1(u,v). \quad (3.30)$$

The other definition of the Lommel function, $V_n(u,v)$, is then used for obtaining equations (3.29) and (3.30).

$$V_n(u,v) = \sum_{s=0}^{\infty}(-1)^s\left(\frac{v}{u}\right)^{n+2s} J_{n+2s}(v). \quad (3.31)$$

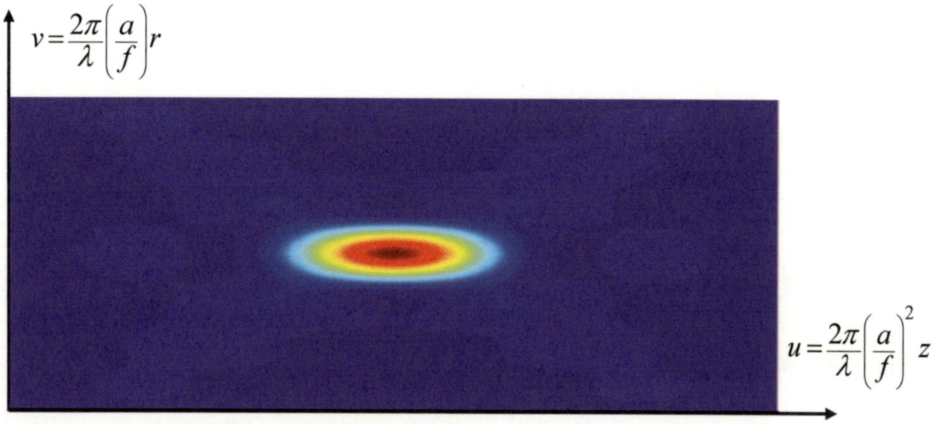

Fig. 7. Intensity distribution near the focal plane calculated using the diffraction integral approach when a flat-top beam profile is focused.

Now, let us calculate the electric field distribution in the focal plane from equations (3.21), (3.25), (3.28), (3.29), and (3.30). First, in the region $|u/v|<1$, we use equations (3.21), (3.25), and (3.28) to calculate the intensity distribution.

$$I(u,v)=4I_0\left[\frac{U_1^2(u,v)+U_2^2(u,v)}{u^2}\right]. \tag{3.32}$$

Figure 7 shows the intensity distribution near the focal plane calculated using equation (3.32) when a flap-top laser beam is focused. In the special case of a focal plane ($u=0$), the intensity distribution is

$$I(0,v)=4I_0\frac{J_1^2(v)}{v^2}. \tag{3.33}$$

Thus, as shown in equation (3.33), the Airy function is obtained when we focus a uniform electric field. If the incoming laser beam has a Gaussian beam profile, then the electric field distribution in equation (3.21) has the form

$$E_1(\rho)=C\times\exp\left[-\frac{a^2\rho^2}{w_0^2}\right]. \tag{3.34}$$

In equation (3.34), we only consider the case of a plane wave (i.e. $R=\infty$). By inserting equation (3.34) into equation (3.21), equation (3.21) becomes

$$\begin{aligned}E(u,v)&=-\frac{2\pi i}{\lambda}\frac{a^2C}{f^2}\exp\left[i\left(\frac{f}{a}\right)^2 u\right]\int_0^1\exp\left[-\frac{a^2\rho^2}{w_0^2}\right]J_0(v\rho)\exp\left[-i\frac{1}{2}u\rho^2\right]\rho d\rho\\&\approx\int_0^1 J_0(v\rho)\exp\left[-\left(\frac{a^2}{w_0^2}+i\frac{u}{2}\right)\rho^2\right]\rho d\rho\end{aligned}. \tag{3.35}$$

Again, in the special case of a focal plane ($u = 0$), the electric field distribution in the focal plane is

$$E(0,v) \approx \int_0^1 J_0(v\rho)\exp\left(-\frac{a^2\rho^2}{w_0^2}\right)\rho d\rho . \tag{3.36}$$

To integrate equation (3.36), we use the following definite integral formula:

$$\int_0^\infty \exp\left(-\alpha x^2\right) I_m(\beta x) J_m(\gamma x) x dx = \frac{1}{2\alpha}\exp\left(\frac{\beta^2-\gamma^2}{4\alpha}\right) J_m\left(\frac{\beta\gamma}{2\alpha}\right). \tag{3.37}$$

The small beam size approximation ($a \gg w_0$) is used to apply the definite integral formula. Then, the final expression for the incoming Gaussian beam is

$$E(0,v) \approx \int_0^1 J_0(v\rho)\exp\left(-\frac{a^2\rho^2}{w_0^2}\right)\rho d\rho = \frac{w_0^2}{2a^2}\exp\left(-\frac{w_0^2 v^2}{4a^2}\right) = \frac{w_0^2}{2a^2}\exp\left(-\frac{r^2}{w_1^2}\right). \tag{3.38}$$

From equation (3.38), the new Gaussian beam size for the focused electric field is obtained as

$$w_1 = \frac{\lambda f}{\pi w_0} . \tag{3.39}$$

Obviously, the new Gaussian beam size for the focused electric field calculated from the Debye integral is exactly the same as that calculated from the geometrical approach. Unlike the above special cases (uniform and Gaussian field profiles), an analytical solution for the

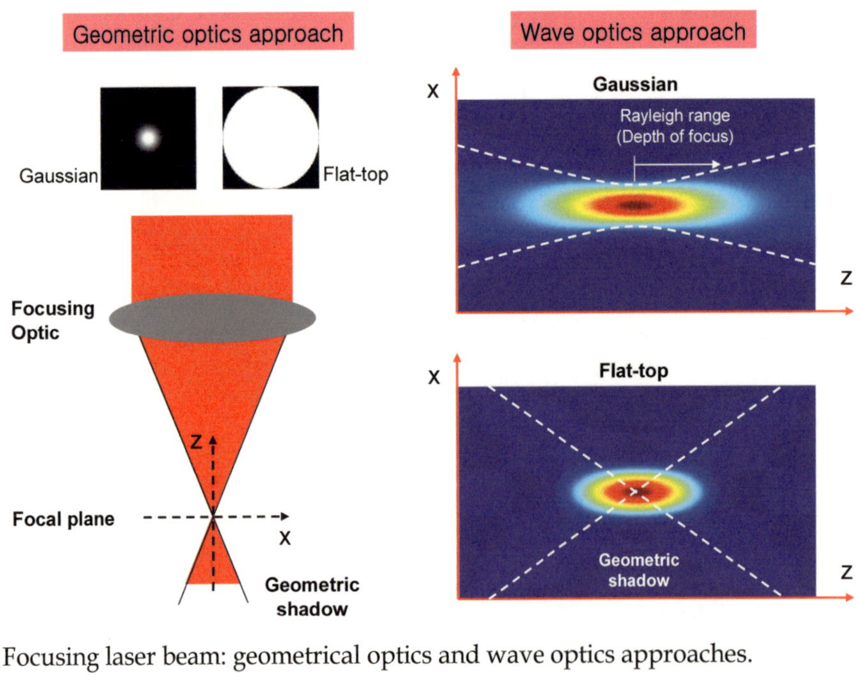

Fig. 8. Focusing laser beam: geometrical optics and wave optics approaches.

focused electric field does not exist for an electric field having an arbitrary magnitude and wavefront. Thus, in case of an arbitrary electric field, it is convenient to use the Fourier transform approach to calculate the focused electric or intensity distribution in and near the focal plane.

3.3 Fourier transform approach

If we assume that a laser beam is focused with an ideal focusing optic having a focal length f , then the electric field distribution at the focal plane can be expressed by

$$E_2(x_2,y_2) \sim \int_{-\infty}^{\infty} E_1(x_1,y_1)\exp\left[ikW(x_1,y_1)\right]\exp\left[i\frac{k}{f}(x_1x_2+y_1y_2)\right]dx_1dy_1 . \tag{3.40}$$

This integral form represents the Fourier transform of the incident electric field having an arbitrary wavefront aberration $W(x_1,y_1)$. Now, let us quickly review the derivation of equation (3.40). Consider that a monochromatic electric field $E_1(x_1,y_1)$ converges by a focusing optic having a focal length f to the axial focal point. Again, the electric field distribution at a point (x_2 , y_2) in the focal plane is given by

$$E(x_2,y_2) = -\frac{i}{\lambda}\frac{e^{-ikf}}{f}\int E_1(x_1,y_1)\frac{e^{iks}}{s}dx_1dy_1 \tag{3.11}$$

where x_1 and y_1 are the coordinates in the aperture plane, and s is the distance from a certain point in the focusing optic to the point (x_2 , y_2). Then, if we express s in the (x, y) coordinate, s is

$$\begin{aligned} s &= \sqrt{(x_1-x_2)^2+(y_1-y_2)^2+z_2^2} = z_2\sqrt{1+\left(\frac{x_1-x_2}{z_2}\right)^2+\left(\frac{y_1-y_2}{z_2}\right)^2} \\ &\approx z_2\left[1+\frac{1}{2}\left(\frac{x_1-x_2}{z_2}\right)^2+\frac{1}{2}\left(\frac{y_1-y_2}{z_2}\right)^2\right] \end{aligned} . \tag{3.41}$$

Because the phase of the electric field varies more quickly than the magnitude, we can approximate s such that $s \approx z_2 \approx f$ in the expression related to the magnitude. Thus, equation (3.11) becomes

$$E(x_2,y_2) = -\frac{i}{\lambda}\frac{e^{-ikf}}{f^2}\exp\left[\frac{ik}{2f}\left(x_2^2+y_2^2\right)\right]\int E_1(x_1,y_1)\exp\left[\frac{ik}{2f}\left(x_1^2+y_1^2\right)\right]\exp\left[-\frac{k}{f}(x_1x_2+y_1y_2)\right]dx_1dy_1 . \tag{3.42}$$

We then derive the expression for electric field distribution in a focal plane when the electric field is focused with a focusing optic having a focal length f . In equation (3.42), one important consideration is the phase delay due to the focusing optic; the phase function including phase delay should be considered in the electric field $E_1(x_1,y_1)$. For this task, we first have to obtain the expression for the phase delay. If we consider a lens having the thickness shown in Fig. 9, then the phase delay after the lens is

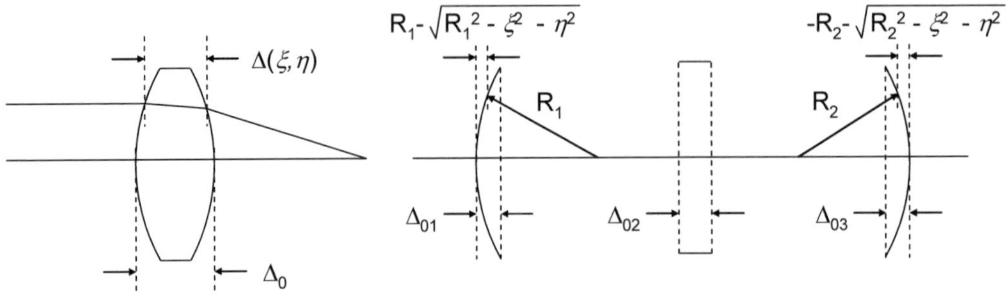

Fi. 9. Calculation of lens thickness function for inducing the phase delay.

$$\phi(x_1,y_1) = nkT(x_1,y_1) + k\left[T_0 - T(x_1,y_1)\right] \tag{3.43}$$

where n is the refractive index of the lens, T_0 is the center thickness of the lens, and $T(x_1,y_1)$ is the thickness function of the lens at a position (x_1, y_1). By considering the geometry of the lens in Fig. 9, the thickness function $T(x_1,y_1)$ can be written as

$$T(x_1,y_1) = T_0 - R_1\left(1-\sqrt{1-\frac{x_1^2+y_1^2}{R_1^2}}\right) + R_2\left(1-\sqrt{1-\frac{x_1^2+y_1^2}{R_2^2}}\right). \tag{3.44}$$

Then, by using a paraxial approximation, we can take only the first two terms in the Taylor series of $\sqrt{\ }$. The thickness function can be simply rewritten as

$$T(x_1,y_1) = T_0 - \frac{x_1^2+y_1^2}{2}\left(\frac{1}{R_1}-\frac{1}{R_2}\right). \tag{3.45}$$

From equation (3.45), the phase function $P(x_1,y_1)$ after a thin lens is

$$P(x_1,y_1) = \exp\left[-i\phi(x_1,y_1)\right] = \exp\left[-i\frac{k}{2f}\left(x_1^2+y_1^2\right)\right]. \tag{3.46}$$

Note that the mathematical formula (3.47) for a lens having radii R_1 and R_2 is used to derive equation (3.46).

$$\frac{1}{f} = (n-1)\left(\frac{1}{R_1}-\frac{1}{R_2}\right). \tag{3.47}$$

Thus, the final expression for the electric field $E_1(x_1,y_1)$ after the lens is

$$E_1(x_1,y_1) = E_0(x_1,y_1)\exp\left[-i\frac{k}{2f}\left(x_1^2+y_1^2\right)\right] \tag{3.48}$$

where $E_1(x_1,y_1)$ means the electric field before the focusing optic. By substituting equation (3.48) into equation (3.42), we quickly obtain equation (3.40). By the same analogy used in equation (3.48), the expression for the wavefront aberration is introduced in equation (3.40).

From equation (3.40), in order to obtain the electric field distribution at the focal plane, an accurate measurement of the wavefront aberration is required.

The wavefront aberration means the deviation of the spatial phase of the laser beam from the reference phase, typically due to imperfections in optics used in the laser system and the beam delivery line and the thermal property of the laser crystal. The wavefront aberration, especially higher-order aberrations, is usually negligible in a small laser system. However, as the size of the laser system (specifically, the beam size) becomes larger, the wavefront aberration of the laser beam can no longer be considered negligible. The wavefront aberration can be expressed using Zernike polynomials $Z_n^m(x,y)$ as follows:

$$W(x,y)=\sum_{n,m} c_{nm} Z_n^m(x,y) \tag{3.49}$$

where n and m are the radial and azimuthal order for the Zernike polynomial, and c_{nm} is the Zernike coefficient. A double-index scheme in the OSA/ANSI standard is generally used to label the Zernike coefficients.

The Shack-Hartmann wavefront sensor is commonly used to measure the wavefront aberration. Figure 10 presents a typical schematic diagram of a Shack-Hartmann wavefront sensor, which consists of a microlens array and a CCD camera. The size of the microlens array ranges from about 100 μm to several hundreds of μm. The microlens array divides the laser beam into hundreds of sub-apertures and focuses them onto the CCD camera. The CCD camera then records the focal spots formed by the microlens array. Next, the locations of focal spots are calculated and compared to the reference position, i.e., when the laser beam has no wavefront aberration. Here, deviations from the reference positions provide the wavefront information. Once the wavefront aberration is obtained, the electric field distribution in the focal plane can be quickly calculated using equation (3.40).

Many other techniques, such as the interferometric method, have been devised to measure the wavefront of the laser beam. Figure 11 describes how to measure the wavefront aberration of a laser beam using a shearing interferometer. Also, the wavefront aberration

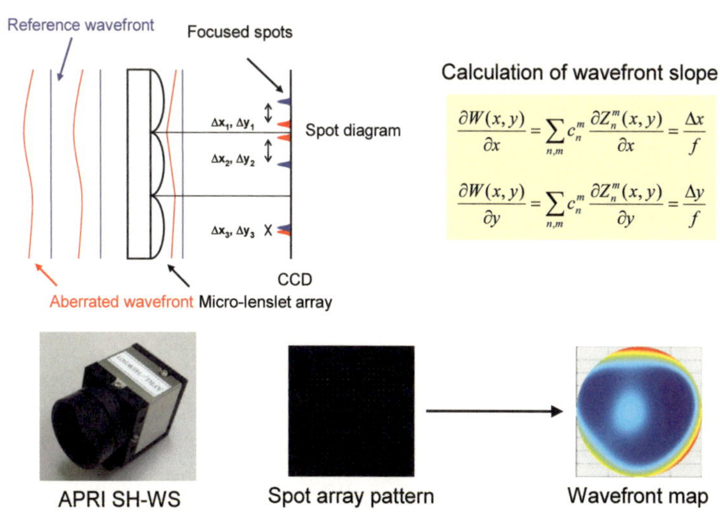

Fig. 10. Principle of Shack-Hartmann wavefront sensing.

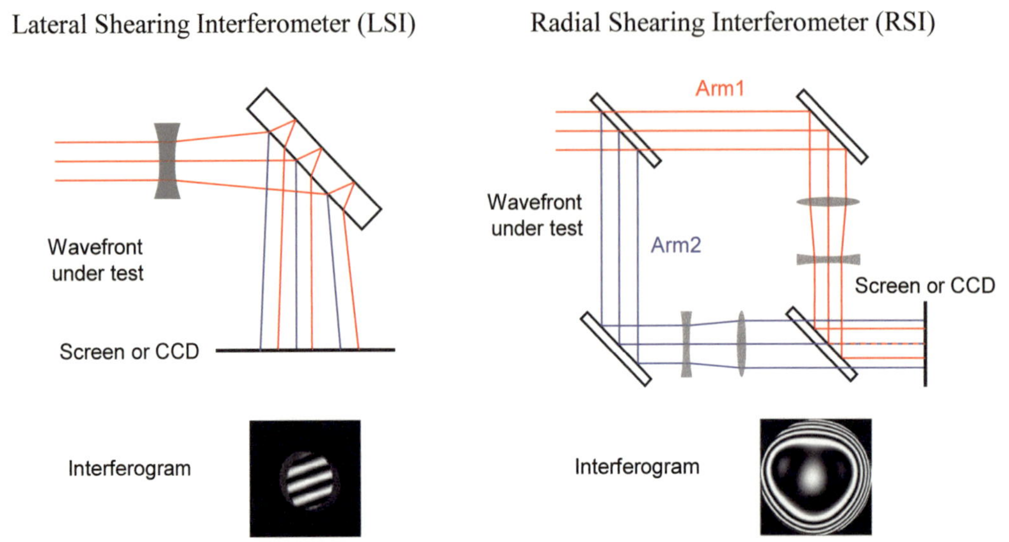

Fig. 11. Lateral and radial shearing interferometers for measuring wavefront of laser beam.

can be reconstructed by directly measuring the focal spot. This technique is called phase retrieval technique. However, the use of this technique is quite limited because the wavefront reconstruction takes time due to the iteration process.

Because the Fourier transform only offers the electric field distribution in the focal plane, a special technique called the focus-shift technique is required in order to calculate the electric field distribution near the focal plane. Figure 12 explains the principle of focus-shift technique. In the figure, if an incident laser beam has no wavefront aberration, the laser beam is then focused at the focal plane of a focusing optic having a focal length of f . If a certain amount (c_{20}) of defocus is intentionally added to the wavefront aberration of the laser beam, the laser beam itself will converge and the focal spot will instead be formed at position d , given by

$$\frac{1}{d}=\frac{4\sqrt{3}c_{20}}{r_0^2} \tag{3.50}$$

where c_{20} is the Zernike coefficient for defocus, and r_0 is the radius of the laser beam. Thus, the added defocus will act as a virtual focusing optic having a focal length of d . Again, if the defocused laser beam is focused by the focusing optic, a new focus will form at position d_{new} due to the dioptric law, such that (see Fig. 12(b)):

$$\frac{1}{d_{new}}=\frac{1}{f}+\frac{1}{d}=\frac{1}{f}+\frac{4\sqrt{3}c_{20}}{r_0^2}\,. \tag{3.51}$$

According to Eq. (3.51), the position of the laser focal spot can be accurately controlled by adding a proper amount of defocus to the wavefront of the incoming laser beam. As mentioned above, this technique is useful in calculating the intensity distribution near the focal plane. Because the Fourier transform of a laser beam always provides the intensity distribution in the focal plane of a focusing optic, the intensity distribution calculated with a different amount of defocus value expresses the intensity distribution of the original laser beam in the plane of z (Fig. 12(b)). In the figure, z is defined by $f - d_{new}$; thus, by introducing various amounts of defocus to the wavefront of an incoming laser beam and performing a finite Fourier transform (FFT), a full description of the three-dimensional intensity distribution near the focal plane is possible for an arbitrary laser beam without having to calculate the diffraction integral in equation (3.11). Of course, the focus-shift technique holds under the thin-lens approximation and under the condition of adding a small amount of defocus. Figure 13 shows the intensity distribution near the focal plane calculated using the Fourier transform approach when a flat-top laser beam profile is focused. As shown in the figure, the overall intensity profile is almost identical to that calculated with the wave optics approach. Also, the Fourier transform approach can be applied to quickly calculate the intensity distribution of a laser pulse near the focal plane.

In order to calculate the intensity distribution of a laser pulse (especially for an ultrashort laser pulse in the femtosecond regime), we have to consider chromatic aberration induced by a focusing optic. The chromatic aberration means the change in defocus values at different wavelengths. Figure 14 shows the calculated and the measured defocus values (c_{20} s), as functions of wavelength, that were induced by the beam expanders in a high power Ti:sapphire laser system. Here, ray-tracing software was used to calculate c_{20} s for the beam expanders at different wavelengths, the values of which are seen to almost linearly

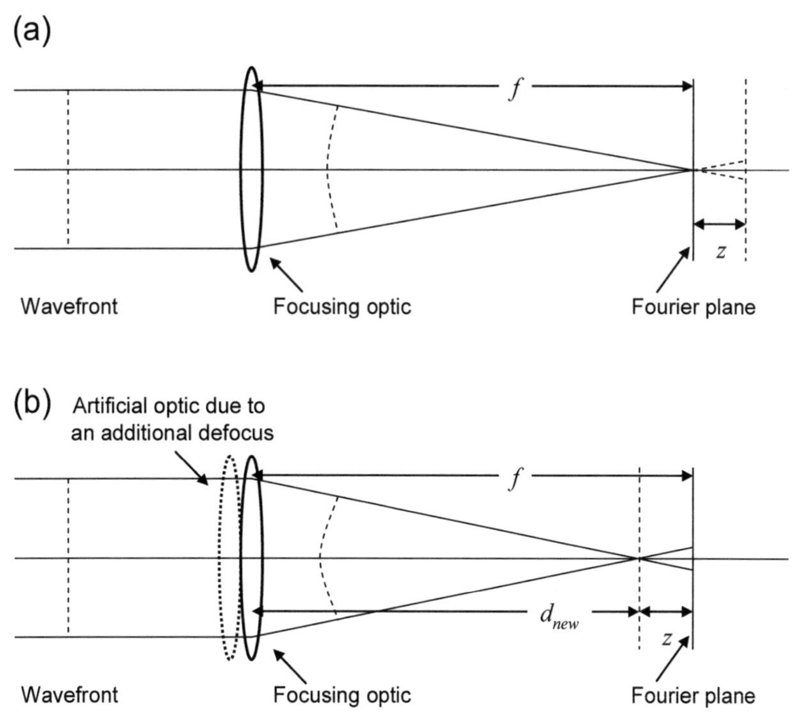

Fig. 12. Explanation of focus-shift technique introducing an arbitrary defocus into laser beam.

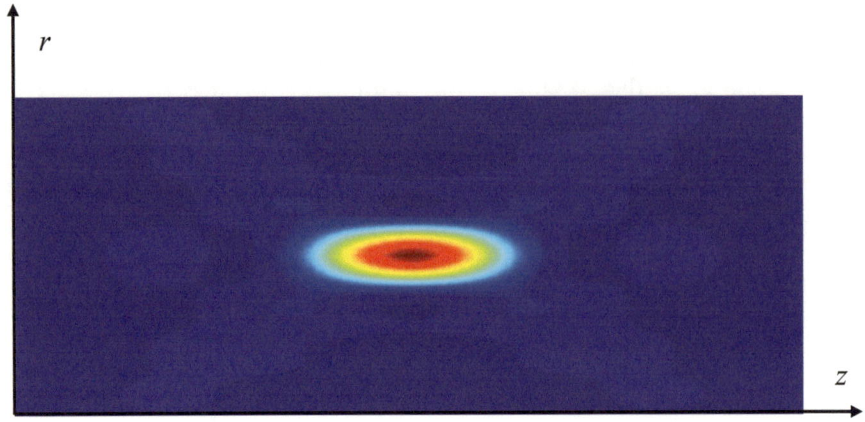

Fig. 13. Intensity distribution near the focal plane calculated using the Fourier transform approach when a flat-top beam profile is focused.

decrease with increasing wavelengths. This decrease implies that, after the beam expanders, the short-wavelength component in the laser spectrum converges while the long-wavelength component diverges. The slope of the decrease was 0.058 μm/10 nm. The change in defocus of the laser pulse in the spectral range was then measured to investigate the chromatic aberration caused by the beam expanders. A wavefront sensor was used to measure the defocus values of the laser pulse at different wavelengths.

Bandpass filters having a bandwidth of 10 nm at wavelengths of 760, 780, 800, 820, and 840 nm, respectively, were placed in front of the wavefront sensor to examine the dependence of the defocus value on the wavelength. The squares in Fig. 14 indicate the measured defocus values, and the error bars are the standard deviations of the measurement. The figure clearly shows that the measured defocuses of the laser pulse agree well with the calculated values for the beam expanders.

Figure 15 shows the spatial and temporal intensity distributions at different positions near the focal plane with and without chromatic aberration. In this calculation, the use of an ideal

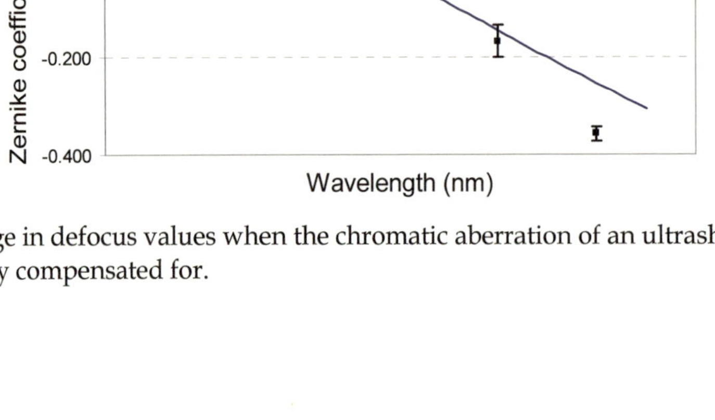

Fig. 14. Change in defocus values when the chromatic aberration of an ultrashort laser beam is not correctly compensated for.

focusing optic having a focal length of 1.5 m is assumed. The change in defocus shown in Fig. 14 was used in the calculations. The monochromatic electric field distributions in the focal plane were calculated in steps of 2 nm over the spectral range, and a focus-shift technique was used to calculate the spatial and temporal intensity distributions at different positions. In Fig. 15, the x-axis of each plot means the time, and the y-axis shows the vertical location at each position. Also, in the figure, by broadening the temporal duration of the focal spot the laser pulse having chromatic aberration showed a very different focusing property than the laser pulse without chromatic aberration. The calculations showed that the peak intensity of the focal spot with chromatic aberration was reduced to about 39% of that of the focal spot without chromatic aberration. The non-symmetric property of the intensity distributions against the focal plane originated from the non-symmetric profile of the laser spectrum. Moreover, with the chromatic aberration, a double-peak structure was observed in several positions near the focal plane. However, the double-peak structure disappeared, recovering to the original pulse duration, when the laser pulse propagated a long distance from the focal plane. Thus, the Fourier transform approach seems to be very useful in calculating the intensity distributions near the focal plane and finally determining the beam propagation properties in both cw and pulsed lasers.

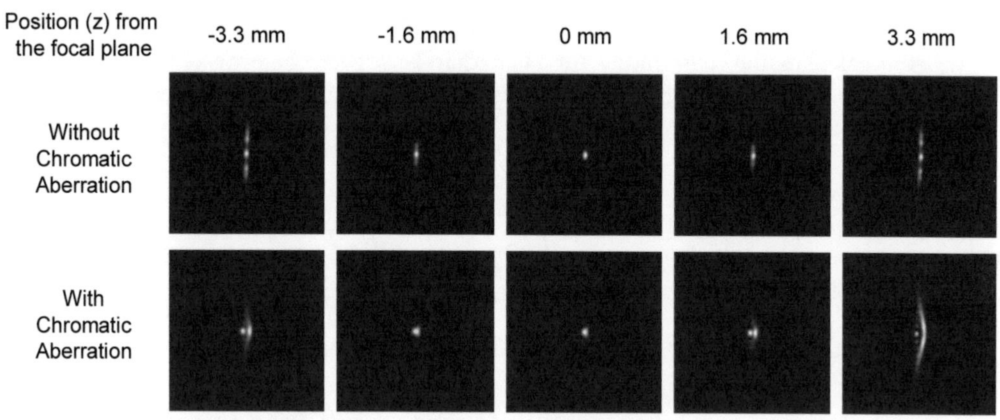

Fig. 15. Spatial and temporal intensity distributions of focal spots at different positions near the focal plane.

4. Beam quality factor

4.1 Definition of beam radius

The propagation of the laser beam can be partially characterized by the beam width and the divergence. In other words, defining the beam quality factor is closely related to the definition of the beam width. For the simple Gaussian laser beam profile, defining the width of the laser beam is trivial. As used in the previous subsections, w_0 and w_1 are used here for the incoming and focused Gaussian beam sizes. However, in some real situations for a complex beam profile with multiple intensity peaks, as shown in Fig. 16, a meaningful and accurate definition of the beam width is not easy. As pointed out by Prof. A. Siegman, there can be many possible definitions of the beam radius for an arbitrary beam profile:

- Width (or half width) at first null

- Transverse knife edge widths
- Width at 1/e or 1/e² in the beam profile
- Radius containing 86% of the entire power
- Second-order moment
- Others

The radius containing 86% of the entire power can be used to calculate the beam propagation of a high-power laser beam in air. The second-order moment is useful to mathematically define the radius of an arbitrary laser beam profile. In the second-order moment, the beam radii for the x- and y-directions are defined by

$$\left\langle w_x^2 \right\rangle = 4\frac{\int_{-\infty}^{\infty}\int_{-\infty}^{\infty} x^2 I(x,y)\,dxdy}{\int_{-\infty}^{\infty}\int_{-\infty}^{\infty} I(x,y)\,dxdy} \text{ and } \left\langle w_y^2 \right\rangle = 4\frac{\int_{-\infty}^{\infty}\int_{-\infty}^{\infty} y^2 I(x,y)\,dxdy}{\int_{-\infty}^{\infty}\int_{-\infty}^{\infty} I(x,y)\,dxdy}. \tag{4.1}$$

In the cylindrical coordinate, the beam radius for a radially symmetrical beam profile is defined by

$$\left\langle w_r^2 \right\rangle = 2\int_0^{\infty}\int_0^{2\pi} r^3 I(r,\varphi)\,drd\varphi \Big/ \int_0^{\infty}\int_0^{2\pi} rI(r,\varphi)\,drd\varphi . \tag{4.2}$$

Now, let us calculate the beam radius for a Gaussian laser beam in terms of the second-order moment. For a Gaussian laser beam having a radius w_0, equation (4.2) becomes

$$\left\langle w_r^2 \right\rangle = 2\frac{\int_0^{\infty} r^3 \exp\left(-\frac{2r^2}{w_0^2}\right)dr}{\int_0^{\infty} r \exp\left(-\frac{2r^2}{w_0^2}\right)dr} = 2\frac{\frac{1!}{2}\frac{w_0^4}{4}}{\frac{0!}{2}\frac{w_0^2}{2}} = w_0^2 . \tag{4.3}$$

And to calculate equation (4.3), we use the definite integral formula

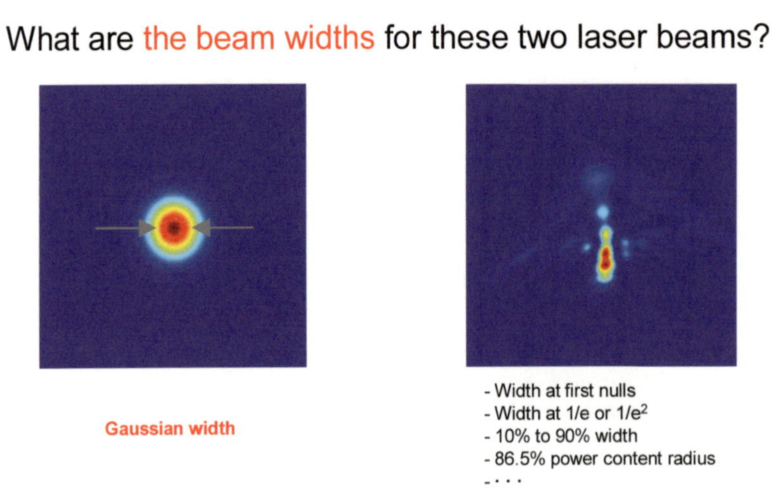

Fig. 16. How to define the beam width for an arbitrary laser beam profile.

$$\int_0^{\infty} x^{2n+1} \exp\left(-px^2\right) dx = \frac{n!}{2p^{n+1}} . \tag{4.4}$$

Thus, for a Gaussian laser beam, we can quickly see the beam radius defined by the second-order moment is again the Gaussian radius w_0. Thus, for a Hermite-Gaussian laser beam of order m and n, equation (4.1) becomes

$$\left\langle w_x^2 \right\rangle = (2m+1) w_0^2 \text{, and } \left\langle w_y^2 \right\rangle = (2n+1) w_0^2 , \tag{4.5}$$

where the following integral formulae are used to obtain the expression for equation (4.5):

$$\int_{-\infty}^{\infty} x^2 H_m\left(\sqrt{2}x/w_0\right) \exp\left(-2x^2/w_0^2\right) dx = 2^m \sqrt{\pi} m! \frac{w_0^3}{2^{3/2}} \left(m + \frac{1}{2}\right), \tag{4.6}$$

$$\text{and } \int_{-\infty}^{\infty} H_m\left(\sqrt{2}x/w_0\right) \exp\left(-2x^2/w_0^2\right) dx = 2^m \sqrt{\pi} m! \frac{w_0}{2^{1/2}} . \tag{4.7}$$

In the same way, we can calculate the beam radius for a Laguerre-Gaussian laser beam having an arbitrary order. For a Laguerre-Gaussian laser beam, for p and l, equation (4.2) becomes

$$\left\langle w_{r,pl}^2 \right\rangle = (2p+l+1) w_0^2 . \tag{4.8}$$

For an Airy beam profile ($I(r) = I_0\left[J_1^2(kar/f)/r^2\right]$), which can be obtained by focusing a flat-top laser beam, equation (4.2) results in

$$\left\langle w_r^2 \right\rangle = 2 \frac{\int_0^{\infty} r J_1^2(kar/f) dr}{\int_0^{\infty} \frac{J_1^2(kar/f)}{r} dr} = \infty . \tag{4.9}$$

The numerator in equation (4.9) continuously grows as the radius increases. Thus, as shown in equation (4.9), the radius is not defined for the Airy beam profile in terms of the second-order moment.

4.2 Theoretical determination of the beam quality factor

As mentioned in Section 4.1, the beam quality factor is closely related to the beam propagation. Thus, knowing the intensity distribution near the focal plane is required. Let us begin with equation (3.2) to derive a useful relation for the beam radius with respect to the distance. As shown in equation (3.2), the general expression of the electric field for a laser beam is

$$E_1(x_1, y_1, z_1) = A(z_1) \times \exp\left[-ik \frac{x_1^2 + y_1^2}{2q(z_1)}\right] \text{, and } \frac{1}{q(z_1)} = \frac{1}{R(z_1)} - i \frac{\lambda}{\pi w_1^2(z_1)} . \tag{3.2}$$

In the plane ($z_1 = z_0$) of the beam waist, $R(z_0) = \infty$ and $\frac{1}{q(z_0)}$ becomes $-i\frac{\lambda}{\pi\left\langle w_0^2(z_0)\right\rangle}$.

Then, after a distance z from z_0, the ABCD matrix is given by

$$\begin{bmatrix} A & B \\ C & D \end{bmatrix} = \begin{bmatrix} 1 & z \\ 0 & 1 \end{bmatrix}. \tag{4.10}$$

Next, by using the relation $\frac{1}{q(z)} = \frac{q(z_0)C + D}{q(z_0)A + B}$, we obtain

$$\frac{1}{q(z)} = \frac{1}{R(z)} - i\frac{\lambda}{\pi\left\langle w^2(z)\right\rangle} = \frac{1}{z + z_R^2/z} - i\frac{\lambda}{\pi\left\langle w_0^2\right\rangle\left[1 + (z/z_R)^2\right]}. \tag{4.11}$$

Again, the definition of the Rayleigh range ($z_R = \pi\left\langle w_0^2\right\rangle/\lambda$) is used to derive equation (4.11). From this equation, the beam radius with respect to the distance from the focal plane, z_0, can be expressed by

$$\left\langle w^2(z)\right\rangle = \left\langle w_0^2\right\rangle\left[1 + \frac{(z - z_0)^2}{z_R^2}\right]. \tag{4.12}$$

Note that equation (4.12) is satisfied for all laser beam modes. From equation (4.12), we know that the change in the size of the beam mode follows a quadratic function with respect to the distance z (after this point, the sign $\langle\ \rangle$ in equation (4.12) will be omitted for convenience). Then, by inserting equations (4.5) and (4.8) into equation (4.12), we can rewrite the beam radii for several laser beams as follows:
For the Hermite-Gaussian beam mode:

$$w_{x,m}^2(z) = W_{0x}^2 + (2m+1)^2\frac{\lambda^2}{\pi^2 W_{0x}^2}(z - z_0)^2 \text{ and } w_{y,n}^2(z) = W_{0y}^2 + (2n+1)^2\frac{\lambda^2}{\pi^2 W_{0y}^2}(z - z_0)^2. \tag{4.13}$$

For the Laguerre-Gaussian beam mode:

$$w_{pl}^2(z) = W_{0,pl}^2 + (2p+l+1)^2\frac{\lambda^2}{\pi^2 W_{0,pl}^2}(z - z_0)^2. \tag{4.14}$$

Here, W_{0x}^2, W_{0y}^2, and $W_{0,pl}^2$ are defined by $(2m+1)w_0^2$, $(2n+1)w_0^2$, and $(2p+l+1)w_0^2$, respectively. In addition, it should be noted that in real situations the laser beam can have arbitrary beam profile. In this case, it is convenient to generalize the beam radius with respect to the distance as follows:

$$w^2(z) = W^2 + M^4\frac{\lambda^2}{\pi^2 W^2}(z - z_0)^2 \tag{4.15}$$

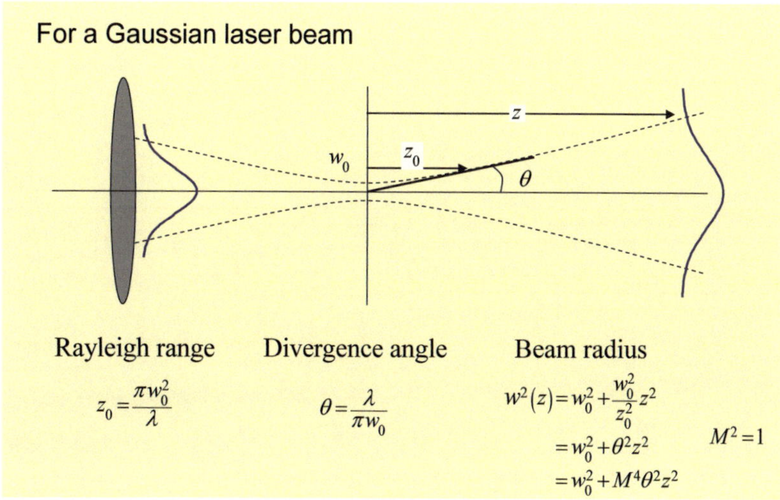

Fig. 17. Beam quality factor, beam radius, divergence angle, and Rayleigh range.

where M^2 is the beam quality factor. From equations (4.13) and (4.14), we can quickly recognize that, for the Hermite-Gaussian and Laguerre-Gaussian beam modes, the beam quality factors are given by $(2m+1)$ and $(2p+l+1)$.

Thus, by using equations (4.1) and (4.2), we can easily derive the property of the radius-divergence product.

$$\left[W^2\theta^2\right] = M^4 \frac{\lambda^2}{\pi^2} \tag{4.16}$$

where θ is the beam divergence defined in Fig. 17. Because the beam quality factor is invariant, it is not affected by the focusing optic once the focusing optic is ideal. In other words, using the invariant property of the beam quality factor, the radius-divergence product remains constant even under different focusing conditions and is given by

$$W_0\theta_0 = W\theta = M^2 \frac{\lambda}{\pi} = \frac{W^2}{Z_0}. \tag{4.17}$$

Finally, from equation (4.17), the beam quality factor for an arbitrary laser beam can be directly calculated by using the second-order moments $W_{x,y}^2$ at the beam waist and the Rayleigh range $Z_{x,y}$:

$$M_x^2 = \frac{\pi W_x^2}{Z_x \lambda} \text{ and } M_y^2 = \frac{\pi W_y^2}{Z_y \lambda}. \tag{4.18}$$

Now, let us calculate the beam quality factor by calculating the beam radii at different positions near the focal plane and the Rayleigh range from the beam radii. In calculating the beam radius, we employ the Fourier transform approach to obtain the intensity distribution, and from the calculated intensity distribution we can calculate the second-order moment as the beam radius. A Gaussian laser beam TEM_{00} is considered first. In this case, the Gaussian

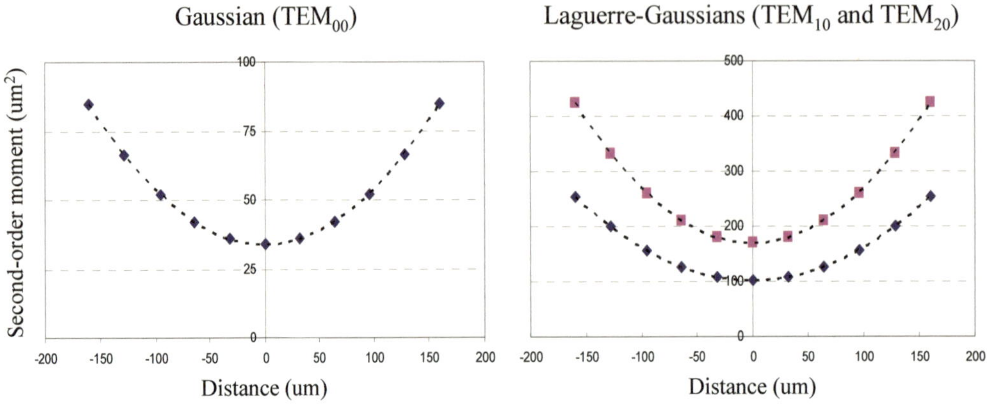

Fig. 18. Calculation of beam radius and the Rayleigh range for Gaussian and higher-order Laguerre-Gaussian (10 and 20) laser beam modes.

width w_1 of the input laser beam was assumed to be 10.337 mm. We also assume that a laser beam having no wavefront aberration is focused by an ideal focusing optic having a focal length of 238 mm, and that the size of the clear aperture of the focusing optic is 70 mm. The large aperture size ensures the extent of the laser beam to be enclosed in the focusing optic. However, the choice of focal length is not important in determining the beam quality factor. The theoretical Gaussian width of the focal spot was about 5.87 μm (calculated from equations (3.10) or (3.39)) at a wavelength of 0.8 μm. In the Fourier transform calculation, the Gaussian width of the focal spot was about 5.82 μm, which is very close to the theoretical width. The calculated second-order moment at the beam waist W^2 and the Rayleigh range Z_0 were about 33.88 μm² and 130.15 μm, respectively. As expected, the second-order moment for the Gaussian beam was given by the square of the Gaussian width. Figure 18 shows the calculated second-order moment for the Gaussian and Laguerre-Gaussian laser beam modes. From the calculated second-order moment and the Rayleigh range, the beam quality factor for a Gaussian beam having no wavefront aberration was calculated as 1.02 using equation (4.18), which is very close to the expected value of 1.00. Subsequent calculations with higher-order Gauss-Laguerre modes showed that the beam quality factors for TEM_{10} and TEM_{20} were 3.04 and 5.08, respectively. The calculated beam quality factors for higher modes were also very close to $(2p+l+1)$, as obtained from the analytical approach. It is thought that the small 1 to 2% error in the beam quality factor comes from calculating the second-order moment.

The beam quality factor for a super-Gaussian beam profile was also investigated because the super-Gaussian beam profile represents the flat-top profile of a real laser beam. Figure 19 shows the normalized intensity profiles and the calculated beam quality factors for Gaussian and super-Gaussian laser beams.

The super-Gaussian orders are 2, 4, 6, 8, 10, 15, 20, 30, and 40 in the calculation. As shown in Fig. 19, the beam quality factor increases with the super-Gaussian order; the beam quality factors for the 30th- and 40th-order super-Gaussian beam profiles are 3.94 and 4.55, respectively. The increase in the beam quality factor with the super-Gaussian order is due to the energy or intensity spread in the focal plane, caused by Airy rings.

With the super-Gaussian intensity profile and a serious wavefront aberration, the intensity distributions at positions -128, -64, 0, 64, and 128 μm from the focal plane are calculated as

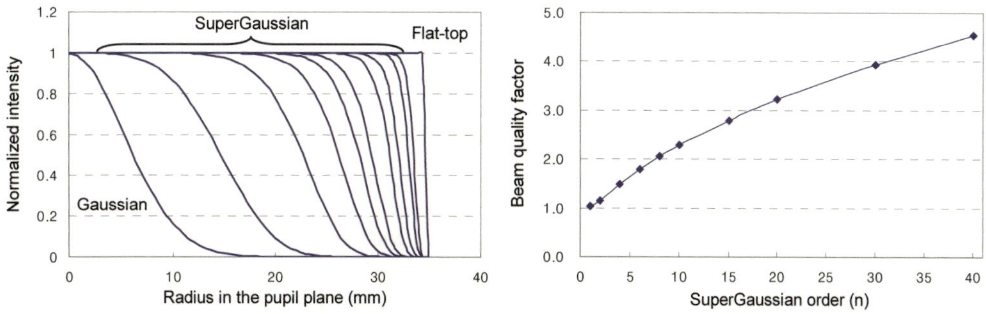

Fig. 19. Intensity profiles of super-Gaussian laser beam modes and the calculated beam quality factors for various super-Gaussian orders.

shown in Fig. 20. The intensity distributions show multiple spikes, and the energy is not well concentrated in a tiny spot because of the wavefront aberration. For these intensity distributions, calculating the second-order moment makes it possible to define the radius of the intensity distribution. In the figure, the blue diamonds and pink squares are the calculated second-order moments $W_x^2(z)$ and $W_y^2(z)$ in the horizontal and vertical directions, respectively. The use of an ideal focusing optic having a focal length of 0.238 m is assumed in this calculation. Again, the dashed line is the second-order polynomial curve fit. The calculated second-order moments for the aberrated laser beam fit the second-order polynomial function well, with the difference between the sagittal and tangential focuses ($z_x - z_y$) being due to astigmatism in the laser pulse. The calculated beam quality factors M_x^2 and M_y^2 for the aberrated laser beam were about 10.1 and 9.9 in the horizontal and the

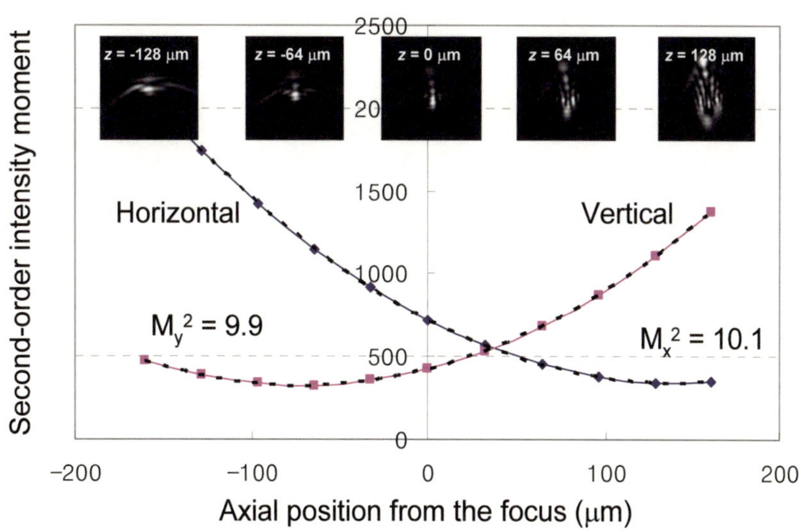

Fig. 20. Calculated intensity distributions and beam radii calculated from the second-order moment. The beam quality factor is then calculated from the beam radii through the second-order polynomial fit.

vertical directions, respectively. The calculated beam quality factors were about 2.6 and 2.5 times higher than those obtained using the same intensity profile having no wavefront aberration. Thus, the beam quality factor can be computationally determined by measuring the intensity of an incoming laser beam and calculating the beam radii at various locations near the focal plane; indeed, calculating the beam radii at various locations is the same procedure as used in the experiment.

4.3 Experimental determination of the beam quality factor

In order to experimentally determine the beam quality factor, measuring the beam radius is first required. For this task, there is a specific instrument, generally called the M-square meter, used to evaluate the beam quality factor. As shown in Fig. 21, the M-square meter basically focuses an incoming laser beam with a relatively long focal length optic. Images at various planes near the focal plane are then recorded by a CCD camera, with beam radii subsequently calculated from the recorded images. Next, from the measured images and the beam radii at various planes, the beam waist and the Rayleigh range can be calculated. Once the beam waist and the Rayleigh range are calculated, the beam quality factor is again given by equation (4.18). Figure 22 shows the image captured from a commercial M-square meter, in which the measured radii at various locations in the M-square meter can be clearly seen. From the measured radii data, the beam quality factor is calculated.

To define the beam diameter, several alternative measurement methods are provided by the International Standardization Office (ISO). In these methods, the beam diameter is defined from the power measurement, instead of the captured image of the laser beam; thus, proper attenuation and accurate calibration of the power meter are necessary. Several measurement methods include: 1) the variable diaphragm method, 2) the moving knife-edge method, and 3) the moving slit method. For the variable diaphragm method, the incoming laser beam can be elliptical but the ratio between the major and the minor axis should be within 1.15:1. The variable diaphragm must be aligned to the beam center to within 10% of the beam diameter. For the knife-edge method, the distance between the energy levels of 10% and 90% should be measured and then multiplied by 1.56. For the moving slit method, a slit having 1/20 width of the laser beam should be used in order to avoid convolution corrections. And, the width between the positions where irradiance has reached 13.5% of the maximum irradiance should be measured.

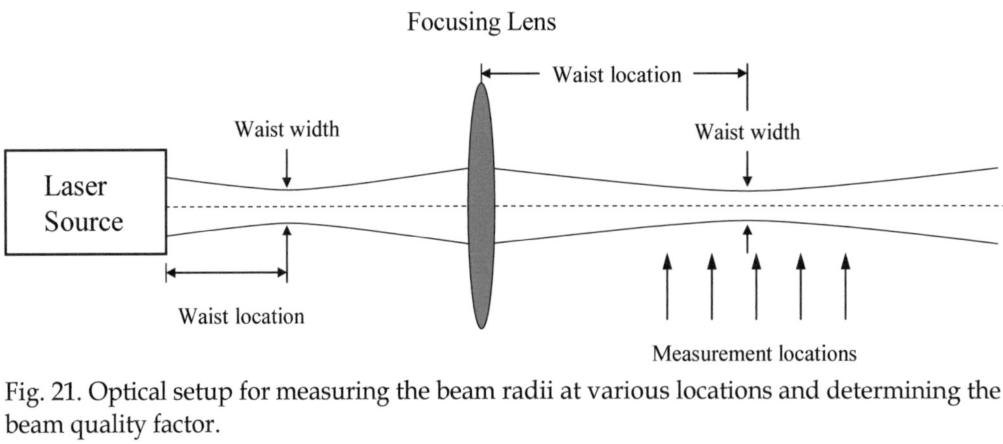

Fig. 21. Optical setup for measuring the beam radii at various locations and determining the beam quality factor.

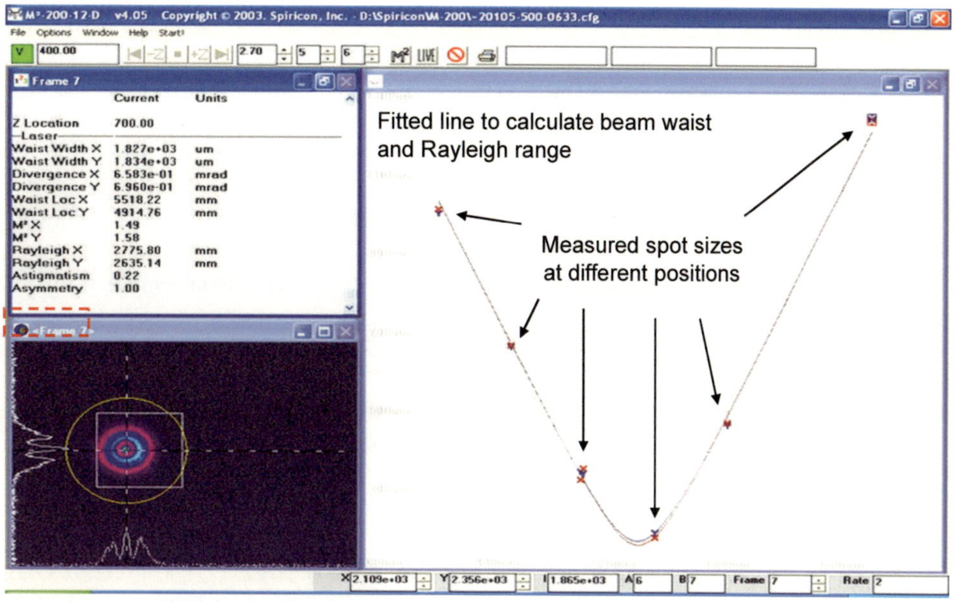

Fig. 22. Measured beam radius and beam quality factor obtained via the M-square meter.

5. Conclusion

The laser pulse has temporal and spatial beam characteristics. The temporal characteristics are related to the pulse duration and the temporal shape of the laser pulse. The spatial characteristics contain the spatial intensity (or electric field) distribution defined by modes and the phase profile of the electric field defined by the wavefront. In order to fully describe the laser pulse in time and space, spatial characteristics such as mode and wavefront should be precisely characterized, in addition to the temporal characteristics. In this Chapter, we describe the intensity distributions as the spatial characteristics of laser beams. In particular, the intensity distribution in the focal plane of a focusing optic or in the vicinity of the focal plane is very important in applications using laser systems because it determines the laser-matter interaction process. We started by deriving the expression for the laser beam intensity and explained the physical background for deriving the mathematical expression of the intensity distribution in the focal plane and in the vicinity of the focal plane. Modification of the intensity distribution with wavefront in the laser beam then followed. Next, the laser pulse was fully characterized in the spatio-temporal domain. The full characterization of the laser pulse quickly introduced a useful concept called the beam quality (or propagation) factor M^2. We also introduced standard methods for experimentally determining the beam quality factor and compared them to an indirect method to determine the beam quality factor based on information pertaining to the intensity and wavefront of the laser pulse. All these topics were discussed under the assumption of the Fraunhofer diffraction region.

6. References

[1] A. E. Siegman, *Laser* (University Science Books, Mill Valley, 1986), Chap. 16.

[2] M. Born and E. Wolf, *Principles of Optics* (Cambridge University Press, Cambridge, 1999), Chap. 8.
[3] P. Milloni and J. H. Eberly, *Lasers* (John Wiley & Sons, New York, 1988), Chap. 14.
[4] O. Svelto and D. C. Hanna, *Principles of Lasers* (Springer, 1998), Chap. 4.
[5] B. E. A. Saleh and M. C. Teich, *Fundamentals of Photonics* (John Wiley & Sons, New York, 1991), Chap. 3.
[6] N. Hodgson and H. Weber, *Laser Resonators and Beam Propagation* (Springer, 2005), Chaps. 2 and 5.
[7] I. S. Gradshteyn and I. M. Ryzhik, *Table of Integrals, Series, and Products* (Academic Press, New York, 2007), Chap. 6.
[8] G. B. Arfken and H. J. Weber, *Mathematical Methods for Physicists* (Academic Press, San Diego, 2001), Chap. 11.
[9] D. Malacara, *Optical Shop Testing* (John Wiley & Sons, New Jersey, 2007), Chaps. 11 and 13.
[10] R. K. Tyson, *Principles of Adaptive Optics* (Academic Press, Boston, 1998), Chap. 5.
[11] A. G. Fox and T. Li, "Resonant Modes in a Maser Interferometer," Bell Sys. Tech. J., 40, 453-488 (1961).
[12] H. Kogelnik and T. Li, "Laser Beams and Resonators," Proc. IEEE, 54, 1312-1329, (1966).
[13] W. H. Carter, "Spot size and divergence for Hermite Gaussian beam of any order," Appl. Opt., 19, 1027-1029 (1980).
[14] J. Durnin, "Exact solutions for nondiffracting beam. I. The scalar theory," J. Opt. Soc. Am. A, 4, 651-654 (1987).
[15] F. Gori, G. Guattari, and C. Padovani, "Bessel-Gauss Beams," Opt. Commun. 64, 491-495 (1987).
[16] D. Wright, P. Greve, J. Fleischer, and L. Austin, "Laser beam width, divergence and beam propagation factor – an international standardization approach," Opt. Quantum Electron. 24, S993-S1000 (1992).
[17] T. M. Jeong, D.-K. Ko, and J. Lee, "Deformation of the Focal Spot of an Ultrashort High-Power Laser Pulse due to Chromatic Aberration by a Beam Expander," Journal of the Korean Physical Society, 52, 1767-1773, (2008).
[18] T. M. Jeong and J. Lee, "Accurate Determination of the Beam Quality Factor of an Aberrated High-Power Laser Pulse," Journal of the Korean Physical Society, 55, 488-494, (2009).

Part 3

Laser-matter Interactions

12

Intensity Effects and Absolute Phase Effects in Nonlinear Laser-Matter Interactions.

Sándor Varró
Research Institute for Solid State Physics and Optics of the Hungarian Academy of Sciences Hungary

1. Introduction

The generation of short pulses of electromagnetic radiations rely in many cases on nonlinear interactions, and, on the other hand, the resulting sources may have very large intensities, and can induce high-order nonlinear processes. Of course, these two faces of light-matter interactions are intimately connected, but they can usually be treated separately, depending on the emphasize put on one side or on the other. The subject of the present chapter belongs to the physics of „strong-field phenomena" taking place in laser-matter interactions, and it partly concerns quantum optics, too. A brief overview of theoretical methods for treating nonlinear light-matter interaction is given, and some characteristic examples are presented. We review the basic classical and quantum approaches in simple terms, by possibly avoiding involved mathematical derivations. We shall attemp to give a clear conceptual framework and illustrative examples, on the basis of which the main characteristics of the intensity and phase effects showing up in various extreme nonlinear light-matter interactions can be understood. We shall limit the discussion mostly to the single-electron picture, and genuine many-body effects will not be considered. On „strong-field phenomena" we mean here those phenomena whose main characteristics are governed by higher powers than linear of the laser intensity. Two typical examples for such processes are the nonlinear photoelectric effect and high-harmonic generation induced by strong laser fields. The former one has been the subject of an extensive research already from the beginning of the sixties of the last century, starting with the works of Keldish (1965) and Bunkin & Fedorov (1965). As for the later developments, see Farkas (1978), Krause et al. (1992), the book by Delone and Krainov (1994), and a recent review on multiphoton surface photoelectric effect by Ferrini et al. (2009). If one photon energy is not enough to ionize an atom (or deliberate an electron from a metal surface), then, in order to have a non-negligible ionization probability per unit time for many-photon absorptions, one needs a high-power source of the radiation. The high-harmonics of an incoming optical radiation stem from the laser-induced nonlinear polarization of the active constituents of matter, which, may for example be bound dipoles of atoms, molecules, solids or free electons in the conduction band of metals or in beams propagating in free space. Another wide class of strong-field phenomena consists of the laser-assisted processes, like, for instance x-ray scattering on atomic electrons in the presence of a high-power laser radiation. Here, in contrast to the induced processes, the primary process can of course take place without any nonlinearity, but the additional interaction modifies the spectrum of the scattered radiation, due to the appearence of side-bands separated by the laser frequency (see e.g. Puntajer & Leubner, 1990). The relative strengths of these side-bands are nonlinear functions of the laser intensity. In all the above-mentioned examples a considerable spectral broadening of the signals may take place, which results in temporal compression, i.e. short pulses are formed. Of course, the appearance and the quality of these pulses crucially depend on the phase relations of its Fourier components. In general, one may say that the generation of short pulses of electromagnetic radiation are based on nonlinear laser-matter interactions. As for the theoretical description of the mentioned, mostly very high-order processes, the greatest challenge is to work out non-pertubative methods, on the basis of which the interactions with the strong fields are taken into account up to arbitrary orders. This is needed because the perturbation theory may break down for large intensities, even to that extent that terms like 'perturbation' or 'higher-order', in fact, simply loose their usual contents. The optical tunneling offers itself as a good example for such a situation, when the electrons within one optical cycle leave the binding potential which is broken down by the high electric field of the laser radiation (Keldish (1965), Bunkin & Fedorov (1965)). Here one encounters with the fundamental question of wave-particle duality of both the electromagnetic radiation and the electron (see e.g. Büttiker & Landauer (1982) and Chiao (1998)). Another conceptually and also practically important question is to what extent can one control the phase of the ultrashort pulses, and, at all, what are the ultimate limits of phase stabilization in a given generating process? The systematic quantitative analysis of these problems is still missing, at least concerning the quantum uncertainties of the phases. One may expect that the usual high-intensity laser fields, being in a highly populated coherent state, can certainly be well represented in a satisfactory manner in the frame of external field approximation, i.e. in terms classical Maxwell fields of definite amplitudes and phases, or in terms of classical stochastic processes. On the other hand, in the generation of extreme pulses, like sub-femtosecond or attosecond pulses, it is an open question whether the fully quantum or the semiclassical description deliver the correct interpretation.

Within the limited space at our disposal, it cannot be our purpose to give an historical overview, and it would be impossible to review the presently ongoing research on strong-field phenomena. Besides studying the important early works, like Ritus & Nikishov (1979), Faisal (1987), Gavrila (1992), Mittleman (1993) and Delone & Krainov (1994), the interested reader can find further references in recent reviews, dealing with various aspects. We refer the reader e.g. to Brabec & Krausz (2000), Ehlotzky (2001), Agostini & DiMauro (2004), Salamin et al. (2006), Mourou et al. (2006) , Krausz & Ivanov (2009), Ehlotzky et al. (2009) and Gies (2009). In fact, the conceptual framework of the theory of strong-field phenomena has not changed very much in the last couple of decades. The main point in the theory has long been the following. In contrast to the usual texbook approach, where the light (or electromagnetic radiation, in general) is treated as a perturbation, the essence of all new methods is just to turn around this scheme, and consider the strong field as the dominating agent in the interaction. Besides direct numerical, *ab initio* calculations, the analytic approaches still play a very important role, because only this can give an intuitive understanding of the mentioned processes, and help to find the really important parameters to be controlled in the experiments. In the meantime, the laser technology has undergone a very fast development, which made it possibile to generate extreme radiation fields in well-controlled ways. Many processes have become accessible for direct experimental basic research, whose results, at the same time, have many potential applications in other branches of science and technology. We shall not touch this aspects, either. In the present chapter we attemp to give a summary of the characteristics of the simplest basic nonlinear processes taking place in strong-field laser-matter interactions. Based partly on our earlier and recent works, we shall present some typical examples, which may serve as guides towards a possibly intuitive insight into the broad field of research on strong-field light-matter interactions.

In Sections 2 and 3, on the occasion of a summary on nonlinear Thomson scatteing, high-harmonic generation and multiphoton ionization, we introduce the basic parameters (dimensionles intensity parameter, the ponderomotive energy shift, and the Keldish-γ or adiabaticity parameter) which govern the characteristics of the processes under discussion. The carrier-envelope phase effect will be studied in simple terms of classical electrodynamics. We shall briefly discuss the multiphoton analogon of the Kramers-Heisenberg dispersion formula. In Section 4 the nonlinear effect of the laser-induced oscillating double-layer potential on metal surfaces and the x-ray generation in the presence of a static homogeneous electric field will be touched. Here we shall introduce a new nonlinearity parameter, that seems to be a key element in the theoretical analysis of many recent experiments showing unexpectedly high laser-induced nonlinearities. We shall briefly discuss the question of optical rectification and the generation of quasi-static wake-fields, whose polarization can be manipulated by changing the carrier-envelope phase difference. Besides, the analogy between the usual Fourier-synthesis of the high-order harmonics stemming from a nonlinear scattering, and the interference of above-threshold electronic de Broglie wave components shall be emphasized. This latter effect may give a possibility for generation high-current attosecond electron pulses. In Section 5 some quantum statistical properties of short electromagnetic radiation pulses will be analysed.

2. Classical description of basic strong field phenomena. Carrier-envelope phase difference effects in ultrashort pulse interactions

In the present section we shall summarize the most important elements of the conceptual framework in which nonlinear laser-matter interactions can be theoretically considered. At the same time, this part serves as an introduction of the important parameters in terms of which the particular natures of these various processes can be quantified. In Table 1. we show the possible combinations of theories in the descriptions of photon-electron interactions. For short, on the word 'photon' we mean electromagnetic radiation in general.

The most fundamental illustrative cases are the nonlinear Thomson and Compton scattering, which in the cleanest way show some general basic characteristics of the nonlinear photon-electron interactions (see e.g. Brown & Kibble (1964), Leubner (1978, 1981), Bergou & Varró (1981b), Gao (2004), Lan et al. (2007), Boca & Florescu (2009), Mackenroth et al. (2010)). According to the description displayed in cell No. 2 in Table 1., consider the classical equation of motion of a point electron of charge $-e$ and rest mass m interacting in vacuum with a linearly polarized plane wave of electromagnetic radiation:

$$m\,d(\gamma\mathbf{v})/dt=-e[\mathbf{E}_0+(\mathbf{v}/c)\times\mathbf{B}_0],\quad \mathbf{E}_0=\varepsilon_0 F(\eta),\quad \eta\equiv t-\mathbf{n}_0\cdot\mathbf{r}(t),\quad \mathbf{B}_0=\mathbf{n}_0\times\mathbf{E}_0\,, \tag{1a}$$

$$F(\eta)=-\partial^2\Pi(\eta)/c^2\partial t^2,\quad \Pi(\eta)=(c/\omega_0)^2F_0 f_0(\eta)\cos(\omega_0\eta+\varphi_0),\quad f_0(\eta)=g_0(\eta)\equiv\exp(-\eta^2/2\tau_0^2), \tag{1b}$$

Electron \ Photon	Trajectory, Rays (Geometric Optics)	Field (Maxwell Theory)	Quantized Field (Photon Picture)
Trajectory, Current (Point Mechanics)	1.	2. Classical Electron Theory	3.Classical Current, Poisson Photon
Field, Current (Wave Mechanics)	4.	5. Semiclassical Description	6.Quantum Current, General Photon
Quantized Field (Negaton-Positon)	7.	8.	9. Quantum Electrodynamics

Table 1. Possible descriptions of photon-electron interactions. The most widely applied descriptions are displayed in cells Nos. 2. and 5. Concerning the electron motion, in each cells the relativistic description is also included. The cells Nos. 1., 4. and 7. are empty, because the direct interaction of point charged particles with rays of radiation are not considered.

where $\mathbf{v}(t)=d\mathbf{r}(t)/dt$ and c are the velocity of the electron and of the light field in vacuum, respectively, and $\gamma=[1-(v/c)^2]^{-1/2}$ is the usual relativistic factor. The amplitude of the electric field strength, the polarization vector and propagation vector are denoted by F_0, $\boldsymbol{\varepsilon}_0$ and $\mathbf{n}_0$, respectively. In Eq. (1b) we have introduced the Hertz potential $\Pi(\mathbf{r},t)=\Pi(\eta)$, and the special case of a quasi-monochromatic wave of mean circular frequency $\omega_0=2\pi/T_0$ having a pulse envelope function $f_0(\eta)$, and the *carrier-envelope phase difference* (CEPD) φ_0. This latter quantity is often simply called *absolute phase*. In general, the pulse envelope function $f_0(\eta)$ has its maximum value 1 at $\eta=0$, and vanishes for $\eta\to\pm\infty$ (as an example, a Gaussian envelope function g_0 of full width at half maximum (FWHM) $2\tau_0(\log 2)^{1/2}$ is shown in the last equation of Eq. (1b)). The first equation of Eq. (1a) is a highly nonlinear second order ordinary differential equation for the electron's position $\mathbf{r}(t)$, because through the variable η the electromagnetic field strengths in their arguments contain this trajectory in a complicated manner. Thus, in general, we do not expect a simple harmonic motion, even if we set the envelope function constant ($f_0=1$), when we are dealing with an ideal monochromatic plane wave. This special case was first studied long ago by Halpern (1924) in the context of a possible description of Compton scattering. Later it turned out that the seemingly complicated equation in Eq. (1a) can be solved analytically for an arbitrary functional form of $f(\eta)$. It can be shown that $d\eta/d\tau=\gamma(1-\mathbf{n}_0\cdot\mathbf{v}/c)$ is a constant of motion (where $d\tau=dt/\gamma$ denotes the proper time element) thus, at the position of the electron the complete argument η of the wave is a linear function of the proper time of the electron. As an example, take an x-polarized wave ($\boldsymbol{\varepsilon}_0=(1, 0, 0)$) propagating in the positive z-direction ($\mathbf{n}_0=(0, 0, 1)$). Then the equation of motion in Eq. (1a) can be brought to the set of equations

$$ma\,d^2x/d\eta^2=-eF(\eta),\quad a\equiv\gamma(1-v_z/c)=\text{const.},\quad d^2z/d\eta^2=(d/2c\,d\eta)(dx/d\eta)^2,\quad d^2y/d\eta^2=0, \tag{2a}$$

$$\gamma mc^2=mc^2+m(dx/d\eta)^2/2. \tag{2b}$$

The first (still exact) equation in Eq. (2a) is completely analogous to a simple Newton equation in a homogeneous time-varying field, which can be immediately integrated, and then, from the third equation, the function $z(\eta)$ can be determined. In Eq. (2b) we have shown the total energy of the electron, which naturally splits to the sum of the rest energy and a term analogous to the nonrelativistic kinetic energy. One has to keep in mind, however, that the trajectory still cannot be expressed by the true time t in a closed form. It is possible to give analytic expressions for $t(\eta)$, $x(\eta)$, $z(\eta)$ as functions of the retarded time η (or, equivalently, of the proper time τ). The „initial data" on the light-like hyperplane η=const. cannot directly correspond to true initial values, because the prescription $t-z/c$=const. cannot account for a true separation of the particles from a real interaction region, since this should be specified on the space-like plane t=const. (in order to secure causality). Concerning the solution of this problem see Varró & Ehlotzky (1992). In the *nonrelativistic limit* ($v/c<<1$, $\gamma\to 1$, $a\to 1$), in *dipole approximation* ($\eta\to t$) and *in a stationary field* we have the simple results

$$md^2x/dt^2=-eF_0\cos(\omega_0 t+\varphi_0),\quad dx/dt=-v_{osc}\sin(\omega_0 t+\varphi_0)+v_0,\quad x=x_{osc}\cos(\omega_0 t+\varphi_0)+v_0 t+x_0, \quad (3a)$$

$$v_{osc}/c=\mu_0,\quad x_{osc}=\mu_0\lambda_0/2\pi,\quad \mu_0\equiv eF_0/mc\omega_0,\quad \mu_0=0.8528\times 10^{-9} I_0[\mathrm{W/cm^2}]^{1/2}\lambda_0[\mu\mathrm{m}], \quad (3b)$$

$$\mu_0= 1.0426\times 10^{-9}\,[I_0/(\mathrm{W/cm^2})]^{1/2}/[h\nu_0/\mathrm{eV}],\quad \mu_0=(I_0/I_c)^{1/2},\quad I_c=cE_c^2/8\pi,\quad E_c=2\pi e/r_0\lambda_0, \quad (3c)$$

where v_{osc} and x_{osc} are the velocity and coordinate amplitudes, respectively, of the electron oscillating in the laser field of intensity I_0 (measured in W/cm^2) and of central wavelength $\lambda_0=2\pi/k_0=2\pi c/\omega_0=c/\nu_0$ (measured in microns, $\mu m=10^{-4}$cm). The *dimensionless intensity parameter* μ_0 introduced in the third equation of Eq. (3b) is the ratio of the amplitude of the velocity oscillation to the velocity of light. Though μ_0 is a classical parameter, in Eq. (3c) we have expressed it in terms of the associated photon energy $h\nu_0$, too, where h denotes the Planck constant. Besides, we have introduced the critical intensity I_c (for which $\mu_0=1$) associated to the critical field strength E_c, where $r_0=e^2/mc^2=2.818\times 10^{-13}$cm is the classical electron radius. The work $eE_c\lambda_C/2\pi=h\nu_0$ done on the electron by this critical electric field (assumed here static) along the Compton wavelength $\lambda_C=h/mc= 2.426\times 10^{-10}$cm just equals the central photon energy $h\nu_0$. According to Eq. (3c), for a laser field of mean photon energy $h\nu_0$=1eV the critical intensity is $I_c=10^{18}$W/cm^2. The parameter μ_0 quantifies the limits of validity of the approximations just applied. If μ_0 approaches unity, then v_{osc} gets close to the velocity of light and the nonrelativistic description is not valid any more. At the same time, the amplitude of the position oscillations becomes comparable with the wavelength, and the dipole approximation breaks down, too, thus the term $\mathbf{n}_0\cdot\mathbf{r}/c$ in the retarded time η cannot be neglected in the equation of motion. Just this kind of term is responsible for the generation of higher-order harmonics by a free electron in the presence of a strong laser field. In the nonrelativistic regime, under the action of the laser field, a pure harmonic oscillation is superimposed on the free inertial motion of the electron. According to classical electrodynamics (Jackson (1962)), for this *simple harmonic* motion the average power radiated per unit solid angle in the nth *higher-harmonic* is:

$$dP_n/d\Omega=(e^2\omega_0^2/2\pi c)(n\tan\theta)^2[\,J_n(\mu_0 n\cos\theta)\,]^2,\quad (e^2\omega_0^2/2\pi c)=\alpha(h\nu_0/T_0),\quad \alpha\equiv 2\pi e^2/hc, \quad (4)$$

where J_n denotes ordinary Bessel function of first kind of order n (see e.g. Gradshteyn and Ryzhik (2000)), and θ is the angle between the propagation direction of the emitted radiation with respect to the direction of oscillation (here: x-axis). In order to make the numerical estimates easier, in Eq. (4) we have also introduced the Sommerfeld fine structure constant $\alpha\approx 1/137$. For instance, for a Ti:Sapphire laser the total power emitted (exclusively to the fundamental frequency) by a single electron becomes ~ $\alpha\mu_0^2(h\nu_0/T_0)\approx\alpha\mu_0^2\times 10^{-4}$ Watt. This is because in the nonrelativistic case under discussion, we have $\mu_0 << 1$, i.e. the argument of the

Bessel function is much smaller than its order n = 1, 2, ..., and the approximation $J_n(nz) \approx (nz/2)^n/n! \approx (z/2e)^{-n}$ is essentially exact. The ratio of the partial powers of the consecutive harmonics is to a good accuracy $(dP_{n+1})/(dP_n) \approx (\mu_0\cos\theta/2e)^2 \lll 1$, thus practically no harmonics are produced, except for the fundamental one. This simple example shows that the higher-harmonic generation on a free electron is an inherently relativistic effect.

In order to illustrate intensity and absolute phase effects in the relativistic regime, we present the solution of the initial value problem of the general equation in Eq. (2) for a stationary field $F(\eta)=F_0\cos(\omega_0\eta+\varphi_0)$. First, let us specify the position and velocity of the electron at the time instant t=0, in the case when $\varphi_0=0$, such that $x(0)=(dx/dt)(0)=z(0)=(dz/dt)(0)=0$, thus $\eta(0)=0$. In accord with an early work by Halpern (1924), on x-ray scattering, we obtain

$$x=\mu_0(\lambda_0/2\pi)[1-\cos(\omega_0\eta)],\quad z=(\mu_0/2)^2(\lambda_0/4\pi)[2\omega_0\eta+\sin(2\omega_0\eta)],\quad 2\omega t=2\omega_0\eta+\beta\sin(2\omega_0\eta), \tag{5a}$$

$$\omega\equiv \omega_0/[1+(\mu_0/2)^2],\quad \beta\equiv(\mu_0/2)^2/[1+(\mu_0/2)^2]<1,\quad z(t)=\beta ct-(\lambda_0/2\pi)\Sigma_k[J_k(k\beta)/k]\sin 2k\omega t. \tag{5b}$$

The summation with respect to k in the last equation of Eq. (5b) runs through all the harmonic indeces k=1, 2, ..., and J_k denotes the ordinary Bessel function of first kind of order k. From the parametric representations of $x(\eta)$ and $z(\eta)$ in Eq. (5a) one can immediately derive the well-know *„figure-8 shape"* for the electron trajectory if one transforms out the uniform motion in the longitudinal z-direction (propagation direction of the ideal laser field). Of course, the modulus of the velocity components are smaller than c for arbitrary high intensities. On the other hand, the energy in Eq. (2b) becomes $\gamma mc^2=mc^2[1+(\mu_0/2)^2(1-\cos(2\omega_0\eta)]$, where the term $(\mu_0/2)^2mc^2\equiv U_p$ is usually called *ponderomotive energy shift*. The velocity of the longitudinal *drift motion* βc approaches c from below in the extreme case when $\mu_0 \ggg 1$, even if the initial velocity is zero. The frequency ω of the electon's fundamental oscillation in the laboratory frame is down-converted by a factor of $[1+(\mu_0/2)^2]$, but, on the other hand, the „renormalized spectrum" becomes very broad for large enough intensities. Really, by applying the asymptotic formula 8.455.1 of Gradshteyn & Ryzhik (2000), $\pi J_n(x) \approx [2(n-x)/3x]^{1/2}K_{1/3}\{[2(n-x)]^{3/2}/3x^{1/2}\}$, where $K_{1/3}$ denotes the modified Hankel function. With the help of this result we have $[J_n(n\beta)]^2 \sim \exp(-n/n_c)$ with $n_c = (3/8)^{1/2}(\mu_0/2)[1+(\mu_0/2)^2] \ggg 1$ is the *critical harmonic number*. As for the longitudinal motion under discussion, the cut-off frequency $\omega_c=n_c\omega$ is of the order of $\mu_0\omega_0$, which may reach the x-uv region of the electromagnetic spectrum. For example, if μ_0=60 we have for the critical number $n_c \approx 16500$, and the spectrum starts from $\omega\approx \omega_0/900$ and ends around $\omega\approx 30\times\omega_0$, covering terahertz and uv radiations as well.

The explicit calculation becomes much more complicated if the phase φ_0 is nonzero. The qualitative difference in this case is that, besides the oscillation around the position x=0, the transverse motion (along the polarization) aquires a drift $x_1(\sin\varphi_0)t$, too. The longitudinal motion is similar to that we have encountered in Eq. (5b), but now the drift part depends on the initial phase:

$$z(t)=\beta c(1+2\sin^2\varphi_0)t-(\lambda_0/2\pi)\Sigma_k(A_k\cos 2k\omega' t+B_k\sin 2k\omega' t), \tag{6}$$

where the Fourier coefficients A_k and B_k are combinations of Bessel functions, whose explicite form is not interesting here. Instead of proceeding further with the analytic formulae, we give in Figs. 1a-d and Figs. 2a-b some typical trajectories of an electron interacting with a strong laser pulse.

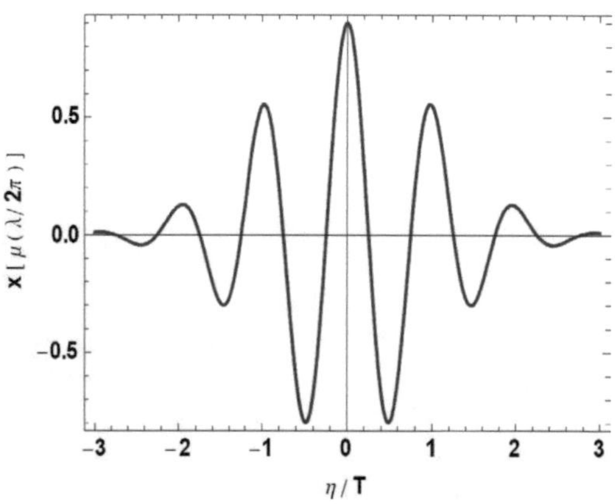

Fig. 1a. Shows the parametric variation of the electron's position along the polarization vector of the strong two-cycle laser field. We have taken a Gaussian pulse displayed in the last equation of Eq. (1b) with φ_0=0, and the intensity parameter was set μ_0=0.9, corresponding to the critical intensity $I_0 \approx 10^{18}$W/cm² for an optical field with $h\nu_0$=1eV.

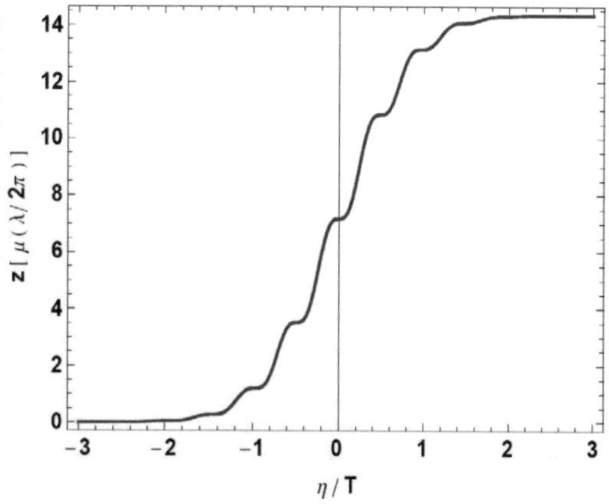

Fig. 1b. Shows the parametric variation of the electron's position along the propagation vector of a strong two-cycle laser field. This figure shows that the longitudinal motion has a systematic drift, whose velocity is essentially βc, as has been given in Eq. (5b). The parameters are the same as in Fig. 1a.

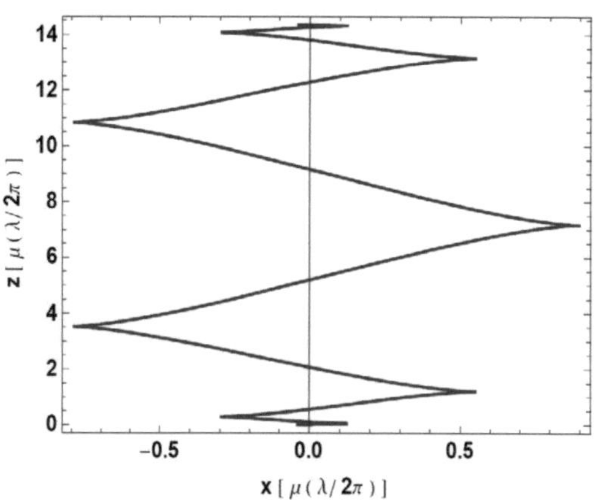

Fig. 1c. Shows the parametric variation of the electron's position along the polarization vector of the strong two-cycle laser field. We have taken a Gaussian pulse displayed in the last equation of Eq. (1b) with φ_0=0, and the intensity parameter was set μ_0=0.9, corresponding to the critical intensity $I_0 \approx 10^{18}$W/cm² for an optical field with $h\nu_0$=1eV.

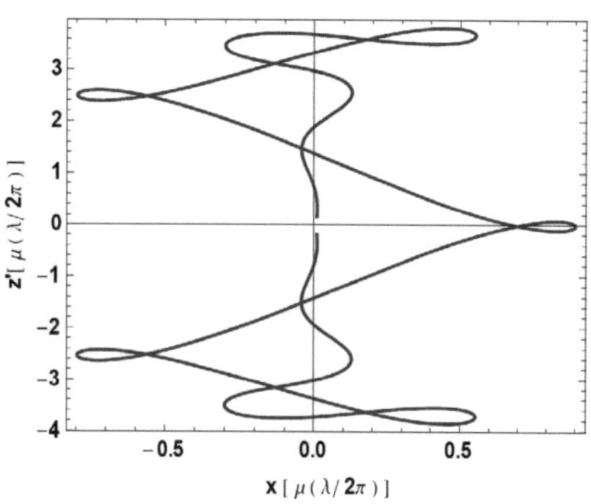

Fig. 1d. Shows the same as Figure 1c, but now the essentially uniform drift motion has been subracted from the longitudinal motion, thus the motion of the electron is viewed from a comoving frame. The coordinate z' has been defined as $z' \equiv z-(\mu_0/2)^2(\lambda_0/4\pi)\times 2\omega_0\eta$, according to Eq. (5a). The curve starts slightly below the origin, towards negative z' values, and then proceeds to positive values and finally gets back to the origin from above.

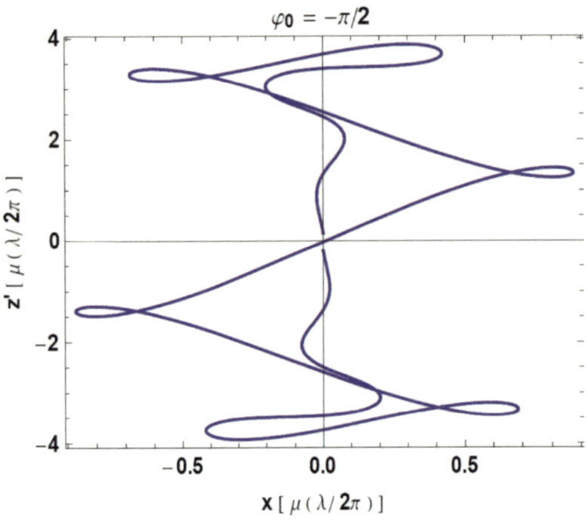

Fig. 2a. Shows the same as Fig. 1d for the reduced electron trajectory, but now the carrier envelope phase difference of the Gaussian pulse has been taken $\varphi_0 = -\pi/2$, i.e. now we are dealing with a sine-pulse. The curve starts slightly below the origin, towards negative z′ values, and then proceeds to positive values and finally gets back to the origin from above.

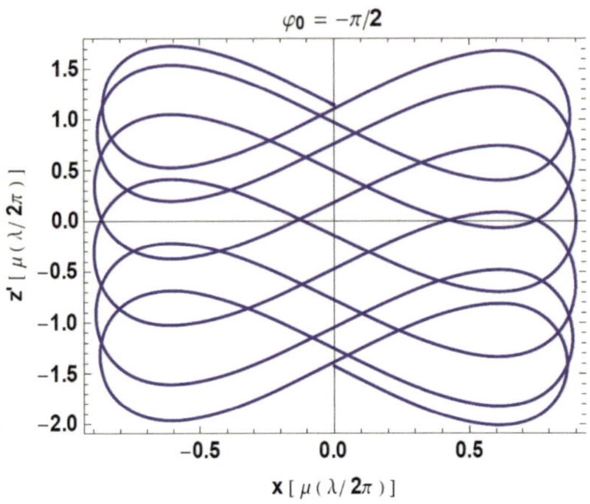

Fig. 2b. Shows the reduced electron trajectory around the maximum of a relatively long (~ 20 cycle) Gaussian sine-pulse (when $\varphi_0 = -\pi/2$). The nearly uniform drift motion has not been transformed out completely, in order to illustrate that the „figure-8 trajectory" is never a stationary lemniscate. The curve starts at the point (0, −1.5) towards negative z′ values, and then proceeds to positive values and finally arrives at the ponit (0,+1.2).

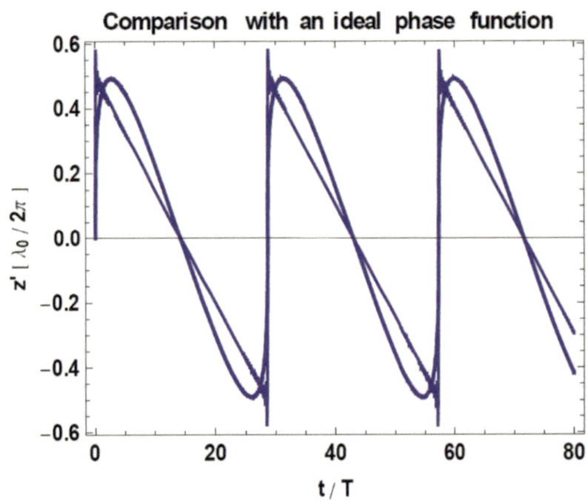

Fig. 3. The real time-dependence of the longitudinal position $z'(t) \equiv -[z(t) - \beta ct]$ of the electron as given in Eq. (5b), where we have subtracted the uniform drift motion βct. The trajectory is compared with the ideal phase function („saw-tooth"), which is given by the famous Fourier series $\Sigma_k(\sin kx)/k = (\pi - x)/2$ on the open interval $(0, 2\pi)$. From the first equation of Eq. (5b), the „saw-tooth" is represented by the sum $\Sigma_k[\sin(2k \times 0.11(t/T))]/k$. which corresponds to the $\mu_0 = 15$ value of the dimensionless intensity parameter.

In order to consider the general role of the ponderomotive energy shift $(\mu_0/2)^2 mc^2=U_p$ introduced after Eq. (5b), we write the relativistic equation of motion in a covariant form, which is equivalent to Eq. (1a) and the relativistic work theorem. From Eq. (1a) we obtain $m d\mathbf{u}/d\tau=(e/c)(u_0\mathbf{E}+\mathbf{u}\times\mathbf{B})$ and $m du_0/d\tau=(e/c)\mathbf{u}\cdot\mathbf{E}$, where $(u_0, \mathbf{u})=\{u^\mu\}=\{dx^\mu/d\tau\}$ is the four-velocity, $\{x^\mu\}=(ct, \mathbf{r})$ is the four-position, and $d\tau=dt/\gamma$ denotes the proper time element of the electron. We adopt the signature $(+ - - -)$, and then the metric tensor g becomes $g_{00}=-g_{ii}=1$ for i=1, 2, 3, and the off-diagonal elements are zero, i.e. $g_{\mu\nu}=0$ for $\mu\neq\nu$, for $\mu, \nu=0, 1, 2, 3$. With the usual summation convention, the four-product of two vectors $\{a^\mu\}$ and $\{b^\nu\}$ reads $a\cdot b\equiv g_{\mu\nu}a^\mu b^\nu=a^\mu b_\mu=a^0b^0-\mathbf{a}\cdot\mathbf{b}$, and the gradient vector is defined as $\partial\equiv\{\partial_\mu\}\equiv\{\partial/\partial x^\mu\}$. With these notations the covariant equation of motion of the electron in the laser field becomes:

$$du_\mu/d\tau=(e/mc)F_{\mu\nu}u^\nu\,,\quad F_{\mu\nu}=\partial_\mu A_\nu-\partial_\nu A_\mu\,,\quad du_\mu/d\tau=-(e/mc)dA_\mu/d\tau+(e/mc)u^\nu(\partial_\mu A_\nu), \tag{7a}$$

$$u_\mu=V_\mu-(e/mc)A_\mu\,,\quad dV_\mu/d\tau=(e/mc)V^\nu(\partial_\mu A_\nu)-(e/mc)^2A^\nu(\partial_\mu A_\nu), \tag{7b}$$

where we have introduced the field tensor $F_{\mu\nu}$, the vector potential A_μ and the part V_μ of the four velocity (soon we shall assume that this is a „slowly-varying" quantity). We emphasize that at this stage Eqs. (7a-b) do not contain any approximations, the above manipulations do not rely on the assumption that V_μ is slowly-varying on any space-time scale. If we take the average over the oscillations of the vector potential in the second equation of Eq. (7b), then, because $\langle V_\mu\rangle=\langle u_\mu\rangle$, we have

$$d\langle V_\mu\rangle/d\tau=-(e/mc)^2\partial_\mu\langle A^\nu A_\nu\rangle\,,\quad m d\langle u_\nu\rangle/d\tau=\partial_\nu[\mu^2(\mathbf{r},t)mc^2/2],\quad m d\langle\mathbf{v}\rangle/dt=-mc^2\nabla\mu^2(\mathbf{r})/2. \tag{7c}$$

In Eq. (7c) we have introduced the generalized intensity parameter μ, in terms of which the ponderomotive energy is expressed as $U_p(\mathbf{r}, t)=\mu_0^2 f(\mathbf{r}, t)(mc^2/4)$, where $\mu_0\equiv eF_0/mc\omega_0$ is the maximum value of the dimensionless intensity parameter, and $f(\mathbf{r}, t)$ is a (slowly-varying) spatio-temporal intensity profile function of the laser beam. The third equation in Eq. (7c) is a (nonrelativistic) special case of the second, more general equation. Here the ponderomotive energy is represented by a conservative (static) potential. It is important to note that, regardless of the sign of the charge of a particle, the ponderomotive potential of the laser field is repulsive towards the lower intensity values (because it is proportional with the square of the charge and with the negative gradient of the intensity distribution). This means, for instance, that an ionized electron experiences a force pushing it out of the interaction region. As a consequence, for instance, the measured energy spectrum of the ionized electrons are different to that they have their „place of birth" inside the laser beam. Since $2mc^2$=1MeV, the maximum energy increase may be of this order if $\mu_0\approx 2$, i.e. if the maximum intensity $I_0\approx 2\times10^{18}$ W/cm^2 for an optical field ($h\nu_0\approx$1eV). This kind of ponderomotive energy increase is still considered as a way towards laser acceleration of charged particles. It is clear that even at nonrelativistic intensities, the ponderomotive energy shift may considerably be larger than the photon energy. For an optical field of 2×10^{13} W/cm^2 intensity the ponderomotive energy shift is $U_{pmax}\approx$10eV.

At the end of the present section we present a simple classical example to illustrate the possibility of linear absolute carrier-envelope phase difference effects. Let us take the Newton equation of a charged particle bound in a linear potential, and driven by a laser pulse represented by a Gaussian envelope, and study the Fourier spectrum of its position:

$$d^2x/dt^2+\Gamma dx/dt+\Omega^2x=-(e/m)F_0\exp(-t^2/\tau^2)\cos(\omega_0t+\varphi_0)=-(e/m)F(t), \tag{8a}$$

$$x(\omega)|^2 = (e/m)^2|F(\omega)|^2/[(\omega^2-\Omega^2)+\Gamma^2], \quad |F(\omega)|^2 = |C(\omega)|^2\,[\cosh(2\tau^2\omega\omega_0)+\cos(2\varphi_0)], \tag{8b}$$

$$M(\omega)\equiv[(|x(\omega)|^2)_{max}-(|x(\omega)|^2)_{min}]/\,[(|x(\omega)|^2)_{max}+(|x(\omega)|^2)_{min}] = 1/\cosh(2\tau^2\omega\omega_0), \tag{8c}$$

where the subsidiary quantity $C(\omega)$ introduced in Eq. (8b) does not depend on φ_0 , its detailed form is unimportant here. In Eq. (8c) we have defined the *modulation function* $M(\omega)$, which is an analogon of the *visibility* used in optics. According to Figs. 4a-b-c, the modulation is negligible except for the very low-frequency part of the spectrum, where it can be considerable if $\omega/\omega_0 < 0.2$ for a three-cycle pulse, for instance. However, this is just that region where the signal (in the present case; the induced dipole) is very small. In the

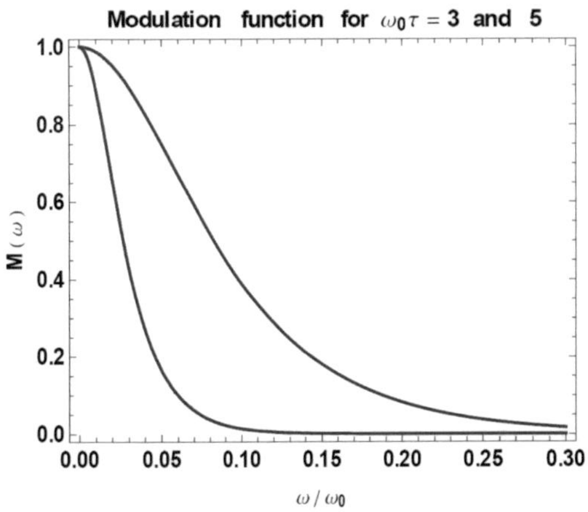

Fig. 4a. Shows the linear modulation function for two few-cycle Gaussian pulses according to the formula given by Eq. (8.c). The upper and the lower curves refer to the values $\omega_0\tau=3$ and $\omega_0\tau=5$, respectively.

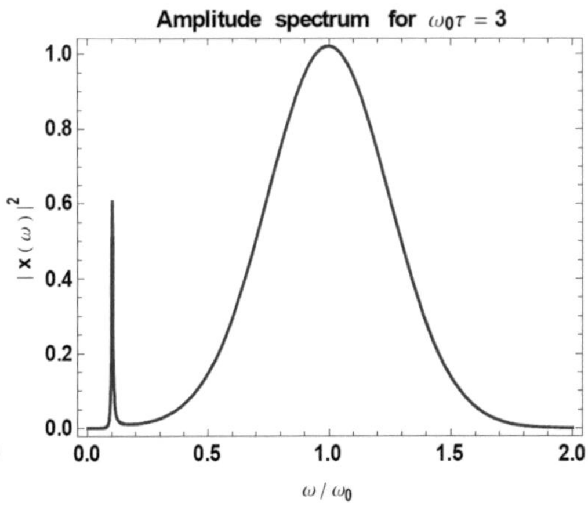

Fig. 4b. Shows the amplitude (dipole) spectrum of the linear oscillator (see Eq. (8b)) with eigenfrequency $\Omega = 0.1\omega_0$ and damping constant $\Gamma = 0.005\omega_0$.

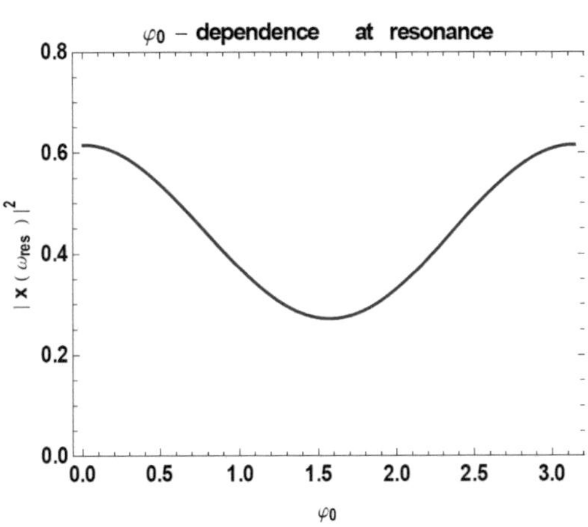

Fig. 4c. Shows the variation of the resonant value of the modulus square of the resonant dipole moment (at frequency $\omega = \Omega = 0.1\omega_0$) as a function of the carrier-envelope phase difference φ_0 , according to Eq. (8b).

linear regime, the spectrum of the response follows the spectrum of the excitation according to this simple model, as is shown in Fig. (4b). If the system has a low-frequency resonance, then both the signal and the modulation become large enough, as can be seen in Figs. 4b-c.

3. Multiphoton ionization and the multiphoton Kramers–Heisenberg formula for high-harmonic generation.

The usual dispersion relations to light scattering are derived in the linear regime, where the incoming radiation is treated as a weak perturbation in comparison of the atomic binding potential. By restricting the present analysis to nonrelativistic intensities, we start with the Schrödinger equation of a single electron interacting jointly with the atomic potential $V(\mathbf{r})$, with the laser field, characterized by the dipole term $e\mathbf{r}\cdot\mathbf{F}(t)$, and with the spontaneously emitted radiation, whose effect is also described by a dipole term of the form $e\mathbf{r}\cdot\mathbf{F}'(t)$:

$$[\mathbf{p}^2/2m + V(\mathbf{r}) + e\mathbf{r}\cdot\mathbf{F}(t) + K(\mathbf{r},t)]\,|\Psi(t)\rangle = i\hbar\partial_t\,|\Psi(t)\rangle, \quad \mathbf{F}(t)=\boldsymbol{\varepsilon}F_0(t)\cos(\omega_0 t+\varphi_0), \tag{9a}$$

$$K(\mathbf{r},t)= e\mathbf{r}\cdot\mathbf{F}'(t), \quad \mathbf{F}'(t)=i\boldsymbol{\varepsilon}'(2\pi\hbar\omega'/L^3)^{1/2}[a\exp(-i\omega' t)-a^{\dagger}\exp(+i\omega' t)], \tag{9b}$$

where $\mathbf{p}=-i\hbar\partial/\partial\mathbf{r}$ is the electron's momentum operator and L^3 is the quantization volume. The frequency, the polarization vector and the quantized amplitude of the emitted photon are denoted by ω', $\boldsymbol{\varepsilon}'$ and a, respectively. The quantized amplitudes satisfy the usual commutation relation $[aa^{\dagger}-a^{\dagger}a]=1$. The system starts at some initial time t_0 in the product state $|\Psi(t_0)\rangle=|\Psi_0\rangle=|\psi_0\rangle|0'\rangle$, and evolves, according to the above Schrödinger equation, to a final state $|\Psi_0(t_1)\rangle$. Here $|\psi_0\rangle$ is the initial electron state, and $|0'\rangle$ symbolizes the vacuum state, where the prime refers to the scattering mode that is initially empty. If we neglect the spontaneous emission during the interaction, then Eq. (9a) can describe multiphoton excitation or ionization, or multiphoton direct or inverse Bremsstrahlung processes. Following the general method due to Varró & Ehlotzky (1993), we can go over to the Kramers-Henneberger frame, and receive a integral equation for the evolution operator of the complete system. By projecting to a final state $\langle\Psi_1|$, we obtain for the S-matrix elements:

$$\langle\Psi_1|\Psi_0(t_1)\rangle=\delta_{10}-(i/\hbar)\int dt\langle\Psi_1|U_0^{+}(t)W(t)U_0(t)|\Psi_0\rangle$$

$$-(i/\hbar)^2\int dt\int dt'\langle\Psi_1|U_0^{+}(t)W(t)U_0(t)\,U_0^{+}(t')W(t')U_0(t')|\Psi_0\rangle+\langle\Psi_1|R_0(t_1)\rangle, \tag{10a}$$

where $U_0(t)$ denotes the unperturbed propagator of the electron, and $|R_0(t_1)\rangle$ represents the third iteration of the integral equation. In deriving Eq. (10a) we have used that $\langle\Psi_1|$ and $|\Psi_0\rangle$ are orthogonal eigenstates of the unpertubed Hamiltonian. The effective interaction represented by the operator $W_\alpha(t)$ has the form

$$W(t)\equiv V_{cl}(\mathbf{r},t)+ K(\mathbf{r}+\boldsymbol{\alpha},t), \quad V_{cl}(\mathbf{r},t)\equiv V(\mathbf{r}+\mathbf{r}_{cl}(t))-V(\mathbf{r}), \quad m d^2\mathbf{r}_{cl}(t)/dt^2=-e\mathbf{F}(t), \tag{10b}$$

where $\mathbf{r}_{cl}(t)$ is the classical position of the electron satisfying the Newton equation. According to Eqs. (3a-c), for a stationary field with $\varphi_0=0$ (or in other words, for smooth switching-on and -off of the laser field) the trajectory simply reads $\mathbf{r}_{cl}(t)=\boldsymbol{\alpha}(t)\equiv\alpha_0\boldsymbol{\varepsilon}_z\sin\omega_0 t$, $\alpha_0\equiv\mu_0\lambda_0/2\pi$. Owing to the periodic modulation of the space-translated atomic potential $V_{cl}(\mathbf{r},t)$, already the first nontrivial term on the right hand side of Eq. (10a) contains all the higher harmonics of the incoming laser radiation:

$$V_{cl}(\mathbf{r},t)=\Sigma_n V_n(\mathbf{r})\exp(-in\omega_0 t), \qquad V_n(\mathbf{r})=\Sigma_l V_n^{(l)}(r)P_l(\cos\theta), \tag{11}$$

where $P_l(\cos\theta)$ is the l-th Legendre polynomial and θ is the polar angle of the electron's position with respect to the polarization vector of the laser field. The detailed form of the „multipole-multiphoton expansion coefficients" $V_n^{(l)}(r)$ have been presented by Varró & Ehlotzky (1993).

Before entering into a short analysis of high-harmonic generation (HHG) in the frame of our semiclassical method, a brief summary may be in order on the famous model by Keldish (1965) on the *multiphoton ionization* and the *optical tunneling* or *cold emission of electrons induced by strong laser fields* (see Bunkin & Fedorov (1965), Farkas et al. (1983,1984), Chin et al. (1983), Farkas & Chin (1984) and Walsh et al. (1994)). Leaving out the interaction term K responsible for the spontaneous emission of radiation, the second term on the right hand side in Eq. (10a) can be approximated in the following way:

$$T_{fi}^{(1)} = -(i/\hbar)\int dt\, \langle \mathbf{p}\,|\,U_p^{\dagger}(t)U_2^{\dagger}(t)U_r^{\dagger}(t)V(\mathbf{r})\,|\,\psi_0\rangle \exp[(i/\hbar)(p^2/2m-E_0)t], \tag{12a}$$

$$U_r(t)=\exp[(ie/\hbar\omega_0)F_0\mathbf{r}\cdot\boldsymbol{\varepsilon}\cos(\omega_0 t)], \quad U_p(t)=\exp[-(i/\hbar)\mathbf{p}\cdot\alpha_0\boldsymbol{\varepsilon}_z\sin\omega_0 t], \tag{12b}$$

$$U_2(t)=\exp[-(i/\hbar)\Delta E_2 t - i(\Delta E_2/2\hbar\omega_0)\sin(2\omega_0 t)], \qquad \Delta E_2=\mu_0^2 mc^2/4, \tag{12c}$$

$$T_{fi}^{(Keldish)} = -(i/\hbar)\int dt \int d^3r\, \psi^*_{\mathbf{p}}(\mathbf{r},t)V(\mathbf{r})\psi_0(\mathbf{r},t) = -2\pi i\Sigma_n T_{fi}^{(n)}\delta[E_p-(E_0-\Delta E_2+n\hbar\omega_0)], \tag{12d}$$

where $\mu_0\equiv eF_0/mc\omega_0=10^{-9}I_0^{1/2}/E_{ph}$ (with $E_{ph}\equiv\hbar\omega_0/\mathrm{eV}$ being the photon energy measured in eV) is the same dimensionless intensity parameter which has already been introduced in Eq. (3c) in the course of the classical analysis. We have also encountered the quantity $\Delta E_2=\mu_0^2 mc^2/4=U_p$, which is just the ponderomotive energy shift, defined after Eq. (5b). In the Keldish model the final state of the electron is taken to be a Volkov state $\psi_{\mathbf{p}}(\mathbf{r}, t)$ of some average energy $E_p= p^2/2m$, which is a dressed state with respect to the interaction with the laser field. In the initial state the interaction with the laser is not incorporated, so it is an unperturbed atomic ground state $\psi_0(\mathbf{r}, t)=\psi_0(\mathbf{r})\exp[-(i/\hbar)E_0 t]$. The explicit form of a nonrelativistic Volkov state reads (where $\mathbf{A}(t)$ is the vector potential of the field $\mathbf{F}(t)$):

$$\psi_{\mathbf{p}}(\mathbf{r},t)=\langle \mathbf{r}\,|\,U_r(t)\,|\,\mathbf{p}\rangle \exp\{-(i/\hbar)\int dt'[\mathbf{p}-(e/c)\mathbf{A}(t')]^2/2m\}, \quad \mathbf{F}(t)=-\partial\mathbf{A}(t)/c\partial t, \tag{13a}$$

$$\exp[-(i/\hbar)\mathbf{p}\cdot\alpha_0\boldsymbol{\varepsilon}_z\sin\omega_0 t]=\exp[-i(\mu_0\mathbf{p}\cdot\boldsymbol{\varepsilon}_z/\hbar k)\sin(\omega_0 t)]=\Sigma_n J_n(\mu_0\mathbf{p}\cdot\boldsymbol{\varepsilon}_z/\hbar k)\exp[-(i/\hbar)(n\hbar\omega_0)t], \tag{13b}$$

where we have explicitely written out the Fourier expansion coefficients of the exponential in Eq. (12b), by using the Jacobi-Anger formula for the generating function of the ordinary Bessel functions (Gradshteyn & Ryzhik (2000)). This is one of the key mathematical formula with the help of which the strength of the *multiphoton side-bands* of additional energies $n\hbar\omega_0$ can be calculated. It is clearly seen that the appearance of the side-bands is a result of the *sinusoidal modulation of the phase of the Volkov state* given in Eq. (13a). The multiphoton ionization process can be viewed as a transition from a side-band to the free electron mass shell corresponding to free propagation. This is expressed by the delta functions in the incoherent superposition of the multiphoton transition amplitudes, describing the conservation of energy:

$$p_n^2/2m = n\hbar\omega_0 - (|E_0| + \Delta E_2) > 0, \quad n = n_{min} + n_{excess} = 1, 2, 3, \ldots, \tag{14}$$

where n_{min} is the minimum number of quanta for the deliberation of the electron from the binding potential, and n_{excess} denotes the number of excess quanta, which may be absorbed additionally in the continuum. This latter phenomenon is called *above-threshold ionization* (ATI). The above-threshold electron spectra of nonlinear photoionization induced by *relatively long laser pulses,* analysed thoroughly e.g. by Krause et al. (1992), Agostini (2001), Paulus and Walther (2001), and recently by Banfi et al. (2005) and Ferrini et al. (2009) for multiphoton surface photoelectric effect, have common features with the corresponding high-harmonic spectra. The initial fall-off, the (occasionally rising) plateau and the sharp cut-off are present in each cases. Just for illustration, in Fig. 5 we show the structure of a typical ATI spectrum, which in many cases very much resembles to the HHG spectrum.

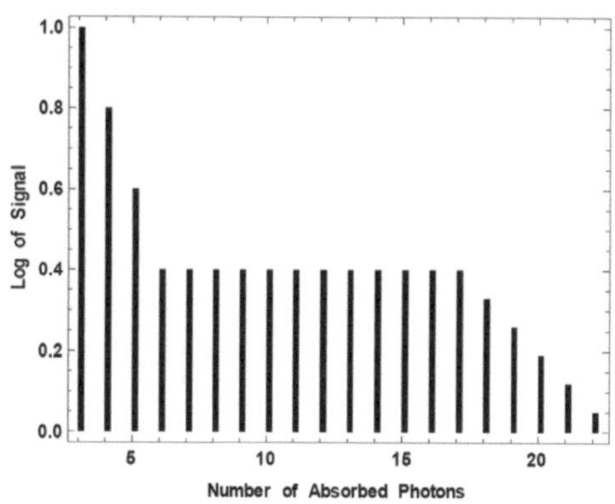

Fig. 5. Shows schematically a typical above-threshold ionization (ATI) electron spectrum. The initial fall-off, the plateau and the cut-off are usually also present in the higher-harmonic generation (HHG) processes (in which case the harmonic order is drawn on the abcissa).

From Eq. (14) it is seen that the ionization potential is increased by the ponderomotive energy shift, which, by the increase of the intensity, may well become even much larger than the photon energy. In such cases the channel corresponding to the minimum number of photons gets closed, and the phenomena of *peak supression* enters in the scene. Keldish (1965) has approximately calculated the *total ionization probability,* by using the method of stationary phase of the classical action in the phase of the electron's Volkov state. On the basis of his method, from Eqs. (12d) and (13a-b), the total probability can be approximated by the formula (see also Landau & Lifshitz (1978)):

$$w \sim \exp[-(2A/\hbar\omega_0)f(\gamma)], \quad f(\gamma) \equiv (1+1/2\gamma^2)\mathrm{Arsh}\gamma - (1+\gamma^2)^{1/2}/2\gamma, \quad \gamma \equiv \omega_0(2mA)^{1/2}/eF_0 \equiv 2\omega_0\tau, \tag{15}$$

where $A = |E_0|$ is the ionization potential (work function), and τ is the tunnel time. *The Keldish γ parameter* defined in Eq. (15) can also be expressed in terms of the dimensionless intensity parameter introduced already in Eq. (3b): $\gamma = 2(A/2mc^2)^{1/2}/\mu_0$. We emphasize that the above approximate formula can be realistic strictly in the case when the ionization energy is much larger than the photon energy, $A = |E_0| \gg \hbar\omega_0$. In the meantime the formula has been considerably refined, in particular by Ammosov, Krainov & Delone (1998), but we shall not enter into these more sophisticated methods. We also note that the tunneling time τ introduced here corresponds to the concept of tunneling time by Büttiker and Landauer (1982). The numerical value of the Keldish γ can be calculated according to the following formula: $\gamma=2\times10^6 E_{ph}(A/I_0)^{1/2}$, where the photon energy E_{ph} and the ionization potential (work function) measured in electron volts, and the peak intensity I_0 measured in W/cm^2. For example for $E_{ph}=1$, $A=25$, and $I_0=10^{12}$ we have $\gamma=10$. On the basis of Eq. (15), in the two extreme cases $\gamma >> 1$ („large frequency and small intensity": „multiphoton regime") and $\gamma << 1$ („small frequency and large intensity": „tunnel regime") we can give a simple physical interpretation of Eq. (15):

$$w \sim [\mu_0^2(mc^2/8A)]^{A/\hbar\omega_0} \sim (I_0)^n \text{ if } \gamma >> 1, \qquad w \sim \exp[-4(2mA^3)^{1/2}/3\hbar e F_0] \text{ if } \gamma << 1, \tag{16}$$

In the first formula of Eq. (16) we mean on the power index n the minimum number of photons needed for the ionization, $n\hbar\omega_0 >\approx A$. The „n-power-law" corresponds to the fall-off regime shown in Fig. 5. The tunnel regime $\gamma < 1$ can be reached with intensities higher than 10^{14} W/cm^2 in the above numerical example. In Figs. 6a and 6b we show in a special case the intensity dependence of the Keldish gamma parameter and the relative probability, respectively.

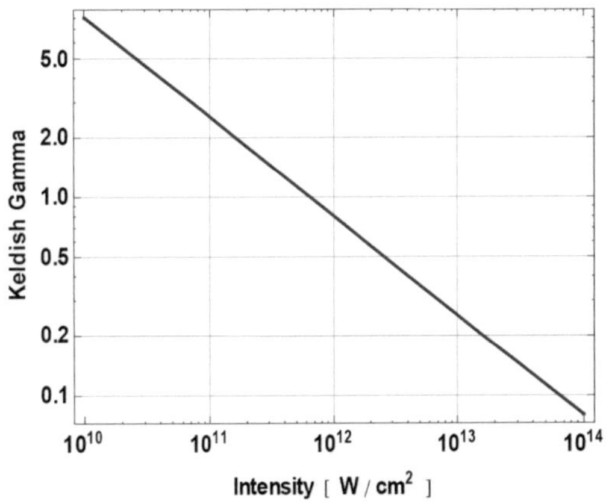

Fig. 6a. Shows the intensity dependence of the Keldish γ parameter, defined in Eq. (15), for the special values A=16eV and $\hbar\omega_0$=0.1eV, which correspond to the ionization of Argon atoms by a CO_2 laser. The transition region is located around the intensity value 10^{12} W/cm^2, where the value of the γ parameter becomes about unity.

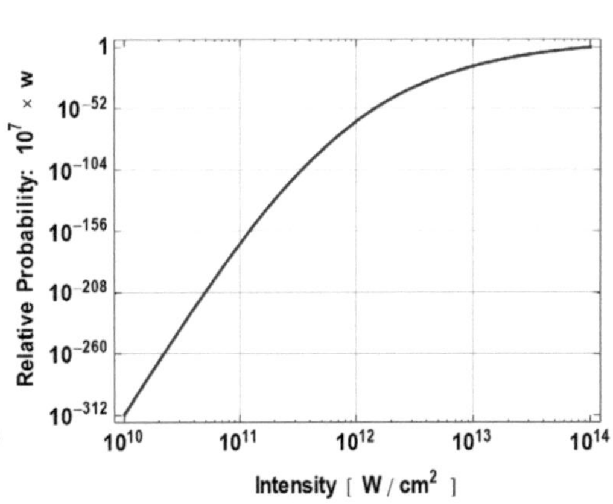

Fig. 6b. Shows the intensity dependence of the Keldish relative probability, according to Eq. (15), for the special values A=16eV and $\hbar\omega_0$=0.1eV, which correspond to the ionization of Argon atoms by a CO_2 laser. Around the intensity value 10^{12} W/cm^2, the dependence changes from the perturbative behaviour to the tunneling regime, as can also be estimated from Eq. (16).

The general equation for the matrix elements for multiphoton transitions, Eqs. (10a-b), can also be used to calculate the production rates of high-harmonics, as has been shown by Varró & Ehlotzky (1993). In contrast to the standard method based on the determination of the laser-induced nonlinear atomic dipole moment, this description relies on scattering theory, and it yields a possible *multiphoton generalization of the Kramers-Heisenberg dispersion relation.* By performing the time integral, we have the energy conservation for an n-th-order process $\hbar\omega'=n\hbar\omega_0-(E_f-E_0)$, thus the formalism describes also the possibility that the final energy E_f of the atomic electron does not coincides with its initial energy E_0. On the basis of Eqs. (10a-b) the differential cross-sections for such processes read

$$d\sigma_n/d\Omega=r_0^2(\omega'/\omega_0)^3(2a_0/\alpha_0)^2|M_n|^2,\quad r_0=e^2/mc^2,\ a_0=\hbar^2/me^2,\quad M_n=(\varepsilon_z\cdot\varepsilon')\delta_{f0}\delta_{n1}+S_1+S_2, \tag{17a}$$

$$S_1\equiv+i\Sigma_k[\langle\psi_f|V_n|E_k\rangle\langle E_k|(\mathbf{r}\cdot\varepsilon')|\psi_0\rangle(E_k-E_0+\hbar\omega'+i\varepsilon)^{-1}], \tag{17b}$$

$$S_2\equiv+i\Sigma_k[\langle\psi_f|(\mathbf{r}\cdot\varepsilon')|E_k\rangle\langle E_k|V_n|\psi_0\rangle(E_k-E_0+\hbar\omega'-n\hbar\omega_0+i\varepsilon)^{-1}], \tag{17c}$$

where in the summation over the intermediate states the positive infinitesimal ε has been introduced. It can be shown that the cross-section given by Eqs. (17a-b-c) reduces to the well-known linear dispersion formula. In Varró & Ehlotzky (1993) we have calculated the matrix elements analytically, and applied to the calculation of the production rates on three kinds of noble gas atoms, and obtained reasonable agreement with the experimental observations.

4. The nonlinear effect of the laser-induced oscillating double-layer potential on metal surfaces. X-ray generation in the presence of a static homogeneous electric field.

In case of multiphoton photoelectric effect of metals, Farkas & Tóth (1990) and Farkas et al. (1998) measured very high-order above-threshold electrons coming from metal targets. The theoretical interpretation of these results has been first given by the present author in Varró & Ehlotzky (1998) and recently in Kroó et al. (2007). In Varró et al. (2010) on the basis of the so-called laser-induced oscillating double-layer potential model the spontaneous emission of radiation by metallic electrons in the presence of electromagnetic fields of surface plasmon oscillations. Our original model belongs to a wider class Floquet-type analyses (see e.g. Kylstra (2001)), and considers the inelastic electron scattering on the *oscillating double-layer potential generated by the incoming laser field at the metal surface. The* model has already been succesfully used to interpret the experimental results on very high order surface photoelectric effect in the near infrared (Farkas & Tóth (1990)) and in the far infrared regime (Farkas et al., 1998). In this description the basic interaction leading to very high nonlinearities is caused by the collective velocity field of the oscillating electrons near the metal surface, within a layer of thickness smaller that the penetration depth δ. Because the quasistatic velocity field is screened inside the metal, the thickness of the layer is taken as the Thomas-Fermi screening length $\delta_s=1/k_F$ where $k_F=(6\pi n_e e^2/E_F)^{1/2}$. The wave function of an electron will then obey the two Schrödinger equations

$$(p^2/2m-V_0-V_D\sin\omega_0 t)\Psi_I=i\hbar\partial_t\Psi_I\ (z<0),\quad (p^2/2m+V_D\sin\omega_0 t)\Psi_{II}=i\hbar\partial_t\Psi_{II}\ (z>0), \tag{18a}$$

where the subscript *I* refers to the interior region (metal) and *II* to the exterior region (vacuum), respectively. Following Varró & Ehlotzky (1998), the amplitude of the collective velocity field

$$V_D=2\pi n_e e^2\alpha_0\delta_s=\mu_0(\omega_p/4\omega_0)(\delta_s/\delta)(2mc^2),\qquad \omega_p=(4\pi n_e e^2/m)^{1/2},\qquad V_D=3\mu_0\times10^4\text{eV}, \tag{18b}$$

where ω_p is the plasma frequency, $\alpha_0 \equiv \mu_0 \lambda_0 / 2\pi$, and δ denotes the skin-depth of the metal taken at the laser frequency ω_0. The interaction with the other agents like surface plasmon fields should in principle still be taken into account, but since their direct effect is much smaller than that of the induced collective velocity field, we have not displayed these direct interaction in Eqs. (7a-b). Thus in Eqs. (7a-b) we only display the Schrödinger equation relevant in the final state interaction, where additional energy redistribution takes place due to the interaction with the induced near field. The outgoing electron current components for which the momenta $p_n=[2m(n\hbar\omega_0 - A)]^{1/2}$ are real ($n \geq n_0=4$ in the case under discussion), corresponding to n-order multiplasmon absorption, have been obtained from the Fourier expansion of Ψ_I and Ψ_{II}. The unknown multiphoton reflection and transmission coefficients, R_n and T_n, respectively, can be determined from the matching equations, i.e. from the continuity of the wave function, $\Psi_I(0,t)=\Psi_{II}(0,t)$ and of its spatial derivative, $\partial_z\Psi_I(0,t)= \partial_z\Psi_{II}(0,t)$ which relation must hold for arbitrary instants of time. The resulting two coupled infinite set of linear algebraic equations for R_n and T_n can be numerically solved without any particular difficulty, moreover, it is possible to derive quite accurate analytic approximate formulas, too. According to these results, the current components normalized to the incoming current can be expressed as

$$j_t(n)=(p_n/q_0)|T_n|^2, \quad |T_n|^2 \approx J_n^2(a) \quad (n \geq n_0), \quad a \equiv 2V_D/\hbar\omega_0, \tag{19}$$

where $q_0=(2mE_F)^{1/2}$ is the average of the initial momenta. For instance, in case of $I_0=2\times10^8 W/cm^2$ we have $2V_D=11eV$, and $a=2V_D/\hbar\omega_0=7$ for a Ti:Sapphire laser of photon energy 1.56eV. We have taken for the ratio $(\delta_s/\delta)=2\times10^{-2}$, i.e. for the $\delta=22.5nm$ skin-depth of gold metal the screening length is about $\delta_s=0.4nm$. Though the standard nonlinearity parameter, $\mu_0=10^{-5}$, is very small at such intensities, the parameter „a" is that large that it can cause 7th order nonlinearities with sizeable probability. This numerical example clearly shows that already at very low intensities used in several experiments the *new nonlinearity parameter* „a" introduced in Eq. (19) has a much larger value than the argument of the Bessel function in Eq. (13b). On the basis of this remarkable quantitative difference, our theory based on introducing the laser-induced near-field is capable of accounting for the basic features of the recently measured unexpectedly broad above-threshold electron spectra. Because of the smallnes of μ_0, in the frame of the standard *nonperturbative* Volkov description there is no chance to interpret several recent and earlier multiphoton experiments. It is interesting to note that these high nonlinearities found in the experiments cannot be accounted for by including the effect of field enhancement due to the generation of surface plasmon oscillations. Though this effect results in a two order of magnitude increase of the intensity parameter, owing to the compression of the radiation field at the metal surface, it is still not enough to have the standard nonperturbative description to work.

The double-layer near-field model has also been applied to consider the generation of x-rays by irradiating metal surfaces with a powerful laser beam in the presence of a static electric field (Varró et al. (1999)). A similar effect has been recently studied by Odžak & Milošević (2005) in their work on high-order harmonic generation in the presence of a static electric field.

Recently, in a series of papers (see Varró (2004) and Varró (2007a-b-c)) the author has studied the scattering of a few-cycle laser pulse on a thin metal nano-layer, and the effect of the carrier-envelope phase difference has been investigated. The analysis has been extended to consider plasma layers, too, in the relativistic regime. Perhaps the most surprising effect which came out from the analysis, was the generation of phase dependent wake-fields. This wake-fields are reflected quasi-static.

By analogy, one may think that if the phases of the *above-threshold electron de Broglie waves* generated at the metal surface are locked (i.e. the difference of the phases of the neighbouring components is a smooth, possibly a constant function of the order, namely the number of absorbed photons), then the Fourier synthesis of these components yields an *attosecond electron pulse train* emanating perpendicularly from the metal surface, quite similarily to the generation of attosecond light pulses from high harmonics (which, on the other hand, are propagating in the specular direction). This expectation is quite natural, because the spacing of the electron peaks in the frequency space is just the optical frequency $h\nu_0/h=\nu_0$, like in the case of high-harmonic generation. The idea has been worked out quite recently by Varró & Farkas (2008a), where further references can be found concerning the general question attosecond electron pulses. Concerning these developments, we refer the reader to the above-mentioned papers.

5. Some quantum phase and other statistical properties of few-cycle and attosecond pulses

It is a natural, and both conceptually and practically important question that to what extent can one control the phase of the ultrashort pulses, and, at all, what are the ultimate quantum limits of phase stabilization in a given generating process? The systematic quantitative analysis of these problems is still missing, at least concerning the quantum uncertainties of the phases. One may expect that the usual high-intensity laser fields, being in a highly populated coherent state, can certainly be well represented in a satisfactory manner in the frame of external field approximation, i.e. in terms classical Maxwell fields of definite amplitudes and phases, or in terms of classical stochastic processes. On the other hand, in the generation of extreme pulses, like sub-femtosecond or attosecond pulses, it is an open question whether the fully quantum or the semiclassical description should be used for the correct interpretation. The reader can judge the importance of such questions by remembering the chapters devoted to the classical theory of coherence in the book by Born & Wolf (2009), for instance. Concerning the general description of quantum coherence and correlations of the electromagnetic radiation, we refer the reader to the books by Klauder & Sudarshan (1968), Scully & Zubairy (1997), Loudon (2000) and to the book by Schleich (2001) on quantum optics in phase space. In this context one has to keep in mind that, after all, though we have been speaking of „photon absorption" or „photon emission" in „multiphoton processes", either the classical or semiclassical framework has been used, in which the word „photon" has simply no meaning. As is general in the study of nonlinear laser-induced processes, the appearance of the side-bands in the electron energy in the phase of the wave function, $\exp[-(i/\hbar)(E_{el}-E'_{el})t-in\omega_0 t]$, is transformed to the mathematically identical form $\exp[-(i/\hbar)(E_{el}-E'_{el}+n\hbar\omega_0)t]$. The delta functions $\delta(E_{el}-E'_{el}+n\hbar\omega_0)$ appearing in the transition probabilities are said to be responsible for the energy conservation $E'_{el}=E_{el}+n\hbar\omega_0$, and then this balance equation is considered as „n-photon absorption" by the electron. On the other hand, in the very sense, no true light quanta has been considered, but rather mere classical Maxwell fields.

Recently we have analysed in quite details the interaction of strong quantized radiation fields with free electron wave-packets (see Varró (2008b-c) and Varró (2010b)), and found that the entangled photon-electron states, developed due to the interaction, are closely related to the number-phase minimum uncertainty states of the photon field. We have also proved (Varró (2010)) that the strong-field interactions necessarily lead to entangled states and the appearance of entropy remnants of a photon-electron system. In a series of papers Fedorov et al. (2006) have shown that the short-pulse and strong-field breakup processes may be considered as a route to study entangled wave-packets. For the other extreme of weak, single-photon fields we have recently presented a method to treat intensity-intensity correlations by simple classical probability theory (Varró (2008c)). The study of the interaction of electrons with quantised radiation field dates back to our earlier works (Bergou & Varró (1981a-b)), in which the multiphoton Bremsstrahlung processes and the nonlinear Compton scattering were described on the basis of an algebraic treatment introduced by us. These analyses has served as a "microscopic foundation" for the semiclassical treatments. Here we mention only one genuine quantal effect, namely that in Compton scattering the frequency ω'_n of the scattered radiation depends on the change in the occupation of the modes of the incoming radiation:

$$\omega'_n=(n\omega_0+\omega_c\mu_0^2\delta/4)/[1+(2n\omega_0/\omega_c+\mu_0^2/2)\sin^2(\theta/2)]\,, \qquad (20)$$

where $\omega_c\equiv mc^2/\hbar$ is the Compton frequency, θ is the scattering angle and $\delta\equiv(N_f-N_i)/N_i$ is the "depletion factor", which is a relative change in the photon occupation number. The "quantum intensity parameter" μ_0 can be defined in a similar manner as in the semiclassical theory, in Section 2, such that in the present definition we *formally* associate an amplitude $A_0=(c/\omega_0)(2\pi\hbar\rho)^{1/2}$ for the vector potential, where $\rho=\langle N_i\rangle/L^3$ is the initial photon density.

In the following we present the simplest quantal description of high-harmonic generation in Thomson (Compton) scattering, where we describe the electron by a classical current density $\mathbf{j}(\mathbf{r},t)$, which is induced by the incoming strong laser field. This part corresponds to the description listed in cell No. 3 in Table 1. According to Eq. (3a) we take a nonrelativistic oscillation caused by a moderately intense field, i.e. $\mathbf{v}(t)=-\varepsilon_0 v_{osc}\sin(\omega_0 t+\varphi_0)$ and $\mathbf{r}(t)=\varepsilon_0 x_{osc}\cos(\omega_0 t+\varphi_0)$, thus we do not consider the effect of the possible drift motion (in fact, for an electron being initially at rest, this assumption does not correspond to any approximation, except that the special case of a smooth switching-on and -off is considered). The equation of motion of the quantized readiation field reads:

$$i\hbar\partial_t|\Phi\rangle=H_{int}|\Phi\rangle,\quad H_{int}=-(1/c)\int d^3r\mathbf{j}(\mathbf{r},t)\cdot\mathbf{A}_Q(\mathbf{r},t),\quad \mathbf{j}(\mathbf{r},t)=-e\mathbf{v}(t)\delta(\mathbf{r}-\mathbf{r}(t)), \qquad (21a)$$

$$\mathbf{A}_Q(\mathbf{r},t)=\Sigma_{ks}[\mathbf{g}(\mathbf{k},s;\mathbf{r},t)a_{ks}+\mathbf{g}^*(\mathbf{k},s;\mathbf{r},t)a^\dagger_{ks}], \qquad (21b)$$

$$\mathbf{g}(\mathbf{k},s;\mathbf{r},t)\equiv c(2\pi\hbar/\omega_k L^3)^{1/2}\boldsymbol{\varepsilon}(\mathbf{k},s)\exp(i\mathbf{k}\cdot\mathbf{r}-i\omega_k t),\quad \omega_k=c|\mathbf{k}|,\quad [a_{ks},a^\dagger_{k's'}]=\delta_{kk'}\delta_{ss'}\,, \qquad (21c)$$

where $\mathbf{A}_Q(\mathbf{r},t)$ is the vector potential of the whole quantized radiation field decomposed into the vector plane wave modes represented by $\mathbf{g}(\mathbf{k},s;\mathbf{r},t)$. The exact solution of Eq. (21a) can be expressed with the help od the displacement operators $D[\alpha_{ks}(t)]$ for each mode, by properly determining the parameters $\alpha_{ks}(t)$, as is shown in the following equation:

$$|\Phi(t)\rangle=\Pi_{ks}D[\alpha_{ks}(t)]\,|\Phi(t_0)\rangle,\quad \alpha_{ks}(t)=(i/\hbar c)\int dt\int d^3r\mathbf{j}(\mathbf{r},t)\cdot\mathbf{g}^*(\mathbf{k},s;\mathbf{r},t). \qquad (22)$$

We summarize some basic properties of the displacement operator $D(\alpha)$ and the coherent states denoted by $|\alpha\rangle$, which are generated from vacuum under the action of the classical current. In order to simplify these formulae we do not display the mode index {**k**, s} (where **k** is the propagation vector and s is the polarization index of the scattered radiation):

$$D(\alpha)\equiv\exp(\alpha a^{\dagger}-\alpha^{*}a), \quad D(\alpha)|0\rangle=|\alpha\rangle=\Sigma_{k}\alpha^{k}/(k!)^{1/2}|k\rangle\exp(-|\alpha|^{2}), \quad a|\alpha\rangle=\alpha|\alpha\rangle, \tag{23}$$

Finally we note that the the displacement property $D^{\dagger}(\alpha)aD(\alpha)=a+\alpha$ means a shift of the quantized amplitudes. By putting explicit form of the electron current in the integral on the right hand side of Eq. (22), and using the Jacobi-Anger formula for the generation of the ordinary Bessel functions, we obtain:

$$\alpha_{ks}(t)=(i/\hbar)ce\mu_{0}(2\pi\hbar/\omega_{k}L^{3})^{1/2}(\varepsilon_{0}\cdot\varepsilon(\mathbf{k},s))\Sigma_{n}\{nJ_{n}[\mu_{0}\lambda_{0}(\varepsilon_{0}\cdot\mathbf{k})]/\mu_{0}\lambda_{0}(\varepsilon_{0}\cdot\mathbf{k})\}\int dt\exp[i(\omega_{k}-n\omega_{0})t]. \tag{24}$$

The time integral for large interaction times give a disctrete sequence of resonant frequencies $\omega_{k}=n\omega_{0}$, which are just the high-harmonic frequencies. In general, if the quantized field is initially on the vacuum state (thus we do not consider induced processes), i.e. $|\Phi(t_{0})\rangle=|0\rangle$, then the quantum state $|\Phi(t)\rangle$, developing due to the interaction with the oscillating electron, will be a multimode coherent state. Because of the frequency condition coming from the resonant time integral in Eq. (24), each harmonic components will be in a coherent state, regardless of their average occupation (which is governed by the size of the Bessel functions). The expectation value of the energy of a particular component is given by the expression:

$$\langle(n\hbar\omega_{0})a^{\dagger}_{n\omega0s}a_{n\omega0s}\rangle=r_{0}^{2}|\mathbf{s}\times\varepsilon_{0}|^{2}n^{4}[2J_{n}(z)/z]^{2}, \quad \mathbf{s}\equiv\mathbf{k}/|\mathbf{k}|, \quad z\equiv n\mu_{0}(\mathbf{s}\cdot\varepsilon_{0}). \tag{25}$$

The formula, Eq. (25) is equivalent with the the classical formula for the high-harmonic production in nonlinear Thomson scattering, however, it contains only the first moment of the photon distribution, so it should be considered as a mean value. Of course, each harmonics of the scattered quantum field is loaded by inherent fluctuations, thus, the phases of the components are uncertain to a lesser or larger extent.

The simplest description of the phase uncertainties of a quantum field can be formulated in terms of the Susskind and Glogower „cosine" and „sine" operators, C and S, respectively, which are defined in analogy with the cosine and sine of a phase of the complex number (see e.g. Varró (2008c)). By introducing the *formal* polar decomposition of the quantized amplitudes, $a=E(a^{\dagger}a)^{1/2}$, $a=(a^{\dagger}a)^{1/2}E^{\dagger}$, we define

$$C\equiv(E+E^{\dagger})/2, \quad S\equiv(E-E^{\dagger})/2i, \quad \Delta C^{2}\equiv\langle(C-\langle C\rangle)^{2}\rangle, \quad \Delta S^{2}\equiv\langle(S-\langle S\rangle)^{2}\rangle, \tag{26}$$

where the variances ΔC^{2} and ΔS^{2} characterize the phase uncertainties. In Fig. 7 we have plotted the sum of this variances as a function of frequency in a Gaussian few-cycle pulse (which may result from high-harmonic generation, resulting in a quantum state like $|\Phi(t)\rangle$ above in Eq. (22)).

The Figure 7 clearly shows that the quantum phase uncertainty is quite steeply increasing as the spectral components are getting off the cental peak, and finally the uncertainty saturates, and takes its maximum value 1. It should be kept in mind that in general the nonlinear sources cannot be described by classical currents. According to our recent study(see Varró (2008b) and Varró (2010))., the joint quantum interaction with localized electron wave packets result in the generation of phase eigenstates of Jackiw type, thus they have very different correlation properties in comparison with coherent states, which are set to be the closest analogons of classical stable fields.

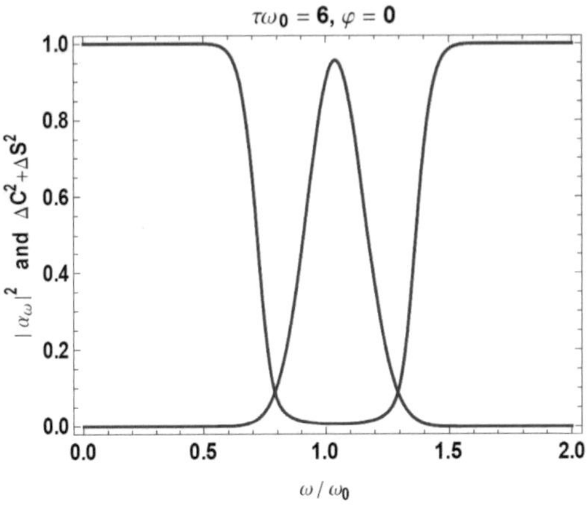

Fig. 7. Shows the normalized quantum amplitude distribution $|\alpha_\omega|^2$ of a 3−cycle femtosecond pulse and the dependencece of the quantum phase uncertainty $\Delta C^2+\Delta S^2$ associated to each spectral components with normalized frequency ω/ω_0. The pulse has been represented by a continuous Gaussian multimode coherent state, like that given in Eq. (22). The spectrum is peaked around ω_0 , where the quantum phase uncertainty has its minimum (it is essentially very close to zero).

6. References

Agostini, P. (2001). Two-color and single-color above threshold ionization. In *Atoms, Solids and Plasmas in Super-intense Laser Fields.* edited by Batani D, Joachain C J, Martellucci S and Chester A N, pp 59-80. (Kluwer Academic/Plenum Publishers, New York, 2001)

Agostini, P. & DiMauro, L. F. (2004). *Rep. Progr. Phys.* 67, 813.

Banfi, F.; Gianetti, C.; Ferrini, G.; Galimberti, G.; Pagliara, S.; Fausti, D. & Parmigiani, F. (2005). *Phys. Rev. Lett.* 94, 037601.

Bergou, J. & Varró, S. (1981a). Nonlinear scattering processes in the presence of a quantised radiation field: I. Non-elativistic treatment. *J. Phys. A: Math. Gen.* 14, 1469-1482.

Bergou, J. & Varró, S. (1981b). Nonlinear scattering processes in the presence of a quantised radiation field: II. Relativistic treatment. *J. Phys. A: Math. Gen.* 14, 2281-2303. (1981).

Boca, M. & Florescu, V. (2009). Nonlinear Compton scattering with a laser pulse. *Phys. Rev. A* 80, 053402 (2009).

Born, M. & Wolf, E. (2009). *The Principles of Optics.* (7th, expanded , fifth printing, Cambridge University Press, Cambridge, 2009).

Brabec, Th. & Krausz, F. (2000). Intense few-cycle laser fields: Frontiers of nonlinear optics. *Reviews of Modern Physics* 72, No. 2, April 2000, pp 545-595. (2000).

Brown, L. S. & Kibble, T. W. B. (1964). Interaction of intense laser beams with electrons. *Phys. Rev.* 133, No. 3A, A705-A719. (1964).

Bunkin, F. V. & Fedorov, M. V. (1965). Cold emission of electrons from the surface of a metal in strong radiation field. *Zh. Exp.Teor. Fiz.* 48, 1341-1646. [*Sov. Phys. JETP* 21, No. 5, November, pp 896-899. (1965)]

Büttiker, M. & Landauer, R. (1982). Traversal time for tunneling. *Phys. Rev. Lett.* 49, No. 23, 1739-1742. (1982).

Chiao, R. Y. (1998). Tunneling time and superluminarity: a tutorial. arXiv:quant-ph/9811019v1. (7. Nov. 1998).

Chin, S. L.; Farkas, Gy. & F. Yergeau, F. (1983). Observation of Kr and Xe ions created by intense nanosecond CO2 laser pulses. *J. Phys. B:At. Mol. Opt. Phys.*, 16, L223-L226.

Delone, N. D. & Krainov, V. P. (1994). *Multiphoton processes in atoms.* (Springer-Verlag, Berlin, 1994).

Ehlotzky, F. (2001). Atomic phenomena in bichromatic laser fields. *Physics Reports* 345, 175-264. (2001)

Ehlotzky, F.; Krajewska, K. & Kaminski, J. Z. (2009). Fundamental processes of quantum electrodynamics in laser fields of relativistic power. *Rep. Prog. Phys.* 72, 046401 (32pp). (2009).

Faisal, F. H. M. (1987). *Theory of multiphoton processes.* (Plenum, New York, 1987).

Farkas, Gy. (1978). *In* Eberly, J. H. & Lambropoulos, P. (Eds.) *Multiphoton Processes.* (New York, London: John Wiley & Sons, Inc., 1978).

Farkas, Gy.; Chin, S. L.; Galarneau, P. & F. Yergeau, F. (1983). A new type of intense CO2 laser induced electron emission from a gold surface. *Optics Comm.*, 48, 275-278.

Farkas, Gy. & Chin, S. L. (1984). Experimental investigation of the possibilities of the optical tunnelling of electron from a metal surface induced by strong CO2 laser pulses, in *Multiphoton Processes.* (Berlin, Springer-Verlag, 1984) pp. 191-199.

Farkas, Gy. Chin, S. L.; Galarneau, P. & Yergeau, F. (1984). First demonstration of ionization of Kr and Xe atoms by intense nanosecond CO2 laser pulses. In: Mandel, L. & Wolf, E. (Eds.) *Coherence and Quantum Optics V.* (Plenum, New York, 1984). pp 843-846.

Farkas, Gy. &Tóth, Cs. (1990). Energy spectrum of photoelectrons produced by picosecond laser-induced surface multiphoton photoeffect. *Phys. Rev. A* 41, 4123-4126.

Farkas, Gy. & Tóth, Cs. (1992). Proposal for attosecond light pulse generation using laser-induced multiple-harmonic conversion processes in rare gases. *Phys. Lett. A* 168, 447-450. (1992).

Farkas, Gy.,Tóth, Cs., Kőházi-Kis, A., Agostini, P., Petite, G., Martin, Ph., Berset, J. M. & Ortega, J. M. (1998). Infrared electron photoemission from a gold surface. *J. Phys. B: At. Mol. Opt. Phys. B* 31, L461-L468. (1998).

Fedorov, M. V.; Efremov, M. A.; Volkov, P. A. & Eberly, J. H. (2006). Short-pulse and strong-field breakup processes: A route to entangled wave-packets. *J. Phys. B: At. Mol. Opt. Phys.* 39, S467-S483. (2006).

Halpern, O. (1924). Zur Theorie der Röntgenstrahlstreuung. *Z. Phys.* 30, 153-172. (1924).

Ferrini, G.; Banfi, F.; Gianetti, C. & Parmigiani, F. (2009). Non-linear electron photoemission from metals with ultrashort pulses. *Nucl. Instr. Meth. Phys. Res. A* 601, 123-131. (2009).

Gaarde, M. B.; Tate, J. L. & Schafer, K. J. (2008). Macroscopic aspects of attosecond pulse generation. *J. Phys. B: At. Mol. Opt. Phys.* 41, 132001 (26pp) (2008).

Gao, J. (2004). Thomson scattering from ultrashort and ultraintense laser pulses. *Phys. Rev. Lett.* 93, 243001 (2004).

Gavrila, M. (1992). (Ed.) *Atoms in intense laser fields.* (Academic Press, New York, 1992).

Gies, H. (2009). Strong field as a probe for fundamental physics. *Eur. Phys. J. D* 55, 311-317.

Glauber, R. J. (1963). The quantum theory of optical coherence. *Phys. Rev.* 131, 2766-2788.

Gradshteyn, I. S. & Ryzhik, I. M. (2000). *Table of Integrals, Series, and Products.* (Sixth Edition, Academic Press, New York, 2000).

Halpern, O. (1924). Zur Theorie der Röntgenstrahlung. *Zeitschrift für Physik* 30, 153-172.

Keldish, L. V. (1964). *Zh. Exp.Teor. Fiz.* 47, 1945-1957. [*Sov. Phys. JETP* 20, 1307-1317 (1964)]

Klauder, J. R. & Sudarshan, E. C. G. (1968). *Fundamentals of quantum optics.* (W. A. Benjamin, Inc., New York, Amsterdam, 1968).

Krause, J. L.; Schafer, K. J. & Kulander, K. C. (1992). Calculation of photoemission from atoms subject to intense laser fields. *Phys. Rev. A* 45, 4998-5010. (1992).

Krausz, F. & Ivanov, M. (2009). Attosecond physics. *Rev. Mod. Phys.* 81, 163-234.

Kroó, N.; Varró, S.; Farkas, Gy.; Oszetzky, D.; Nagy, A. & Czitrovszky, A. (2007). Quantum metal optics. *Journal of Modern Optics* 54, 2679-2688. (2007).

Kroó, N,; Farkas, Gy.; Dombi, P. & Varró, S. (2008). Nonlinear processes induced by the enhanced , evanescent fields of surface plasmons excited by femtosecond laser pulses. *OPTICS EXPRESS* 16, 21656-21661. (2008).

Kylstra, N., Joachain, C. J. & Dörr M. (2001). Theory of multiphoton ionization of atoms. In *Atoms, Solids and Plasmas in Super-intense Laser Fields*, edited by Batani D, Joachain C J, Martellucci S and Chester A N, pp 15-36. (Kluwer Academic/Plenum Publishers, New York, 2001)

Landau, L. D. & Lifshitz, E. M. (1978). *Teoreticheskaya Fizika. III. Kvantovaya Mehanika.* (Izdatelstvo Nauka, Moscow, 1978). The author has used the the Hungarian translation by P. Boschán (Tankönyvkiadó, Budapest, 1978), based on the third, exteded edition, edited by L. P. Pitayevskii. The Keldish-method is discussed in details in the 3. excersize of § 77. on the pages 355-357 of this textbook. The application of the method of stationary phase for determining the ionization probabilty is worked out.

Lan, P.; Lu, P.; Cao, W. & Wang, X. (2007). Nonlinear Thomson scattering in the few-cycle regime. J. Phys. B: At. Mol. Opt. Phys. 40, 403-411. (2007).

Leubner, C. (1978). The spectrum of nonlinear inverse Compton radiation by the method of steepest descents. Astron. Astrophys. 69, 149-154. (1978).

Leubner, C. (1981). Spectral and angular distribution of synchro-compton radiation in a linearly polarized wacuum wave of arbitrary intensity. Astron. Astrophys. 96, 373-379. (1981).

Lewenstein, M., Balcou, Ph., Ivanov, M. Yu., L'Huillier, A. & Corkum, P. B. (1994). Theory of high-harmonic generation by low-frequency laser fields. *Phys. Rev. A* 49, 2117-2132.

Loudon, R. (2000). *The Quantum Teory of Light.* (University Press, Oxford, 2000)

Mackenroth, F.; Di Piazza, A. & Keitel, C. H. (2010). Determining the carrier-envelope phase of intense few-cycle laser pulses. Phys. Rev. Lett. 105, 063903 (2010).

Mittleman, M. H. (1993). *Introduction to the theory of laser-atom interactions.* (Plenum, New York, 1993).

Mourou, G. A.; Tajima, T. & Bulanov, S. V. (2006). Relativistic optics. *Rev. Mod. Phys.* 78, 309-371.

Odžak, S. & Milošević, D. B. (2005). High-order harmonic generation in the presence of a static electric field. *Phys. Rev. A* 72, 033407 (2005).

Paulus, G. G. & Walther, H. (2001). The classical and the quantum face of above-threshold ionization. In *Atoms, Solids and Plasmas in Super-intense Laser Fields.* edited by Batani D, Joachain C J, Martellucci S and Chester A N, pp 285-300. (Kluwer Academic/Plenum Publishers, New York, 2001)

Puntajer, A. K. & Leubner, C. (1990). Classical versus semiclassical predictions for harmonic generation in laser—free electron scattering under experimentally realizable conditions. J. Appl. Phys. 67, (3), 1607-1609. (1990).

Ritus, V. I. & Nikishov, A. I. (1979). Quantum electrodynamics of phenomena in intense fields. *Works of the Lebedev Physical Institute* 111, 5-278. (in Russian)

Salamin, Y. I.; Hu, S. X.; Hatsagortsian, K. Z. & Keitel, Ch. H. (2006). Relativistic high-power laser-matter interactions. *Physics Reports* 427, 41-155 (2006)

Schleich, W. P. (2001). *Quantum optics in phase space* (Wiley-VCH, Weinheim, 2001)

Scully, M. O. & Zubairy, M. S. (1997). *Quantum Optics* (University Press, Cambridge, 1997)

Varró, S. & Ehlotzky, F. (1992). Thomson scattering in strong external fields. *Zeitschrift für Physik D – Atoms, Molecules and Clusters* 22, 619-628 (1992).

Varró, S. & Ehlotzky, F. (1993). A new integral equation for treating high-intensity multiphoton processes. *IL Nuovo Cimento* 15 D, No. 1., 1371-1396 (1993).

Varró, S. & Ehlotzky, F. (1998a). High-order multiphoton ionization at metal surfaces by laser fields of moderate power. *Phys. Rev. A* 57, 663-666. (1998).

Varró, S. & Ehlotzky, F. (1998b). Redistribution of electron energies at the interface between laser-radiation-filled space and vacuum. *J. Phys. B : At. Mol. Opt. Phys.* 31, 2145-2151. (1998).

Varró, S.; Farkas, Gy. & Ehlotzky, F. (1999). Generation of x-rays by irradiating metal surfaces with a powerful laser beam in the presence of a strong static electric field. *Optics Communication* 172 , 47-53. (1999).

Varró, S. (2004). Scattering of a few-cycle laser pulse on a thin metal layer : the effect of the carrier-envelope phase difference. *Laser Physics Letters* 1, Nos. 1-2, 42-45. (2004).

Varró, S. (2007a). Reflection of a few-cycle laser pulse on a metal nano-layer: generation of phase dependent wake-fields. *Laser Physics Letters*, 4 , 138-144. (2007).

Varró, S. (2007b). Scattering of a few-cycle laser pulse by a plasma layer : the role of the carrier-envelope phase difference at relativistic intensities. *Las. Phys. Lett.* 4, 218-225. (2007).

Varró, S. (2007c). Linear and Nonlinear Absolute Phase Effects in Interactions of Ulrashort Laser Pulses with a Metal Nano-Layer or with a Thin Plasma Layer. *Laser and Particle Beams* 25, No. 3, 379-390. (2007). E-print: arxiv.org : physics/0610266

Varró, S. & Farkas, Gy. (2008a). Attosecond electron pulses from interference of above-threshold de Broglie waves. *Laser and Particle Beams* 26, 9-19. (2008). E-print: arXiv: 0705.0212v1 [physics.plasm-ph]

Varró, S. (2008b). Entangled photon-electron states and the number-phase minimum uncertainty states of the photon field. *New Journal of Physics* 10, 053028 (35pp).

Varró, S. (2008c). Correlations in single-photon experiments. *Fortschritte der Physik—Progress of Physics*, 56, No. 1, 91-102. (2008). E-print: arXiv: 0707.1305v1 [quant-ph]

Varró, S. (2010). Entangled states and entropy remnants of a photon-electron system. *Physica Scripta* 82, in press (8pp)

Varró, S.; Kroó, N.; Farkas, Gy. & Dombi, P. (2010). Spontaneous emission of radiation by metallic electrons in the presence of electromagnetic fields of surface plasmon oscillations. *Journal of Modern Optics* 57, 80-90. (2010). E-print: arXiv: 0705.0212v1 [physics.plasm-ph]

Walsh, T. D. G. ; Ilkov, F. A. & Chin, S. L. (1994). *J Phys B:At Mol Phys*, 27, 3767 (1994).

13

Multiphoton Selective Excitation and Analytical Control of Small Molecules in Intense Laser Fields: an Algebraic Model

Yujun Zheng
School of Physics, Shandong University, Jinan 250100
China

1. Introduction

The advent of high-power and short-pulse laser technology has led to the study of atomic and molecular multiphoton processes because of its importance in many areas, including photochemistry, fluorescence imaging, and photoionization (Denk et al., 1990; Blackwell et al., 1997; Constant et al., 1996; Paramonov, 2005; Windhorn et al., 2002; Oomens et al., 2004). The quantum control of problem by laser is also an active subject both in the theoretical and experimental area (Tannor & Rice, 1985; Brumer & Shapiro, 1986; Judson & Rabitz, 1992; Levis et al., 2001; Zou et al., 2006). The studies of the quantum control are motivated by fundamental interest in quantum properties of light, and the possible applications in physics, chemistry, and quantum computation, cryptology etc.. The comprehensive reviews could be found in Refs. (Gordon & Rice, 1997; Rice & Zhao, 2000; Brumer & Shapiro, 2003; Kral et al., 2007). One of the considerable attention in this field is multiphoton selective (vibrational) excitation and dissociation, which is important to control the chemical reaction and the quantum states on demand. Various approaches are suggested, for example, the pump-dump laser, optimal control theory, phase control etc.. Some theoretical and computational methods have been developed for improvements in experiment, such as the Coulter transformation method, the Floquet theory method, and dressed molecule picture etc. (Colgan et al., 2001; Chang et al., 1985; Chu & Telnov, 2004; Leforestier & Wyatt, 1985; 1983; Dibble & Shirts, 1991). Much of the work, however, has studied the problem by numerically solving the time-dependent Schrödinger equation (TDSE) as an initial value problem in Hilbert space. The development of generalized Floquet formalisms allows the reduction of the periodical or quasiperiodical time-dependent Schrödinger equation into a set of time-independent coupled equations or Floquet matrix eigenvalue problem. The Floquet method has also been employed to study the problem of molecules in intense laser fields, but it has usually been used to discuss the atomic problems (Leasure et al., 1981; Burke et al., 2000; Colgan et al., 2001; Chu & Telnov, 2004).

In this chapter we present the algebraic approach on a different background. In the algebraic framework we can obtain the explicit expression of the time-evolution operator directly, which avoiding the complex process of solving the time-dependent Schrödinger equation numerically. The time-evolution operator can be expressed a product of a finite number of exponential operators if the operators in Hamiltonian close under commutation. The parameters in time-evolution operator satisfy a set of differential equations. For the past years the application of time-dependent problems of the dynamical Lie algebraic approach are advanced. Algebraic methods have been extensively used to study problems in nuclear physics, molecular physics and quantum optics etc., for example, the problems of vibrational excited states and potential energy surfaces for small ployatomic molecules have been solved successfully (Iachello, 1981; van Roosmalen et al., 1984; Benjamin et al., 1983; Ding & Zheng, 1999; Zheng & Ding, 2001; 1999; Benjamin et al., 1983; Lu & Kellman, 1995; 1997; Kellman, 1995; Zheng & Ding, 2001). This approach is recently used to study the interacting quantum system, the dynamical symmetries and its breaking in nanophysics, Tavis-Cummings problem and the decoherence in a general three-level system, the dynamical entanglement of vibrations etc.. (Batista & Ortiz, 2004; Kikoin et al., 2004; Vadeiko & Miroshnichenko, 2003; Rau & Zhao, 2005; Zheng, 2006; Hou et al., 2006; Dai et al., 2002; Feng et al., 2007a;b; Liu et al., 2008; Feng et al., 2010). In the present work, both the multiphoton selective excitation and analytical control of small molecules (diatomic and triatomic molecules) in intense laser fields are studied in the algebraic framework. The influences of laser pulse frequencies and shapes on control are discussed.

In sec. 2 we derived the time-evolution operator of the system by using algebraic method for the diatomic and triatomic molecules. The analytical expression of the transition probability is obtained via the time-evolution operator. In sec. 3 we discussed the selective vibrational excitation of OH and OD by shaped and chirped laser pulses and the multiphoton excitation problem. For triatomic molecules, we calculated the selective bond excitation of HCN and DCN molecules in the linear chirped laser pulses with different shapes. The problem of intramolecular vibrational redistribution (IVR) is also discussed in this section. The chapter ends with some concluding remarks in sec. 4.

2. Theoretical background

2.1 The general considerations

We consider the problem of small molecules (diatomic and triatomic molecules) in the laser field, the more ambitious case of including other environment are not pursued at this time. The Hamiltonian of the system considered here can be written as

$$\mathcal{H} = \mathcal{H}_{mol} + \mathcal{V}, \tag{1}$$

where $\mathcal{H}_{mol}$ is the Hamiltonian of an unperturbed molecule, and $\mathcal{V}$ is its interaction with a laser field. As usually, we here consider the interaction between the molecule and the laser field within the electric dipole approximation,

$$\mathcal{V} = -\boldsymbol{\mu} \cdot \mathcal{E}(t), \tag{2}$$

where $\mathcal{E}(t)$ is the external laser field, and $\boldsymbol{\mu}$ is the molecular dipole moment.

It is suitable to consider this problem in *the interaction representation*. The Hamiltonian (1) could be splited into an $\mathcal{H}_0$ part and into an additional " interaction " $\mathcal{H}'$, namely, the Hamiltonian (1) can be rewritten as

$$\mathcal{H} = \mathcal{H}_0 + \mathcal{H}'(t). \tag{3}$$

In the interaction representation, the system Hamiltonian is written as

$$\mathcal{H}_I(t) = e^{i\mathcal{H}_0 t/\hbar}\mathcal{H}'(t)e^{-i\mathcal{H}_0 t/\hbar}, \tag{4}$$

and the time-evolution operator is given by

$$i\hbar\frac{\partial}{\partial t}\mathcal{U}_I(t) = \mathcal{H}_I(t)\mathcal{U}_I(t), \tag{5}$$

with the initial condition $\mathcal{U}_I(t=0)=1$. Once we know the evolution operator of the system, we can extract all the information of the system from the evolution operator. Hence, the influences of laser pulse frequencies and shapes can be investigated and take these quantum control on demand. Specially, here we are interested in the transition probability of the system from the initial state $|v_i\rangle$ to the final state $|v_f\rangle$,

$$\mathcal{P}_{v_i,v_f}(t) = \left|\langle v_f|\mathcal{U}_I(t)|v_i\rangle\right|^2. \tag{6}$$

The corresponding long-time average probability is defined as

$$\langle\mathcal{P}_{v_i,v_f}\rangle = \lim_{T\to\infty}\left\{\frac{1}{T}\int_0^T \mathcal{P}_{v_i,v_f}(t)dt\right\}. \tag{7}$$

Also, the long-time-averaged absorption energy spectra are obtained using

$$\langle\varepsilon\rangle = \sum_f \langle\mathcal{P}_{v_i,v_f}\rangle\varepsilon_{v_i,v_f}, \tag{8}$$

and the averaged number of photons absorbed by the molecule can be calculated by

$$\langle n(t)\rangle = \sum_f \frac{\varepsilon_{v_i,v_f}}{\hbar\omega_L}\mathcal{P}_{v_i,v_f}(t), \tag{9}$$

where ε_{v_i,v_f} is the energy difference between the states $|v_f\rangle$ and $|v_i\rangle$, ω_L is the frequency of the external laser field. Here we neglect the obvious labels of the various aspects of the laser field in Eqs. (6) and (7). In the following sections we would obtain the obvious expressions of Eqs. (6) $\sim$ (9) in the theoretical framework.

2.2 Theoretical framework for diatomic molecules

We consider here the one-dimensional vibrational motion of a diatomic molecule. For completeness, we brief review the well-known results for the $U(2)$ dynamical symmetry of diatomic molecules. The $U(2)$ algebra is successfully applied to describe vibrations of diatomic molecule, and $U(2)$ possesses two dynamical symmetry chains (Levine, 1983; 1982; Kellman, 1985; Cooper, 1997; 1998):

$$\begin{aligned} U(2) &\supset U(1), \\ U(2) &\supset O(2). \end{aligned} \tag{10}$$

The generators of $U(2)$ could be realized in terms of two boson creation and annihilation operators:

$$t^\dagger t, s^\dagger s, t^\dagger s, s^\dagger t. \tag{11}$$

The new four generators could be constructed using these four bilinear products, familiar with Schwinger,

$$\begin{aligned} \hat{Q} &= (1/2N)^{1/2}(t^\dagger s + s^\dagger t), \\ \hat{P} &= (1/2N)^{1/2}(t^\dagger s - s^\dagger t), \\ \hat{I}_0 &= N^{-1}(s^\dagger s - t^\dagger t), \\ \hat{E}_0 &= N^{-1}(t^\dagger t + s^\dagger s), \end{aligned} \tag{12}$$

which N is the eigenvalues of the number operator $\hat{N}(=t^\dagger t + s^\dagger s)$.
The Hamiltonian of a non-rotating Morse oscillator could be written as

$$\mathcal{H}_{mol} = \hbar\omega_0(\frac{\hat{P}^2}{2} + \frac{\hat{Q}^2}{2}), \tag{13}$$

where ω_0 is the frequency of the anharmonic oscillator, $\hat{P}$ and $\hat{Q}$ are the bilinear products constructed by the generators of $U(2)$ algebra in Eq. (12) (Levine, 1983; 1982).
An equivalent expression of Morse oscillator Hamiltonian Eq. (13) is

$$\mathcal{H}_{mol} = \hbar\omega_0(\hat{A}^\dagger\hat{A} + \frac{\hat{I}_0}{2}), \tag{14}$$

where the creation ($A^\dagger$) and annihilation (A) operators are defined as (Levine, 1983),

$$\hat{A}^\dagger = \frac{1}{\sqrt{2}}(\hat{Q} - i\hat{P}), \quad \hat{A} = \frac{1}{\sqrt{2}}(\hat{Q} + i\hat{P}), \tag{15}$$

which obey the commutation relations

$$[A, A^\dagger] = I_0, \quad [I_0, A] = -2\chi_0 A, \quad [I_0, A^\dagger] = 2\chi_0 A^\dagger. \tag{16}$$

It should be noted that I_0 is an operator and tends to the identity operator in the harmonic limits, and $\chi_0 \to 0$ in this limits. The anharmonic correction is given in order of $\chi_0 = \frac{1}{N}$.
In our numerical calculation we would take the replace of N by $N+1$. According to Cooper *et al* (Cooper, 1997; 1998) this modification can lead to quite accurate fits to the vibrational energy levels of diatomic molecules, potential function including the zero-point energy and achieves the known value of dissociation energy.
The molecular dipole moment $\boldsymbol{\mu}(x)$ could be expanded in a series at the equilibrium, and we keep the first order in our present work, as in Refs.(Hay & Dunning, 1976; Walker & Preston, 1977; Tung & Yuan, 1987; Chelkowski, 1995; Dai et al., 2002), i.e.

$$\boldsymbol{\mu}(x) = \boldsymbol{\mu}_0 x. \tag{17}$$

The interaction between the molecule and the laser field Eq. (2), by using Eq. (15), can be rewritten as

$$\mathcal{V} = -\frac{d(t)}{2}\left(A^\dagger + A\right), \tag{18}$$

where the time-dependent parameter $d(t)$ is

$$d(t) = \boldsymbol{\mu}_0 \cdot \mathcal{E}(t)\frac{1}{\alpha}\sqrt{\frac{\hbar\omega_0}{D}}. \tag{19}$$

Here we choose $\mathcal{H}_0 = \mathcal{H}_{mol}$ and $\mathcal{H}' = \mathcal{V}$, the system Hamiltonian, in *the interaction representation* from Eq.(4), reads

$$\begin{aligned} \mathcal{H}_I(t) &= e^{i\mathcal{H}_0 t/\hbar}\mathcal{V}e^{-i\mathcal{H}_0 t/\hbar} \\ &= e^{i\omega_0\chi_0 t\hat{A}^\dagger\hat{A}}e^{i\omega_0 t I_0/2}d(A^\dagger + A)e^{-i\omega_0 t I_0/2}e^{-i\omega_0\chi_0 t A^\dagger A} \\ &= de^{i\omega_0\chi_0 t}e^{i\omega_0 I_0 t}A^\dagger de^{i\omega_0\chi_0 t}e^{-i\omega_0 I_0 t}A \\ &\equiv \gamma_+ A^\dagger + \gamma_- A. \end{aligned} \tag{20}$$

Here we have taken the operator I_0 as an identity approximately in the exponential term since the anharmonic correction is of order χ_0. It is often of the order of 1% or less for realistic molecules (Levine, 1983).
Since the generators $A^{\dagger}$, A and I_0 form a closed Lie algebra (see Eq. (16)), the evolution operator can be written as (Wei & Norman, 1964; Alhassid & Levine, 1978)

$$\begin{aligned} \mathcal{U}_I &= e^{-\frac{i}{\hbar}\mu_0 \hat{I}_0} e^{-\frac{i}{\hbar}\mu_+ \hat{A}^{\dagger}} e^{-\frac{i}{\hbar}\mu_- \hat{A}} \\ &\equiv \prod_r U_r(t), \text{ and } U_r(t) = e^{-\frac{i}{\hbar}\mu_r(t) X_r}, \end{aligned} \tag{21}$$

where the coefficients $\mu_r(t)$ $(r = 0, +, -)$ are known as the Lagrange parameters (Alhassid & Levine, 1978). The operators in Eq. (21) are $X_0 \equiv I_0$, $X_+ \equiv A^{\dagger}$ and $X_- \equiv A$.
The equations of $U_r(t)$, from the Eqs.(5) and (20), are as

$$i\hbar \frac{\partial U_r(t)}{\partial t} = \frac{\partial \mu_r(t)}{\partial t} X_r U_r(t), \tag{22}$$

or the equations of the Lagrange parameters $\mu_r(t)$, by using the Baker-Hausdorff expansion, are given by

$$\begin{aligned} \dot{\mu}_0 &= -\frac{i}{\hbar}\mu_+ \gamma_- e^{\frac{i}{\hbar}2\chi_0\mu_0}, \\ \dot{\mu}_+ &= \gamma_+ e^{-\frac{i}{\hbar}2\chi_0\mu_0} - \frac{\chi_0}{\hbar^2}\mu_+^2 \gamma_- e^{\frac{i}{\hbar}2\chi_0\mu_0}, \\ \dot{\mu}_- &= \gamma_- e^{\frac{i}{\hbar}2\chi_0\mu_0}, \end{aligned} \tag{23}$$

with the initial conditions

$$\mu_r(t=0) = 0, \quad (r = 0, \pm). \tag{24}$$

The probability from the initial state $|v_i\rangle$ to the final state $|v_f\rangle$, from Eq. (6), is

$$\mathcal{P}_{v_i,v_f}(t) = |\langle v_f | \mathcal{U}_I(t) | v_i \rangle|^2 = |\lambda(t)|^2 \delta_{v_f, v_i - m + n}, \tag{25}$$

where

$$\begin{aligned} \lambda(t) &= \exp\{ -\frac{i}{\hbar}\mu_0[1 - 2\chi_0(v_i - m + n)]\} \\ &\times \sum_{n=0}^{\infty} \frac{1}{n!}(-\frac{i}{\hbar}\mu_+)^n \sqrt{\prod_{n'=0}^{n} [1 - \chi_0(v_i - m' + n' - 1)](v_i - m' + n')} \\ &\times \sum_{m=0}^{\infty} \frac{1}{m!}(-\frac{i}{\hbar}\mu_-)^m \sqrt{\prod_{m'=0}^{m} [1 - \chi_0(v_i - m')(v_i - m' + 1)]}. \end{aligned} \tag{26}$$

The analytic expression of vibrational transition probability is obtained and we can tackle many concrete examples using this expression.

2.3 Theoretical framework for triatomic molecules

For the case of triatomic molecule, its dynamical symmetry group is

$$U_1(2) \otimes U_2(2). \tag{27}$$

For stretching vibrations in a triatomic molecule, by using Eq. (15), the algebraic Hamiltonian reads

$$\mathcal{H}_m = \hbar\omega_{01}(\hat{A}_1^\dagger \hat{A}_1 + \frac{\hat{I}_{01}}{2}) + \hbar\omega_{02}(\hat{A}_2^\dagger \hat{A}_2 + \frac{\hat{I}_{02}}{2}) - \lambda(\hat{A}_1^\dagger \hat{A}_2 + \hat{A}_2^\dagger \hat{A}_1), \tag{28}$$

where ω_{01} and ω_{02} are the angular frequencies of the bond 1 and bond 2 of triatomic molecule, respectively. λ is the coupling coefficient, we include here the kinetic and potential linear couple terms.

The interaction between the triatomic molecule and the laser field can be expressed, similar to the diatomic molecule, as:

$$\mathcal{V} = d_1(\hat{A}_1^\dagger + \hat{A}_1) + d_2(\hat{A}_2^\dagger + \hat{A}_2), \tag{29}$$

where

$$d_1 = -\frac{1}{2\alpha_1}\sqrt{\frac{\hbar\omega_{01}}{D_1}}\boldsymbol{\mu}_1 \cdot \mathcal{E}(t), \text{ and } d_2 = -\frac{1}{2\alpha_2}\sqrt{\frac{\hbar\omega_{02}}{D_2}}\boldsymbol{\mu}_2 \cdot \mathcal{E}(t). \tag{30}$$

Here the linear dipole moment form of $\boldsymbol{\mu}_1$ and $\boldsymbol{\mu}_2$ (Umeda et al., 1994) is employed for the molecule.

In the interaction picture, the Hamiltonian of the system for the triatomic molecule is

$$\begin{aligned} \mathcal{H}_I(t) &= e^{i\mathcal{H}_0 t/\hbar}\mathcal{H}' e^{-i\mathcal{H}_0 t/\hbar} \\ &= d_1(\kappa_{1+}\hat{A}_1^\dagger + \kappa_{1-}\hat{A}_1) + d_2(\kappa_{2+}\hat{A}_2^\dagger + \kappa_{2-}\hat{A}_2) \\ &\quad -\lambda(\kappa_{1+}\kappa_{2-}A_1^\dagger A_2 + \kappa_{1-}\kappa_{2+}A_2^\dagger A_1), \end{aligned} \tag{31}$$

where

$$\kappa_{j+} = e^{i\omega_{0j}(x_{0j}+I_{0j})t}, \text{ and } \kappa_{j-} = e^{i\omega_{0j}(x_{0j}-I_{0j})t}, \ (j = 1,2). \tag{32}$$

In the calculation we have chosen

$$\begin{aligned} \mathcal{H}_0 &= \hbar\omega_{01}(\hat{A}_1^\dagger \hat{A}_1 + \frac{\hat{I}_{01}}{2}) + \hbar\omega_{02}(\hat{A}_2^\dagger \hat{A}_2 + \frac{\hat{I}_{02}}{2}), \\ \mathcal{H}' &= -\lambda(\hat{A}_1^\dagger \hat{A}_2 + \hat{A}_2^\dagger \hat{A}_1) + d_1(\hat{A}_1^\dagger + \hat{A}_1) + d_2(\hat{A}_2^\dagger + \hat{A}_2). \end{aligned} \tag{33}$$

We partition the Hamiltonian (31), for the simplicity of algebraic structure and calculation, into two parts

$$\mathcal{H}_I(t) = \mathcal{H}_I^{(0)} + \mathcal{H}_I^{(1)}(t), \tag{34}$$

where we let

$$\mathcal{H}_I^{(0)} = d_1(\kappa_{1+}\hat{A}_1^\dagger + \kappa_{1-}\hat{A}_1) + d_2(\kappa_{2+}\hat{A}_2^\dagger + \kappa_{2-}\hat{A}_2), \tag{35}$$

and

$$\mathcal{H}_I^{(1)}(t) = -\lambda(\kappa_{1+}\kappa_{2-}A_1^\dagger A_2 + \kappa_{1-}\kappa_{2+}A_2^\dagger A_1). \tag{36}$$

The evolution equation of the evolution operator $\mathcal{U}_I^{(0)}(t)$, corresponding to Hamiltonian (35), is as follows (Landau & Lifshitz, 1977)

$$i\hbar\frac{\partial}{\partial t}\mathcal{U}_I^{(0)}(t) = \mathcal{H}_I^{(0)}\mathcal{U}_I^{(0)}(t), \tag{37}$$

with the initial condition $\mathcal{U}_I^{(0)}(0) = 1$.

If $\mathcal{U}_I^{(0)}(t)$ is known, the evolution operator $\mathcal{U}_I(t)$ describing the whole system (corresponding total Hamiltonian (31)) will be given by (Landau & Lifshitz, 1977)

$$\mathcal{U}_I(t) = \mathcal{U}_I^{(0)}(t) \cdot \mathcal{U}_I^{(1)}(t), \tag{38}$$

where $\mathcal{U}_I^{(1)}(t)$ satisfies the evolution equation

$$i\hbar \frac{\partial}{\partial t} \mathcal{U}_I^{(1)}(t) = \mathcal{H}_I^{(1)\prime}(t) \mathcal{U}_I^{(1)}(t), \tag{39}$$

with the initial condition $\mathcal{U}_I^{(1)}(0) = 1$. The Hamiltonian $\mathcal{H}_I^{(1)\prime}(t)$ is given by

$$\mathcal{H}_I^{(1)\prime}(t) = \mathcal{U}_I^{(0)\dagger}(t) \mathcal{H}_I^{(1)}(t) \mathcal{U}_I^{(0)}(t). \tag{40}$$

It is easily to show that the operator in Hamiltonian (35) close under the commutation (16). The time-evolution operator $\mathcal{U}_I^{(0)}$ can be written as (Wei & Norman, 1964; Alhassid & Levine, 1978)

$$\mathcal{U}_I^{(0)} = e^{-\frac{i}{\hbar}\mu_{01} I_{01}} e^{-\frac{i}{\hbar}\mu_{1+} A_1^\dagger} e^{-\frac{i}{\hbar}\mu_{1-} A_1} e^{-\frac{i}{\hbar}\mu_{02} I_{02}} e^{-\frac{i}{\hbar}\mu_{2+} A_2^\dagger} e^{-\frac{i}{\hbar}\mu_{2-} A_2}. \tag{41}$$

Following the similar procedures to those carried out in preceding section, the Lagrange parameters are satisfied

$$\begin{aligned}
\dot{\mu}_{01} &= -\frac{i}{\hbar} d_1 \kappa_{1-} \mu_{1+} e^{\frac{i}{\hbar} 2\chi_{01}\mu_{01}}, \\
\dot{\mu}_{1+} &= d_1 \kappa_{1+} e^{-\frac{i}{\hbar} 2\chi_{01}\mu_{01}} - \frac{\chi_0}{\hbar^2} d_1 \kappa_{1-} \mu_{1+}^2 e^{\frac{i}{\hbar} 2\chi_{01}\mu_{01}}, \\
\dot{\mu}_{1-} &= d_1 \kappa_{1-} e^{\frac{i}{\hbar} 2\chi_{01}\mu_{01}}, \\
\dot{\mu}_{02} &= -\frac{i}{\hbar} d_2 \kappa_{2-} \mu_{2+} e^{\frac{i}{\hbar} 2\chi_{02}\mu_{02}}, \\
\dot{\mu}_{2+} &= d_2 \kappa_{2+} e^{-\frac{i}{\hbar} 2\chi_{02}\mu_{02}} - \frac{\chi_{02}}{\hbar^2} d_2 \kappa_{2-} \mu_{2+}^2 e^{\frac{i}{\hbar} 2\chi_{02}\mu_{02}}, \\
\dot{\mu}_{2-} &= d_2 \kappa_{2-} e^{\frac{i}{\hbar} 2\chi_{02}\mu_{02}}
\end{aligned} \tag{42}$$

The initial conditions are $\mu_r(t = 0) = 0$, $(r = 01, 1\pm, 02, 2\pm)$.
The Hamiltonian $\mathcal{H}_I^{(1)\prime}(t)$ can be then expressed as

$$\begin{aligned}
\mathcal{H}_I^{(1)\prime}(t) &= (\mathcal{U}_I^{(0)})^{-1} \mathcal{H}_I^{(1)}(t) \mathcal{U}_I^{(0)} \\
&= \beta_0 \hat{I}_{01} \hat{I}_{02} + \beta_1 \hat{I}_{02} \hat{A}_1^\dagger + \beta_2 \hat{I}_{02} \hat{A}_1 + \beta_3 \hat{I}_{01} \hat{A}_2^\dagger + \beta_4 \hat{I}_{01} \hat{A}_2 \\
&\quad + \beta_5 \hat{A}_1^\dagger \hat{A}_2 + \beta_6 \hat{A}_1 \hat{A}_2^\dagger + \beta_7 \hat{A}_1^\dagger \hat{A}_2^\dagger + \beta_8 \hat{A}_1 \hat{A}_2,
\end{aligned} \tag{43}$$

where

$$\beta_0 = \frac{\eta_1}{\hbar^2} \mu_{1-} \mu_{2+} (1 - \frac{\chi_{02}}{\hbar^2} \mu_{2+} \mu_{2-}) + \frac{\eta_2}{\hbar^2} \mu_{1+} \mu_{2-} (1 - \frac{\chi_{01}}{\hbar^2} \mu_{1+} \mu_{1-}),$$

$$\beta_1 = -\frac{i}{\hbar} \eta_1 \mu_{2+} (1 - \frac{\chi_{02}}{\hbar^2} \mu_{2+} \mu_{2-}) + \frac{i}{\hbar^3} \eta_2 \chi_{01} \mu_{1+}^2 \mu_{2-},$$

$$\beta_2 = -\frac{i}{\hbar^3} \eta_1 \chi_{01} \mu_{1-}^2 \mu_{2+} (1 - \frac{\chi_{02}}{\hbar^2} \mu_{2+} \mu_{2-}) + \frac{i}{\hbar} \eta_2 \mu_{2-} (1 - \frac{\chi_{01}}{\hbar^2} \mu_{1+} \mu_{1-})^2,$$

$$\beta_3 = \frac{i}{\hbar^3}\eta_1\chi_{02}\mu_{1-}\mu_{2+}^2 - \frac{i}{\hbar}\eta_2\mu_{1+}(1-\frac{\chi_{01}}{\hbar^2}\mu_{1+}\mu_{1-}),$$

$$\beta_4 = \frac{i}{\hbar}\eta_1\mu_{1-}(1-\frac{\chi_{02}}{\hbar^2}\mu_{2+}\mu_{2-})^2 - \frac{i}{\hbar^3}\eta_2\chi_{02}\mu_{1+}\mu_{2-}^2(1-\frac{\chi_{01}}{\hbar^2}\mu_{1+}\mu_{1-}),$$

$$\beta_5 = \eta_1(1-\frac{\chi_{02}}{\hbar^2}\mu_{2+}\mu_{2-})^2 + \frac{\eta_2}{\hbar^4}\chi_{01}\chi_{02}\mu_{1+}^2\mu_{2-}^2,$$

$$\beta_6 = \frac{\eta_1}{\hbar^4}\chi_{01}\chi_{02}\mu_{1-}^2\mu_{2+}^2 + \eta_2(1-\frac{\chi_{01}}{\hbar^2}\mu_{1+}\mu_{1-})^2,$$

$$\beta_7 = \eta_1\frac{\chi_{02}}{\hbar^2}\mu_{2+}^2 + \eta_2\frac{\chi_{01}}{\hbar^2}\mu_{1+}^2,$$

$$\beta_8 = \eta_1\frac{\chi_{01}}{\hbar^2}\mu_{1-}^2(1-\frac{\chi_{02}}{\hbar^2}\mu_{2+}\mu_{2-})^2 + \eta_2\frac{\chi_{02}}{\hbar^2}\mu_{2-}^2(1-\frac{\chi_{01}}{\hbar^2}\mu_{1+}\mu_{1-})^2. \quad (44)$$

The coefficients η_1 and η_2, in Eq. (44), are defined as

$$\eta_1 = -\lambda\kappa_{1+}\kappa_{2-}e^{\frac{i}{\hbar}2(\chi_{02}\mu_{02}-\chi_{01}\mu_{01})}, \text{ and } \eta_2 = -\lambda\kappa_{1-}\kappa_{2+}e^{\frac{i}{\hbar}2(\chi_{01}\mu_{01}-\chi_{02}\mu_{02})}. \quad (45)$$

The time-evolution operator $\mathcal{U}_I^{(1)}(t)$, corresponding to Hamiltonian Eq. (43), can be obtained, by using the Magnus approximation,

$$\mathcal{U}_I^{(1)}(t) = \exp\left[\sum_{m=1}^{\infty}\Gamma_m(t)\right], \quad (46)$$

where Γ_m denotes the integrals of m-fold multiple commutators (Klarsfeld & Oteo, 1989). The total time-evolution operator $\mathcal{U}_I(t)$ of the whole system can be calculated by using Eq. (38). The transition probability from state $|v_{1i}, v_{2i}\rangle$ to state $|v_{1f}, v_{2f}\rangle$ is

$$P_{if}(t) = |\langle v_{1f}, v_{2f}|\mathcal{U}_I(t)|v_{1i}, v_{2i}\rangle|^2, \quad (47)$$

The explicit expression of the vibrational transition probability can be found in Ref. (Feng et al., 2007a). The average energy in bonds can also be obtained by using the time-evolution operator.

3. Numerical results and discussions

3.1 Diatomic molecules

In this section, we study the multiphoton transition of the diatomic molecules. For the comparison with previous study, we use the local OH and OD bond of the HOD molecule in our numerical model. These models are extensively studied in previous study (Korolkov & Paramonov, 1997; Stranges et al., 2002; Amstrup & Henriksen, 1992; Elghobashi et al., 2003). All parameters, in atomic units, are taken from Refs. (Korolkov & Paramonov, 1997; Stranges et al., 2002; Amstrup & Henriksen, 1992; Elghobashi et al., 2003), namely, ω_0=0.01664a.u., χ_0=0.02323a.u., D=0.1614a.u., α=1.156a.u. for OH and ω_0=0.0122a.u., χ_0=0.01645a.u., D=0.1636a.u., α=1.142a.u. for OD, respectively.

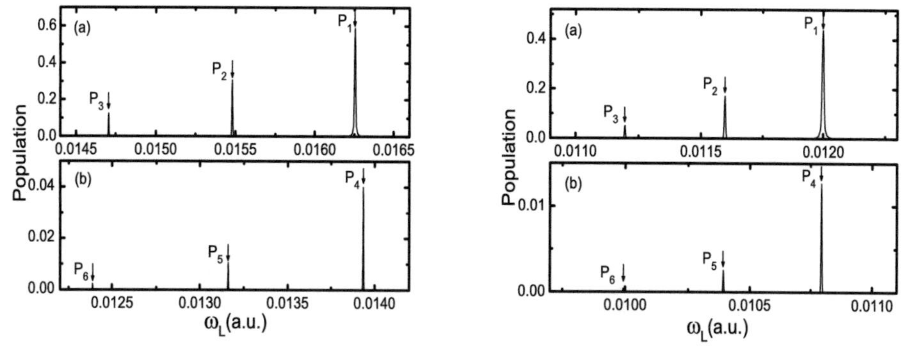

Fig. 1. The long-time average populations of OH (left) and OD (right) as a function of laser frequency. (a)The transitions from the ground state to the first, second and third excited states. (b)The transitions from the ground state to the fourth, fifth and sixth excited states. P_i $(i = 1,2,3,4,5,6)$ denotes the transition $|v = 0\rangle \rightarrow |v = i\rangle$.

3.1.1 The multiphoton vibrational transition

We first consider the simple case: *the multiphoton vibrational transition*. In this case we choose the laser field as follows

$$\mathcal{E}(t) = \mathcal{E}_0 sin(\omega_L t). \tag{48}$$

The initial state of the molecules are set on the ground state. The frequency of the infrared laser field is adjusted to the resonant frequency ω_r.

Multiphoton transitions are said to be in resonance if the energy of one or several quanta of laser photons is close to the transition energy from the initial to an intermediate excited state or to that between the intermediate. The transition is called multiphoton resonance transition, if the following condition is satisfied (Jakubetz et al., 1990; Lin & Fujimura, 1984)

$$\omega_r \cong \omega_n = \frac{\varepsilon_f - \varepsilon_0}{n\hbar}, \tag{49}$$

where $\varepsilon_f - \varepsilon_0$ is the energy gap between the ground state and the fth excited state.

Fig. 1 gives the long-time average probabilities of OH and OD from the ground state to the different excited states using Eq. (7). We can obtain the resonance frequency ω_r corresponding to the peaks of the transition probabilities. The orders of multiphoton transitions could be determined using *the formal intensity law* (Lin & Fujimura, 1984),

$$\ln P_n = n \ln I + C, \tag{50}$$

where I is the intensity of the laser field, P_n is the transition probability for the n-photon process, and C is a constant.

If no saturation occurs, one can determine the order of the multiphoton transition from the slope of Eq. (50). The order of the n-photon transition could be known from the slopes of the log-log plots of the transition probabilities as a function of laser intensity I, which correspond to the transition probabilities in Fig. 1, respectively. The detail results are summarized in Tab. 1.

As shown in Tab. 1, the transitions from the ground state to the first, second, third, fourth, fifth and sixth excited states belong to one-, two-, three-, four-, five-, six-photon transitions. However, the deviation between the transition frequency ω_r and the frequency

		P_1	P_2	P_3	P_4	P_5	P_6
	ω_r	0.016253	0.015480	0.014707	0.013934	0.013161	0.012388
OH	n	1.002	2.004	3.005	4.008	5.003	5.984
	ω_n	0.016221	0.015835	0.015453	0.015063	0.014698	0.014359
	$\lvert\omega_n - \omega_r\rvert$	0.000032	0.000355	0.000746	0.001129	0.001537	0.001971
	ω_r	0.011999	0.011598	0.011197	0.010796	0.010395	0.009994
OD	n	1.003	2.014	3.011	3.973	5.002	6.020
	ω_n	0.011963	0.011716	0.011556	0.011474	0.011192	0.010959
	$\lvert\omega_n - \omega_r\rvert$	0.000036	0.000118	0.000359	0.000678	0.000797	0.000965

*All parameters are atomic units except n.

Table 1. Multiphoton vibrational transitions of OH and OD

ω_n is becoming larger. This deviation denotes the non-resonant transition appears when the vibrational state goes up.

From Fig. 1, we can also see that the resonance frequencies ω_r are little shifting when the system reaches the different finial states. Namely, the resonant transition frequency decreases with the final energy level going up. And because of the higher dissociation energy in deuterated case the transition probabilities of OD are all smaller than those of OH at the same laser intensity. The value of the interval between the transition peaks in left panel of Fig. 1 (for OH) is about $0.000773a.u.$, and the value in right of Fig. 1 (for OD) is about $0.000401a.u.$. However, the interval of the transition peaks of OH is larger than that of OD. We attribute it to the difference of anharmonicity parameter of OH and OD, i.e., the anharmonic parameters of OH is larger than that of OD. The ratio of the anharmonicity parameters of OH and OD is 1.4 which is comparable to the ratio of 0.000773 to 0.000401 (≈ 1.9). Furthermore, we can see the deviations ($\omega_r - \omega_n$) of OH are also larger than those of OD in Tab. 1. These results show the anharmonicity of molecular vibrations has an important influence on the resonant transition frequency.

3.1.2 The multiphoton selective vibrational transition of OH

Sugimori *et al.* investigated the multiphoton absorption of OH using numerical simulation (Sugimori et al., 2005). For the comparison, we here take the same molecular parameters as in Ref. (Sugimori et al., 2005). We make compare all the calculations for the 1-photon, 2-photon, 3-photon and 4-photon absorption in Ref. (Sugimori et al., 2005) with our analytical results. In this comparison, the laser field is employed with the same with Ref. (Sugimori et al., 2005)

$$\begin{aligned} \mathcal{E}(t) &= \mathcal{E}_0 g(t)\sin(\omega_L t), \\ g(t) &= exp[-(t-T_0)^2/\sigma^2], \end{aligned} \tag{51}$$

where $T_0 = 100ps$, $\sigma = 30ps$ and the laser frequency ω_L is calculated using Eq.(49). In Tab. 2 a comparison is made between numerical results in Ref. (Sugimori et al., 2005) and our analytical results. A good agreement is found, indicating that our algebraic model is excellent approximation.

Since there exist a little deviation of the resonant transition frequency for the different transition(see details in Tab. 1), we then take the calculated frequency as the laser frequency and study the multiphoton selective vibrational transition of OH. We find the selectivity of the vibrational transition will decrease as the value of the laser frequency has a little change (about $0.000001a.u. \approx 6.6 \times 10^3 MHz$). The little changes on the laser frequency may cause

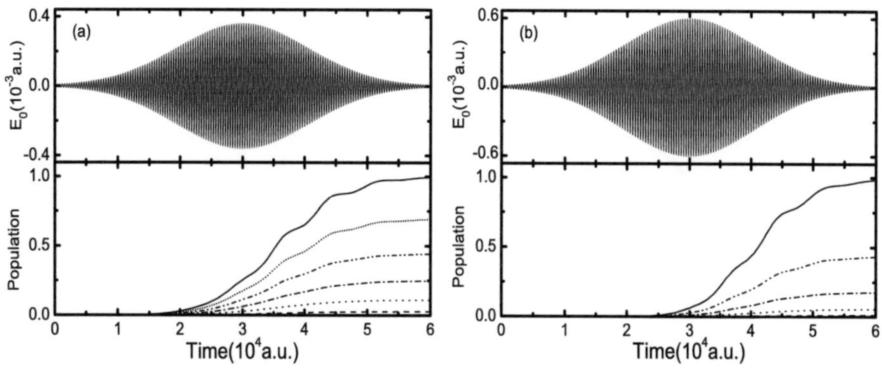

Fig. 2. Time dependence of the laser field and population: (a)1-photon transition between $|0\rangle$ and $|1\rangle$, (b)2-photon transition between $|0\rangle$ and $|2\rangle$, by changing the laser intensity with the fixed pulse duration at the non-chirped laser pulse. The laser parameters are given in sec.3.1.3.

	OH^*		OH	
	I_{max}	Ratio(%)	I_{max}	Ratio(%)
1-photon	4.74×10^6	100.00	4.68×10^6	99.99
2-photon	5.68×10^8	99.98	5.57×10^8	99.49
3-photon	5.58×10^9	99.88	5.42×10^9	99.23
4-photon	2.36×10^{10}	94.51	2.27×10^{10}	95.31

* Numerical solution from Ref. (Sugimori et al., 2005).

Table 2. The comparison of maximum laser intensity $I_{max}(W/cm^2)$ and the ratio of the target state population

the less selectivity and the stronger laser intensity. This denotes the selectivity of vibrational transition is sensitive to the laser frequency.

3.1.3 Control of vibrational excitation of OH and OD molecules

We have obtained the resonant transition frequencies of OH and OD from the ground state to some excited states in previous section. In this section, we use these resonant frequencies to study the selective vibrational transition of OH and OD. The laser field is

$$\begin{aligned} \mathcal{E}(t) &= \mathcal{E}_0 f(t) \cos \Phi(t), \\ f(t) &= \exp[-(t-\tau/2)^2/(\tau/4)^2], \end{aligned} \tag{52}$$

where the Gaussian pulse envelop $f(t)$ is employed since it is extensively employed in experiment (Zou et al., 2006). The frequency of the field is chosen constant for non-chirped laser pulses(ω_r), or time-dependent for chirped ones, namely, the phase of the laser field pulse is chosen as

$$\Phi(t) = \begin{cases} \omega_r t, & \text{(non-chirped pulse)} \\ \Phi_0 + \int_0^t \left[\triangle\omega_1 + \triangle\omega_2 e^{-(t'/\tau)^2}\right] dt' & \text{(chirped pulse)} \end{cases}$$

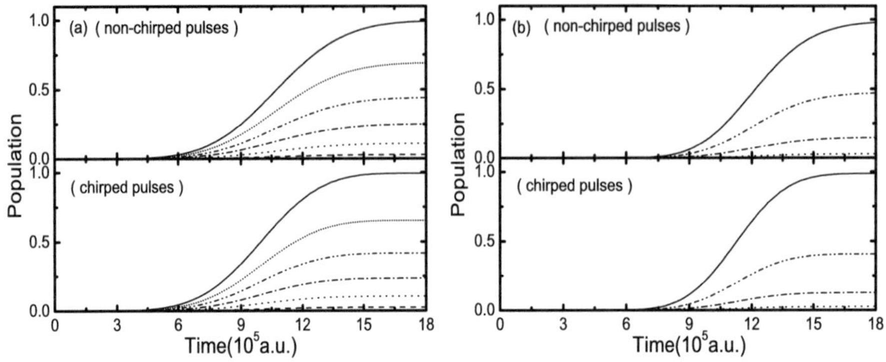

Fig. 3. Comparison of populations between the chirped by changing the laser intensity with the fixed pulse duration for OH molecule. (a)1-photon transition between $|0\rangle$ and $|1\rangle$, the value of the laser field strength from the lowest curve to the highest one in the non-chirped pulse is from $\mathcal{E}_0 = 0.000001 a.u.$ to $\mathcal{E}_0 = 0.000011 a.u.(\approx 4.25 \times 10^6 W/cm^2)$, and in the chirped pulse is from $\mathcal{E}_0 = 0.0000017 a.u.$ to $\mathcal{E}_0 = 0.0000102 a.u.(\approx 3.59 \times 10^6 W/cm^2)$. (b)2-photon transition between $|0\rangle$ and $|2\rangle$, the value of the laser field strength from the lowest curve to the highest one in the non-chirped pulse is from $\mathcal{E}_0 = 0.00004 a.u.$ to $\mathcal{E}_0 = 0.00012 a.u.(\approx 5.05 \times 10^8 W/cm^2)$, and in the chirped pulse is form $\mathcal{E}_0 = 0.00003 a.u.$ to $\mathcal{E}_0 = 0.00011 a.u.(\approx 4.29 \times 10^8 W/cm^2)$.

That is, the two types of the laser pulse are used to study the selective vibrational transition: *the non-chirped laser pulse* and *the chirped laser pulse*.

The selective excitations of 1-photon and 2-photon of OH are studied by the non-chirped laser pulse. The pulsewidth is set to be $\tau = 60000 a.u.(\approx 1.4 ps)$ as in Refs. (Just et al., 1992; Korolkov et al., 1996). We adjust the laser intensity to get maximum populations from $\mathcal{E}_0 = 0.000055 a.u.$ to $\mathcal{E}_0 = 0.00033 a.u.(\approx 0.4 \times 10^{10} W/cm^2)$ for 1-photon, and from $\mathcal{E}_0 = 0.00013 a.u.$ to $\mathcal{E}_0 = 0.00065 a.u.(\approx 0.15 \times 10^{11} W/cm^2)$ for 2-photon. In Fig. 2 we plot the time dependence of the laser field and population with different intensity. From Fig. 2 we see that the oscillations appear in the population with the laser intensity increases under the condition of the laser pulse duration is relatively small. To avoid the oscillations, we adjust the laser pulse duration and we find the oscillations disappear completely at the pulsewidth $\tau = 18 \times 10^5 a.u.(\approx 40 ps)$. This value is consentient to the pulsewidth in Ref. (Sugimori et al., 2005). Then we study the changes of the transition probabilities when the non-chirped laser pulse is turned to the chirped one. The transitions from the ground state to the first and the

	OH			*OD*		
	$\Delta\omega_1(a.u.)$	$\Delta\omega_2(a.u.)$	$I_{max}(W/cm^2)$	$\Delta\omega_1(a.u.)$	$\Delta\omega_2(a.u.)$	$I_{max}(W/cm^2)$
1-photon	0.016253	0.0000042	3.59×10^6	0.011999	0.0000040	4.80×10^6
2-photon	0.015480	0.0000040	4.29×10^8	0.011598	0.0000033	5.61×10^8
3-photon	0.014707	0.0000038	4.25×10^9	0.011197	0.0000025	5.05×10^9
4-photon	0.013934	0.0000036	1.76×10^{10}	0.010796	0.0000018	1.91×10^{10}

Table 3. The optimal laser parameters of multiphoton selective vibrational excitation of *OH* and *OD*

second excited states at the different laser intensity are given in Fig. 3. The laser intensity begin to increase from a relative small value till the populations get the maximum values at the end of the non-chirped or chirped pulse (see Fig. 3 caption). For the chirped pulse, we take the frequency $\triangle\omega_1$ as the resonant frequency, namely, we let $\triangle\omega_1 = \omega_r$. The chirped term $\triangle\omega_2$ is adjusted to get a good selectivity at a certain laser intensity. From Fig. 3, we can see that the populations begin to increase and get the maximum earlier by the chirped pulse than the non-chirped one. It shows the chirped pulse gives a better control of vibrational excitation of molecules. Finally, we give the selective excitation of 1-photon, 2-photon, 3-photon and 4-photon of OH and OD. We can obtain complete selective vibrational excitation by a Gaussian shaped and chirped laser pulse. The optimal parameters of the laser pulse are given in Tab. 3. We find the values of the chirped term of OH are larger than those of OD which could be also relevant to the anharmonicity of molecular vibrations. In the same laser duration, the laser frequency of OH decrease faster than that of OD, but the laser intensity of OD is higher than that of OH.

From the above discussion, we see that the optimal laser frequency must decrease as a function of time in order to achieve a good selectivity of excitation. Furthermore, the chirping of OH is faster than that of OD.

3.2 Triatomic molecules

In this section we study the control of triatomic molecules. As for triatomic molecules, there exist richer information of selective laser excitation, for example, there is the phenomenon of intramolecular vibrational redistribution (IVR). The IVR can interfere with selective excitation. Here we investigate the direct resonant excitation of the intramolecular bond for linear triatomic molecule HCN and DCN.

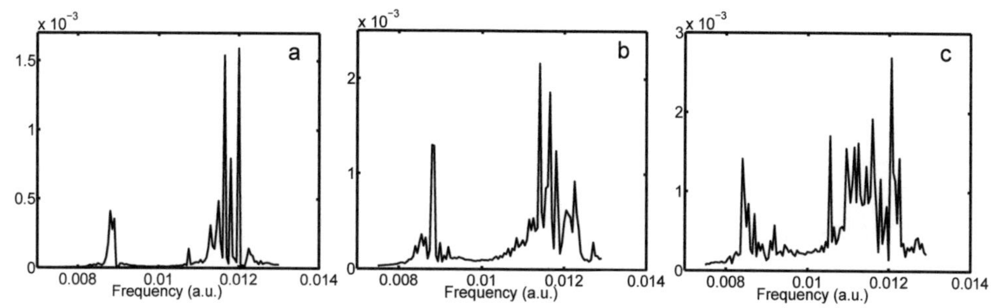

Fig. 4. Long-time-averaged energy absorption spectra of DCN at different laser intensities. (a) E_0=0.003a.u. (I≈0.32TW/cm^2), (b) E_0=0.009a.u. (I≈2.84TW/cm^2), (c) E_0=0.015a.u. (I≈7.89TW/cm^2).

The Hamiltonian (28) is used to calculate the stretch vibrational levels of HCN and DCN molecules successfully, and it can reproduce the real situations of the molecules (Feng & Ding, 2007). The molecular parameters of HCN and DCN are given in Table 1 of Ref. (Feng & Ding, 2007).

We investigate the effects of linearly chirped pulses with different shapes on the molecules. The linear chirped pulses is successfully used to study vibrational excitation and dissociation of the molecules (Liu et al., 1995; 1999; Lin et al., 1998). Furthermore, a linearly chirped pulse can be readily be prepared in the laboratory (Witte et al., 2003; Balling et al., 1994). The laser

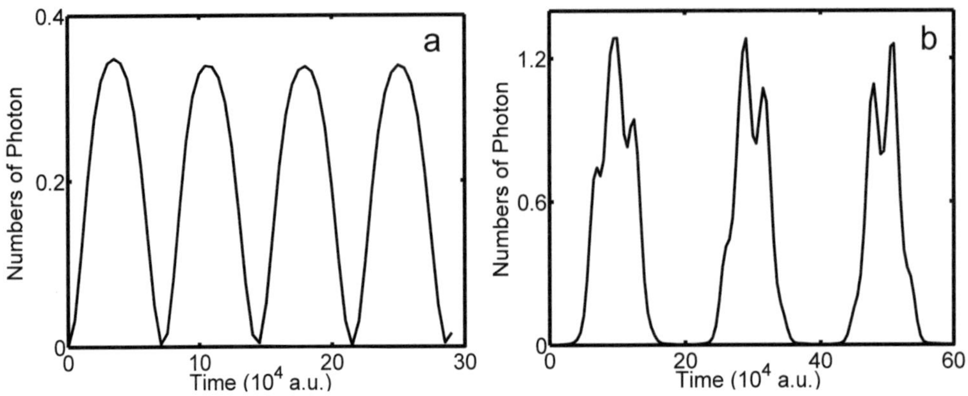

Fig. 5. Time-dependent average number of photons absorbed by DCN with E_0=0.003a.u. (I≈0.32TW/cm^2). (a) ω_L=0.012a.u.(2630cm^{-1}), the single-photon resonance, (b) ω_L=0.01075a.u.(2360cm^{-1}), the seven-photon resonance.

field is as follows

$$\begin{aligned} E(t) &= E_0 f(t)\cos\Phi(t), \\ \Phi(t) &= \Omega_0 t(1 - \frac{\alpha_c t}{2\tau}). \end{aligned} \tag{53}$$

In our numerical calculation, Ω_0 is taken as the anharmonic frequency of the selected bond, namely, we let $\Omega_0 = \omega_{01}$ or ω_{02}. The instantaneous frequency is taken as $\Omega_{ins}(t) = \Omega_0[1 - \alpha_c(\frac{t}{\tau})]$, the amount of chirping is then determined by the parameters α_c. The target excitation state is chosen $v = 10$ which determines the value of the parameters α_c. Researches as in Ref. (Gong et al., 2005; Dion et al., 1999) show the ionization threshold is estimated to be at $10^{14}W/cm^2$, the laser intensities chosen in this investigation are all far below this value and other parameters are similar to Ref. (Brezina & Liu, 2004) and given in figure captions. We here consider three kinds of laser shapes: *rectangular, Gaussian and Sech-shaped.*

3.2.1 Resonant excitation probability

The dipole moment of DCN molecule is given by using Eq. (17) with experimental values (Hyde & Hornig, 1952). We assume the molecule to be in the ground state at $t = 0$. Fig. 4 gives the long-time-averaged absorption energy spectra of DCN molecule at different laser intensities. Fig. 5 gives the averaged number of photons absorbed by the molecule.

Based upon Eq. (49), n-photons transition resonant frequencies can be found in the curves of the long-time-averaged vibrational transition probabilities as a function of external field frequency. The frequency of the infrared laser field is adjusted to obtain the highest peaks of the transition probabilities from $\omega_L = 0.007$a.u. to $\omega_L = 0.013$a.u.

Fig. 4 shows that as the laser intensity increases, both the long-time-averaged absorption energies and the resonant excitation peaks increase. Calculation of the vibrational transition probabilities shows that the highest resonance excitation peak at different laser intensities belongs to different multiphoton resonances, which means an efficient multiphoton resonance can be achieved only under certain laser intensity. This result is consistent with experiment (Borsella et al., 1983).

The average number of photons absorbed by the molecule have been studied at E_0=0.003a.u. At ω_L=0.012a.u. and ω_L=0.01075a.u., the single-photon and the seven-photon resonant

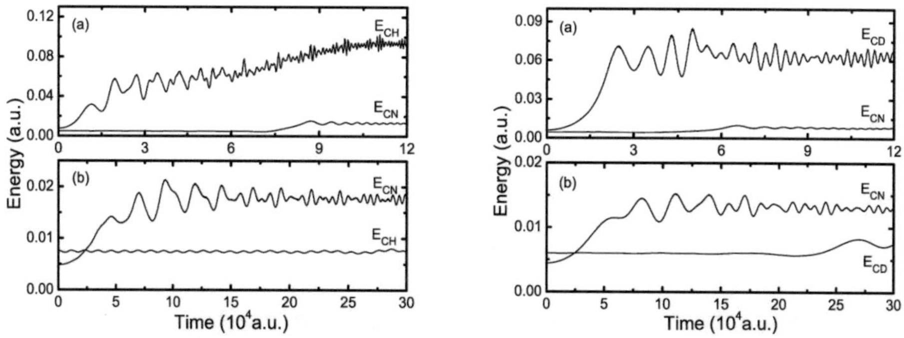

Fig. 6. The selective excitations of C-H bond and C-N bond for HCN (left column) and C-D bond and C-N bond for DCN (right column) to the $v = 10$ level by a resonance linearly chirped pulse with a rectangular shape. (a)Time-dependence of energy in C-H bond (E_{CH}) (left) or C-D bond (E_{CD}) (right) and in C-N bond(E_{CN}). The pulse parameters are $I = 0.9 \times 10^{12} W/cm^2$; $\tau \approx 2.8ps$. $\alpha_c = 0.38$ for HCN and $\alpha_c = 0.26$ for DCN. (b)Time-dependence of energy in C-H bond (E_{CH}) (left) or C-D bond(E_{CD}) (right) and in C-N bond (E_{CN}). The pulse parameters are $I = 0.5 \times 10^{13} W/cm^2$; $\tau \approx 7.2ps$. $\alpha_c = 0.13$ for HCN and $\alpha_c = 0.11$ for DCN.

excitations are found; the corresponding curves of the time-dependent average number of photons absorbed by the molecule are given in Fig. 5. The curve follows a simple Rabi oscillation, and the vibrational period is about 150 optical cycles (about 464 a.u. of one optical cycle) in the single-photon resonance. Within the Rabi cycle, the single-photon curves have cone-shaped peaks and broad bases. However, the average number of photons absorbed of the seven-photon resonance has split peaks and a long periodical behavior, the period being about 430 optical cycles. The multiphoton resonant excitation is a long-time process. This result also agrees with other relevant research (Walker & Preston, 1977).

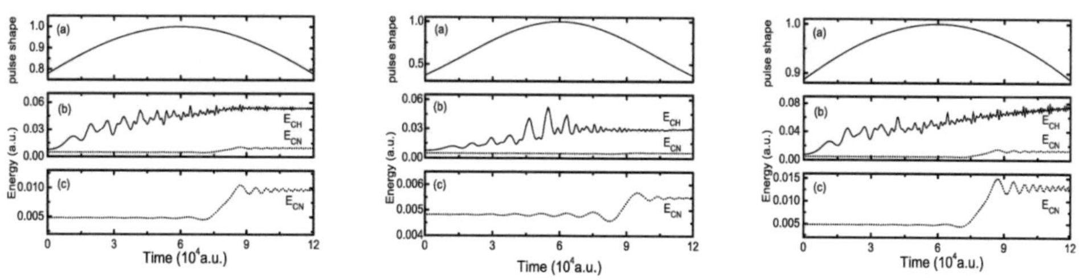

Fig. 7. The selective excitation of C-H bond of HCN to the $v = 10$ level by a resonance linearly chirped pulse with a Gaussian shape. (a)Time dependence of Gaussian pulse shape $f(t)$. (b)Time-dependence of energy in C-H bond(—E_{CH}) and in C-N bond($\cdots E_{CN}$). (c)The enlargement of C-N bond($\cdots E_{CN}$) energy in panel (b). The pulse parameters are $I = 0.9 \times 10^{12} W/cm^2$; $\alpha_c = 0.38$; $\tau \approx 2.8ps$. The left column is for $f(t) = exp[-(t-\tau/2)^2/\tau^2]$, the central column is for $f(t) = exp[-(t-\tau/2)^2/(\tau/2)^2]$, and the right column is for $f(t) = sech[(t-\tau/2)/\tau]$.

3.2.2 The selective bond excitation via the rectangular laser shape

The selective vibrational excitation of C-H (C-D) and C-N bonds in HCN and DCN molecules can be achieved by using the rectangular laser shape as shown in Fig. 6. The selective target bond picks up a higher energy but the energy of the other bond has a relatively smaller values at the end of the pulse. The selective bond excitation in molecule HCN is plotted in the left column of Fig. 6. In the top-left panel of Fig. 6, we can see the energy in C-N bond does not oscillate near the initial energy but it has a pronounced increase after the energy in C-H bond grows. There are two possibilities about this phenomenon. One is due to the intramolecular vibrational redistribution(IVR) as energy leaks from the highly excited C-H bond; the other is due to an accidental resonance with the pulse frequency which is near the C-N frequency(ω_{02}) at the end of the laser pulse. From the bottom-left panel of Fig. 6, we see the selective vibrational excitation of the C-N bond needs longer pulse duration and more higher laser intensity. The C-N bond is much heavier and stronger than the C-H bond, so the C-H bond may break firstly before the C-N bond is excited to the higher excitation level or dissociation and the selectivity of the C-N bond decreases. The results are consonant with the other study (Brezina & Liu, 2004).

The selective vibrational excitation of DCN is given in the right column of Fig. 6 and the similar case is observed. However, from top-right panel of Fig. 6, we can see that the energy in C-D bond does not increase always but begin to oscillate at some value, and the energy in C-N bond begin to increase earlier than that of HCN. In bottom-right panel of Fig. 6 when the C-N bond is excited, the C-D bond picks up energy at the end of the pulse. Compared to the left column of Fig. 6, energies in the selective C-D bond or the selective C-N bond are smaller than those of HCN. These results represent IVR (intramolecular vibrational redistribution) in DCN molecule is faster and the selectivity of DCN is much less than that of HCN which consist with the reference (Chelkowski & Bandrauk, 1991).

3.2.3 C-H bond excitation in HCN molecule by a Gaussian and Sech-shaped laser pulses

In this subsection we show the influences of laser pulse on the energy oscillations in a bond of HCN molecule. Fig. 7 (the left and center columns) shows that the oscillations of energy in C-H bond decreased at the end of the Gaussian shaped laser pulse and the pulsewidth of the Gaussian shape have a deep influence on oscillations of energy. But the energy in C-H bond also decreases in the Gaussian shaped laser pulse and the value is smaller than in the rectangular laser shape. In contrast, using the Sech-shaped laser pulses, the value of energy becomes large but still smaller than in the rectangular laser shape as shown in the right column of Fig. 7.

From the figures, we find the values of energy in the bond excitation is decided by the pulse area, and the oscillations of energy is related to the pulse shape. In order to get small oscillation and high energy, we take a Super-Gaussian shape laser pulse and give the C-H and C-N bond selective excitation in Fig. 8. As we expected, the energy in the C-H bond has a larger values and smaller oscillations compared to the other shaped cases. The C-N bond excitation has also a better selectivity than the rectangular case at the same laser intensity.

4. Conclusion

We have studied selective vibrational excitation for small molecules in an algebraic model. It is achieved state-selective excitation of diatomic molecules and bond selective excitation of triatomic molecules successfully. The results are in good agreement with the other researches. In this investigation, we find the optimal laser frequency must decrease as a function of time

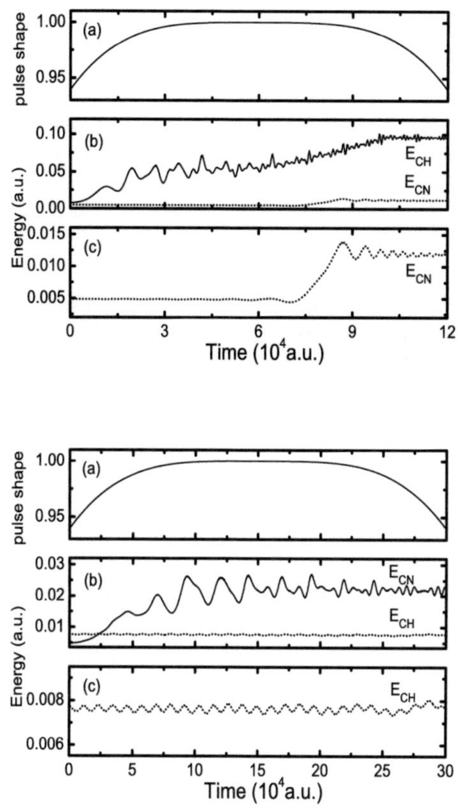

Fig. 8. The selective excitation of C-H bond of HCN to the $v = 10$ level by a resonance linearly chirped pulse with a Super-Gaussian shape. (a)Time dependence of Super-Gaussian pulse shape: $f(t) = exp[-(t-\tau/2)^4/\tau^4]$. (b)Time-dependence of energy in C-H bond(—E_{CH}) and in C-N bond($\cdots E_{CN}$). (c)The enlargement of C-N bond($\cdots E_{CN}$) energy in panel (b). The pulse parameters for left column are $I = 0.9 \times 10^{12} W/cm^2$; $\alpha_c = 0.38$; $\tau \approx 2.8ps$. The pulse parameters for right column are $I = 0.5 \times 10^{13} W/cm^2$; $\alpha_c = 0.13$; $\tau \approx 7.2ps$.

and the transition probability is very sensitive to the laser frequency to the control state excitation of a diatomic molecule. For the triatomic molecule, although the intramolecular vibrational redistribution makes selective excitation difficult, selective excitation to a specific vibrational bond can be achieved by using the appropriate chirped and shaped pulse. To get a good selectivity of bond excitation, the pulse area and the pulse shape are of the same importance. The pulse area decides the value of energy in bonds and the proper laser shape decreases the oscillations of energy. In our cases, a Super-Gaussian shape laser can be used to obtain a better selectivity of bond excitation. Furthermore, the bending motions and rotations can be taken into account and this method can also be extend to study the tetra-atomic molecular case.

More fundamental extensions are possible. The dynamical entanglement of realistic small molecules in the external laser fields can be investigated based on the algebraic models. The quantum control of the von Neumann entropy, and geometric phase of realistic molecules could be a natural investigation in the future.

5. Acknowledgments

The author thanks Profs. Shiliang Ding and Weiyi Ren, Dr. Hairan Feng and Ms. Yan Liu for their contributions to the work reviewed here. This work summarized herein was supported by the National Science Foundation of China (grant no. 10874102), and National Basic Research Program of China (973 Program, Grant No. 2009CB929404). Partial financial supports from the Research Fund for the Doctoral Program of Higher Education (Grant No. 200804220004) is acknowledged.

6. References

Alhassid, Y. & Levine, R. D. (1978). Connection between the maximal entropy and the scattering theoretic analyses of collision processes. *Phys. Rev.A*, Vol.18, No.1, 89-116.

Amstrup, B. & Henriksen, N. E. (1992). Control of HOD photodissociation dynamics via bond-selective infrared multiphoton excitation and a femtosecond ultraviolet laser pulse. *J. Chem. Phys.*, Vol.97, No.11, 8285-8295.

Balling, P.; Mass, D. J. & Noordam, L. D. (1994). Interference in climbing a quantum ladder system with frequency-chirped laser pulses. *Phys. Rev. A*, Vol.50, No.5, 4276-4285.

Batista, C. D. & Ortiz, G. (2004). Algebraic approach to interacting quantum systems. *Adv. Phys.*, Vol.53, No.1, 1-82.

Benjamin, I.; Levine, R. D. & Kinsey, J. L. (1983). High-lying levels of ozone via an algebraic approach. *J. Phys. Chem.*, Vol.87, No.5, 727-729.

Blackwell, M. ; Ludowise, P. & Chen, Y. (1997). Control of chemical reaction pathways by femtosecond ponderomotive forces: Time-resolved multiphoton ionization spectroscopic study of OCIO photodissociation. *J. Chem. Phys.*, Vol.107, No.1, 283-286.

Borsella, E.; Fantoni, R. ; Giardini-Guidoni, A.; Adams, D. R. & Cantrell, C. D. (1983). Resonance structure of multiple-photon excitation in symmetric and asymmetric tops. *Chem. Phys. Lett.* , Vol.101, No.1, 86-92.

Brezina, R. & Liu, W. K. (2004). Control of bond excitation and dissociation in HCN using laser pulses. *J. Phys. Chem. A*, Vol.108, No.41, 8852-8859.

Brumer, P. & Shapiro, M. (1986). Control of unimolecular reactions using coherent light. *Chem. Phys. Lett*,, Vol.126, No.6, 541-546.

Brumer, P. & Shapiro, M. (2003) *Coherent Control of Atomic and Molecular Process*, Wiley, New York.

Burke, P. G.; Colgan, J.; Glass, D. H. & Higgins, K. (2000). R-matrix-Floquet theory of molecular multiphoton processes . *J. Phys. B: At. Mol. Opt. Phys.*, Vol.33, No.2, 143-167.

Chang, J.; Ding, S. L. & Wyatt, R. E. (1985). Theory of multiphoton excitation of methane in intense laser fields.I.A purely vibrational treatment. *J. Chem. Phys.*, Vol.83,No.7, 3244-3254.

Chelkowski, S.; Bandrauk, A. D. & Corkum, P. B. (1995). Efficient molecular dissociation by a chirped ultrashort infrared laser pulse. *Phys. Rev. Lett.*, Vol.65, No.19, 2355-2358.

Chelkowski, S. & Bandrauk, A. D. (1991). Control of vibrational excitation and dissociation of small molecules by chirped intense inflared laser pulses. *Chem. Phys. Lett.*, Vol. 186,No.2-3, 264-269.

Chu, S.-I & Telnov, D. A. (2004). Beyond the Floquet theorem: generalized Floquet formalisms and quasienergy methods for atomic and molecular multiphoton processes in intense laser fields. *Physics Reports*, Vol.390, No.1-2, 1-131.

Colgan, J.; Glass, D. H.; Higgins, K. & Burke, P. G. (1998). The calculation of molecular multiphoton processes using the R-matrix-Floquet method. *Comput.Phys.Commun.*, Vol.114, No.1-3, 27-41.

Colgan, J.; Glass, D. H.; Higgins, K. & Burke, P. G. (2001). R-matrix Floquet theory of molecular multiphoton processes: II. Multiphoton ionization of H_2. *J. Phys. B: At. Mol. Opt. Phys.*, Vol.34, No.11, 2089-2106.

Constant, E.; Stapelfeldt, H. & Corkum, P. B. (1996). Observation of Enhanced Ionization of Molecular Ions in Intense Laser Fields. *Phys. Rev. Lett.*, Vol.76, No.22, 4140-4143.

Cooper, I. L. & Gupta, R. K. (1997). Application of $U(2) \supset O(2)$ and $U(2) \supset U(1)$ dynamical symmetries to diatomic molecular vibrations. *Phys. Rev. A*, Vol.55, No.6, 4112- 4119.

Cooper, I. L. (1998). Diatomic Potential Functions Derived from Accurate Fits to Vibrational Energy Levels Using Algebraic Theory. *J. Phys. Chem. A*, Vol.102, No.47, 9565-9569.

Dai, Y.; Geng, Z. & Ding, S. (2002). Multiphoton selective dissociation of a HF molecule in intense laser fields by the Lie-algebra approach. *Phys. Rev. A*, Vol.66, No.4, October-2002, 043415(6).

Denk, W.; Strickler, J. H. & Webb, W. W. (1990). Two-photon laser scanning fluorescence microscopy. *Science*, Vol.248, No.4951, 73-76.

Dibble, B. G. & Shirts, R. B. (1991). Theoretical prediction of ultrahigh vibrational excitation using infrared picosecond pulse trains: Coherent absorption of several photons each of a different frequency. *J. Chem. Phys.* Vol.94, N0.5, 3451-3467.

Ding, S. & Zheng, Y. (1999). Lie algebraic approach to potential energy surface for symmetric triatomic molecules. *J. Chem. Phys.*, Vol.111, No.10, 4466-4471.

Dion, C. M.; Keller, A.; Atabek, O. & Bandrauk, A. D. (1999). Laser-induced alignment dynamics of HCN: Roles of the permanent dipole moment and the polarizability. *Phys. Rev. A*, Vol.59, No.2, 1382-1391.

Elghobashi, N.; Krause, P.; Manz, J. & Oppel, M. (2003). $IR + UV$ laser pulse control of momenta directed to specific products: Quantum model simulations for $HOD^* \to H + OD$ versus $HO + D$. *Phys. Chem. Chem. Phys.*, Vol.5, 4806-4813.

Feng, H. & Ding, S. (2007). Lie algebraic approach to multiphoton selective excitation of linear triatomic molecules in intense infrared laser fields. *J. Phys. B: At. Mol. Opt. Phys.*, Vol.40, No.1, 69.

Feng, H.; Liu, Y.; Zheng, Y.; Ding, S. & Ren, W. (2007). Analytical control of small molecules by intense laser pulses in an algebraic model. *Phys. Rev. A*, Vol.75, No.6, 063417(12).

Feng, H.; Zheng, Y. & Ding, S. (2007). Study on infrared multiphoton excitation of the linear triatomic molecule by the Lie-algebra approach. *Eur. Phys. J. D*, Vol.42, No.2, 227-234.

Feng, H.; Li, P.; Zheng, Y. & Ding, S. (2010). Lie algebraic approach to dynamical entanglement of vibrations in triatomic molecules. *Prog. Theor. Phys.*, Vol.123, No.2, 215-225.

Gong, J.; Ma, A. & Rice, S. A. (2005). Isomerization and dissociation dynamics of HCN in a picosecond infrared laser field: A full-dimensional classical study. *J. Chem. Phys.*, Vol.122, No.14, 144311(10).

Gordon, R. J. & Rice, S. A. (1997). Active control of the dynamics of atoms and molecules. *Annu. Rev. Phys. Chem.*, Vol.48, 601-641.

Hay, P. J. & Dunning, T. H. (1976). Polarization CI wavefunctions: the valence states of the NH radical. *J. Chem. Phys.*, Vol.64, No.12, 5077-5087.

Hou, X.; Chem, J. & Ma, Z. (2006). Dynamical entanglement of vibrations in an algebraic model. *Phys. Rev. A*, Vol.74, No.6, 062513(6).

Hyde, G. E. & Hornig, D. F. (1952). The Measurement of Bond Moments and Derivatives in

HCN and DCN from Infrared Intensities. *J. Chem. Phys.*, Vol.20, No.4, 647-652.
Iachello, F. (1981). Algebraic methods for molecular rotation-vibration spectra. *Chem. Phys. Lett.*, Vol.78, No.3, 581-585.
Jakubetz, W.; Just, B.; Manz, J. & Schreier, H. J. (1990). Mechanism of state-selective vibrational excitation by an infrared picosecond laser pulse studied by two techniques: fast Fourier transform propagation of a molecular wave packet and analysis of the corresponding vibrational transitions. *J. Phys. Chem.*, Vol.94, No.6, 2294-2300.
Judson, R. S. & Rabitz, H. (1992). Teaching lasers to control molecules. *Phys. Rev. Lett.*, Vol.68, No.10, 1500-1503.
Just, B.; Manz, J. & Trisca, I. (1992). Chirping ultrashort infrared laser pulses with analytical shapes for selective vibrational excitations. Model simulations for $OH(v = 0) \rightarrow OH(v \leq 10)$. *Chem. Phys. Lett.*, Vol.193, No.5, 423-428.
Kellman, M. E. (1985). New raising and lowering operators for highly anharmonic coupled oscillators. *J. Chem. Phys.*, Vol.82, No.7, 3300-3308.
Kellman, M. E. (1995). *in Molecular Dynamics and Spectroscopy by Stimulated Emission Pumping*, World Scientific, Singapore.
Kikoin, K.; Avishai, Y. & Kiselev, M. N. (2004). Dynamical symmetries in nanophysics. *arXiv:cond-mat/0407063*.
Klarsfeld, S. & Oteo, J. A. (1989). Recursive generation of higher-order terms in the Magnus expansion. *Phys. Rev. A*, Vol.39, No.7, 3270-3273.
Kral, P. ; Thanopulos, I. & Shpiro, M. (2007). Colloquium: Coherently controlled adiabatic passage. *Rev. Mod. Phys.*, Vol.79, No.1, 53-77.
Korolkov, M. V.; Paramonov, G. K. & Schmidt, B. (1996). State-selective control for vibrational excitation and dissociation of diatomic molecules with shaped ultrashort infrared laser pulses. *J. Chem. Phys.*, Vol.105, No.5, 1862-1879.
Korolkov, M. V. & Paramonov, G. K. (1997). Ultrafast laser-pulse control for selective excitation of high vibrational states and dissociation of diatomic molecules in an environment. *Phys. Rev. A*, Vol.56, No.5, 3860- 3869.
Landau, L. D. & Lifshitz, M. E. (1977). *Quantum Mechanics*, Pergamon Press.
Leasure, S. C. ; Milfeld, K. F. & Wyatt, R. E. (1981). Quantum molecular dynamics in intense laser fields: Theory and applications to diatomic molecules. *J. Chem. Phys.*, Vol.74, No.11, 6197-6211.
Leforestier, C. & Wyatt, R. E. (1983). Optical potential for laser induced dissociation. *J. Chem. Phys.*, Vol.78, No.5, 2234-2344.
Leforestier, Claude & Wyatt, Robert E. (1985). Role of Feshbach resonances in the infrared multiphoton dissociation of small molecules. *J. Chem. Phys.*, Vol.82, No.2, 752-757.
Levine, R. D. (1982) *Intramolecular Dynamic*, Reidel, Dordrecht.
Levine, R. D. (1983). Representation of one-dimensional motion in a morse potential by a quadratic hamiltonian. *Chem. Phys. Lett.*, Vol.95, No.2, 87-90.
Levis, R. J.; Menkir, G. M. & Rabitz, H. (2001). Selective Bond Dissociation and Rearrangement with Optimally Tailored, Strong-Field Laser Pulses. *Science*, Vol.292, No.709.
Lin, J. T.; Lai, T. L.; Chuu, D. S.& Jiang, T. F. (1998). Quantum dynamics of a diatomic molecule under chirped laser pulses. *J. Phys. B*, Vol.31, No.4, L117-L126.
Lin, S. H. & Fujimura, Y. (1984). *Multiphoton Spectroscopy of Molecules*, academic press, INC.London.
Liu, W. K.; Wu, B. & Yuan, J. M. (1995). Nonlinear dynamics of chriped pulse excitation and dissociation of diatomic molecules. *Phys. Rev. Lett.*, Vol.75, No.7, 1292-1295.

Liu, W. K.; Yuan, J. M. & Lin, S. H. (1999). Classical dynamics of multiphoton excitation and dissociation of diatomic molecules by infrared laser pulses. *Phys. Rev. A*, Vol.60, No.2, 1363-1370.

Liu, Y.; Zheng, Y.; Ren, W. & Ding, S. (2008). Dynamical entanglement of vibrations in small molecules through an analytically algebraic approach. *Phys. Rev. A*, Vol.78, No.3, 032523(9).

Lu, Z. M. & Kellman, M. E. (1995). Assigning vibrational spectra of chaotic molecules. *Chem. Phys. Lett.*, Vol.247, No.3, 195-202.

Lu, Z. M. & Kellman, M. E. (1997). Phase space structure of triatomic molecules. *J. Chem. Phys.*, Vol.107, No.1, 1-15.

Oomens, J.; Moore, D. T. ; Meijer, G. & von Helden, G. (2004). Infrared multiple photon dynamics and spectroscopy of cationic PABA and its dehydroxylated fragment ion. *Phys. Chem. Chem. Phys.*, Vol.6, 710-718.

Paramonov, G. K. (2005). Ionization and dissociation of simple molecular ions in intense infrared laser fields: Quantum dynamical simulations for three-dimensional models of HD^+ and H_2^+ . *Chem. Phys. Lett.*, Vol.411, No.4-6, 350-356.

Philip, P. & Light, J. C. (1966). On the exponential form of time-displacement operators in quantum mechanics. *J. Chem. Phys.*, Vol.44, No.10, 3897-3912.

Rau, A. R. & Zhao, W. (2005). Time-dependent treatment of a general three-level system. *Phys. Rev. A*, Vol.71, No.6, 063822(6).

Rice, S. A. & Zhao, M. (2000). *Optical Control of Molecular Dynamics*, Wiley, New York.

Stranges, S.; Rithcer, R. & Alagia, M. (2002). High-resolution inner-shell photoabsorption of the OH and OD free radicals. *J. Chem. Phys.*, Vol.116, No.9, 3676-3680.

Sugimori, K.; Ito, T.; Nagao, H. & Nishikawa, K. (2005). Theoretical study of multiphoton processes in diatomic molecules. *Int. J Quantum Chem.*, Vol.105, No.6, 596-604.

Tannor, D. J. & Rice, S. A. (1985). Control of selectivity of chemical reaction via control of wave packet evolution. *J. Chem. Phys.*, Vol.83, No.10, 5013-5018.

Tung, M. & Yuan, J. M. (1987). Dissipative quantum dynamics:Driven molecular vibrations. *Phys. Rev. A*, Vol.36, No.9, 4463-4473.

Umeda, H.; Sugawara, M.; Fujimura, Y. & Koseki, S. (1994). Multiphoton dissociation dynamics of hydrogen cyanide in nonstationary laser fields: important role of dipole moment function. *Chem. Phys. Lett.*, Vol.229, No.3, 233-238.

van Roosmalen, O. S.; Benjamin, I. & Levine, R. D. (1984). A unified algebraic model description for interacting vibrational modes in ABA molecules. *J. Chem. Phys.*, Vol.81, No.12, 5986-5997.

Vadeiko, I. P. & Miroshnichenko, G. P. (2003). Algebraic approach to the Tavis-Cummings problem. *Phys. Rev. A*, Vol.67, No.5, 053808(12).

Walker, R. B. & Preston, R. K. (1977). Quantum versus classical dynamics in the treatment of multiple photon excitation of the anharmonic oscillator. *J. Chem. Phys.*, Vol.67, No.5, 2017-2028.

Wei, J. & Norman, E. (1964). On global representations of the solutions of linear differential equations as a product of exponentials. *Proc. Am. Math. Soc.*, Vol.15, No.2, 327-334.

Windhorn, L.; Witte, T.; Yeston, J. S.; Proch, D.; Motzkus, M.; Kompa, K. L. & Fuss, W. (2002). Molecular dissociation by mid-IR femtosecond pulses. *Chem. Phys. Lett.*, Vol.357, Mo.1-2, 85-90.

Witte, T.; Hornung, T.; Windhorm, L. & Proch, D. (2003). Controlling molecular ground-state dissociation by optimizing vibrational ladder climbing. *J. Chem. Phys.*, Vol.118, No.5,

2021-2024.

Wu, G. Z. (1998). Classification and assignment of eigenstates of highly excited vibrational manifolds via broken constants of motion. *Chem. Phys. Lett.*, Vol.292, No.4-6, 369-378.

Zheng, Y. (2006). Comment on "Time-dependent general quantum quadratic Hamiltonian system". *Phys. Rev. A*, Vol.74, No.6, 066101(3).

Zheng, Y. & Ding, S. (1999). Vibrational spectra of HCN and OCS from second-order expansion of the $U_1(4) \otimes U_2(4)$ algebra. *Phys. Lett. A*, Vol.256, No.2-3, 197-204.

Zheng, Y. & Ding, S. (2001). Saddle points of potential-energy surfaces for symmetric triatomic molecules determined by an algebraic approach. *Phys. Rev. A*, Vol.64, No.3, 032720(5).

Zheng, Y. & Ding, S. (2001). Saddle points of potential-energy surfaces for symmetric triatomic molecules determined by an algebraic approach. *Phys. Rev. A*, Vol.64, No.3, 032720(5).

Zheng, D. S. & Wu, G. Z. (2003). Coset potential approach to the vibrational dissociation: the case study of DCN and HCN. *Chem. Phys.*, Vol.290, No.1, 121-127.

Zou, S.; Ren, Q.; Balint-Kurti, G. G. & Manby, F. R. (2006). Analytical control of molecular excitations including strong field polarization effects. *Phys. Rev. Lett.*, Vol.96, No.24, 243003(4).

14

Laser Plasma Accelerators: Towards High Quality Electron Beam

Victor Malka
Laboratoire d'Optique Appliquée, ENSTA Paris Tech, CNRS, Ecole Polytechnique Paris Tech, UMR 7639, 91761 Palaiseau France

1. Introduction

In 1979 Tajima and Dawson, Tajima & Dawson (1979), on the basis of theoretical work and simulations, have shown that relativistic plasma waves are suitable for the development of compact accelerators. In the proposed schemes, electrons were injected externally and were accelerated through the very high electric field (GV/m) sustained by relativistic plasma waves driven by lasers, in the laser beatwave regime or in the laser wakefield regime. Several experiments have been performed following their idea and injected electrons at the few MeV level have indeed been accelerated by electric fields in the GV/m range in a plasma medium using the beat wave and laser wakefield schemes. With the development of more powerful lasers, much higher electric fields were achieved, from few GV/m to more than 1 TV/m, Malka et al. (2002). This fast increase in the laser power has allowed physicists (i) to produce electron beams directly from the interaction of an intense laser with an underdense plasma, i.e. without any external injection and (ii) at the same time to explore new theoretical schemes: forward Raman instability, self-modulated laser wakefield, direct laser acceleration, and forced laser wakefield. In all those experiments, the distribution of the electron was Maxwellian and the beam quality was far from what is required for some applications. These first beams did not compare well to beams produced by conventional accelerators. To control or to shape the electron beam distribution, one has to reduce the injection to a very limited volume of phase space. For external injection, this means that the injected electron beam must have a duration much shorter that the plasma period, i.e. less than ten of femtoseconds. Note that longer electron bunches can be injected and compressed by the plasma wave when they are injected before the laser pulse, Khachatryan et al. (2004). For internal injection, the control of injection can be done either in the bubble/blowout regime using one single laser beam or by colliding two or three laser pulses. In 2004, a major breakthrough came with the experimental demonstration of the bubble/blowout regime when three groups produced for the first time an electron beam with a quasimonoenergetic distribution, Mangles et al. (2004); Geddes et al. (2004); Faure et al. (2004). The bubble regime is reached when the laser power is high enough and when the laser pulse length and waist match the plasma wavelength. When these conditions are met, the laser ponderomotive force expels the plasma electrons radially and leaves a cavitated region behind the pulse. Electrons are progressively injected at the back of this cavity forming a dense electron beam in the cavity. The increasing charge of the forming electron beam progressively reduces the electric value and the injection process eventually stops, leading to the formation of a quasimonoenergetic electron beam.

A continuous transition between the self-modulated laser wakefield, forced laser wakefield, and bubble regimes has been measured by changing the electron density, i.e., by changing the ratio of the pulse length to the plasma wavelength. The use of two laser beams instead of one offers more exibility and enables one to separate the injection from the acceleration processes. The first laser beam the pump beam (the injection beam) is used to heat electrons during its collision with the pump beam. During the beating, electrons can gain enough momentum to catch the relativistic plasma wave and consequently be trapped and efficiently accelerated. In 2006, stable and tunable quasimonoenergetic electron beams were measured by using two laser beams in the colliding scheme with a counterpropagating geometry, Faure et al. (2006).

2. Theory and evolution of electron acceleration using laser-plasma interaction

2.1 Maxwell equations

When a laser pulse propagates in a medium, its corresponding electromagnetic wave evolution is described by the Maxwell equations :

$$\begin{cases} \vec{\nabla}.\vec{E} = \dfrac{\rho}{\varepsilon_0} & \vec{\nabla}.\vec{B} = 0 \\ \vec{\nabla}\wedge\vec{E} = -\dfrac{\partial \vec{B}}{\partial t} & \vec{\nabla}\wedge\vec{B} = \dfrac{1}{c^2}\dfrac{\partial \vec{E}}{\partial t} + \mu_0 \vec{j} \end{cases} \tag{1}$$

where $\vec{E}$ and $\vec{B}$ are the electric and magnetic fields respectively, ρ and $\vec{j}$ describe the medium and are the local density and the current density respectively. ε_0 and μ_0 are the permittivity and permeability of vacuum. c is the celerity of light. The operator $\vec{\nabla}$ corresponds to the spatial partial derivative vector

$$\vec{\nabla} = \left(\frac{\partial}{\partial x}, \frac{\partial}{\partial y}, \frac{\partial}{\partial z}\right) \tag{2}$$

From this set of equations, one obtains the equation of propagation of the electric field. In vacuum ($\rho = 0$ and $\vec{j} = \vec{0}$), this is written :

$$\nabla^2 \vec{E}(\vec{x},t) - \frac{1}{c^2}\frac{\partial^2 \vec{E}}{\partial t^2}(\vec{x},t) = \vec{0} \tag{3}$$

In the frequency domain (for the temporal variable), it becomes :

$$\nabla^2 \vec{E}(\vec{x},\omega) + \frac{\omega^2}{c^2}\vec{E}(\vec{x},\omega) = \vec{0} \tag{4}$$

In a more general case of a propagation in an isotropic medium, one introduces the refractive index η in the equation of propagation, which contains the response of the medium

$$\nabla^2 \vec{E}(\vec{x},\omega) + \frac{\omega^2}{c^2}\eta^2(\omega)\vec{E}(\vec{x},\omega) = \vec{0} \tag{5}$$

2.2 Laser parameters

In general, one uses potentials to describe these fields. There exist a vector potential $\vec{A}$ and a scalar potential Φ that verify :

$$\begin{cases} \vec{E} = -\vec{\nabla}\Phi - \dfrac{\partial \vec{A}}{\partial t} \\ \vec{B} = \vec{\nabla}\wedge\vec{A} \end{cases} \tag{6}$$

These potentials are not described uniquely. It is possible to find other solutions with a gauge transformation. We usually work in Coulomb gauge $\vec{\nabla}.\vec{A} = 0$.
In the following, we will use the normalized vector potential $\vec{a}$, defined by :

$$\vec{a} = \frac{e\vec{A}}{m_e c} \tag{7}$$

where e is the electron charge and m_e its mass.
One also introduces the intensity I, which is the average of the Poynting vector over an optical cycle :

$$I = c^2 \varepsilon_0 \left\langle \vec{E} \wedge \vec{B} \right\rangle_t \tag{8}$$

where brackets design the temporal average on one optical cycle.

2.3 Gaussian beams

Short laser pulses delivered by laser systems have a broad spectrum which contains many modes locked in phase. This spectrum is usually described simply by a gaussian envelope, thus also leading to an gaussian temporal envelope, which is close to reality. In the same way, the spatial profile of the laser pulse at the focal plane is also represented by a gaussian function. The electric field has the following form for a linearly polarized pulse :

$$\vec{E}(r,z,t) = \frac{E}{2} f(r,z) g(t,z) \exp\left[-i(k_0 z - \omega_0 t)\right] \vec{e_x} + \vec{cc} \tag{9}$$

Equation (9) contains a carrying envelope with wave number k_0 and frequency ω_0 and spatial and temporal information contained in $f(r,z)$ and $g(t)$ respectively. The following gaussian expressions (10) verify the equation of propagation of the electric field in vacuum in the paraxial approximation. These expressions reproduce accurately the electric field of the laser when the focusing optics have small aperture.

$$\begin{aligned} g(t,z) &= \exp\left[-2\ln 2 \left(\frac{t - z/c}{\tau_0}\right)^2\right] \\ f(r,z) &= \frac{\mathrm{w}_0}{\mathrm{w}(z)} \exp\left[-\frac{r^2}{\mathrm{w}^2(z)} - i\frac{k_0 r^2}{2R(z)}\right] \exp i\phi(z) \end{aligned} \tag{10}$$

where τ_0 is the pulse duration at full width at half maximum (FWHM), w_0 is the waist of the focal spot (the radius at $1/e$ of the electric field in the focal plane $z = 0$). $\phi(z)$ is the Gouy phase. Functions $\mathrm{w}(z)$ and $R(z)$ represent respectively the radius at $1/e$ of the electric field and the radius of curvature of the wave front. These functions take the following form :

$$\mathrm{w}(z) = \mathrm{w}_0 \sqrt{1 + \frac{z^2}{Z_r^2}} \tag{11}$$

$$R(z) = z\left(1 + \frac{Z_r^2}{z^2}\right) \tag{12}$$

$Z_r = \pi \mathrm{w}_0^2 / \lambda_0$ is the Rayleigh length. This physical parameter represents the length where the laser intensity on axis has dropped by a factor 2 compared to the intensity in the focal plane $(z = 0)$.

Starting from this expression of the electric field, the following relation exists between the maximal intensity I_0 and the power P :

$$I_0 = \frac{2P}{\pi w_0^2} \tag{13}$$

with $P = 2\sqrt{\frac{\ln 2}{\pi}}\frac{U}{\tau_0} \sim \frac{U}{\tau_0}$, where U is the energy contained in the pulse.
Then, the following relation lies the maximal intensity I_0 and the maximum of the normalized vector potential a_0

$$a_0 = \left(\frac{e^2}{2\pi^2\varepsilon_0 m_e^2 c^5}\lambda_0^2 I_0\right)^{1/2} \tag{14}$$

When a_0 exceeds unity, the oscillations of an electron in the laser field become relativistic. In laser plasma accelerators the motion of the electrons is mostly relativistic[1].

2.4 Plasma parameters

A plasma is a state of matter made of free electrons, totally or partially ionized ions and neutral atoms or molecules, the whole medium being globally neutral. Let's assume an initially uniform, non-collisional plasma in which a slab of electron is displaced from the equilibrium position. The restoring force which applies on this electron slab, drives them towards the equilibrium position. For the time scale corresponding to the electron motion, one neglects the motion of the ions because of the inertia. This gives in the end oscillations around the equilibrium position at a frequency called the electron plasma frequency ω_{pe}

$$\omega_{pe} = \sqrt{\frac{n_e e^2}{m_e \varepsilon_0}} \tag{15}$$

where n_e is the unperturbed electron density.
This frequency has to be compared to the laser frequency : if $\omega_{pe} < \omega_0$ then the characteristic time scale of the plasma is longer than the optical period of the incoming radiation. The medium can't stop the propagation of the electromagnetic wave. The medium is said to be transparent or under-dense. On the opposite, when $\omega_{pe} > \omega_0$ then the characteristic time scale of the electrons is fast enough to adapt to the incoming wave and to reflect totally of partially the radiation. The medium is said to be overdense.
These two domains are separated at frequency ω_0, which corresponds to the critical density[2] $n_c = \omega_0^2 m_e \varepsilon_0 / e^2$

2.4.1 Electric field of the plasma wave

One considers now a periodic sinusoidal perturbation of the electron plasma density in a uniform ion layer. Mechanisms responsible for the excitation of the plasma wave will be described in the following section. The density perturbation δn is written :

$$\delta n = \delta n_e \sin(k_p z - \omega_p t) \tag{16}$$

where ω_p and k_p are the angular frequency and the wave number of the plasma wave.

[1]For a visible laser light intensity $I_0 = 3 \times 10^{18}$ W/cm^2, to which corresponds a $a_0 = 1.3$.
[2] For an wavelength $\lambda_0 = 820$ nm, one obtains a critical density of $n_c = 1.7 \times 10^{21}$ cm^{-3}

This density perturbation leads to a perturbation of the electric field $\delta\vec{E}$ via the Poisson equation:

$$\vec{\nabla}.\delta\vec{E} = -\frac{\delta n\, e}{\varepsilon_0} \tag{17}$$

This gives

$$\delta\vec{E}(z,t) = \frac{\delta n_e\, e}{k_p \varepsilon_0} \cos(k_p z - \omega_p t)\vec{e_z} \tag{18}$$

Because we want to describe the electron acceleration to relativistic energies by a plasma wave, we consider now a plasma wave with a phase velocity is close to the speed of light $v_p = \omega_p / k_p \sim c$. Let $E_0 = m_e c \omega_{pe} / e$. The electric field becomes :

$$\delta\vec{E}(z,t) = E_0 \frac{\delta n_e}{n_e} \cos(k_p z - \omega_p t)\vec{e_z} \tag{19}$$

One notice that the electric field is dephased by $-\pi/4$ with respect to the electron density.

2.4.2 Lorentz's transform

Let's now describe what happens to an electron placed in this electric field. The goal is to obtain the required conditions for trapping to occur. The following variables are introduced to describe the electron in the laboratory frame : z the position, t the associated time, β the velocity normalized to c, $\gamma = 1/\sqrt{1-\beta^2}$ the associated Lorentz's factor. In the frame of the plasma wave, let z', t' ,β' and γ' represent the equivalent quantities.
The frame linked to the plasma wave is in uniform constant translation at speed $v_p = \beta_p c$. One writes γ_p the Lorentz's factor associated to this velocity. The Lorentz's transform allows to switch from the laboratory frame to the wave frame :

$$\begin{cases} z' = \gamma_p(z - v_p t) \\ t' = \gamma_p(t - \frac{x}{c}) \\ \gamma' = \gamma\gamma_p(1 - \vec{\beta}.\vec{\beta_p}) \end{cases} \tag{20}$$

In this new frame, without magnetic field, the electric field remains unchanged $\delta\vec{E}'$

$$\delta\vec{E}'(z') = \delta\vec{E}(z,t) = E_0 \frac{\delta n_e}{n_e} \cos(k_p z' / \gamma_p)\vec{e_z} \tag{21}$$

Consequently, in terms of potential, the electric field is derived from potential Φ' defined by

$$\vec{F} = -e\delta\vec{E}' \equiv -\vec{\nabla}'\Phi' \tag{22}$$

This leads to

$$\Phi'(z') = mc^2 \gamma_p \frac{\delta n_e}{n_e} \sin(k_p z' / \gamma_p) \equiv mc^2 \phi'(z') \tag{23}$$

Finally, one writes the total energy conservation for the particle in this frame compared to the initial energy at the injection time (labelled with subscript 0) :

$$\gamma'(z') + \phi'(z') = \gamma'_0(z'_0) + \phi'_0(z'_0) \tag{24}$$

Equation 24 gives the relation between the electron energy and its position in the plasma wave. Figure 1 illustrates the motion of an electron injected in this potential. Finally, we perform the reverse Lorentz's transform to give this energy in the laboratory frame.

For $\beta' > 0$, the scalar product in eq. 20 is positive

$$\gamma = \gamma'\gamma_p + \sqrt{\gamma'^2 - 1}\sqrt{\gamma_p^2 - 1} \tag{25}$$

For $\beta' < 0$, scalar product in eq. 20 is negative

$$\gamma = \gamma'\gamma_p - \sqrt{\gamma'^2 - 1}\sqrt{\gamma_p^2 - 1} \tag{26}$$

2.4.3 Electron trajectories

Figure 1 represents an example of electron trajectory in a plasma wave. In this phase space, the closed orbits correspond to trapped particles. Open orbits represent untrapped electrons, either because the initial velocity is too low, or to high. The curve which separates these two regions is called the separatrix.

This separatrix gives the minimum and maximum energies for trapped particles. This is comparable to the hydrodynamic case, where a surfer has to crawl to gain velocity and to catch the wave. In terms of relativistic factor, γ has to belong to the interval $[\gamma_{min};\gamma_{max}]$ with :

$$\begin{cases} \gamma_{min} = \gamma_p(1+2\gamma_p\delta) - \sqrt{\gamma_p^2 - 1}\sqrt{(1+2\gamma_p\delta)^2 - 1} \\ \gamma_{max} = \gamma_p(1+2\gamma_p\delta) + \sqrt{\gamma_p^2 - 1}\sqrt{(1+2\gamma_p\delta)^2 - 1} \end{cases} \tag{27}$$

where $\delta = \delta n_e / n_e$ is the relative amplitude of the density perturbation.

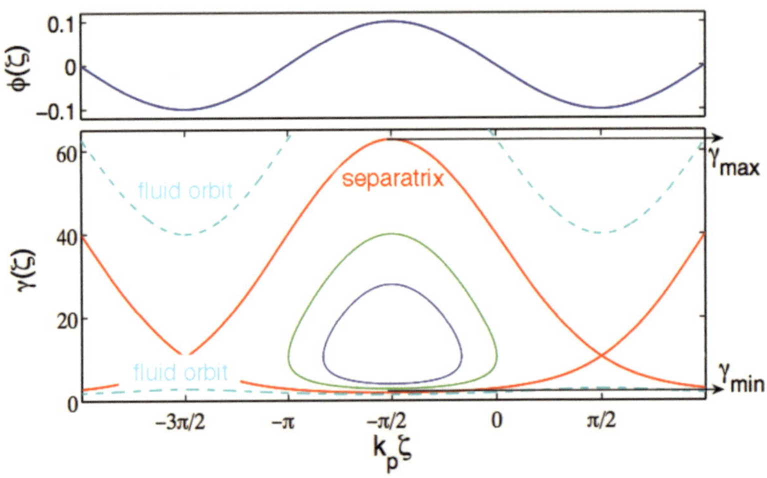

Fig. 1. Up: Potential in phase space. Down: Trajectory of an electron injected in the potential of the plasma wave in the frame of the wave with the fluid orbit (dashed line), the trapped orbit and in between in red the separatrix.

One deduces that the maximum energy gain ΔW_{max} for a trapped particle is reached for a closed orbit with maximum amplitude. This corresponds to the injection at γ_{min} on the separatrix and its extraction at γ_{max}. The maximum energy gain is then written

$$\Delta W_{max} = (\gamma_{max} - \gamma_{min})mc^2 \tag{28}$$

For an electron density much lower than the critical density $n_e \ll n_c$, one has $\gamma_p = \omega_0/\omega_p \gg 1$ and

$$\Delta W_{max} = 4\gamma_p^2 \frac{\delta n_e}{n_e} mc^2 \tag{29}$$

For electron travelling along the separatrix, the time necessary to reach maximal energy is infinite because there exist a stationnary point at energy γ_p. On other closed orbits, the electron successively gains and looses energy during its rotation of the phase space. In order to design an experiment, one needs an estimation of the distance an electron travels before reaching maximal energy gain. This length, which is called the dephasing length L_{deph}, corresponds to a phase rotation of $\lambda_p/2$ in the phase space. In order to have a simple analytical estimation, one needs to assume that the energy gain is small compared to the initial energy of the particle and that the plasma wave is relativistic $\gamma_p \gg 1$, then the dephasing length is written

$$L_{deph} \sim \gamma_p^2 \lambda_p \tag{30}$$

This concept of dephasing length in a 1D case can be refined in a bi-dimensional case. Indeed, if one also takes into account the transverse effects of the plasma wave, this one is focusing or defocusing for the electrons along their acceleration, Mora (1992). Because these transverse effects are shifted by $\lambda_p/4$ with respect to the pair acceleration/deceleration, the distance over which the plasma wave is both focusing and accelerating is restricted to a rotation of $\lambda_p/4$ in phase space, which decreases by a factor 2 the dephasing length from eq. 30.

$$L_{deph}^{2D} \sim \gamma_p^2 \lambda_p/2 \tag{31}$$

In these formulas, one has considered a unique test electron, which has no influence on the plasma wave. In reality, the massive trapping of particles modifies electric fields and distorts the plasma wave. Finally, this linear theory is difficult to apply to highly non-linear regimes which are explored experimentally. Some non-linear effects concerning short pulses are described in the next section. Nonetheless, these formulas are usefull to scale the experiments.

2.5 Non-linear effects

2.5.1 Ponderomotive force

Let's take a non-relativistic electron for a short while. In a laser field with a weak intensity, the average position of an electron is constant. If one only keeps linear terms in fluid equation there remains, Kruer (1988):

$$\frac{\partial \vec{v_e}^{(l)}}{\partial t} = -\frac{e}{m_e}\vec{E} \tag{32}$$

The electron directly varies with the electric field. Let's consider now a laser pulse slightly more intense, so that the electron velocity becomes slightly non linear $\vec{v_e} = \vec{v_e}^{(l)} + \vec{v_e}^{(nl)}$ with $\|\vec{v_e}^{(nl)}\| \ll \|\vec{v_e}^{(l)}\|$. The second order terms satisfy the following equation

$$\frac{\partial \vec{v_e}^{(nl)}}{\partial t} = -(\vec{v_e}^{(l)}.\vec{\nabla})\vec{v_e}^{(l)} - \frac{e}{m_e}(\vec{v_e}^{(l)} \wedge \vec{B}) \tag{33}$$

By keeping the low frequency component of the equation of motion, i.e. by averaging over an optical cycle, one obtains

$$m_e \frac{\partial \left\langle \vec{v_e}^{(nl)} \right\rangle_t}{\partial t} = -\frac{\vec{\nabla} I}{2cn_c} \equiv \vec{F_p} \tag{34}$$

$\vec{F}_p$ is called the ponderomotive force. This force repels charged particles from regions where the laser intensity gradient is large. This ponderomotive force[3] derives from a ponderomotive potential which is written as follow

$$\phi_p = \frac{I}{2cn_c} = \frac{e^2E^2}{4m_e\omega_0^2} \tag{35}$$

2.5.2 Laser self-focusing

For a laser intensity above 10^{19} W/cm^2, the motion of an electron in an intense laser field becomes relativistic. In this case, local properties of the medium vary as function of the laser intensity. In particular, the refractive index in the equation of propagation (eq. 5) depends on laser intensity $\eta(I) = \eta_0 + \eta_2 I$. The plasma medium acts as a focusing lens for the electromagnetic field of the laser. If one considers only the relativistic contribution, the critical power for self-focusing P_c for a linearly polarized laser pulse[4] is written, Sprangle et al. (1987) :

$$P_c = \frac{8\pi\varepsilon_0 m_e^2 c^5}{e^2}\frac{n_c}{n_e} \tag{36}$$

This formula doesn't account for other phenomena which also modify the refractive index : the plasma wave, the ponderomotive effect on the electrons, the ion channel created by a long prepulse. For instance, the plasma wave tends to defocus the laser pulse, which might prevent the pulse from self-focusing at P_c, Ting et al. (1990). Then, because of an electron density bump at the front of the plasma wave, the laser field in the first plasma bucket can't self-focus , Sprangle & Esarey (1992). Consequently, the laser pulse tends to erodes by the front. In particular, this theory predicts that it's not possible for a laser pulse shorter than the plasma wavelength to remain self-focused.
In reality, current experiments use very intense laser pulses $a_0 \gg 1$ and density perturbations are not linear anymore. Then, consequences on the self-focusing of very short laser pulses are less obvious.

3. Acceleration mechanisms

At first glance, the electromagnetic field associated to the laser doesn't seem a good solution to accelerate electrons: the electric field is mainly transverse to the propagation of the wave and its direction alternates every half period of the oscillation. Acceleration mechanisms presented here require an intermediary : the plasma wave. This one is excited by the laser pulse and allows to create a longitudinal electrostatic field favourable to the acceleration of electrons. The general diagram is represented on Fig. 2.
In section 2.4.1, a simple model of the electron acceleration in a plasma wave has been presented. Now, the link between the electromagnetic field of the laser and the plasma wave has to be described. Several mechanisms have been developed to excite a large-amplitude plasma wave. These acceleration mechanisms have evolved as the laser pulse duration shortened and maximal intensity increased. Initially, the acceleration was well described by linear formulas. Then, as the intensity increased, non-linear mechanisms have appeared (Raman instability ,Drake et al. (1974), relativistic self-focusing, Mori et al. (1988), relativistic

[3] For an intensity $I_0 = 1 \times 10^{19}$ W/cm^2 and a wavelength 1 μm, one obtains a ponderomotive potential of $\phi_p = 1$ MeV
[4] For an electron density $n_e = 10^{19}$ cm^{-3}, for a laser wavelength $\lambda_0 = 1\mu$m, one obtains a critical power $P_c = 2$ TW

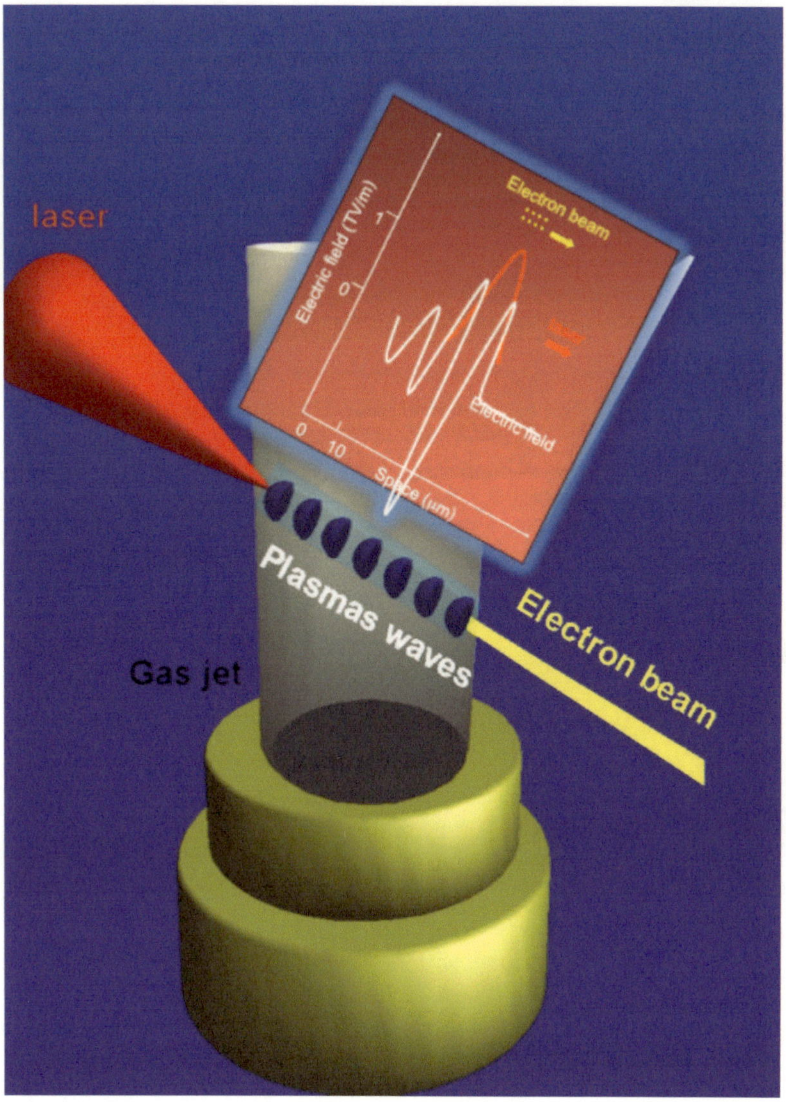

Fig. 2. Principle of laser-plasma acceleration : from the interaction of an intense laser pulse with a gas jet, one obtains an electron beam at the output.

self-modulation ,McKinstrie & Bingham (1992)) which allowed to reach even higher electric fields and particle beams with unique properties.

3.1 Linear regime

3.1.1 Laser wakefield

Acceleration in a laser wakefield has been introduced by Tajima and Dawson, Tajima & Dawson (1979). The perturbed electron density driven by the laser pulse is favourable to the acceleration of particles. The electron density profile obtained behind a gaussian laser pulse has been reported for $a_0 \ll 1$, Gorbunov & Kirsanov (1987). For a linearly polarized laser pulse with full width at half maximum (FWHM) $\sqrt{2\ln 2}\,L$ (in intensity), the normalized vector potential is written[5] :

$$a^2(z,t) = a_0^2 \exp\left[-\left(\frac{k_0 z - \omega_0 t}{k_p L}\right)^2\right] \tag{37}$$

In this case, the associated electric field is

$$\vec{E}(z,t) = E_0 \frac{\sqrt{\pi} a_0^2}{4} k_p L \exp(-k_p^2 L^2/4) \cos(k_0 z - \omega_0 t) \vec{e_z} \tag{38}$$

Equation 38 explicitly shows the dependence of the amplitude of the wave with the length of the exciting pulse. In particular, the maximal value for the amplitude is obtained for a length $L = \sqrt{2}/k_p$ (see Fig. 3).

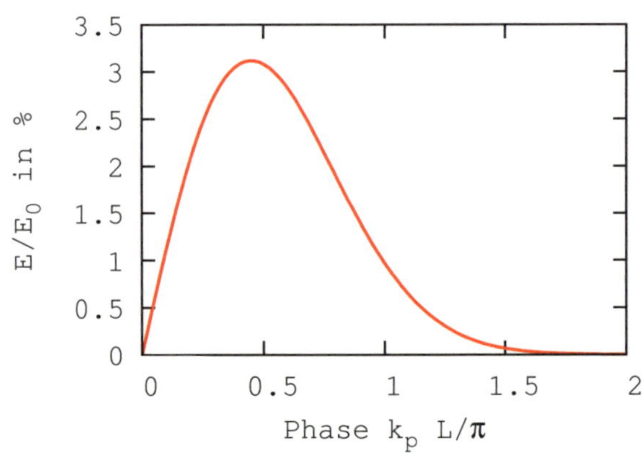

Fig. 3. Amplitude of the electric field as function of the length of a gaussian laser pulse for a normalized vector potential $a_0 = 0.3$.

In figure 4 the density perturbation and the corresponding electric field produced by a 30 fs laser pulse at low intensity $I_{laser} = 3 \times 10^{17}$ W/cm^2 are shown. One can note that in the linear regime the electric field has sinusoidal shape and reach maximal values of a few GV/m.

[5] For an electron density $n_e = 10^{19}$ cm^{-3}, the optimal pulse duration equals $L = 2.4\ \mu$m (equivalent to a pulse duration $\tau = 8$ fs). For $a_0 = 0.3$, the maximal electric field is $E = 10$ GV/m

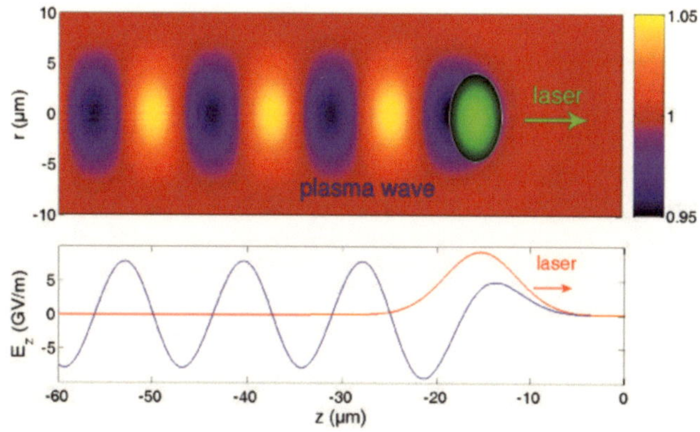

Fig. 4. density perturbation (top) electric field (bottom) produced in the linear regime.

3.1.2 Non linear regime

Thanks to the development of laser systems with a high power and a short pulse duration non linear plasma waves can be produced. In the non linear regime the laser pulse excites at resonance plasma wave with much higher amplitude to which corresponds electric field 100 times larger than in the linear regime. One can notice on figure 5 that the radial density perturbation has a horse shoe behavior with bent wakes. As a_0 grows, wakes become steeper and the wave front becomes curved due to the relativistic shift of the plasma frequency.

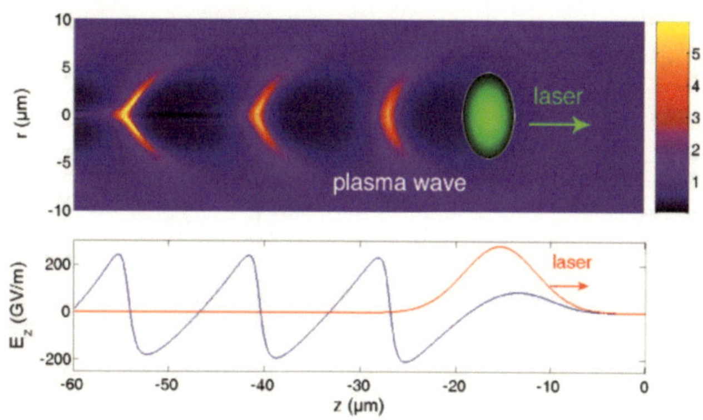

Fig. 5. density perturbation (top) electric field (bottom) produced in the non linear regime.

3.2 Self-injection

3.2.1 Self modulated wakefield

When the laser power exceeds the critical power for relativistic self-focusing, it temporal shape can be modulated during the propagation in the plasma medium. For laser pulse longer than the plasma wavelength the pulse can be "sausaged" into shorter pulses which excite in a resonant way the relativistic plasma waves. These effects that have been predicted of the basis of numerical simulations, Andreev et al. (1992); Antonsen & Mora (1992); Sprangle & Esarey (1992) are illustrated on Fig. 6. This mechanism, which is very similar to Forward Raman Scattering instability, can be described as the decomposition of an electromagnetic wave into a plasma wave an a frequency shifted electromagnetic wave.

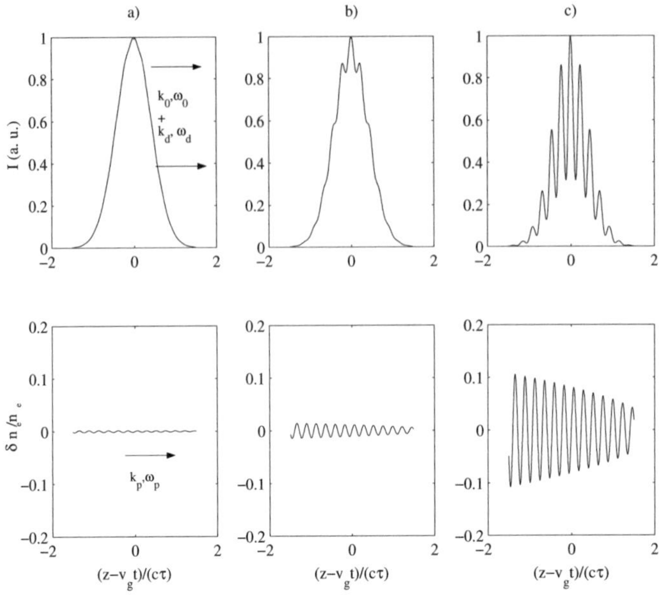

Fig. 6. Self-modulation of the laser envelope and coupling with the plasma wave amplitude. Initially, the laser propagates on a plasma density perturbation (a). This modulates the laser envelope, which increases the coupling with the plasma wave, the amplitude of which increases (b). Finally, the self-modulation mechanism generates a train of laser pulses spaced by a plasma wavelength, which resonantly excites a large amplitude plasma wave (c).

During experiments carried out in England in 1994, Modena et al. (1995), the amplitude of the plasma waves reached the wavebreaking limit, where electrons initially belonging to the plasma wave are self-trapped and accelerated to high energies. The fact that the external injection of electrons in the wave is no longer necessary is a major improvement. Electron spectrum extending up to 44 MeV have been measured during this experiment. This regime has also been reached for instance in the United States at CUOS, Umstadter et al. (1996), at NRL, Moore et al. (2004). However, because of the heating of the plasma by these relatively "long" pulses, the wave breaking occurred well before reaching the cold wave breaking limit, which limited the maximum electric field to a few 100 GV/m. The maximum amplitude of the plasma wave has also been measured by Thomson scattering to be in the range 20-60 %, Clayton et al. (1998).

3.2.2 Forced wakefield

These unique properties of laser-plasma interaction at very high intensity, previously explored only on very large infrastructures, became accessible for smaller systems, fitted to university laboratories. These laser systems, also based on chirped pulse amplification, Strickland & Mourou (1985) and using here Titanium Sapphire crystals, fit in a room of several tens of meters square and deliver on-target energy of 2-3 J in 30 fs. This corresponds to 100 TW-class laser systems which can deliver an intensity of a few 10^{19} W/cm^2 after focusing. Many publications have shown that these facilities which deliver a modest energy and operate at a high repetition rate, can produce energetic electron beams with a quality higher than

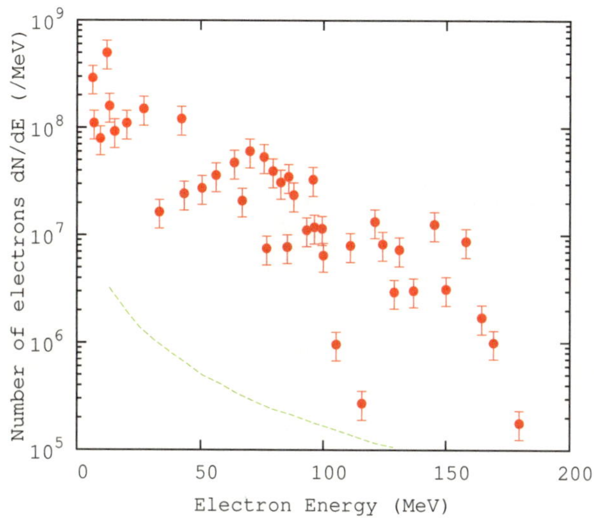

Fig. 7. Typical electron spectrum obtained at $n_e = 7.5 \times 10^{18}$ cm^{-3} with a 1J-30fs laser pulse focused down to a waist of $w_0 = 18\mu$m. The dashed line corresponds to the detection threshold

larger facilities. For instance, using the laser from "Salle Jaune" at LOA, electrons have been accelerated to 200 MeV in 3 mm of plasma, Malka et al. (2002). The mechanism involved is called forced laser wakefield to distinguish it from the self-modulated regime.
Indeed, thanks to short laser pulses, the heating of the plasma in the forced laser wakefield is significantly lower than in the self-modulated wakefield. This allows to reach much higher plasma wave amplitudes and also higher electron energies. Thanks to a limited interaction between the laser and the accelerated electrons, the quality of the electron beam is also improved. The measurement of the normalized transverse emittance has given values comparable to those obtained with conventional accelerators with an equivalent energy (normalized rms emittance $\varepsilon_n = 3\pi$ mm.mrad for electrons at 55 ± 2 MeV), Fritzler et al. (2004). Electron beams with maxwellian spectral distributions (exponential decay, see Fig. 7), generated by ultra-short laser pulses, have been produced in many laboratories in the world : at MPQ in Germany, Gahn et al. (1999), at LOA in France, Malka et al. (2001), at NERL in Japan, Hosokai et al. (2003), and at LBNL in USA, Leemans et al. (2004) for instance.

3.2.3 Bubble regime

More recently, theoretical work based on 3D PIC simulations have shown the existence of a robust acceleration mechanism called the bubble regime, Pukhov & Meyer-ter-Vehn (2002). In this regime, the dimensions of the focused laser are shorter than the plasma wavelength in longitudinal and also transverse directions. Thus, the laser pulse looks like a ball of light with a radius smaller than 10 μm. If the laser energy contained in this volume is high enough, the ponderomotive force of the laser expels efficiently electrons from the plasma radially, which forms a cavity free from electrons behind the laser, surrounded by a dense region of electrons. Behind the bubble, electronic trajectories intersect each other. A few electrons are injected in the cavity and accelerated along the laser axis, thus creating an electron beam with radial and longitudinal dimensions smaller than those of the laser (see Fig. 8).

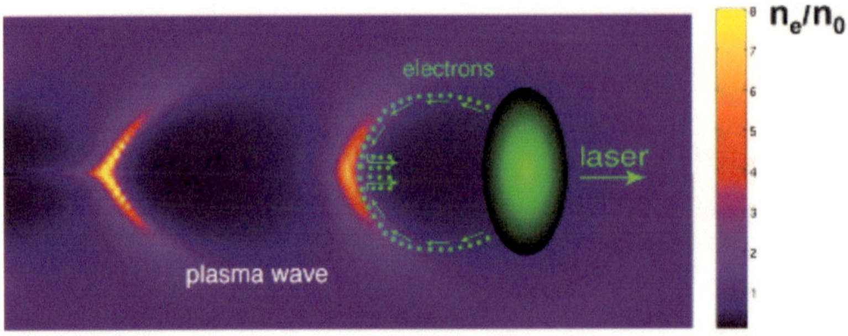

Fig. 8. Acceleration principle in the bubble regime. Electrons circulated around the cavitated region before to be trapped and accelerated at the back of the laser pulse

The signature of this regime is a quasi monoenergetic electron distribution. This contrasts with previous results reported on electron acceleration using laser-plasma interaction. This properties comes from the combination of several factors :

- The electron injection is different from that in the self-modulated or forced regimes. Injection doesn't occur because of the breaking of the accelerating structure. It is localized at the back of the cavity, which gives similar initial properties in the phase space to injected electrons.
- The acceleration takes place in a stable structure during propagation, as long as the laser intensity is strong enough.
- Electrons are trapped behind the laser, which suppresses interaction with the electric field of the laser.
- Trapping stops automatically when the charge contained in the cavity compensates the ionic charge.
- The rotation in the phase-space also leads to a shortening of the spectral width of the electron beam, Tsung et al. (2004).

Several laboratories have obtained quasi monoenergetic spectra : in France, Faure et al. (2004) with a laser pulse shorter than the plasma period, but also with pulses longer than the plasma period in England, Mangles et al. (2004), in the United States, Geddes et al. (2004), then in Japan, Miura et al. (2005) and in Germany, Hidding et al. (2006). The interest of such a beam is important for applications : it is now possible to transport and to refocus this beam by magnetic fields. With a maxwellian-like spectrum, it would have been necessary to select an energy range for the transport, which would have decreased significantly the electron flux. Electrons in the GeV level were also observed in this regime using in a uniform plasma, Hafz et al. (2008) or in plasma discharge, i.e, a plasma with a parabolic density profile, Leemans et al. (2006) with a more powerful laser which propagates at high intensity over a longer distance.

3.2.4 Colliding laser pulses scheme

The control of the parameters of the electron beam (such as the charge, energy, and relative energy spread) is a crucial issue for many applications. In the colliding scheme successfully demonstrated at LOA, it has been shown that not only these issues were addressed but also that a high improvement of the stability was achieved. In this scheme, one laser beam is used to create the relativistic plasma wave, and a second laser pulse which when it collides with

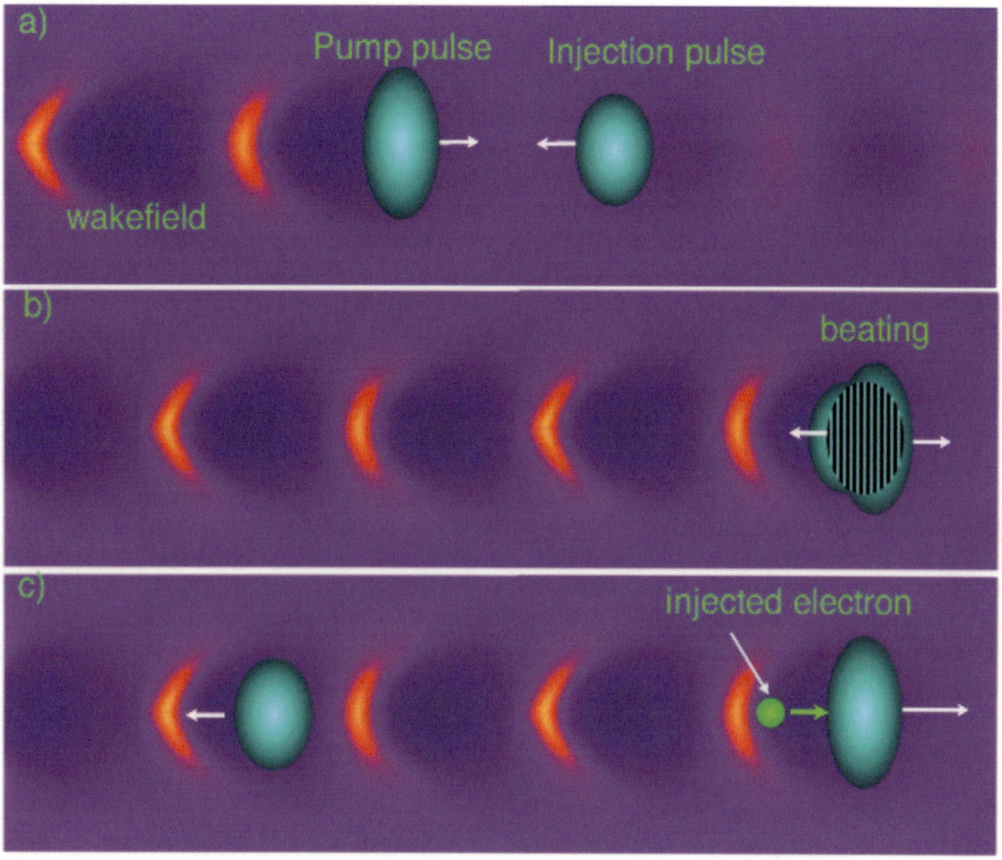

Fig. 9. Principle of injection in the counterpropagating colliding pulse scheme. (1) The two laser pulses have not collided yet; the pump pulse drives a plasma wake (2) The pulses collide and their interference sets up a beatwave that preaccelerates electrons. (3) preaccelerated electrons are trapped and further accelerated in the wake.

the main pulse, creates a standing wave which heats locally electrons of the plasma. The scheme of principle of the colliding laser pulses is shown in Fig 9. The control of the heating level gives not only the number of electrons which will be trapped and accelerated but also the volume of phase space, or in other words, the energy spread of the injected electrons bunch. In the pioneer work of E. Esarey *et al.*, Esarey et al. (1997), a fluid model was used to describe the evolution of the plasma wave whereas electrons were described as test particles. Electron trajectories in the beatwave as well as their energy gain were derived analytically from theory in the case of laser pulses with circular polarization. It has been shown that this approach fails to describe quantitatively the physics occurring at the pulse collision, Rechatin et al. (2007). In the fluid approach, the electron beam charge has been found to be one order of magnitude greater than the one obtained in PIC simulations. For a correct description of injection, one has to describe properly (i) the heating process, e.g. kinetic effects and their consequences on the dynamics of the plasma wave during the beating of the two laser pulses, (ii) the laser pulse evolution which governs the dynamics of the relativistic plasma waves,

Davoine et al. (2008). New unexpected feature have shown that heating mechanism can be achieved when the two laser pulses are crossed polarized. The stochastic heating can be explained by the fact that for high laser intensities, the electron motion becomes relativistic which introduces a longitudinal component through the $\mathbf{v} \times \mathbf{B}$ force. This relativistic coupling makes it possible to heat electrons. Thus, the two perpendicular laser fields couple through the relativistic longitudinal motion of electrons. The heating level is modified by tuning the intensity of the injection laser beam or by changing the relative polarization of the two laser pulses. This consequently changes the volume in the phase space and therefore the charge and the energy spread of the electron beam. When the pulses have the same polarization, electrons are trapped spatially in the beatwave and can not sustain the collective plasma oscillation inducing a strong inhibition of the plasma wave which persists after the collision. When the polarizations are crossed, the motion of electrons is only slightly disturbed compared to their fluid motion, and the plasma wave is almost unaffected during the collision, which tends to facilitate trapping. In addition to enhance stability, tuning the electron beam energy can be achieved by adjusting the position of the collision in the gas jet, Faure et al. (2006). The collision point can be modied by simply changing the delay between the two laser pulses. If the lasers collide at the entrance of thegas jet, electrons will be injected at an early stage and they can be accelerated over the whole gas jet length (2 mm). Thus, their energy will be high. On the contrary, injection at the exit of the gas jet will limit the acceleration length and will lead to a low energy beam. The robustness of this scheme has also allow to carry out very accurate studies of the dynamic of electric field in presence of high current electron beam. This beam loading effect has been used to reduce the relative energy spread of the electron beam. We have shown that there is an optimal load can flattened the electric field in such suitable way that a very small, 1%, relative energy spread electron beam can outcome form the target as shown on figure 10 . In this case, the more energetic electrons are slightly slow down and accelerated at the same energy that the slower one. In case of lower charge, this effect doesn't play any role and the energy spread depend mainly of the heating volume. For higher charge, the load is too high and the most energetic electrons slow down too much that they get energies even smaller that the slower one, Rechatin et al. (2009), increasing the relative energy spread. The optimal load was observed experimentally in agreement with full 3D PIC simulations, its corresponds to a current in the 20-40 kA. The decelerating electric field due to the electron beam was found to be in the GV/m/pC .

4. Future of the laser-based acceleration

Conventional accelerator technology has progressed through a long road paved by scientific challenges. A recent example is the development of superconductivity for high current acceleration in RF cavity, which has required tens of years of theoretical investigations and experimentations to understand the physical processes and finally to control the technology which has been successfully used in accelerators such as LEP/LHC (CERN), or HERA (DESY-Hamburg). Laser plasma accelerator researches follow the same road paved with many successful (and unsuccessful) experiments. Thanks to this pioneering works and judging from the incredible results achieved over the last three years, the time has come where a technological approach has to be considered. Two stages laser plasma accelerators schemes should allow the development of few GeV electron beam with a small relative energy spread and emittance, Malka et al. (2006). In parallel, fundamental and experimental research should of course be pursued to explore new regimes and to validate theories and numerical codes. The improvement of the laser plasma interaction with the evolution of short-pulse laser

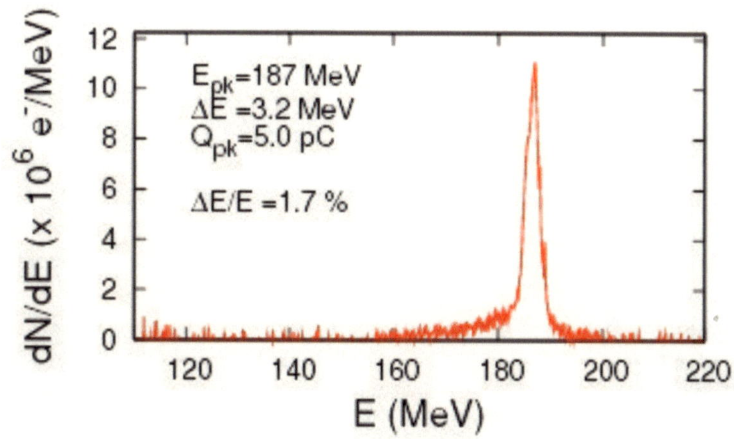

Fig. 10. Typical spectrum obtained in the colliding laser pulse scheme showing a 1% relative energy spread.

technology, a field in rapid progress, will still improve this new and very promising approach which potential societal applications in material science, medicine, chemistry and radiobiolgy, Malka et al. (2008). The ultra short duration (few fs) of the electron beam, Rechatin et al. (2010), and consequently his very high current (few kA) comparable to the one delivers at SLAC for LCLS experiment, where very bright X rays beam was produced by saturating the gain of their free electron laser, indicate that laser plasma accelerators should play a significant role in the compactness of free electron laser design and achievement.

5. Acknowledgements

I acknowledge warmly J. Faure, Y. Glinec, C. Rechatin, X. Davoine, E. Lefebvre, and A. Lifschitz who have largely contributed during this last decade to the work I presented in this book chapter. I also acknowledge the support of the European Research Council for funding the PARIS ERC project (contract number 226424), of the European Community-Research Infrastructure Activity under the FP6 and of European Community "Structuring the European Research Area" program (CARE, contract number RII3-CT-2003-506395), of the European Community-New and Emerging Science and Technology Activity under the FP6 "Structuring the European Research Area" program (project EuroLEAP, contract number 028514), and of the French national agency ANR-05-NT05-2-41699 "ACCEL1".

6. References

Andreev, N. E., Gorbunov, L. M., Kirsanov, V. I., Pogosova, A. A. & Ramazashvili, R. R. (1992). Resonant excitation of wakefields by a laser pulse in a plasma, *JETP Lett* 55: 571.

Antonsen, Jr., T. M. & Mora, P. (1992). Self-focusing and Raman scattering of laser pulses in tenuous plasmas, *Phys. Rev. Lett.* 69(15): 2204–2207.

Clayton, C. E., Tzeng, K.-C., Gordon, D., Muggli, P., Mori, W. B., Joshi, C., Malka, V., Najmudin, Z., Modena, A., Neely, D. & Dangor, A. E. (1998). Plasma wave generation in a self-focused channel of a relativistically intense laser pulse, *Phys. Rev. Lett.* 81(1): 100.

Davoine, X., Lefebvre, E., Faure, J., Rechatin, C., Lifschitz, A. & Malka, V. (2008). Simulation of quasimonoenergetic electron beams produced by colliding pulse wakefield acceleration, *Phys. Plasmas* 15(11): 113102.

Drake, J. F., Kaw, P. K., Lee, Y. C., Schmidt, G., Liu, C. S. & Rosenbluth, M. N. (1974). Parametric instabilities of electromagnetic waves in plasmas, *Phys. Fluids* 17(4): 778.

Esarey, E., Sprangle, P., Krall, J. & Ting, A. (1997). Self-focusing and guiding of short laser pulses in ionizing gazes and plasmas, *IEEE J. Quant. Electron.* 33(11): 1879–1914.

Faure, J., Glinec, Y., Pukhov, A., Kiselev, S., Gordienko, S., Lefebvre, E., Rousseau, J.-P., Burgy, F. & Malka, V. (2004). A laser-plasma accelerator producing monoenergetic electron beams, *Nature* 431: 541–544.

Faure, J., Rechatin, C., Norlin, A., Lifschitz, A., Glinec, Y. & Malka, V. (2006). Controlled injection and acceleration of electrons in plasma wakefields by colliding laser pulses., *Nature* 444: 737–739.

Fritzler, S., Lefebvre, E., Malka, V., Burgy, F., Dangor, A. E., Krushelnick, K., Mangles, S. P. D., Najmudin, Z., Rousseau, J.-P. & Walton, B. (2004). Emittance measurements of a laser-wakefield-accelerated electron beam, *Phys. Rev. Lett.* 92(16): 165006.

Gahn, C., Tsakiris, G. D., Pukhov, A., Meyer-ter-Vehn, J., Pretzler, G., Thirolf, P., Habs, D. & Witte, K. J. (1999). Multi-MeV electron beam generation by direct laser acceleration in high-density plasma channels, *Phys. Rev. Lett.* 83(23): 4772–4775.

Geddes, C. G. R., Tóth, C., van Tilborg, J., Esarey, E., Schroeder, C. B., Bruhwiler, D., Nieter, C., Cary, J. & Leemans, W. P. (2004). High quality electron beams from a laser wakefield accelerator using plasma-channel guiding, *Nature* 431: 538–541.

Gorbunov, L. M. & Kirsanov, V. I. (1987). Excitation of plasma waves by an electromagnetic wave packet, *Sov. Phys. JETP* 66: 290–294.

Hafz, N., Jeong, T. M., Choi, I. W., Lee, S. K., Pae, K. H., Kulagin, V. V., Sung, J. H., Yu, T. J., Hong, K.-H., Hosokai, T., Cary, J. R., Ko, D.-K. & Lee, J. (2008). Stable generation of GeV-class electron beams from self-guided laserplasma channels, *Nature Photonics* 2: 571–577.

Hidding, B., Amthor, K.-U., Liesfeld, B., Schwoerer, H., Karsch, S., Geissler, M., Veisz, L., Schmid, K., Gallacher, J. G., Jamison, S. P., Jaroszynski, D., Pretzler, G. & Sauerbrey, R. (2006). Generation of quasimonoenergetic electron bunches with 80-fs laser pulses, *Phys. Rev. Lett.* 96(10): 105004.

Hosokai, T., Kinoshita, K., Zhidkov, A., Nakamura, K., Watanabe, T., Ueda, T., Kotaki, H., Kando, M., Nakajima, K. & Ueseka, M. (2003). Effect of a laser prepulse on a narrow-cone ejection of mev electrons from a gas jet irradiated by an ultrashort laser pulse, *Phys. Rev. E* 67: 036407.

Khachatryan, A. G., van Goor, F. A., Boller, K.-J., Reitsma, A. J. W. & Jaroszynski, D. A. (2004). Extremely short relativistic-electron-bunch generation in the laser wakefield via novel bunch injection scheme, *Phys. Rev. ST Accel. Beams* 7(12): 121301.

Kruer, W. L. (1988). *The physics of laser plasma interactions*, Addison-Wesley, New-York.

Leemans, W. P., Nagler, B., Gonsalves, A. J., Tòth, C., Nakamura, K., Geddes, C. G. R., Esarey, E., Schroeder, C. B. & Hooker, S. M. (2006). GeV electron beams from a centimetre-scale accelerator, *Nature Physics* 2: 696–699.

Leemans, W. P., van Tilborg, J., Faure, J., Geddes, C. G. R., Tóth, C., Schroeder, C. B., Esarey, E., Fubiani, G. & Dugan, G. (2004). Terahertz emission from laser accelerated electron bunches, *Phys. Plasmas* 11(5): 2899.

Malka, V., Faure, J., Gauduel, Y. A., Lefebvre, E., Rousse, A. & Ta Phuoc, K. (2008). Principles and applications of compact laserplasma accelerators, *Nature Physics* 44: 447–453.

Malka, V., Faure, J., Marqus, J.-R., Amiranoff, F., Rousseau, J.-P., Ranc, S., Chambaret, J.-P., Najmudin, Z., Walton, B., Mora, P. & Solodov, A. (2001). Characterization of electron beams produced by ultrashort (30 fs) laser pulses, *Phys. Plasmas* 8: 0.

Malka, V., Fritzler, S., Lefebvre, E., Aléonard, M.-M., Burgy, F., Chambaret, J.-P., Chemin, J.-F., Krushelnick, K., Malka, G., Mangles, S. P. D., Najmudin, Z., Pittman, M., Rousseau, J.-P., Scheurer, J.-N., Walton, B. & Dangor, A. E. (2002). Electron acceleration by a wake field forced by an intense ultrashort laser pulse, *Science* 298: 1596–1600.

Malka, V., Lifschitz, A., Faure, J. & Glinec, Y. (2006). Staged concept of laser-plasma acceleration toward multi-gev electron beams, *Phys. Rev. ST Accel. Beams* 9(9): 091301.

Mangles, S., Murphy, C. D., Najmudin, Z., Thomas, A. G. R., Collier, J. L., Dangor, A. E., Divall, A. J., Foster, P. S., Gallacher, J. G., Hooker, C. J., Jaroszynski, D. A., Langley, A. J., Mori, W. B., Nooreys, P. A., Viskup, R., Walton, B. R. & Krushelnick, K. (2004). Mono-energetic beams of relativistic electrons from intense laser plasma interactions, *Nature* 431: 535.

McKinstrie, C. J. & Bingham, R. (1992). Stimulated Raman forward scattering and the relativistic modulational instability of light waves in rarefied plasma, *Phys. Fluids B* 4(8): 2626.

Miura, E., Koyama, K., Kato, S., Saito, S., Adachi, M., Kawada, Y., Nakamura, T. & Tanimoto, M. (2005). Demonstration of quasi-monoenergetic electron-beam generation in laser-driven plasma acceleration, *Appl. Phys. Lett.* 86: 251501.

Modena, A., Dangor, A., Najmudin, Z., Clayton, C., Marsh, K., Joshi, C., Malka, V., Darrow, C., Neely, D. & Walsh, F. (1995). Electron acceleration from the breaking of electron plasma waves, *Nature* 377: 606–608.

Moore, C., Ting, A., Krushelnick, K., Esarey, E., Hubbard, R., Hafizi, B., Burris, H., Manka, C. & Sprangle, P. (2004). Electron trapping in self-modulated laser wakefields by raman backscatter, *Phys. Rev. Lett.* 79: 3909.

Mora, P. (1992). Particle acceleration in a relativistic wave in the adiabatic regime, *Phys. Fluids B* 4(6): 1630–1634.

Mori, W. B., Joshi, C., Dawson, J. M., Forslund, D. W. & Kindel, J. M. (1988). Evolution of self-focusing of intense electromagnetic waves in plasma, *Phys. Rev. Lett.* 60(13): 1298–1301.

Pukhov, A. & Meyer-ter-Vehn, J. (2002). Laser wake field acceleration: the highly non-linear broken-wave regime, *Appl. Phys. B* 74: 355–361.

Rechatin, C., Davoine, X., Lifschitz, A., Ismail, A. B., Lim, J., Lefebvre, E., Faure, J. & Malka, V. (2009). Observation of beam loading in a laser-plasma accelerator, *Phys. Rev. Lett.* 103(19): 194804.

Rechatin, C., Faure, J., Davoine, X., Lundh, O., Lim, J., Ben-Ismal, A., Burgy, F., Tafzi, A., Lifschitz, A., Lefebvre, E. & Malka, V. (2010). Characterization of the beam loading effects in a laser plasma accelerator, *New Journal of Physics* 12(4): 045023.

Rechatin, C., Faure, J., Lifschitz, A., Malka, V. & Lefebvre, E. (2007). Plasma wake inhibition at the collision of two laser pulses in an underdense plasma, *Phys. Plasmas* 14(6): 060702.

Sprangle, P. & Esarey, E. (1992). Interaction of ultrahigh laser fields with beams and plasmas, *Phys. Fluids B* 4(7): 2241.

Sprangle, P., Tang, C.-H. & Esarey, E. (1987). Relativistic self-focusing of short-pulse radiation beams in plasmas, *IEEE Trans. Plasma Sci.* PS-15(2): 145–153.

Strickland, D. & Mourou, G. (1985). Compression of amplified chirped optical pulses, *Opt.*

Comm. 56: 219–221.
Tajima, T. & Dawson, J. M. (1979). Laser electron accelerator, *Phys. Rev. Lett.* 43(4): 267.
Ting, A., Esarey, E. & Sprangle, P. (1990). Nonlinear wake-field generation and relativistic focusing of intense laser pulses in plasmas, *Phys. Fluids B* 2(6): 1390.
Tsung, F. S., Narang, R., Mori, W. B., Joshi, C., Fonseca, R. A. & Silva, L. (2004). Near-gev-energy laser-wakefield acceleration of self-injected electrons in a centimeter-scale plasma channel, *Phys. Rev. Lett.* 93(18): 185002.
Umstadter, D., Chen, S.-Y., Maksimchuk, A., Mourou, G. & Wagner, R. (1996). Nonlinear optics in plasmas and laser wake field acceleration of electrons, *Science* 273: 472.

15

Mechanisms of Nanoparticle Formation by Laser Ablation

Tatiana Itina[1] and Karine Gouriet[2]
[1]*CNRS/University of Lyon,*
[2]*CNRS/Ecole des Mines de Saint-Etienne,*
France

1. Introduction

During the last decade, numerous experiments have been performed, clearly demonstrating that an interaction of a short (several nanoseconds or shorter) laser pulse with a solid target leads to the formation of nanoparticles. This technique, known as pulsed laser ablation (PLA), has become a promising method of the synthesis of nanoclusters for photonics, electronics and medicine (Movtchan et al., 1995; Yamada et al., 1996; Makimura et al., 1996; Geohegan et al., 1998; Albert et al., 2003). The PLA method has several advantages compared to more traditional techniques. In particular, it was shown that this method provides a possibility for chemically clean synthesis, which is difficult to achieve under more conventional nanoparticle production conditions. In addition, several experimental studies indicated that the cluster size distribution could be controlled in PLA by carefully choosing the laser irradiation parameters and properties of the background gas. Furthermore, laser ablation allows for an easy production of colloidal metal nanoparticles for biological and medical applications.

Despite the large number of the experimental results, the theoretical understanding of the physical and chemical mechanisms leading to the formation of nanoparticles during the PLA is still lacking. The number of theoretical studies of these mechanisms remains limited because both the continuum hydrodynamic models and the classical nucleation theory become inapplicable under the typical PLA conditions. Under these conditions, laser plume expansion and all collisional processes inside the plume occur so rapidly that equilibrium conditions are not attained. In addition, fast laser energy deposition may induce an explosive volume ejection rather than an equilibrium surface evaporation. Nevertheless, the laser ablation process is often described by a thermal desorption model, which considers the ablation as a rather slow layer-by-layer evaporation of monomers from the target surface (Anisimov, 19968). The presence of nanoparticles in the laser plume is then explained by using a Zel'dovich-Raizer condensation model (Zeldovich, 1966; Luk'yanchuk et al., 1998; Gusarov et al., 2000). This approach is appropriate for interpretation of the experimental findings obtained in laser ablation with hundred of nanoseconds and longer laser pulses, and in the presence of a background gas (Ohkubo et al., 2003; Boldarev et al., 2001). In the PLA with shorter laser pulses, however, clusters can be ejected directly from the target as a result of the target disintegration by laser-induced explosion-like process (Bulgakov, 2004; Amoruso et al., 2004; Zhigilei, 2003). In this case, the common thermal desorption and condensation model is insufficient and only a detailed molecular-level simulation can provide a complete description of the nanoparticle formation process. One of the possible approaches for such simulations is to combine the molecular dynamics technique (MD) with the direct simulation Monte Carlo method (DSMC) (Zhigilei & Garrison, 2000; Bird, 1994; Itina et al., 2002; Zeifman et al., 2002; Mizuseki et al., 2001; Briehl & Urbassek, 1999). This combination allows one to properly account for both the processes of cluster ejection and their following evolution during the laser plume expansion as a result of the gas-phase collisions.

So far, only several theoretical models were proposed to explain the growth of nanoparticles. For instance, such models as the model of Rice Ramsperger and Kassel (RRK) (Malakhovskii & Ben-Zion, 2002) and the Classical Nucleation Theory (CNT) were used in many studies (Kinjo et al. 1999; Zhong et al., 2006; Schenter et al., 1999). An alternative approach that can be successfully used for the investigation of the condensation-evaporation processes in small systems is based on the molecular dynamics (MD) method. In this method, molecular movement and interactions are simulated directly and one can easily study system dynamics (Kinjo et al. 1999; Zhong et al., 2006; Frenkel & Smit, 1996). Therefore, MD was also used to study the formation and evolution processes of clusters of many materials, and the advantages of this approach were demonstrated (Malakhovskii & Ben-Zion, 2002; Zhong et al., 2006).

Here, first we study the evolution of clusters in the presence of the same gas species. The study focuses on the influence of the surrounding gas and cluster parameters on the time evolution of clusters and small molecules formed in typical laser ablation experiments. The cluster parameters, such as cluster temperature and size, are chosen based on the cluster properties found in the ablation calculations. In the same way, the corresponding characteristics of the surrounding environment, such as gas density and temperature, are set to represent the ablated plume. Calculations are performed for two different materials, molecular matrix and metal. The roles of both simulation parameters and material properties are investigated. The applicability of the RRK theory is verified.

Then, we present the numerical results obtained with a hybrid MD-DSMC model of PLA for different laser pulses and fluences. The objective of this study is to bridge a gap between the previous molecular dynamics simulations and typical ablation experiments. In fact, the proposed combination of methods allows us to attain both time and length scales typical for ablation experiments. By using the combined model, we investigate the time evolution of clusters in the plume. We examine different types of collisions leading to cluster growth and decay. As a result, we determine the major collisional processes responsible for the observed changes in cluster distributions. These results are related to the recent experimental results on the ablation plume dynamics.

2. Simulation details

2.1 MD calculations of cluster evolution

In the first part of our calculations, classical molecular dynamics (Frenkel & Smith, 1996 and Zhigilei et al., 1998) is used to study a system composed by a nickel cluster surrounded by a gas of the same atoms. In many applications, metal nanoparticles are used due to their optical (plasmonic), and chemical (sensing) properties (Amoruso et al., 2007). The effective Embedded Atom Model (EAM) potential (Daw & Baskes, 1984) is used in these calculations. This force field contains two terms: a pairwise additive interaction ϕ of the atomic cores shielded at short interatomic distances and secondly a contribution F(ρ) of delocalized electrons to the energy, which is modeled by multi-body term assuming a linear superposition ρ_i of atomic electron densities ρ_{at}. The energy of an atom i can be written as

$$E_i = F_i\left[\rho_i\right] + \frac{1}{2}\sum_{i \neq j}\phi_{ij}\left(r_{ij}\right) \tag{1}$$

The subscripts i and j run over the number of atoms N in the system. A cut-off radius of 0.53 nm is used for the electron density and the pair potential contribution to the Ni-Ni interaction.

First, we form a cluster of N particles with the initial temperature T_0. Then, a background gas is generated based on the corresponding gas number density and temperature. The volume of the cell is constant and equal to 100×100×100 nm³. Periodic boundary conditions in all directions are imposed. After generating the gas, we heat it, so that each gas particles has a velocity based on the Maxwell-Boltzmann distribution with a given gas temperature. Then the cluster is located in the center of the cell. Thus, we can vary four basic parameters: two for the cluster (initial temperature and size) and two for the gas (temperature and density).

In the calculations, we monitor the following processes (i) simple collision

$$A_i + A \xleftrightarrow{\nu_{coll}} A_i + A$$

with the collision rate ν_{coll}; and (ii) evaporation and condensation processes

$$A_i + A \xleftrightarrow{\nu_e,\nu_c} A_{i+1}$$

with reaction rates ν_e and ν_c respectively. The ν_e (or ν_c) rate is calculated as the ratio of the number of evaporated cluster molecules (or condensated gas molecules on the cluster) with the time step δt. This time step is sufficiently larger than the MD calculation time step (1 fs for nickel) and in order to 50 ps. Calculations typically continue until about 3 ns. This time delay is long enough for the establishment of a dynamic equilibrium between the evaporation and condensation processes. This condition corresponds to ~2 ns in the present calculations.

To determine the number of cluster molecules, we used the neighbor method. Before starting the simulation, we define differently the both cluster and gas molecules. We calculate the neighbor list at each 25 fs, and the cluster molecule list at each 50 ps. If a molecule own the cluster at the last step is not in the new list, this molecule is evaporated and we increased the number of evaporated molecules N_{eva}; and inversely. To investigate the influence of each parameter on the others, we ran several simulations with 3 fixed parameters.

The cluster evaporation process can be described as unimolecular thermal decomposition process by using the RRK theory (Malakhovskii & Ben-Zion, 2002 21). In the RRK theory, the evaporation and the condensation of clusters are controlled by the sublimation energy $D_n = E_n - E_{n-1}$, where E_n is the absolute binding energy of the cluster with n molecules. The evaporation rate is defined as following

$$\nu_e^{RRK} = \nu_0 g_n \left[\frac{E_n - D_n}{E_n}\right]^{s_n - 1} \tag{2}$$

where ν_0 is the typical molecular vibrational frequency, ν_n is the number of vibrational degrees of freedom (1 for dimers and 3N-6 for larger clusters), g_n is the number of surface molecules and D_n is the molecular evaporation energy. For clusters larger than 50 molecules in size, the evaporation rate can be approximate by

$$\nu_e^{RRK} = \nu_0 g_n \exp\left[-\frac{D_n}{E_n}\right] \tag{3}$$

For large clusters, we use the following approximations (Malakhovskii & Ben-Zion, 2002)

$$g_n = (36\pi)^{1/3}\left(N^{1/3} - 1\right)^2 \tag{4}$$

In addition, we can approximate the condensation rate by the collision rate of monomers with the cluster

$$\nu_c = \bar{c}\pi\left(r_0 + r_n\right)^2 \rho \tag{5}$$

The average thermal velocity of the monomers in the gas is $\bar{c} = \left(8k_B T_g / \pi m\right)$, ρ is the monomer density, k_B is Boltzmann's constant, T_g is the gas temperature and m is the molecular mass, r_0 and r_n are the hard sphere equivalent radius of a monomer and a cluster. In this expression for the condensation rate, we approximate the cross section by the surface area of a sphere of radius r_n and assume unit sticking probability upon collision. The hard sphere equivalent radius is calculated based on the gyration radius of the cluster r_{gyr} (Arunachalam et al., 1999).

$$r_n = \sqrt{\frac{5}{2}} r_{gyr} + r_0 \tag{6}$$

To compare the calculation results with the RRK theory, we calculate the evaporation and condensation rates for a set of simulation parameters (cluster size and temperature and gas density and temperature and interatomic potentials) that corresponds to the typical laser applications, such as laser ablation experiments (Handschuh et al., 1999; Vitiello et al; 2005).

2.2 Combined MD-DSMC numerical model

A combined MD-DSMC model is developed for the calculation of the laser plume formation and long-term evolution. In the model, two different numerical methods are used (i) molecular dynamics for target material disintegration and ejection of a mixture of clusters and monomers (14); and (ii) direct simulation Monte Carlo for the calculation of large-scale three-dimensional plume evolution (Itina et al., 1997). Here, as in the previous MD calculations (Zhigilei et al., 1998; Zeifman et al., 2002), the breathing sphere model is adopted to model laser pulse absorption and relaxation processes in the MD part. The calculation parameters in the MD part are the same as in these calculations for a molecular solid target. The pulse duration was set to be 15 ps.

The transition between the DSMC and MD calculations is performed by using MD results obtained at 1 ns after the laser pulse for the design of the initial conditions for the subsequent DSMC calculation. The collision probabilities in DSMC calculations are parameterized to describe the material properties in a gas phase, such as viscosity coefficient and velocity dependence of the equivalent collision cross sections.

To couple the two models, all plume species are divided into five groups depending on their size. The groups are chosen as follows: 1 - monomers, 2 - clusters of 2 to 15 molecules, 3 - clusters of 16 to 100 molecules, 4 - clusters of 101 to 1000 molecules, and 5 - clusters of more than 1000 molecules. The calculations are performed with laser pulse duration of 15 ps and an absorbed laser fluence of 61.38 J/m^2. Cluster size is sampled in each group according to the size distributions obtained in the MD simulation.

In the present study, we assume that the spatial laser beam profile is top hat and the radial expansion of the ablation plume is small during the first MD stage. Therefore, the radial distributions of the simulation parameters are assumed to be uniform within the laser spot. The axial distributions of the initial DSMC parameters are determined by fitting the axial distributions of density, velocity, radial and axial temperatures for different groups obtained from the MD calculations. To fit the axial density distributions, we use normal distribution functions

$$f_i(z) = \frac{d_i}{\sigma_i\sqrt{2\pi}} \exp\left[-(z-\mu_i)^2 / 2\sigma_i^2\right] \tag{7}$$

where z is the distance from the initial target surface. The parameters d_i, μ_i, and σ_i are calculated from numerical fits to the MD results performed for each group, and i=1-5. The linear dependencies

$$g_i(z) = A_i z - B_i \tag{8}$$

are used to fit axial velocity distributions, where A_i and B_i are fitting parameters for each component (Zhigilei, 2003). The distributions of the axial and radial temperatures as well as the internal temperatures of the clusters are fitted by six-order polynomial dependencies. More details of the procedure used for the statistical description of the output of MD simulations will be presented elsewhere.

2.3 Modification of the DSMC method used for the calculation of ablation flows with clusters

In the DSMC simulations, the laser plume is modeled by an ensemble of simulated particles (here, about $8{\cdot}10^6$). The simulated particles are introduced based on the gas-dynamic parameters obtained from the MD simulations of the initial stage of the plume formation. The physical space is divided into a network of cells with dimensions smaller than the local mean free path. The time is incremented by a small time step Δt. The calculations consist in the repetition of three main steps: (i) indexing of the simulated particles; (ii) calculation of a representative set of collisions and chemical reactions in each cell; (ii) movement of the simulated particles and calculation of the interactions with the boundaries (Bird, 1994). In the present work, collisions are calculated for all particles together based on their probabilities by using a NTC technique (Bird, 1994).

The initial and boundary conditions are set as follows. At the beginning, simulated particles are introduced according to the distributions calculated by MD as described in Section 2.2 above. The axis r = 0 is set to be the axis of radial symmetry. At the target surface (Z = 0), a partial diffuse reflection and re-deposition of particles is sampled by using an accommodation coefficient. The accommodation coefficient of the target surface is set to be

0.5 in this work. The simulated volume is increased several times during the calculation to cover the plume expansion region.
The calculation of collisions and reactions in the present DSMC includes the treatment of the following processes

i. elastic collisions;
ii. inelastic non-reactive collisions;
iii. sticking reactions in cluster - molecule collisions;
iv. aggregation reactions in cluster - cluster collisions;
v. cluster evaporation reactions.

The non-reactive collisions are calculated based on the conservation laws of energy and momentum. For clusters, the reference cross-sections are computed based on their effective radii $r_n = an^{1/3} + b$, where a=3.97 Å, b=1.59 Å, n is the number of monomers in the cluster. A soft-spheres collision model is used in the present modeling calculation. As we described above, the initial values of the internal energy, E_n, are assigned to each cluster based on the internal temperature distribution obtained from the MD simulation and the number of degree freedom in the cluster. To describe the internal energy modification in clusters and small molecules, Larsen-Borgnakke statistical model (Bird, 1994) is used. Sticking process is included for collisions in which the relative translational energy before the collision is smaller than the binding energy of the final cluster. In this case, the sticking of two clusters containing n_1 and n_2 molecules occurs with the probability (Hittema & McFeaters, 1996)

$$p_1 = p_0 \times (n_1 n_2)^{-x} \tag{9}$$

where x is a parameter (x = 1/12 in the present simulation (33)), and p_0 is a constant. The parameters in (3) are selected based on a separate MD study of cluster evolution in a gas (to be published in a forthcoming paper). The binding energy in each cluster is given by a combination of volume and surface terms

$$E_b = a_V n - a_s n^{2/3} \tag{10}$$

where a_V and a_S are material constants and n is the number of molecules in the cluster. To ensure the energy conservation in the sticking reaction, the relative translational energy of the monomer and cluster was transferred into the internal energy of the resulting cluster, whereas the center-of-mass velocity was conserved.
The rate of evaporation from a cluster of n atoms with internal energy E_n is calculated according to a simple exponential decay (Brady et al., 1979)

$$\ln N / N_0 = -k_{ev} t , \tag{11}$$

where k_{ev} is the decay constant, or the rate coefficient (Jarold, 1994).

3. Results and discussion

3.1 Cluster evolution

A series of calculation are performed for a cluster with different initial size, but with the same initial cluster temperature and gas density and temperature. Calculation results demonstrate that both evaporation [Fig. 1(a)] and condensation [Fig. 1(b)] rates increase with initial cluster size. The size dependencies are fitted by a polynomial function with the same order of the size function in theoretical rates. We obtain a reasonable agreement between the theoretical fits and our results. The evaporation and condensation rates are larger than 1.2 10^{11} s^{-1} for clusters containing around 500 atoms and larger than 2.0 10^{11} s^{-1} for N>1200 atoms.

Furthermore, the initial cluster temperature is varied, keeping other initial parameters equal to those in the previous series of calculations. Simulations results demonstrate that the evaporation rate (Fig. 2(a)) decreases exponentially with the inverse of cluster temperature, while the cluster temperature dependency of the condensation rate is not significant (Fig.2(b)). In addition, one can see that the calculated dependency of evaporation rate can be described by the RRK theory. In the RRK theory, the evaporation rate exponentially decreases with cluster temperature and the coefficient in the exponential (sublimation energy) corresponds to the minimal kinetic energy required to obtain a significant evaporation. Our results show that we can determine the sublimation energy D_N for the clusters. For instance, for nickel cluster with 604 atoms, we found D_N = 1.725 eV.

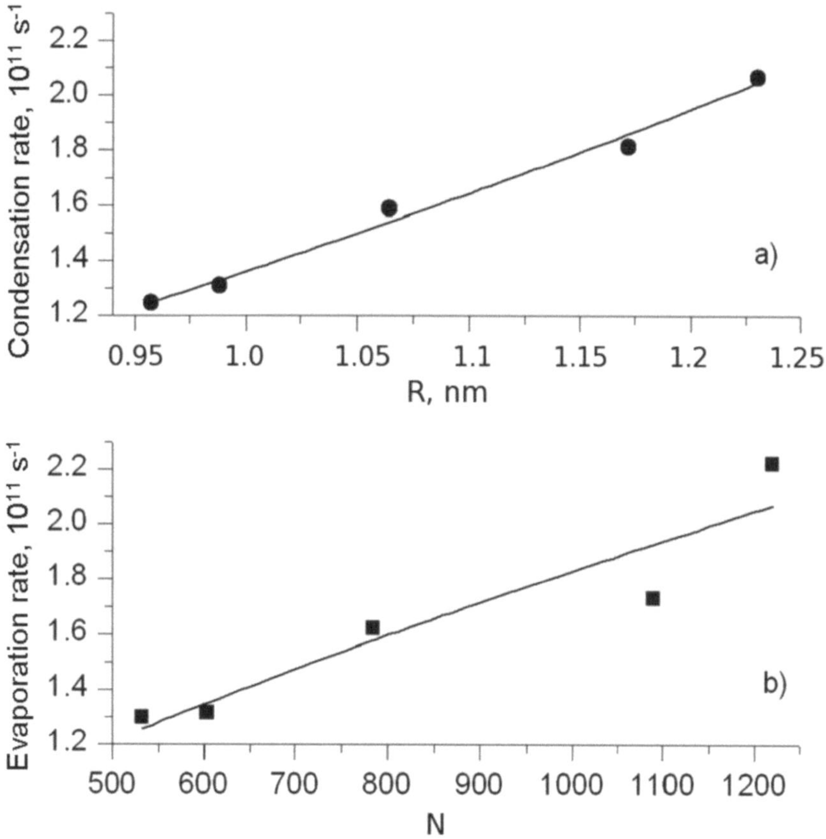

Fig. 1. (a) Condensation rate as a function of R cluster radius of nickel. The condensation rate is fitted by a function given by the collision rate such as A + $B(r_1+R)^2$, where A=-1.973 1010 s^{-1} and B=1.181 10^{29} $s^{-1}m^{-2}$. (b) Evaporation rate as a function of the number of atoms, N, in a nickel cluster. The evaporation rate is fitted by a function given by the RRK evaporation rate. These results are obtained for clusters of different sizes heated initially to 2400 K in a gas with the number density of 10^{19} cm^{-3} at 5036 K.

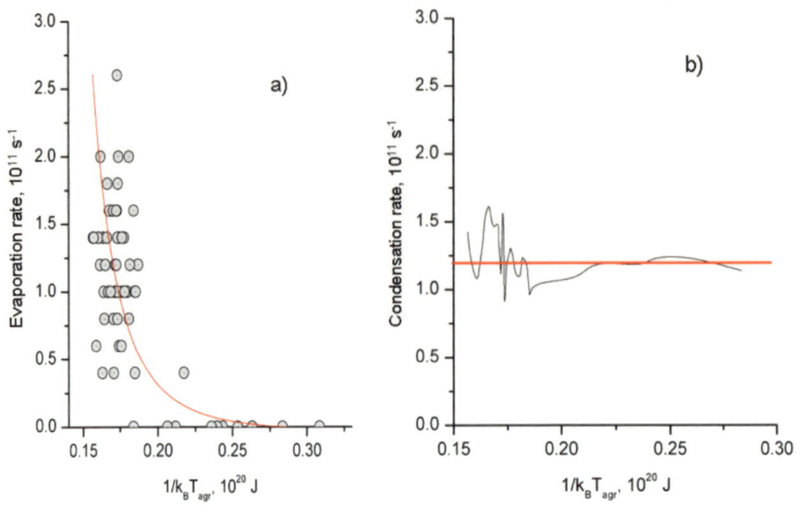

Fig. 2. (a) Evaporation rate as a function of cluster temperature of nickel. The evaporation rate is fitted by a function given by the RRK evaporation rate such as $A\ exp(-D_N/k_BT)$, where A=2.57 10^{12} s^{-1} and D_N=1.725 eV. (b) Condensation rate as a function of cluster temperature (nickel). The results are obtained for the cluster containing initially 604 molecules with different temperatures in a background gas with number density of 10^{19} cm^{-3} at 5036 K.

Then, cluster evaporation and condensation processes are investigated by running our MD simulations with the same cluster parameters (material, initial size and temperature) but with different gas properties (density and temperature). Interestingly, the obtained calculation results show that the evaporation and condensation rates are proportional to the gas density (Fig. 3) and the square of temperature (Fig. 4). In particular, our simulations indicate a dramatic influence of both gas density and temperature on the evaporation process. For a nickel cluster with initially 604 atoms, calculations show that the evaporation rate is larger than 1.0 10^{11} s^{-1}, if the initial gas temperature is larger than the boiling point (T_{boil}=3190 K at P=0 Pa, for nickel) and its density is larger than 1.0 10^{19} cm^{-3}. So, under the given conditions, for both gas temperature and density, the evaporation process definitely prevails over the condensation one. Interestingly, an intersection is obtained at 4000 K. This effect can be explained by interplay between the condensation and evaporation processes. At these temperatures, each process balances the other process, and the cluster size does not change significantly.

One can therefore expect that small clusters (N<50 molecules) evaporate rapidly in the ablated plume regions where the density and the temperature of monomers are large ($T>T_{boil}$ and n>1.0 10^{19} cm^{-3}). These conditions typically correspond to the beginning of the laser plume formation and its expansion in laser ablation.

Thus, the calculation results reveal the role of gas-cluster collisions in both evaporation and condensation processes. The performed MD simulations confirm that the RRK theory and the collision rate can be used to define the evaporation and condensation rates. Similar results for both studied materials are obtained. However, we can modify the RRK solution for the definition of evaporation and calculate the sticking probability for the condensation rate estimation based on our MD simulations. For this we propose analytical expressions for the evaporation ν_e and condensation ν_c rates, for large clusters of nickel and molecular material as follows

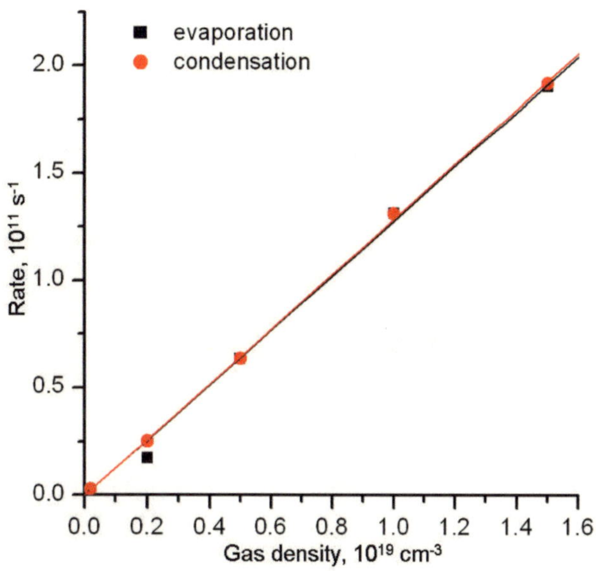

Fig. 3. Evaporation rates and condensation rates for a nickel cluster containing initially 604 molecules at 2490 K as a function of monomers gas density in a gas at 5000 K. Linear fits are added to the dependencies.

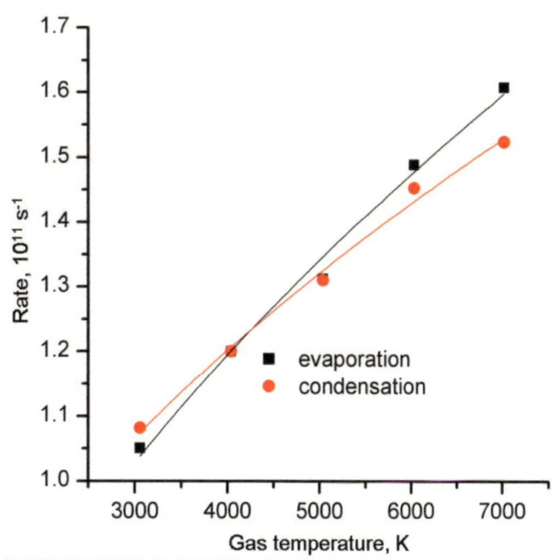

Fig. 4. Evaporation rates and condensation rates for a nickel cluster containing initially 604 molecules at 2490 K as a function of gas temperature. Here, the background gas number density is 10^{19} cm^{-3}.

$$\nu_e = v_e^0 \bar{c} \rho \, g_n \exp\left[-\frac{D_n}{E_n}\right] \tag{12}$$

$$\nu_c = v_c^0 \bar{c} \rho \left(r_n + r_0\right)^2 \tag{13}$$

where $\nu_e^0 = 9.1272 \cdot 10^{-18}$ and $\nu_c^0 = 5.4201$, the average thermal velocity of the monomers in the gas $\bar{c} = \left(8k_B T_g / \pi m\right)$, ρ is the monomer density, k_B is Boltzmann's constant, T_g is the gas temperature and m the molecular mass, r_0 and r_n the hard sphere equivalent radius of a monomer and a cluster. D_n is the molecular evaporation energy and E_n is the internal energy.

3.2 Plume dynamics and composition

A series of calculations are performed with the combined MD-DSMC model. First, we consider the laser plume dynamics.

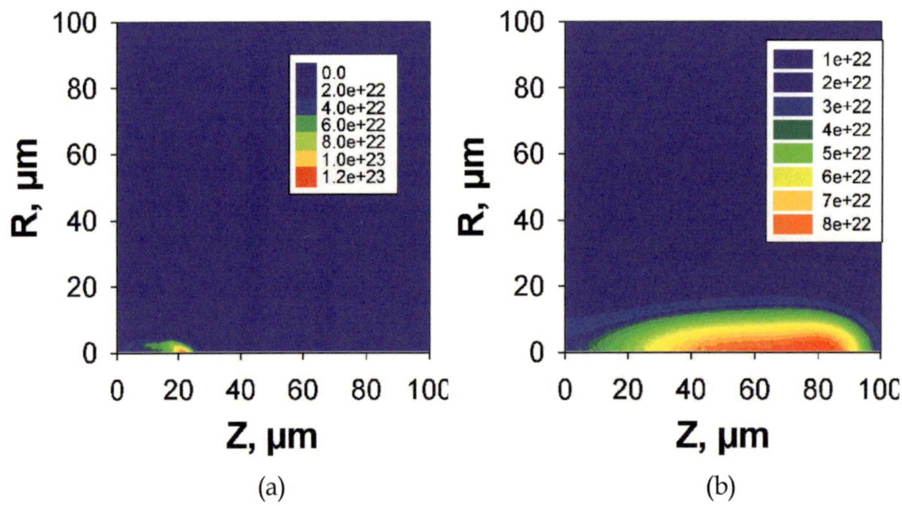

Fig. 5. Two-dimensional distributions of (a) cluster number density and (b) monomer density at a time delay of 50 ns. The results are obtained in a MD-DSMC simulation performed for a pulse duration of 15 ps, laser fluence of 61 J/m^2, and a laser spot radius of 10 μm.

Distributions of the cluster and monomer densities shown in Fig. 5 illustrate the laser plume expansion away from the irradiated target. Monomers constitute the fastest plume component, whereas clusters are moving slower. In addition, we find that the larger are clusters the lower is their average axial velocity. Quantitatively, the center-of mass velocity of the monomer component of the plume is about 1560 m/s, whereas the one of clusters is as small as 555 m/s. Here, clusters bigger than 10 molecules were considered.

Furthermore, the changes in the plume composition and the influence of evaporation/condensation processes are examined. In order to provide a reliable description of the evaporation/condensation processes in DSMC, the cluster evolution is first investigated in a separate 5 ns long MD simulation performed for a gas box containing one cluster of a given initial size and temperature. These calculations demonstrate that, for temperatures observed by the end of MD simulations of laser ablation, small cluster evaporation prevailed over growth within the considered time of 5 ns. The parameters used in the DSMC part are adjusted to these results.

The evolution of the plume composition in the MD-DSMC simulation of plume expansion is shown in Fig. 6 (a). One can see that at the beginning and until about 10 ns, the number of clusters decreases, whereas the number of monomers rises. After that, the situation changes and the number of clusters in the plume steadily increases. At a delay of 200 ns, the number of clusters reaches the one that was in the plume at 1 ns after the beginning of laser pulse. If one looks at the cluster abundance distribution [Fig. 6 (b)], which also varies with time, one can see that the number of small clusters decreases, while bigger clusters become more abundant.

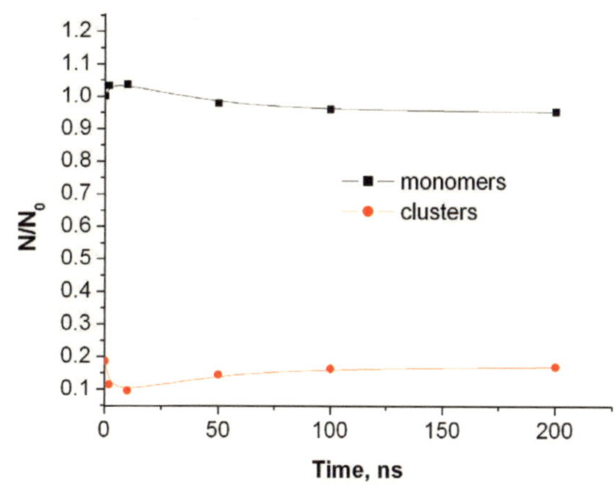

(a)

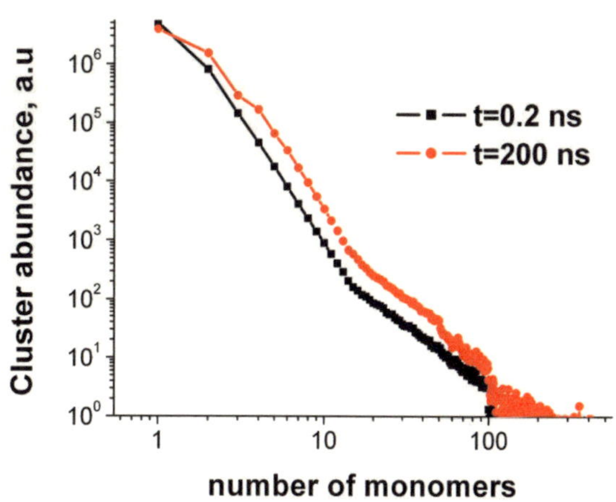

(b)

Fig. 6. (a) Time evolution of the number of monomers and clusters in the plume. (b) Cluster abundance distribution. The results are obtained in a MD-DSMC simulation performed for a pulse duration of 15 ps, laser fluence of 61 J/m² and a laser spot radius of 10 μm.

As we have already seen from the MD calculations, the cluster size distribution obtained soon after the ejection is a decaying function. During the following plume expansion, numerous collisions take place in the plume and affect the distribution. The indication of these changes obtained in the MD-DSMC is displayed in Fig. 6(b). One can see the decrease in the number of small clusters and the growth in the number of average ones. The distribution shape agrees with the experimental findings. These results will be discussed in more details elsewhere.

The calculations may help explaining the results of the experiments on plume imaging and spectroscopy obtained for short and ultra-short laser pulses. Results of Noël et al (Noël and Hermann, 2009) show typical plume images recorded in our laboratory at different times in ultra-short laser ablation experiments with pulse width of 100 fs at 400 nm wavelength. One can clearly observe two different plumes: the faster one and the slower one. Based on our simulation results and on additional spectroscopic study, we can conclude that the first component is composed of atoms and small clusters and the slower one is composed of bigger clusters.

4. Conclusion

We have studied the evaporation and condensation processes of a molecular cluster in the presence of a background gas composed of the same molecules as the cluster for two different materials: (i) molecular material, which is modeled by the breathing sphere model; and (ii) metallic material (nickel), which is modeled by the embedded atom model. We have analyzed the calculated evaporation rate by comparing the MD calculation results with the rates given by the RRK theory. The condensation rate is analyzed by using collision rate. As a result, we have observed that both gas density and temperature have strong influence on the evaporation process. We have shown that, in the presence of a gas, evaporation is a collision-driven process, just like condensation is. Therefore, the evaporation rate depends on gas density and temperature. As a result, both evaporation and condensation rates depend on the gas parameters.

The obtained MD results are in agreement with a slightly modified RRK collision rate expression. The corresponding modifications are proposed based on the MD calculation results. In particular, our calculation results have demonstrated that gas-cluster collisions affect cluster evaporation. The condensation rate has been also explained by using the cluster-gas collision rate with a sticking probability determined by our MD simulations.

Both the calculations performed with the combined MD-DSMC model and the experiments clearly demonstrate the role of two following channels of the cluster production by short-pulsed laser ablation

i. direct cluster ejection upon the laser-material interaction;
ii. collision-affected condensation and evaporation in the ablated plume flow.

The first process results from the fast energy absorption by the target, which leads to the explosive decomposition of a substantial volume of the rapidly heated target material. For short laser pulses, material decomposition might occur through either "phase-explosion" (Garrison et al., 2003), photomechanical spallation (Zhigilei & Garrison 2000), or fragmentation (Perez & Lewis, 2002; Glover, 2003) mechanisms. These processes take place in volume rather than on the surface and lead to the ejection of a mixture of clusters and monomers.

The second mechanism is due to gas-phase collisions and evaporations. These processes are similar to the phenomena taking places in aggregation sources (Briehl & Urbassek, 1999; Haberland, 1994). The major advantage of short and ultra-short laser pulses for cluster synthesis is the presence of the laser-ejected small molecules and clusters in the ablated flow. As a result, the formation of diatomic molecules in three-body collisions, which represents a "bottleneck" for cluster formation in common aggregation sources, is not crucial for cluster synthesis by short laser ablation.

5. References

Albert, O.; Roger, S.; Glinec, Y.; Loulergue, J. C.; Etchepare, J.; Boulmer-Leborgne, C.; Perriere, J. & Millon, E. (2003). *Appl. Phys. A: Mater Sci. Process*, 76, 319

Amoruso, S. ; Bruzzese, R. ; Spinelli, N. ; Velotta, R. ; Vitello, M. & Wang, X. (2004). *Europhys. Lett.* 67, 404

Amoruso, S.; Ausaniob, G. ; Bruzzese, R. ; Campana, C. & Wang, X (2007). *Appl. Surf. Sci.*, 254(4), p.1012

Anisimov, S. I. (1968). *Zh. Eksp. Teor. Fiz.* 54, 339 (*Sov. Phys. JETP*, 27, 182 (1968))

Arunachalam, V. ; Lucchese, R. R. & Marlaow, W. H. (1999). *Phy. Rev. E* 60, p 2051

Bird, G. A. (1994). *Molecular Gas Dynamics and the Direct Simulation of Gas Flows*, Clarendon, Oxford.

Boldarev, A. S. ; Gasilov, V. A. ; Blasco, F. ; Stenz, C.; Dorchies, F.; Salin, F.; Faenov, A. Ya.; Pikuz, T. A.; Magunov, A. I. & Skobelev, I. Yu (2001). *JETP Letters*, 73, 514

Brady, J. W.; Doll, J. D. & Thompson, D. L. (1979). *J. Chem. Phys.*, 71, 2467

Briehl, B. & Urbassek, H. M. (1999). *J. Vac. Sci. Technol. A* 17, 256

Bulgakov, A. V. ; Ozerov, I.; Marine, W. (2004). *Appl. Phys. A* 79, 1591

Daw M. S. & Baskes, M. I (1984). *Phys. Rev. B*, 29, p. 6443

Frenkel, D. & Smit, B. (1996). *Understanding molecular simulation*, Academic Press

Garrison, B. J.; Itina, T. E. & Zhigilei, L. V. (2003). *Phys. Rev. E*,68, 041501

Geohegan, D. B. ; Puretzky, A. A. ; Dusher, G. & Pennycook, S. J. (1998). *Appl. Phys. Lett.*, 72, 2987

Glover, T. E. (2003). *J. Opt. Soc. Am. B*, 20, 125

Gusarov, A. V. ; Gnedovets, A. V. & Smurov, I. (2000). *J. Appl. Phys.* 88, 4362

Haberland, H. (1994). *Clusters of Atoms and Molecules*, ed. by H. Haberland (Springer, Berlin), p. 205

Handschuh, M. ; Nettesheim, S. & Zenobi, R. (1999). *Appl. Surf. Sci.*, 137(1-4) p. 125–135

Hittema H. & McFeaters, J. S. (1996). *J. Chem. Phys.* 105, 2816

Itina, T. E.; Tokarev, V. N.; Marine, W. & Autric, M. (1997). *J. Chem. Phys.* 106, 8905

Itina, T.E.; Hermann, J. ; Delaporte, P. & Sentis, M. (2002). *Phys. Rev. E*, 66, 066406

Jarold, M. F. (1994). *Clusters of Atoms and Molecules*, ed. by H. Haberland (Springer, Berlin,), p. 163.

Kinjo, T.; Ohguchi, K. ; Yasuoka, K. & Matsumoto, M. (1999). *Computational Materials Science, 14*, 138-141

Luk'yanchuk, B. ; Marine, W. & Anisimov, S. (1998). *Laser Phys.* 8, 291

Makimura, T. ; Kunii, Y. & Murakami, K. (1996). *Jpn. J. Appl. Phys. Part 1*, 35, 4780

Malakhovskii, A. V. & Ben-Zion, M. (2001). *Chem. Phys, 264*, 135-143

Mizuseki, H. ; Jin, Y. ; Kawazoe, Y. & Wille, L. T. (2001). *Appl. Phys. A* 73, 731

Movtchan, I. ; Dreyfus, R. W. ; Marine, W. ; Sentis, M. ; Autric, M. & Le Lay, G. (1995). *Thin Solid Films* 255, 286
Noël, S.; Hermann, J. (2009)*Applied Physics Letters*, 94:053120
Ohkubo, T.; Kuwata, M. ; Luk'yanchuk, B. ; Yabe, T. (2003). *Appl. Phys. A.*, 77, 271
Perez, D. & Lewis, L. J. (2002). *Phys. Rev. Lett.*, 89, 25504
Schenter, G. K.; Kathmann, S. M. & Garett, B. C. (1999). *Phys. Rev. Letters, 82*, 3484-3487
Vitiello, M.; Amoruso, S.; Altucci, C. ; de Lisio, C. & Wang, X. (2005).*Appl. Surf. Sci.*, 248(1-4), p. 404
Yamada, Y.; Orii, T.; Umezu, I. ; Takeyama, S. & Yoshida, Y. (1996). *Jpn. J. Appl. Phys. Part 1,* 35, 1361
Zeldovich, Y. B. & Raizer, Yu. P. (1966). *Physics of Shock Waves and High Temperature Hydrodynamic Phenomena* Academic Press, London
Zeifman, M. I. ; Garrison, B. J. & Zhigilei, L. V. (2002). *J. Appl. Phys.* 92, 2181
Zhigilei, L. V. ; Kodali, P. B. S. & Garrison, B. J. (1998). *J. Phys. Chem. B*, 102, 2845-2853
Zhigilei, L. V. & Garrison, B. J. (2000). *J. Appl. Phys.* 88, (3), 1281
Zhigilei, L. V. (2003). *Appl. Phys. A* 76, 339
Zhong, J. ; Zeifman, M. I. & Levin, D. A. (2006). *J. Thermophysics and Heat Transfer, 20*, 41-45

16

Ablation of 2-6 Compounds with Low Power Pulses of YAG:Nd Laser

Maciej Oszwaldowski, Janusz Rzeszutek and Piotr Kuswik
Poznan University of Technology, Faculty of Technical Physics
Poland

1. Introduction

The 2-6 compounds and their mixed crystals are important semiconductor materials with practical applications in various areas of solid state electronics and optoelectronics. Most of the applications need these materials in a thin film form. One of the most versatile methods of obtaining thin films and their composed structures is the Pulsed Laser Deposition (PLD) method. That method has been used many times to the deposition of the 2-6 compound films and the result of the investigations of both the target ablation process and the obtained films physical properties were published in numerous publications. The earlier publications have been summarized in several excellent reviews (Cheung & Sankur, 1988; Christley & Hubler, 1994; Dubowski, 1991). In spite of the existing broad experimental material, optimum technological conditions for obtaining 2-6 compound layers by PLD with pre-defined properties has not as yet been determined. This is because the layers properties depend on the Pulsed Laser Ablation (PLA) process of the target material. The ablation depends on such parameters of the process as: the energy and duration of the laser pulse, pulse repetition frequency and the angle of incidence, target preparation method and some others. Therefore, the PLA is a multi-parameters process.

It has been recently shown (Rzeszutek et al., 2008a,b) that pulsed laser ablation of CdTe target with low - power pulses of YAG:Nd laser can be an effective method for the deposition of high quality CdTe thin films. The advantages of using low-power pulses of YAG:Nd laser for the CdTe ablation are following. The YAG:Nd laser is as such a very stable and environmentally harmless laser that can be very easy handled. Because the thermal evaporation of CdTe results in nearly congruent vaporisation of Cd and Te (Ignatowicz S. & A. Koblendza 1990), it may be expected that the low - power pulsed laser ablation should be a very effective method of the deposition of CdTe thin films. However, the most important reason that ablation is performed in the low-power regime, realized by long pulse duration of 100 μs, is to minimize the splashing effect that is the effect of emitting of macroscopic particularities from the target (Cheung & Sankur, 1988). That degrades the quality of the thin films obtained by the laser ablation. Therefore, the type of the laser and its pulse duration time are dictated by practical reasons.

In this chapter we summarize our earlier experiments on the ablation of CdTe and add new results on the ablation of CdSe and ZnTe not as yet published. Like CdTe, these two latter compounds, are rather volatile materials, and the use of the low-power YAG:Nd laser ablation for their thin film preparation can be substantiated largely in the same way as it is done above for CdTe. Therefore, in the following we will present and discuss the results on the PLA of a group of the 2-6 compounds, which allows on some generalization of the conclusions.

However, our main goal is not the presentation of the physical properties of the 2-6 thin films obtained in the low-power regime of the YAG:Nd PLA. It is rather the ablation process itself and its dependence on the parameters of the process. Our main points of interest are: the dependence of the ablation process of the 2-6 compounds on the target preparation method and laser pulse energy and the effect of these factors on the velocity distribution of emitted particles.

The chapter's material is organized in the following way. In Sec. 2 the experimental procedures are described. Here a general experimental set-up for performing PLA is given together with the description of the Time-Of-Fly measurement method. Sec. 3 is devoted to the dependence of the pulsed laser ablation of the 2-6 compounds on target preparation method. In particular, the vapour stream intensity and the chemical composition and their mutual evolution with time are investigated with the help of a quadrupole mass spectrometer. These studies are performed for three kinds of targets: a target made of CdTe bulk crystal (BC target), a target made of CdTe fine powder pressed under the pressure of 700 atms (PP target), and a target made of loose (non-pressed) CdTe powder (N-PP target). Results obtained for PP targets made of CdSe and ZnTe are also presented. Sec. 4 deals with the velocity distribution of emitted particles. It starts with a theoretical background and continues with experimental velocity distribution of particles and comparison with the theory. The velocity distribution is determined by the time-of-fly (TOF) spectrometry performed by a quadrupole mass spectrometer. This section deals also with the angular distribution of particles. In Sec. 5 final conclusions are drawn.

2. Experimental procedures

2.1 Apparatus for pulsed laser ablation of semiconductor materials

The pulsed laser ablation of the 2-6 materials has been performed in an apparatus for pulsed laser deposition of semiconductor thin films described earlier (Oszwałdowski et al., 2003). A general scheme of the main part of the apparatus is shown in Fig. 1. Important elements of the apparatus are:

Laser. A typical neodymium doped yttrium–aluminum–garnet (YAG:Nd) laser is used. It has the following parameters: wavelength ,1.064 μm; maximum pulse energy, 0.5 J; instability of the pulse energy, 6%; pulse duration in the free generation mode, 100 μs; pulse duration in the *Q*-switched mode, 10 ns; repetition time, 10–50 Hz; beam divergence, 3 mrad; and beam diameter, 7 mm.

In the further described experiments the Nd:YAG laser operates at 25 Hz or 35 Hz pulse frequency. The pulse energy is changed from 0.13 J to 0.25 J; however most of the experiments are performed with the energy of 0.16 J. The laser spot on the target has the effective (roughly FWHM) diameter of 0.2 cm, thus the surface density of the energy is changed from 4 J/cm^2 to 8 J/cm^2, and the most frequently applied energy density is 5 J/cm^2. For the applied laser pulse duration of 100 μs, the pulse power is changed from $1.3 \cdot 10^3$ W to $2.5 \cdot 10^3$ W, and the most frequently applied power is $1.6 \cdot 10^3$ W. Therefore, the applied laser pulse powers fall into the low power regime (Cheung &.Sankur, 1988; Christley &Hubler, 1994). The low pulse power and the relatively large laser spot are chosen to diminish the splashing.

Optical path of laser beam. The laser radiation beam falls onto a focusing mirror having the focal length of 80 cm. This mirror is attached to a guide that enables to shift the mirror, and thereby its focal point in relation to the targets plane. As a rule, in order to decrease the radiation surface density power with the aim of avoiding splashing, the mirror focal plane is shifted from the target plane. The extent of the off focusing depends on the target material.

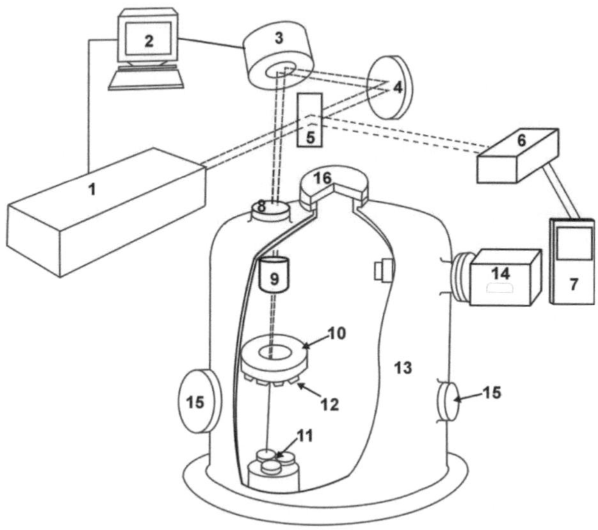

Fig. 1. Sketch of the apparatus for PLD of semiconductor thin films.
1. YAG:Nd laser, 2. Computer controlled system of laser beam monitoring, 3. Device switching laser beam between targets (optical deflector) , 4. Focusing mirror, 5. Quartz plate, 6. Photodiode system for measurement laser beam intensity, 7. Meter or oscilloscope, 8. Optical port of laser beam, 9. Heater of internal optical port, 10. Substrate holder and heater, 11. targets, 12. Substrates, 13. Vacuum chamber, 14. QMS at first port, 15. Peep holes, 16. Second QMS port.

The concave mirror reflects the beam onto a flat mirror of the optical deflector, which directs the beam through an opening in the substrate holder/heater onto a surface of one of the targets.
Other details of the apparatus construction less important for the present studies can be found in the source article (Oszwałdowski et al., 2003)
Quadrupol Mass Spectrometer (QMS) The apparatus is supplied with a quadrupole mass spectrometer (QMS, HALO 301, Hiden Analytical) equipped with a pulse ion counter. The action of the QMS is synchronized with the laser action by a specially designed electronic device. With this improvement, the vapour cloud ejected from the target by a laser pulse arrives at the spectrometer head in a proper time to be recorded and analysed on its chemical composition. Determination of chemical composition of the vapour stream and the velocity distribution of emitted particles are main functions of QMS in present investigations.

2.2 Time-Of-Fly experiments

An important part of the present investigations is performed with Time-Of-Fly (TOF) experiments. They are carried out with the use of the quadrupole mass spectrometer equipped with a pulse ion counter (PIC) as an ion detector. Here, the option of the measurements of the delay times between the electric pulse triggering the laser shot and the detection of the ionized particles by the ion detector is exploited for the determination of the particle delay time distribution (Rzeszutek et al., 2008b). From that, the velocity distribution of particles in the vapour stream is determined. The total particle delay time is composed of the following partial delay times: the laser pulse generation time, the particle emission time, the particle TOF between the target and the orifice of the ionizer, the particle arrival to PIC time, the PIC reaction time (given by the manufacturer to be between 30 ns and 50 ns). The time of the laser pulse generation is determined with the help of an electronic circuit equipped with a photodiode as a radiation detector. The time of the laser pulse generation is assumed to be the FWHM of the signal shown, which is about of 70 μs, and thus is close to the value of the 100 μm given by the QMS manufacturer.

The sketch of the configuration applied in the measurement of TOF is shown in Fig. 2. The TOFs are measured with the substrate heater, *H* removed from its position (10) in the vacuum chamber shown in Fig. 1. The sum of the remaining delay times is determined from the difference in the total delay times t_1 and t_2, measured at two different distances l_1 = 43 cm and l_2 = 24 cm between the target and the ionizer entry. For this purpose, the particle velocity $v = (l_1 - l_2)/(t_1 - t_2)$ was determined in the first step. Then, from the knowledge of v, t_1 and t_2 the sum of the remaining delay times is determined to be 0.12 ms. In the subsequent measurements, the TOFs were measured only for the distance l_1 and the TOF velocity was determined from the equation: $v = l_1 / (t_1 - 0.12)$, where t_1 is in milliseconds. The measured values of $(t_1 - 0.12)$ were in the range from 0.4 to 4 ms, whereas the range of measurable delay times was from 0.1 to 100 ms. Thus, the system was capable to measure the TOFs of all particles that appeared at the ionizer.

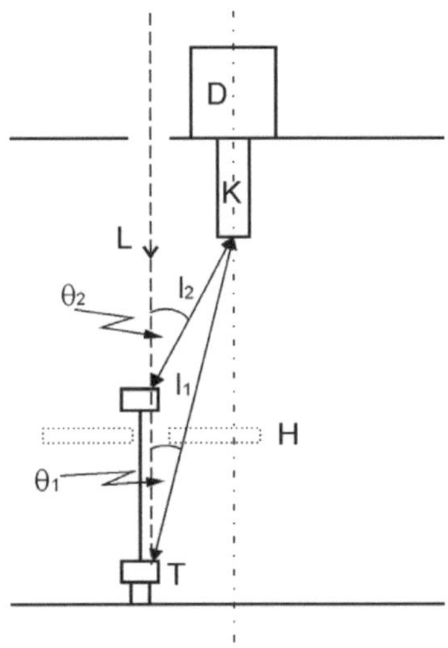

Fig. 2. Sketch of configuration applied for TOF and angular distribution measurements. (*D*) particle detector (PIC), (*K*) particle ionizer and quadrupole, (*T*) target, (*L*) laser beam, l_1 & l_2 distances between lower and upper position of target, respectively, (*H*) substrate heater, removed from position for TOF measurements.

The target holder is a rotating copper cup having the inner diameter ϕ = 2 cm. The angular velocity of the cup can be changed.

2.3 Target preparation method

In the present experiments three types of targets are used: a target made of powdered material poured directly into the holder cup (non-pressed powder target, N-PP target), a pellet made of a fine powder pressed at a pressure of 700 atm (pressed powder target, PP target), and a slice cut off from a CdTe bulk crystal (bulk crystal target, BC target). The diameter of the targets made of the powder is 2 cm and the diameter of the bulk CdTe crystal target is 1 cm. The ablation runs, lasting 9-14 minutes are performed at a constant laser power. The quadrupole system of the QMS head is directed roughly towards the target. The orifice of the head is lightly shifted parallel to the target surface in such a way that the line joining the orifice centre with the target centre makes an angle of 19° with the target normal. The distance between the orifice and the target surface is 43 cm. During the ablation process the pressure in the vacuum chamber is about 10^{-6} torr.

3. Pulsed laser ablation of 2-6 compounds: Dependence on target preparation method

3.1 Dependence of vapour stream intensity on pulse

The study of the dependence of the vaporisation intensity of CdTe, CdSe and ZnTe on the laser pulse energy is performed on the PP targets. The vapour stream intensity for each compound is deduced from two different and independent measurements. In the first measurement method total amount of the mass ablated by the action of 10000 laser pulses of a given energy is measured by weighing the pellet before and after the ablation and evaluating the difference. From these data the average mass ablated by a single pulse is determined. In the second measurement method, the total number of counts is registered, by the QMS, in the same ablation process for the isotope ^{110}Cd in the case of CdTe and CdSe and the isotope ^{66}Zn in the case of ZnTe. From that, the average number of counts for a single pulse is determined. Thus, in both measurements, a magnitude proportional to the vapour stream density is determined. The measurement results for the laser pulse energies ranging from 130 mJ to 250 mJ are shown in Fig. 3.

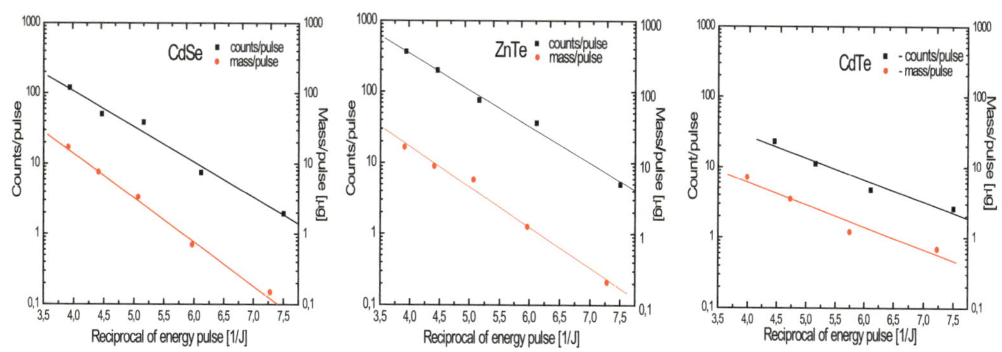

Fig. 3. Dependence of the vapour stream intensity from CdSe, ZnTe and CdTe PP targets on the inverse of the laser pulse energy ranging from 130 mJ to 250 mJ. The left scale shows the number of QMS counts per a single laser shot. The right scale shows the mass ablated by a single laser shot

It is seen in the figure that for both measurement methods, there is a good agreement as for the character of the dependence of the vaporisation intensity on the pulse energy. This dependence is linear in the scale logarithm of the vapour stream density versus the inverse of the pulse energy. In the applied range of the energy density, the mass evaporated by a single laser shot is between 0.6 μg and 8 μg for CdTe, and between 0.1 μg and 20 μg for CdSe and ZnTe. Thus the effectiveness of evaporation for CdSe and ZnTe is higher.
The presented studies of the dependence of the vapour stream density on pulse energy for the targets made of CdTe, CdSe and ZnTe are performed in the power density range (4-8)*10^4 W/cm^2 that is for the densities smaller than 10^6 W/cm^2, which is a region known as the low power density range (Cheung & Sankur, 1988). In this range the particle emission is expected to have the thermal character, in which the stream density S depends on the thermal energy kT, acquired from the pulse energy E according to the relationship:

$$S \propto \exp\left(-\frac{\Delta H}{kT}\right) \tag{1}$$

where ΔH is the heat of vaporisation. Fig. 3 shows that the results for the PP targets comply with Eq. (1) under the assumption that the kT is proportional to the pulse energy. However, it should be pointed out that in the case of materials having a high vapour pressure, the ablation with laser pulses in the low power density regime does not mean a low particle stream density (Kelly & Miotello, 1994).
Since CdTe, CdSe and ZnTe show in Fig. 3 a linear dependence of the stream density on the energy pulse reciprocal, it is possible to calculate the slopes of the curves. They should be roughly proportional to the heats of vaporisation (enthalpies of sublimation). The determined curve slopes for CdSe, ZnTe and CdTe respectively are: -1.33 μg J/pulse, -1.24 μg J/pulse and -0.78 μg J/pulse. The respective enthalpies of sublimation for CdSe, ZnTe and CdTe are: 1.7 10^6 J/kg (Bardi et al., 1988), 1.6 10^6 J/kg (Nasar & Shamsuddin, 1990) and 1.2 10^6 J/kg (Bardi et al., 1988). Comparing the absolute values of the curve slopes with the values of the enthalpies of sublimation, we find some correlation between them. Namely, they decrease in the same order and the values for CdSe and ZnTe are very close, whereas corresponding values for CdTe are distinctly smaller. The correspondence between the curve slopes and the sublimation enthalpies seems to further confirm the thermal nature of the ablation process.
Each change in the pulse energy has an effect on the surface appearance of the ablated area. A similar effect has the degree of the spatial overlapping of two consecutive laser shots on the target. The surface appearance of a PP target made of CdTe and ablated with laser shots having the energy of 160 mJ is shown in Fig. 4. The shown in the figure detail is a fragment of a 2 mm wide circular track carved by the laser beam on the target surface. The left-hand side of the figure marked a) shows an area ablated with 20000 laser shots, of which spots did not overlapped. After moving the laser spot along the target radius towards the target centre and reducing the angular speed of the target to a half of its initial value, the consecutive laser spots overlapped. The result of the ablation performed with overlapping spots is shown on the right-hand side of the figure, marked b). The left- and the right-hand side of the figure are separated by a narrow and smooth part of the target surface, marked c) that was not laser ablated. It is seen in the figure that the laser ablation results in formation of a surface structure consisting of granular forms. However, the topography of the part

ablated with the overlapping spots is richer and shows higher roughness. The ablation with overlapping spots increases local temperature of the ablated area formed by 2-3 consecutive shots. This effect, called further overheating, is equivalent to the increase in the energy of the laser pulse.

The effect of a genuine increase of the laser pulse energy is shown in Fig. 5.

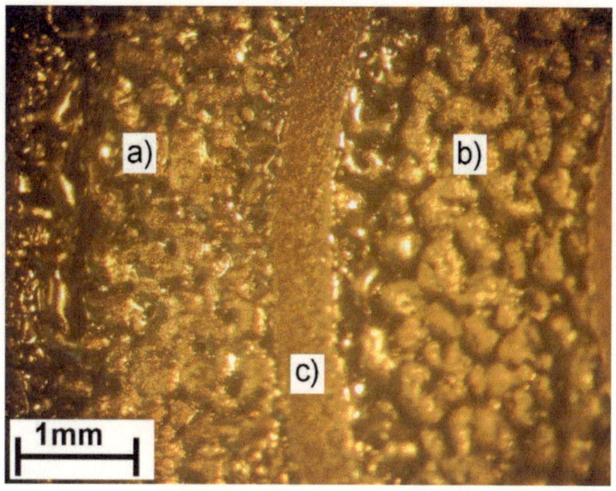

Fig. 4. Surface appearance of CdTe PP target ablated with 160 mJ laser pulses (optical microscope).

Part a) shows fragment of circular track ablated with 20000 non-overlapping laser shots. Part b) shows fragment of circular track ablated with 20000 overlapping laser shots. Both parts are separated by narrow circular strip (c) of the material that was not laser ablated.

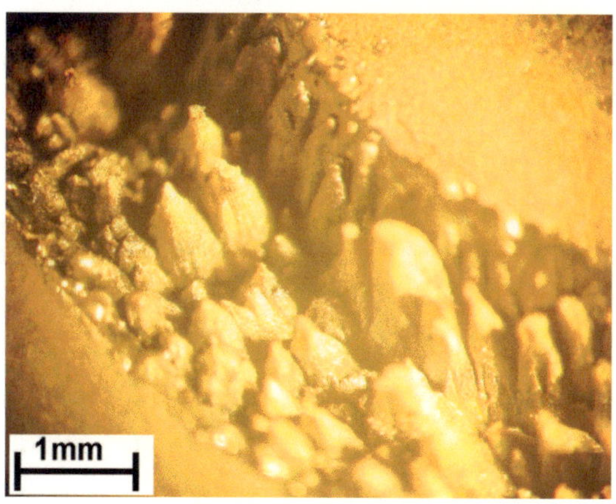

Fig. 5. Surface appearance of CdTe PP target ablated with 250 mJ laser pulses (optical microscope). Surface structure is obtained after 30000 laser shots.

The observed fragment of the circular track is a result of the ablation with non-overlapping shots, 30000 in number, with the pulse energy of 250 mJ. Comparing Fig. 4 and Fig. 5, it can be seen that the increase in the shot energy leads to a more developed surface structure showing considerably higher roughness. This is quite a general observation for all studied materials, as may be concluded from Fig. 6 that shows the results for CdSe and ZnTe PP targets.

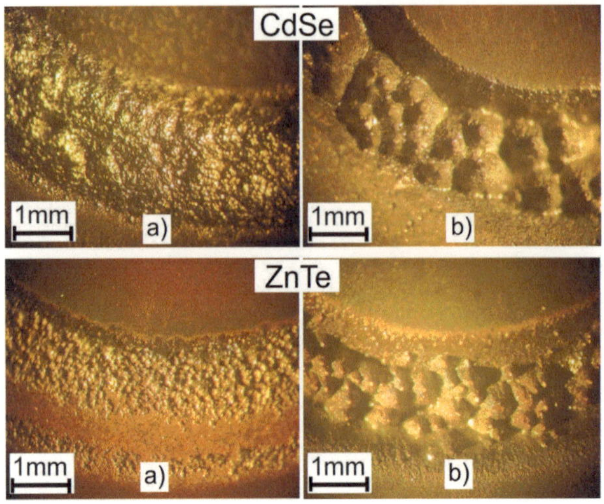

Fig. 6. Surface morphology of CdSe and ZnTe PP targets after ablation with laser pulse energy: 130 mJ, a) and 250 mJ, b).

It is clearly seen that the fragments of the targets ablated with the 130 mJ pulses is much smoother that the fragments ablated with the 250 mJ pulses. In Figs 4, 5, and 6 one can observe than at the higher pulse energy (250 mJ) the formation of characteristic conical forms occurs in the ablated material. This formation can be associated with the granular nature of the PP targets.

3.2 Vapour chemical composition and its time dependence

In order to perform the stream intensity measurements with QMS, it is necessary to choose a proper isotope of each element of the compounds. The number of isotopes of Cd, Zn, Se and Te respectively is: 8, 5, 6 and 7. For the monatomic species we have chosen the following isotopes: ^{110}Cd, ^{66}Zn, ^{78}Se, and ^{128}Te. For the diatomic species we have chosen: $^{256}Te_2$ resulting from the sum (pairing) of the monatomic species: ^{128}Te + ^{128}Te and ^{126}Te + ^{130}Te. For $^{156}Se_2$ we have chosen resulting from the sum of the monatomic species: ^{78}Se + ^{78}Se, ^{76}Se + ^{80}Se, ^{74}Se + ^{83}Se. This choice of the masses is an optimum from the point of view of the measurement convenience. With this choice, the QMS signals from all the chosen masses have comparable amplitudes. That enables their convenient observation on the screen in the same signal scale.

In the case of CdTe, all three forms of the target are investigated. Prior to the investigations of the vapour streams generated by the laser, we studied the vaporisation of CdTe powder by the normal thermal vaporisation from a heated quartz crucible. We were particularly interested in the ratio of the vapour streams of the monatomic and the diatomic forms, $J(Te)/J(Te_2)$. In the investigations we have found that at relatively slow thermal vaporisation of CdTe powder, the ratio of the QMS signals from the masses 128 and 256 is 0.25 and shows tendency to increase to about 0.5 at a fast vaporisation. Hence, taking into account the species abundances we obtained that purely thermal evaporation of CdTe gives at least a 20 % participation of monatomic Te in the stream.

The investigations of the chemical composition of the vapour stream generated by the laser pulses are performed both with overlapping and non-overlapping laser shots. The ablations are carried out with 160 mJ pulses and the frequency of 35 Hz. A typical ablation time is 9 minutes, and that corresponds to 20000 laser pulses. Results obtained for the BC target are shown in Fig. 7. Results obtained for non-overlapping spots are shown in the left-hand side panels, and those for overlapping laser shots are shown in the right-hand side panels. The panels a1 and a2 show the time dependence of the QMS signals from the ^{110}Cd and the ^{128}Te isotopes as well as for the $^{256}Te_2$ molecules.

It should be noted that the particle emission starts with some delay, which amounts to about one minute, counting from the beginning of the ablation. Such a delay for the start of the ablation was also observed in earlier works for various target materials. In the case of CdTe it was explained by the existence of an energy threshold for the ablation. The existence of the threshold was explained with a complicated mechanism, in which a two-phonon mechanism is employed at the initial stage to heat-up the material to the level sufficient for the generation of numerous structural defects and a decrease in the energy gap that enable much more effective single-phonon absorption in the final stage (Dubowski, 1991). At that moment the target material shows very high absorption for the laser radiation.

Comparing the dependences shown in the panels a1 and a2, it is seen that the stream intensity at the beginning of the ablation is higher for the ablation with overlapping spots, but it decreases considerably with time. No such a decrease is observed for the non-overlapping spots. The higher stream intensity observed for the ablation with the overlapping spots is clearly due to the higher local temperature at the spot area that results from a cumulative heat effect of the overlapping laser shots. This is an outcome of the poor heat conductivity of CdTe.

It may be seen in the panels a1 and a2 of Fig, 7 that the time dependence of the stream intensity for all masses has the same character, except for the first two minutes. This means that after the first two minutes the masses are emitted from the target congruently.

To make it more clear, we determine the time dependence of the relative signal intensities $S(^{110}Cd)/S(^{256}Te_2)$ and $S(^{110}Cd)/S(^{128}Te)$ from the data shown in the panels a1 and a2. The results are shown in the panels b and c respectively. It is seen that if the first two minutes are skipped, the signal ratios do not show any clearly marked dependence on time. That means that the total vapour stream has stoichiometric composition corresponding to the CdTe compound. It is also observed in Fig. 7 that the signal ratio $S(^{128}Te)/S(^{256}Te_2)$ increases with time during the first two minutes from the value of about 1.5 to the value of about 2.5, and then tends to saturate at this value. Taking into account the particle abundances (Rzeszutek et. al., 2008a), it results that the total particle stream ratio $J(Te)/J(Te_2)$ increases from 1.2 to 2.0. The latter value is by about an order of magnitude higher than the 0.25 obtained for the pure thermal evaporation. Therefore, in the case of the bulk crystal target and the ablation without the laser spot overlapping, the vapour stream contains twice more monatomic Te particles than diatomic Te_2 particles. This is quite different from the case of the same target, but ablated with the laser spot overlapping. As seen in the panel b2, in that case the signal ratio $S(^{128}Te)/S(^{256}Te_2) = 0.5$, and it is constant from the beginning of the ablation. The value 0.5 corresponds to the particle stream ratio $J(Te)/J(Te_2)$ amounting to 0.41 and, as mentioned earlier, this is also the value characteristic for a fast thermal evaporation at high temperatures.

The panels c1 and c2 of Fig. 7 show that the signal ratio $S(^{110}Cd)/S(^{128}Te)$ decreases during the first two minutes of the ablation, and then saturates. It is clear that in the case of the non-overlapping laser shots, this initial decrease is in fact due to a relatively smaller participation

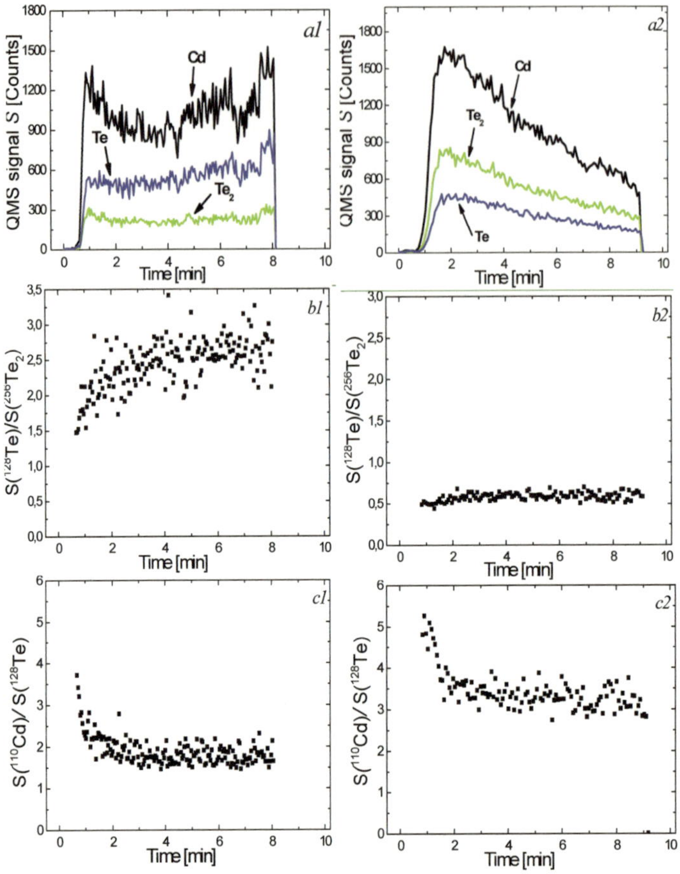

Fig. 7. Time dependence of QMS signals obtained during laser ablation of CdTe BC target. Left panels show dependence for non-overlapping laser shots, and right panels show dependence for overlapping laser shots. Panels a1 and a2 show time dependence of QMS signals from Cd, Te and Te_2. Symbols $S(^{128}Te)/S(^{256}Te_2)$ or $S(^{110}Cd)/S(^{128}Te)$ in remaining panels mean ratio of signals S from Te and Te_2 or from Cd and Te respectively. Ablation is performed with laser frequency of 35 Hz and pulse energy of 160 mJ.

of the monatomic Te species in comparison with the diatomic Te_2 species in the total tellurium stream (panel b1). The time dependence of the ratio $S(^{110}Cd)/S(^{256}Te_2)$ (not shown) does not reveal any such a decrease. On contrary, in the case of the overlapping laser shots, that initial decrease, as may be concluded from the panel b2, cannot be associated with any change in the participation of the monatomic Te species in the total tellurium stream. Thus, the initial decrease is associated with an initial excess of cadmium in the vapour stream.

Investigations of the ablation process of the PP targets are performed in the way analogous to those described for the bulk CdTe target. These investigations reveal that at the same ablation conditions the erosion of the target made of pressed power is considerably larger than that of the BC target. Since the initial smoothness of both types of the targets was similar, this may confirm that the much larger roughness of the pressed powder target is associated with its granular character. The investigation results for the CdTe PP targets are shown in Fig. 8.

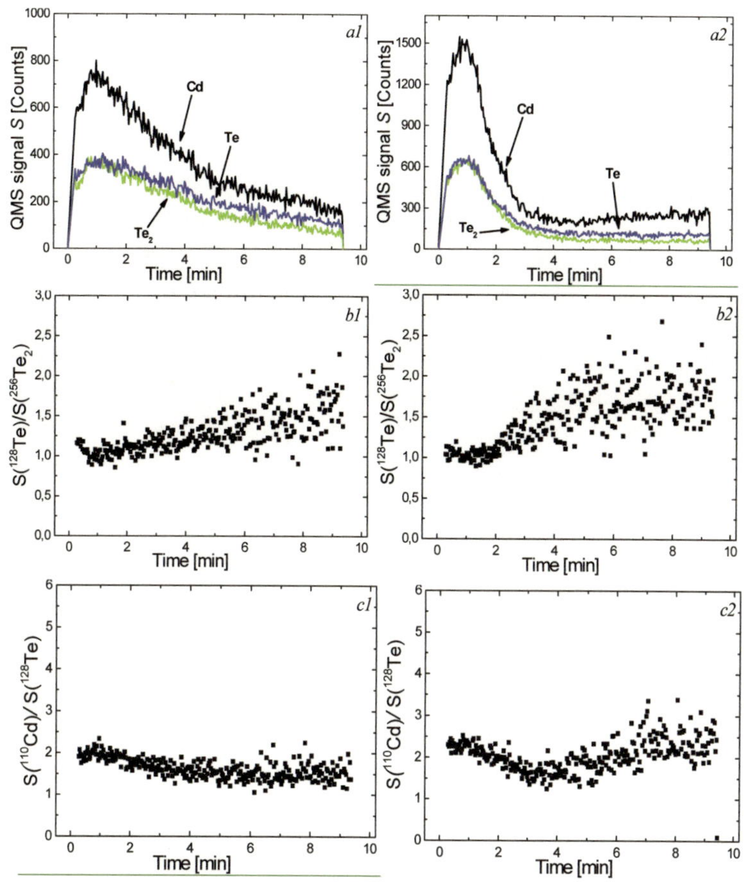

Fig. 8. Time dependence of QMS signal obtained during laser ablation of CdTe PP target. Left panels show dependence for non-overlapping laser spots, and right panels show dependence for overlapping laser spots. Panels a1 and a2 show time dependence of QMS signals from Cd, Te and Te_2.. Symbols $S(^{128}Te)/S(^{256}Te_2)$ or $S(^{110}Cd)/S(^{128}Te)$ in remaining panels, mean ratio of signals S from Te and Te_2 or from Cd and Te respectively. Ablation is performed with laser frequency of 35 Hz and pulse energy of 160 mJ.

It is seen in the panels a1 and a2 that in contrast to the BC target, in the present case the particle emission process commences immediately after the start of the laser action. This means that the powdered CdTe has sufficiently large number of structural defect to be strongly absorbent for the laser radiation. In the case of the non-overlapping laser spots, the magnitude of the QMS signal is slightly lower at the ablation beginning, as compared to the signal magnitude from the bulk crystal CdTe target, and further decreases with time. On the other hand, in the case of the overlapping laser spots, the magnitude of the QMS signal is comparable with that from the bulk crystal CdTe target at the ablation beginning, but its further decrease with time is stronger. Comparing the results for the overlapping and non-overlapping spots in the panels a1 and a2 of Fig.8, it is seen that the signal decrease is distinctly stronger in the case of the overlapping spots. In contrast to the ablation of the bulk crystal CdTe target, in the present case during the first 1-2 minutes of the intense particle emission, sporadic splashing is observed. Approximately after that time, a crust is formed on the ablated surface of the target. The crust is located in a groove formed by the laser action. At that time, the target surface becomes increasingly rough. The crust is expected to be formed in the process of melting and subsequent freezing of the powder.

Panels b1 and b2 in Fig. 8 show the ratio of the signals for the masses 128 and 256. During the first 1-2 minutes this ratio is roughly constant, and next it increases with the ablation time from the value of 1 to about 1.7. These values correspond to the vapour stream ratio J(Te)/J(Te_2) from 0.82 to 1.4. The signal ratio S(^{110}Cd)/S(^{128}Te) slightly decreases during the first three minutes of the ablation, and then slightly increases with time. This behaviour is more clearly seen in the case of the overlapping laser spots (panel c2). The increase with time and simultaneous increase in the S(^{128}Te)/S($^{256}Te_2$) ratio can be understood under the assumption that in addition to the direct laser ablation, there is also a contribution from purely thermal evaporation of the target material associated with its local overheating resulting from the low thermal conductivity of the pressed powder. In such a case, one could expect the thermal evaporation component would show an excess of cadmium. This assumption is supported by the fact that the increase in the ratio S(^{110}Cd)/S(^{128}Te) is more pronounced in the case of the overlapping laser shots, which cause a larger overheating.

The characteristic features of the ablation process of the N-PP target can be presented by a comparison of the experimental results obtained for the N-PP target with those obtained for the BC and the PP targets. In the case of the N-PP target, the signal magnitudes both for the ablation with and without overlapping of the laser shots are markedly higher than those for the PP target, and also higher than those for the BC target. During the ablation of an N-PP target splashing is observed and that is particularly intense during the first 1-2 minutes. Moreover, during the ablation, a glowing tail is formed and it follows the laser spot in its travel around the moving target. The glowing part of the target has to be the source of purely thermal evaporation of the target material. Like in the case of the BC and PP targets, the ablation with overlapping laser shots is more effective. In comparison with the PP target, the decrease with time of the stream intensity is markedly smaller, and resembles that occurring for the bulk crystal target, however, with the exception that the particle emission starts immediately after the laser action onset. The composition of the tellurium vapour stream is dominated by the thermal evaporation from the glowing spot (Rzeszutek et al., 2008a).

As in the case of the PP target, the laser ablation leads to the formation of a crust on the top of the ablated powder. The ablated surface roughness of the N-PP target is considerably higher than that of the PP target. Also the laser carved groove is considerably deeper.

The investigations of the chemical composition of the vapour stream for ZnTe and CdSe are performed on pressed powder targets. The ablations are carried out with 25 Hz pulses in time of 9 minutes that corresponds to about 10000 laser pulses. The pulse energy is 250 mJ for CdSe, and 220 mJ for ZnTe. Results obtained for ZnTe are shown in the left-hand side panels, and those for CdSe are shown in the right-hand side panels of Fig. 9.

Panel a1 shows the time dependence of the QMS signals from ^{66}Zn and ^{128}Te isotopes as well as from $^{256}Te_2$ molecules. On the other side, panel a2 shows the time dependence of the QMS signals from ^{110}Cd and ^{78}Se isotopes and from $^{156}Se_2$ molecules. It may be observed in both panels that at an initial stage of the ablation, the signal intensity increases. This may be associated with gradual heating up the target by the laser action. If this is the case, comparison of Figs. 7 and 9 leads to the conclusion that the heating up is much faster for CdTe than that for ZnTe. The reason for that is unknown.

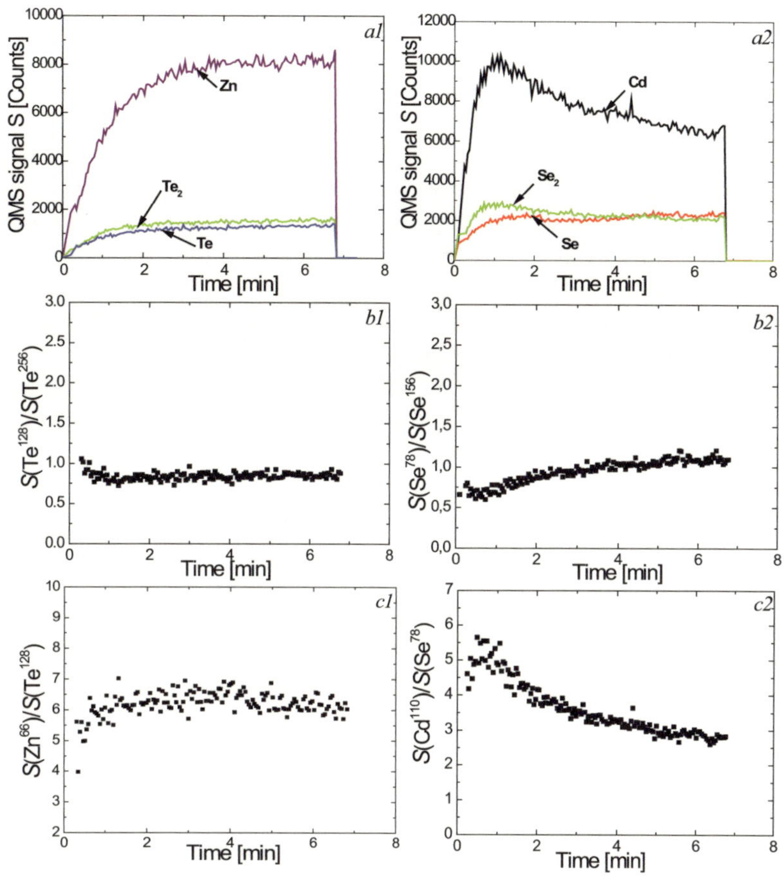

Fig. 9. Time dependence of QMS signal obtained during laser ablation of PP targets made of ZnTe, left panels and CdSe, right panels. Panel a1 shows time dependence of QMS signals from Cd, Te and Te_2, and panel a2 shows time dependence of QMS signals from Cd, Se and Se_2. In remaining panels are shown ratios of signals S from various particle streams. Ablations are performed with laser non-overlapping spots, frequency of 25 Hz and pulse energy of 220 mJ, for ZnTe, and 250 mJ, for CdTe.

Panels a1 and a2 show also that time dependences of the signals from all masses are more or less the same. A more precise investigation of this observation can be done by the determination of the ratios of signals from various masses. This is done in further panels. Panel b1 shows the time dependence of the ratio of the signals from the masses 128 and 256, and panel b2 shows the time dependence of the ratio of the signals from the masses 156 and 78. As it is seen, with the exception for the initial stage of the ablation, both ratios are time independent in the first approximation. It is interesting that the signal ratio $S(^{128}Te)/S(^{256}Te)$ is close to unity and weakly time dependent both for ZnTe and CdTe (Fig. 7). Thus the signal ratio is weakly dependent of the chemical composition.

Panel c1 shows the time dependence of the ratio of the signals from the masses 66 and 128. The ratio is time independent in the first approximation.

Panel c2 shows the time dependence of the ratio of the signals from the masses 110 and 78. This ratio decrease with time. That may be the result of the increase with time of the stream of the mass 78 observed in panel b2.

The observed dependence of the particle emission magnitude for CdTe, CdSe and ZnTe on the target properties can in a large part be explained with the difference in the target heat conductance. For example, the fact that the stream intensity emitted from the N-PP target is higher than the intensities found for BC and PP targets at the same ablation conditions can be explained by the poor heat conductance of the N-PP target that leads to a local temperature increase, and finally results in a more intensive particle emission.

It is observed that for all targets, the intensity of the stream of the emitted particles is initially higher for the overlapping spots, but the intensity strongly decreases with time, and after first 3-4 minutes the intensity decreases below the level observed for the non-overlapping spots. The higher initial vapour stream intensity in the case of the overlapping spots can also be explained in terms of the local temperature increase. However, the strong decrease in the vapour intensity with time needs additional explanation. First of all, it is observed that the decrease in the vapour intensity is strictly correlated with the increase in the surface roughness. The latter is expected to result in a decrease in the angular dependence of the emitted particles. It is known that the laser ablation of a flat and smooth target surface leads to a very directional particle emission. Preferred is the "forward emission", i.e. emission in the direction normal to the surface. In the case of a rough surface the local forward emission can be directed quite differently than the direction of the normal to the target surface as a whole. Thus, it is expected that emission from a rough target will be much less directional than that from a smooth target. In such a case the number of particles reaching the QMS detector, of which the angular position is close to the target normal, will strongly decrease with the increase in the surface roughness. Therefore, it is possible that the observed decrease in the measured by QMS stream intensity is associated with the structural changes (roughness) of the target surface resulting in changes in the stream angular dependence, rather than reflects any real changes in the total stream intensity.

Cheung in his PLA studies of CdTe (Cheung, 1987) found that the stream intensity ratio $J(Te)/J(Te_2)$ strongly increases with the laser pulse power. In some qualitative agreement with that, we find that in the case of the typical thermal evaporation, this ratio increases with temperature, from a value of about 0.2 to a value of about 0.4, and ratios $J(Te)/J(Te_2$ higher than 0.5 can be reached in the typical laser ablation only. Therefore, one can assume that the increase in the ratio $J(Te)/J(Te_2)$ is a measure of a departure from the purely thermal evaporation to the typical laser ablation. In the result, one can conclude that the most typical in character pulsed laser ablation has the ablation of the BC target with no spatial overlapping of the consecutive laser shots. The steady increase in the $J(Te)/J(Te_2)$ ratio with time to the final value of 2.0 suggests that in a later time the target temperature still increases, which is equivalent to an increase in the laser pulse power. The same thermal effects can be responsible for the increase with time of the $J(Te)/J(Te_2)$ ratios for the remaining targets. The very low $J(Te)/J(Te_2)$ ratio in the case of the bulk crystal target ablated with overlapping shots can be related to a considerably larger spot overlapping (larger than in the case of the powder targets) resulting from the smaller diameter of that target. Due to the local overheating of the target, the particle emission has the chemical composition that is closer to a pure thermal evaporation than to a typical laser ablation.

3.3 Dependence on target preparation method: Summary and conclusions

The main goal of this Section is to explain the effect of the target preparation method on the ablation process. The basic investigation method is the determination of the vapour stream intensity and chemical composition, as well as their real time evolution with the help of a quadrupole mass spectrometer synchronized with the laser action. The studies are performed for three kinds of targets made of CdTe, CdSe and ZnTe. Such investigations are carried out for the first time. They revealed that the ablation has a thermal character; i. e. the process is thermally activated through an energy barrier (Kelly & Miotello, 1994). In spite of the fact that he ablation process is performed in the low power regime, it is very effective because of the high volatility of the 2-6 compounds. The thermal ablation is concluded on the basis of Fig. 3, which shows that for all compounds the stream density is proportional to the pulse energy reciprocal. Hence, the experimental results can be described by the relationship of Eq. (1), if one assumed that the thermal energy kT is gained from the laser pulse energy.

It may be observed in Fig. 3 that for CdSe and ZnTe the evaporated masses per pulse are comparable, and for CdTe they are smaller. Nevertheless, for the pulse energy of 1/6 J (=166 mJ) the emitted mass is about 1 μg/pulse for all the compounds. This means that a target prepared from a homogeneous mixture of powders of these compounds should under the laser ablation emit the vapour of chemical composition corresponding to the target average composition. Thus homogeneous thin films of mixed crystals of the compounds could be obtained.

In the PLA experiments, the target erosion is large and is characterized by a considerable surface roughness, especially in the case of the powder targets. The surface roughness increases with the laser pulse power and frequency and also with the degree of spatial overlapping of subsequent laser pulse spots on the target surface. With the development of the roughness the ablation effectiveness decreases. It seems that the apparent decrease in the particle emission is connected with a change in the angular distribution of the emitted particles introduced by the roughness. The less forward directed particle emission from the target, confirmed further on by the studies of the stream angular distribution, could be important for the practical application of PLD as a thin film deposition method because the typical very directional particle evaporation is a drawback of PLD.

In the present experiments, large surface roughness is as a rule formed in the conditions of target overheating, which results in an enhanced target vaporisation. In this case the vapour stream consists of a component typical for the true laser vaporisation and a component typical for the thermal vaporisation associated with the target overheating. These two components can be distinguished in the investigations of the ratio of the number of the monatomic VI group particles (VI = Te or Se) to the number of the diatomic particles VI_2 (Te_2., Se_2). In the case of CdTe, the stream ratio $J(Te)/J(Te_2) < 0.5$ is typical for pure thermal vaporisation and the $J(Te)/J(Te_2) > 0.5$ is typical for true laser ablation. The highest value of the ratio is found during the ablation of the CdTe BC target without any spot overlapping. Only in this case the vapour stream has the stoichiometric composition (Cd/Te = 1) from the beginning of the ablation. In the other ablation processes, the vapour stream has some excess of Cd during the first minutes of the ablation. This period can be prolonged in the ablation of the powder targets with the spot overlapping. These effects are understood in terms of the differences in the thermal conductivity between the targets.

The obtained experimental results can be a basis of a comparison of the targets from the point of view of their practical value. The most useful chemical composition of the vapour stream in the low power regime can be obtained with the target made of a slice of a crystalline 2-6 compound ablated with well specially separated subsequent laser shots. The stream composition is in this case stoichiometric from the beginning and has a high particle ratio J(VI)/J(VI_2). The latter is advantageous for obtaining high quality epitaxial films of the 2-6 compounds (Cheung, 1987). A substantial drawback of that target, when ablated with the Nd:YAG laser, is the energetic threshold for the ablation and the (irregular) delay in the ablation process above the threshold.

On contrary, the main advantage of the powder targets over the BC target is the lack of any ablation threshold or delay. Also the costs of preparation of the powder targets are substantially lower. However, the smaller effective heat conductance of these targets poses the problem of the target overheating even in a more demanding form. To avoid the generation of the thermal component of the vapour stream, the subsequent laser shots should be well spatially separated. In addition to this, the laser frequency and/or improved heat sink may have to be applied. The target prepared from the compressed powder is a reasonable choice for the pulsed laser deposition of thin films of the 2-6 compounds, provided that the initial ablation, in which the more intense splashing and the non-stoichiometric composition of the vapour take place, is eliminated. This can be done by the use of a typical shutter that is a standard equipment of most vacuum evaporation plants.

The N-PP target is the simplest form of the target that can be prepared instantly. The main advantage of that target is that due to the very low thermal conductivity, it can be ablated effectively with very small laser powers. However, owing to the splashing, the large thermal component, and the non-stoichiometry of the vapour stream, its applicability to the pulsed laser deposition of the thin films of the II-VI compounds is rather limited to some preliminary deposition tests.

4. Pulsed laser ablation of 2-6 compounds: Velocity distribution of emitted particles

In this section we continue the investigation of the ablation process through the determination of the emitted particle velocity and angle distribution. The velocity distribution is determined by the time-of-fly spectrometry. It is known that the TOF spectrometry is a very useful characterisation method of the PLA process; however, the interpretation of results obtained needs a lot of care (Kelly & Miotello, 1994).

The information on the distribution of particle velocities during the laser ablation of the 2-6 compounds is rather limited. The PLA of CdTe was performed by Cheung (Cheung, 1987). He used 100 μm pulses of a Nd:YAG laser and found that the particle velocity distribution is the full Maxwell-Boltzmann. Namiki et al. (Namiki et al., 1986) also observed the full M-B distribution of particles for CdS. However, as shown further on, our velocity distributions obtained from the TOF studies are much narrower than the classic M-B distribution, which means that ablation takes place in the condition of the formation of the Knudsen layer (KL) followed by adiabatic expansion (KL-AE). These studies are carried out for all three types of targets and from them the velocities of the stream and the most probable velocities in the centre-of-mass system are determined.

The formation of KL-AE leads to a specific angular distribution of the emitted particles, which is strongly forward directed. The investigation of those distributions for the three types of the target is also the subject of our studies.

4.1 Theoretical velocity distribution

The laser ablation process, in particular PLA, of various solid state materials both organic and inorganic, liquids and suspensions is a point of interest for many researchers working in a wide variety of scientific domains. This diversity may be the reason of the existing confusion on the nomenclature, definitions and assumptions encountered in the literature devoted to the physical description of the gas (plasma) generated in vacuum by laser pulses. In this connection, it seems appropriate to introduce the definitions and formulae that are used in the presentation and interpretation of the experimental results presented in the further part of this contribution.

The velocity distribution of a perfect gas at temperature, T is described by the Maxwell-Boltzmann distribution function

$$f(v) \propto v^2 \exp\left(-\frac{v^2}{v_0^2}\right) \tag{2}$$

where v is the particle velocity and

$$v_o = \sqrt{2kT/m}\,, \tag{3}$$

is the most probable particle velocity in the gas. In Eq. (3), T is the gas temperature, and m is the particles mass. In the Cartesian coordinate system, in which the gas mass centre is resting, $v^2 = v_x^2 + v_y^2 + v_z^2$, and $-\infty < v_x, v_y, v_z < +\infty$. The same functional form has the velocity distribution of particles emitted during the thermal ablation of a target, provided that the particle concentration is sufficiently small for the emission process to be collisionless (Kools et al., 1992; Anisimov, 1968). If the target normal is directed along the z-axis, the movement along this direction is limited by the condition, $v_z > 0$, which means that the centre-of-mass of the particle stream drives away from the target surface. Therefore, the gas is described by the distribution defined by Eq. (2) with $v_z > 0$. This so-called "half-range" M-B distribution can be the basis of the interpretation of the TOF measurements, provided that the following measurement conditions are fulfilled (Kelly & Dreyfus, 1988):

a. Both the detector sensitive area and the emission area are small in comparison with the distance between them,
b. The detector is on-axis, i.e., the detector area and the emission area are parallel and have a common normal,
c. The particle emission time is much smaller than TOF.

The specific geometry of TOF measurements causes the TOF distribution is different from that given by Eq. (2) by a pre-exponential factor v^n. In the case of the TOF distribution we have

$$f_{\mathrm{TOF}}(v) \propto v^4 \exp\left(-\frac{v^2}{v_0^2}\right). \tag{4}$$

It should be mentioned that this form of the TOF distribution function with $n = 4$ applies to the particle detector sensitive to particle concentration. This is the case when a QMS is applied (Kelly & Dreyfus, 1988). The v in Eq. (4) is the velocity measured within a small solid angle around the target surface normal, i. e., $v \cong v_z$.

It is sometimes argued that the pre-exponential factor in Eq. (4) should be v^3 (Kools et al., 1992) instead of v^4. We are, however, more convinced by the arguments given by Kelly (Kelly, 1992) for the pre-exponential factor v^4, and will thus use it in the following. However, in view of the dominant role of the exponential term, the difference between those two pre-exponential factors does not have any meaningful effect on the general interpretation of the presented experimental results.

Particle emission with collisions is accompanied by the formation of the Knudsen layer at the target surface, where the collisions occur. In such a case, Eq. (2), with the condition, $v_z > 0$, applies at the closest vicinity of the target surface only. Moreover, owing to the collisions, the gas cloud evolves to the distribution called *shifted M-B* or *M-B on stream velocity*. In this distribution v_z in Eq. (2) has to be replaced by $(v_z - u)$, where u is the so-called *stream velocity* or *centre-of mass-velocity*. In the result, when the Knudsen layer is formed one obtains:

$$f(v) \propto v^2 \exp\left(-\frac{(v-u)^2}{v_0^2}\right). \tag{5}$$

It is important to note that, owing to collisions, for this distribution we have: $-\infty < v_x, v_y, v_z < +\infty$. For $u = 0$, we are in the COM system, and the distribution becomes the classic M-B distribution of Eq. (2). Now, v_0 is the most probable velocity among the velocities, v in the COM system, and Eq. (3) defines temperature in the COM system. In the laboratory system, to the distribution described by Eq. (5) belongs the following most probable velocity:

$$v_{0,r} = \frac{u}{2} + \sqrt{\left(\frac{u}{2}\right)^2 + v_0^2}\,. \tag{6}$$

It is observed that for $u >> v_0$, from Eq. (6) one obtains $v_{0,r} \cong u$.

With the formation of the Knudsen layer, Eq. (4) evolves to the form

$$f_{\mathrm{TOF}}(v) \propto v^4 \exp\left(-\frac{(v-u)^2}{v_0^2}\right). \tag{7}$$

Again, v is the velocity measured within a small solid angle around the target surface normal ($v \cong v_z$), and v_0 is given by Eq. (3). From the extremum condition for this distribution function, one obtains the following most probable velocity in the laboratory system:

$$v_{0,TOF} = \frac{u}{2} + \sqrt{\left(\frac{u}{2}\right)^2 + 2v_0^2} \tag{8}$$

Since for $u = 0$, Eq. (5) reduces to Eq. (4), where v_0 is the most probable velocity of the classic M-B distribution in the COM coordinate system.

For very high density of emitted particles, the number of collisions in the KL increases leading to the adiabatic expansion. In that case the velocity distribution is still described by Eq. (5) (Kools et al., 1992). However, it becomes more "narrow", i. e. its FWHM decreases.

4.2 Experimental velocity distribution and its comparison with theoretical predictions

The reliability of the obtained TOF experimental data depends on the fulfilment of the conditions (a)-(c) listed in Sec. 4.1. In an earlier paper (Rzeszutek et al., 2008b) it was shown that the described in Sec. 2.2 equipment for TOF measurements fulfils the conditions. Thus the equations of Sec. 4.1 are applicable to the interpretation of the performed TOF experiments.

The TOF measurements are performed for all atomic and diatomic species emitted from the investigated targets. In the case of CdTe the TOF measurements are carry out for all three sorts of target, and in the case of CdSe and ZnTe they are carry out on PP targets only.

The measurement results of the TOF velocity distribution for a CdTe BC target ablated with 220 mJ laser pulses are shown in Fig. 10.

The solid lines being theoretical fits to experimental points are obtained with the use of Eq. (7). The theoretical curve fits in Fig. 10, as well as those in Figs. 11–13 are performed with the computer application Origin 7.0 with the option of the least squares method. The values of the velocities u and v_0 obtained from those fitting are given in the inserts to the figures. In

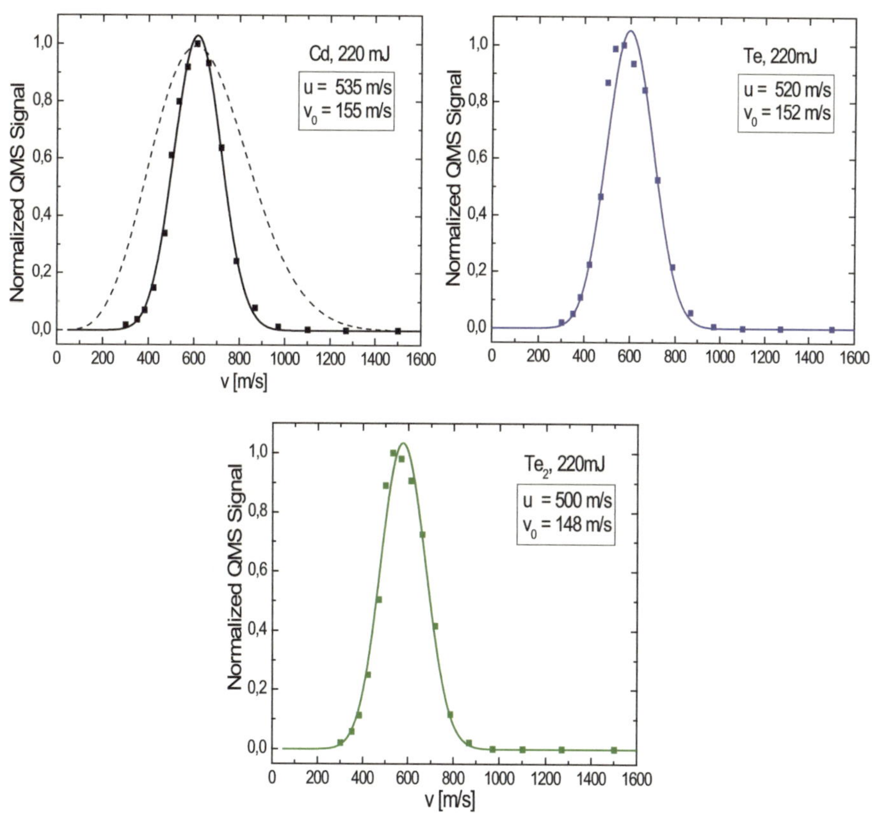

Fig. 10. TOF velocity distributions for atoms of Cd and Te and diatomic molecules, Te_2. Results are for CdTe BC target ablated with 220 mJ laser pulses with frequency of 25 Hz. Solid curves are obtained from theoretical fit to the experimental points with help of Eq. (7). Fitting parameters u and v_0 are given in inserts. For Cd fit with Eq. (4) is also made and shown as dashed curve for comparison.

the case of Cd, the fit with $u = 0$ is also given for comparison. This is in effect a fit to the TOF distribution of the classic M-B distribution (Eq. (4)). It is clear from the comparison that the experimental distributions are rather narrow, and thereby have to be described by the TOF shifted M-B distribution (Eq. (7)). It is seen in Fig. 10 that the fit with Eq. (7) is rather good, and therefore the values of the velocities u and v_0 are determined with a fairly good accuracy. The description of the experimental distributions with Eq. (7) means that the Knudsen layer is formed, hence the particles encounter collisions.

The velocity distribution for the CdTe PP target is determined for two laser pulse energies: 280 mJ and 160 mJ. The results are shown in Fig. 11.

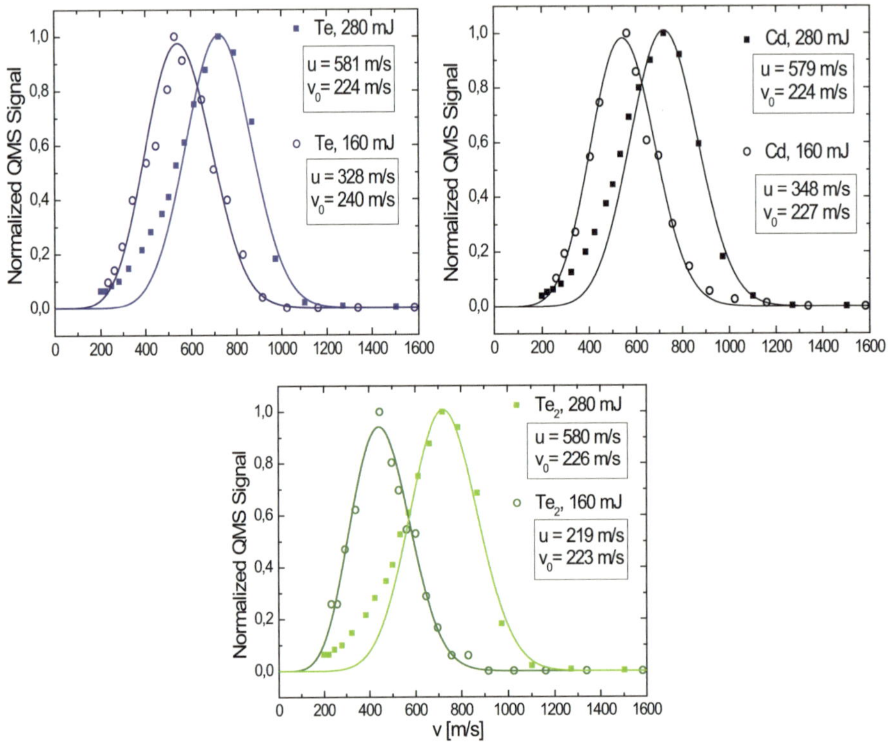

Fig. 11. TOF velocity distributions for atoms of Cd and Te and diatomic molecules Te_2. Results are for CdTe PP target ablated with 280 mJ and 160 mJ laser pulses of 25 Hz frequency. Solid curves are obtained from theoretical fit to experimental points with the help of Eq. (7). Fitting parameters u and v_0 are given in the inserts.

It is seen that the velocity distributions for the 280 mJ pulses are shifted to higher velocities, as may be expected. It is also observed that the theoretical fit to the experimental points is generally good, with the exception of the low velocity side for the 280 mJ pulses. The existence of the "low velocity tail" for the 280 mJ pulses can be ascribed to the target overheating, which results in the emission of the low energy particles of the heat origin, as described in Sec. 3. In the case of the 280 mJ pulses we observe, as in the BC target case, a close similarity between all particle distributions, leading to very close values of the most probable velocity. On contrary, in the case of 160 mJ pulses, the most probable velocity of the diatomic molecules is clearly smaller than those of the monatomic particles.

The results of the TOF velocity distribution measurements for the CdTe N-PP target ablated with 160 m J pulse energies are shown in Fig. 12.

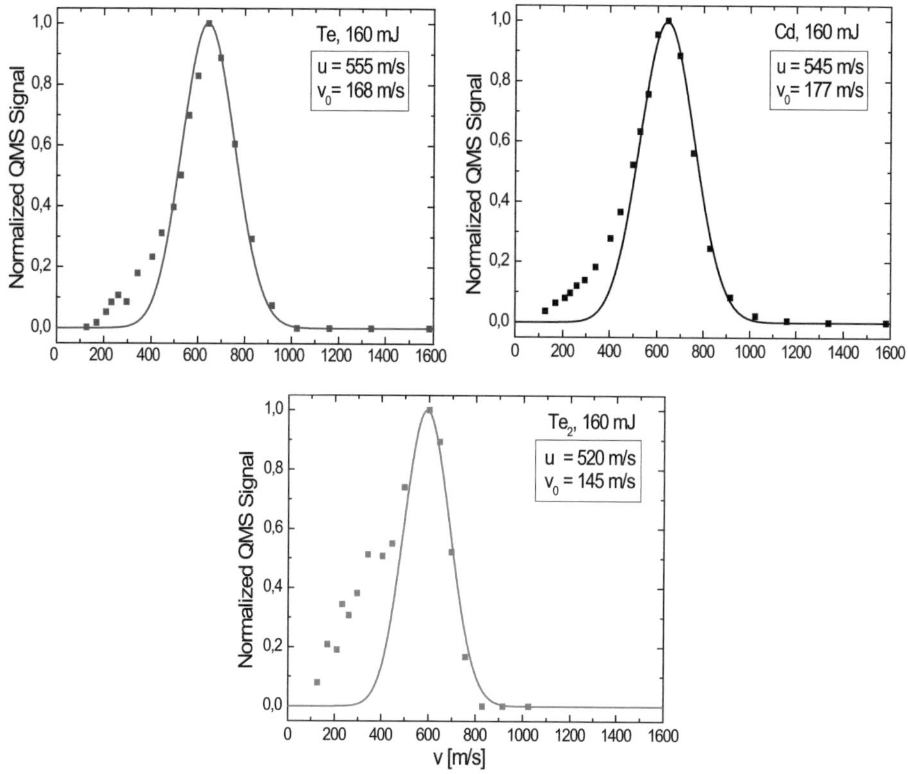

Fig. 12. TOF velocity distributions for atoms of Cd and Te and diatomic molecules Te_2. Results are for CdTe N-PP target ablated with 160 mJ laser pulses of 25 Hz frequency. Solid curves are obtained from theoretical fit to experimental points with the help of Eq. (7). Fitting parameters u and v_0 are given in the inserts.

It may be seen that even for this relatively small pulse energy, the velocity distributions are affected by the low temperature tail. This low temperature tail is particularly clearly observed in the case of Te_2 molecules. This may be understood, because the Te_2 diatomic molecules are larger in number than the monatomic Te in the particle stream emitted thermally from an overheated target. The distributions for the CdTe N-PP target are similar to those of CdTe PP target of the CdTe PP target ablated with 280 mJ pulses. This means that due to a smaller thermal conductivity of the CdTe N-PP target, the effective temperature of the target ablation is for that target higher. This also confirms the conclusion of Sec. 3 that at the same laser pulse energy the effectiveness of the laser ablation of a N-PP target is higher in comparison with a PP target.

The TOF velocity distributions for CdSe and ZnTe are measured for PP targets with 160 mJ laser pulse energies and the frequency of 25 Hz. They are shown in Fig. 13.

Like in the case of the CdTe PP target, the particle velocity distributions for CdSe and ZnTe are narrow and show "the low velocity tail", which is more clearly seen for ZnTe. Therefore, the general features of the distributions are dependent on the compound chemical composition.

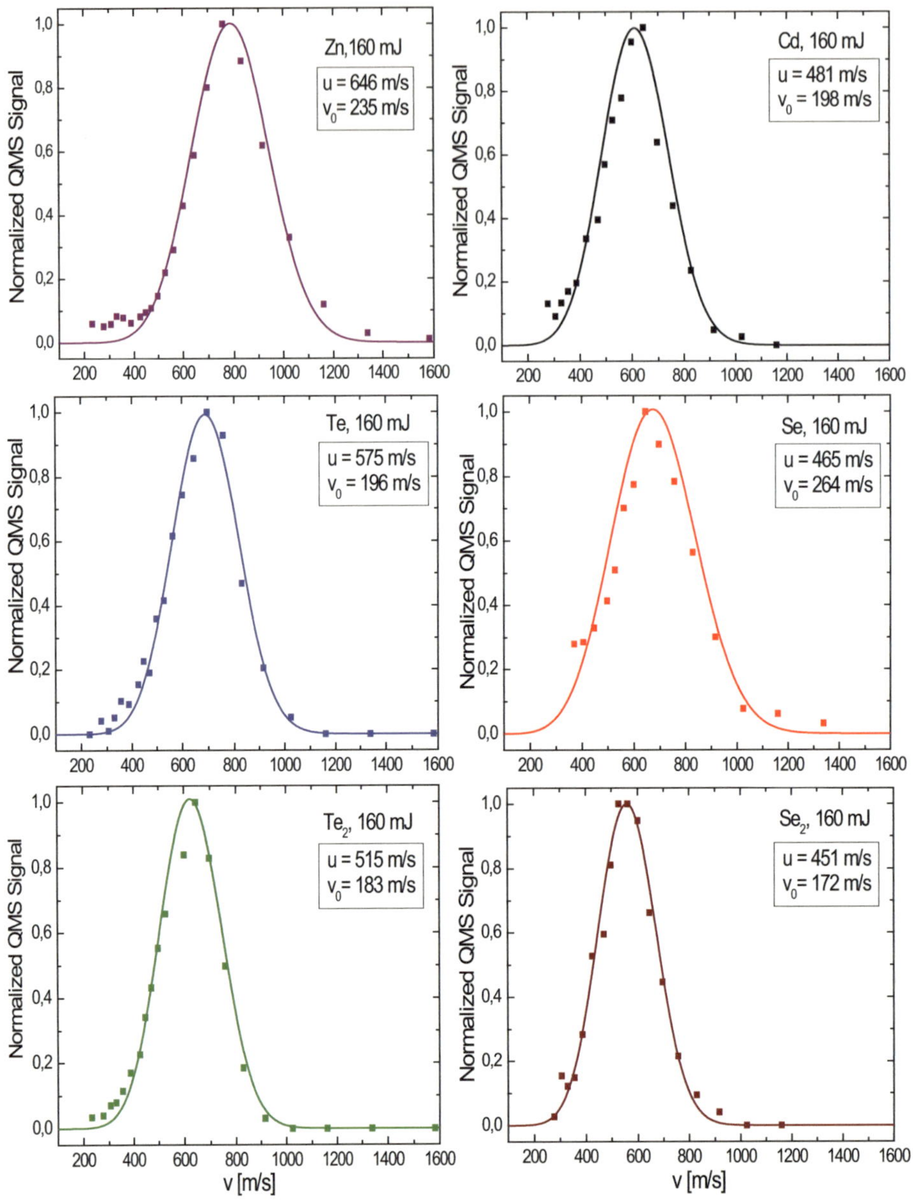

Fig. 13. TOF velocity distributions for CdSe PP and ZnTe PP targets ablated with 160 mJ laser pulses of frequency of 25 Hz Solid curves are obtained from theoretical fit to experimental points with the help of Eq. (7). Fitting parameters *u* and v_0 are given in the inserts.

4.3 Particle velocity distribution: Discussion and conclusions

In the present studies is observed substantial narrowing of the velocity distributions for all types of targets. The physical parameters of the gas phase, as determined form the best fit of the data in Figs. 10 – 13, are compiled in Table 1.

Target	energy pulse [mJ]	particle	v_0 [m/s]	u [m/s]	c [m/s]	T [K]	$v_{0,r}$ [m/s]	T_r [K]	M [-]
		Cd	155	535	141	159	577	2200	3.8
BC,	220	Te	152	520	139	178	561	2424	3.7
CdTe		Te_2	148	500	135	337	541	4499	3.7
		Cd	224	579	204	332	656	2843	2.8
PP,	280	Te	224	581	204	386	657	3326	2.8
CdTe		Te_2	226	580	206	786	658	6660	2.8
		Cd	227	348	207	341	460	1400	1.7
PP,	160	Te	240	328	215	443	455	1592	1.5
CdTe		Te_2	223	219	311	766	358	1973	0.7
		Cd	177	545	162	207	597	2361	3.4
N-PP	160	Te	168	555	153	217	602	2789	3.6
CdTe		Te_2	145	520	132	324	558	4789	3.9
		Zn	235	646	215	219	722	2072	3.0
PP,	160	Te	196	575	179	296	635	3109	3.2
ZnTe		Te_2	183	515	167	516	573	5062	3.1
		Cd	198	481	181	259	552	2016	2.7
PP,	160	Se	264	466	241	327	585	1606	1.9
CdSe		Se_2	172	451	157	278	509	2432	2.9

Table 1. Particle stream parameters emitted from various targets.

The first column in the table shows the target preparation method and material, the second column gives the energy of the laser shot at which the target was ablated, and the third column specifies the emitted particles. The fourth and the fifth columns present respectively the values of the most probable particle velocity v_0 in the frame of COM system and the COM velocity, u in the laboratory system. In the eighth column values of most probable velocity in the laboratory system $v_{0,r}$ are given. If $u > v_0$ as it is in the present case, the most probable velocities given by Eqs. (8) and (6) are rather close. In the present case, $v_{0,r}$ is smaller than $v_{0,TOF}$ by about 20%. This may be verified by comparing the values of $v_{0,r}$ in the table with the values of $v_{0,TOF}$ read out from the distribution maxima in Figs 10 – 13. We prefer to present in the table the values of $v_{0,r}$ because they are not directly associated with the measurement method (TOF).

To $v_{0,r}$ one can formally ascribe temperature T_r according to the equation analogous to the relationship described by Eq. (3):

$$kT_r = \frac{mv_{0,r}^2}{2}, \tag{9}$$

where T_r is a certain measure of the average kinetic energy, in the laboratory system, of particles subjected to the adiabatic expansion. Because it describes the gas phase after its

thermalization in the collisional process, T_r is smaller than the target surface temperature (Kools et al., 1992) The calculated values of T_r are shown in the ninth column. It may be seen that for each 2-6 compound, irrespective of the target preparation method, T_r increases with the particle mass, which means that the emitted particles form independent energetic systems.
The values of the most probable velocities v_0 in the COM system are considerably smaller than the values of $v_{0,r}$. The temperatures in COM system calculated from these values of v_0, obtained by using Eq. (3) are displayed in the seventh column of the table. It may be seen that they are generally low, and can be even lower than room temperature. To these temperatures can be ascribed sound velocity, c according to the classic rule applicable to perfect gas:

$$c = \sqrt{\gamma kT / m} = v_0\sqrt{\gamma / 2}\,, \tag{10}$$

where $\gamma = (j+5)/(j+3)$ is the specific heat ratio (adiabatic index) for perfect gas, and j is the number of internal degrees of freedom accessible at T. For the monatomic species, j = 0, and for the diatomic species maximum j = 3. The values of the sound velocities are given in the sixth column of the table. For the diatomic particles we present data for j = 0 as the number of occurring degrees of freedom in the present experimental conditions is not known. In the last column we present the expansion Mach number $M = u/c$, which is the measure of the velocity distribution narrowing.
Table 1 reveals the following interesting trends in the behaviour of the emitted particles:

1. The velocity $v_{0,r}$ depends on the pulse energy and the target material and preparation method. For the CdTe PP target, the velocities are higher for 280 mJ pulses than for 160 mJ pulses. On the other hand, for 160 mJ pulses, the values of $v_{0,r}$ are higher for the CdTe N-PP target than those for the CdTe PP target. This is obviously due to the smaller thermal conduction of the N-PP target resulting in higher evaporation temperature. In turn, the values of $v_{0,r}$ for the N-PP target are comparable with those of the BC target, in spite of the smaller pulse energy of the former. This is also due to the difference in the thermal conduction of the targets.
2. In general, for a given target material and preparation method the values of $v_{0,r}$ for monatomic and diatomic particles are comparable Nevertheless, some differences in the dependence on the mass can be noted. For example, for a smaller effective pulse energy (the case of CdTe BC and CdTe PP-160 mJ), the velocity $v_{0,r}$ decreases with the mass and for higher effective pulse energy (the case of CdTe PP-280 mJ), the velocity $v_{0,r}$ becomes roughly independent of the mass. The case of CdTe N-PP target does not follow that rule, which may be associated with a disturbance in the TOF spectrum by the thermal component of particles in the Te_2 stream. Among the PP targets ablated with 160 mJ laser pulses the highest values of $v_{0,r}$ shows the ZnTe target, and smaller values shows CdTe target.
3. One could expect that as v_0 is the most probable velocity in COM system, described by the classic M-B distribution, the most probable velocity, v_0 of various species will scale with the mass in the same way as it occurs in the perfect gas, that is as $m^{-0.5}$. This is obviously based on the assumption that temperature T is the same for all species. Actually, as may be seen in the table, the velocities for the same ablation process do, in general, decrease with the mass, but the decrease is considerably smaller than that described by the $m^{-0.5}$ dependence. This generally weaker mass dependence seems to

show that the multicomponent gas does not attain the thermal equilibrium in COM system. This presumably occurs because of an insufficient number of collisions. The lack of equilibrium reflects temperature T that increases with the increase in mass. That means that different species form different energetic subsystems.

4. The COM velocity u shows a weak dependence on the particle mass, with the exception of the CdTe PP target ablated with 0.16 J pulses. For the same ablation conditions, the value of u decreases with mass for less energetic particles, and becomes nearly independent of mass for more energetic particles. A clear example of this behaviour is the result obtained for the CdTe PP target ablated with 160 mJ and 280 mJ pulses. Therefore, it is observed that for a more energetic ablation, the levelling of the value of the COM velocity u takes place for all masses.

With the exception for the CdTe PP target ablated with 160 mJ pulses, for which M considerably decreases with mass, the Mach numbers weakly depend on the mass. Since M is a measure of the degree of the stream forward directionality and increases with the number of collisions, it is understood that the Mach numbers are higher in the case of the CdTe PP target ablated with 280 mJ pulses than in the case of the same target ablated with 160 mJ pulses. However, it is not clear why the Mach numbers of PP target ablated with 280 mJ pulses are smaller than those of CdTe BC and CdTe N-PP targets. The Mach numbers for the ZnTe PP target are high, and those for the CdSe are smaller.

The peculiarities of the velocity distributions observed in our experiments and their comparison with the data published for materials other than the 2-6 compounds (Rzeszutek et al., 2008b) show that the stream behaviour is governed by a number of ablation parameters. Unfortunately, our results cannot be compared with results obtained by other researchers for 2-6 compounds, because publications on such data practically do not exist. Nevertheless, it seems to be clear from the available data that the number of collisions prior to the gas expansion plays an important role, especially in the process of levelling the stream velocities for different masses. It should be pointed out that the velocity levelling has no satisfactory theoretical explanation yet. This is obviously due to the complexity of the processes occurring in the gas phase. Kelly and Dreyfus (Kelly & Dreyfus, 1988) suggest that in a two-component system the levelling may take place due to the formation of KL at the solid-gas boundary.

In view of the lack of any detailed interpretation of the velocity equalization, it may be worthy to consider a similar effect in the mesoscopic scale that takes place in the matrix-assisted laser ablation of molecular solids. In this technique, large molecules called the analytes are incorporated into a matrix of much smaller, but still large molecules that readily absorb the incident laser ablation.

Computer simulations performed with the molecular dynamics method show that the ablation of such molecular systems reproduces most of the characteristic features that are observed in the laser ablation of the typical solids (Zhigilei, et al., 1998; Zhigilei et al., 2003). This includes the fluence threshold for ablation, cluster formation, and the velocity distribution, in particular the forward peaking and the levelling. As for the latter, it is interpreted as a simple entrainment of the large molecules by the matrix molecules in the stream. Thus, it is the model of dragging suspended particles in flowing fluid. If this model were adopted to the microscopic scale of the small molecules and atomic species, most of the above discussed peculiarities of the velocity distribution could be explained. In particular, for sufficient fluence that results in a large number of collisions in the gas phase, the larger particles with larger masses have to be entrained and thereby they level their velocities to the common stream velocity.

4.4 Angular distribution of particles

Particle emission with velocity distribution described by Eq. (7) is expected to be strongly forward directed (Saenger, 1994). In this respect, the measurements of the angular distribution of the particles are auxiliary to TOF measurements.

For the determination of the angular distribution we use a typical method of measuring the profile of a deposited thin film. For this purpose, glass plates are glued with liquid indium to the surface of the circular substrate heater, shown schematically in Fig. 2 by the dotted lines. The substrate plates are aligned along several concentric circles that are centred vertically above the laser spot on the target. The film thickness is measured with an interference optical microscope. This measurement procedure is described in more detail in (Rzeszutek et al., 2008b).

Having determined the distribution of the film thickness along the radius, the dependence of the film thickness on the polar angle θ can be obtained. Such dependence for all kinds of the CdTe targets ablated with 160 mJ laser pulses with frequency of 35 Hz is shown in Fig. 14.

For the interpretation of the experimental results, it is assumed that the theoretical dependence of the film thickness d on θ is of the form (Saenger, 1994):

$$d = d_0 \cos^m \theta \tag{11}$$

where d_0 is the thickness in the circle centre, i. e. corresponding to $\theta = 0$. The dependence of the relative film thickness d/d_0 on θ, predicted by Eq. (11) for several values of m is shown in Fig. 14.

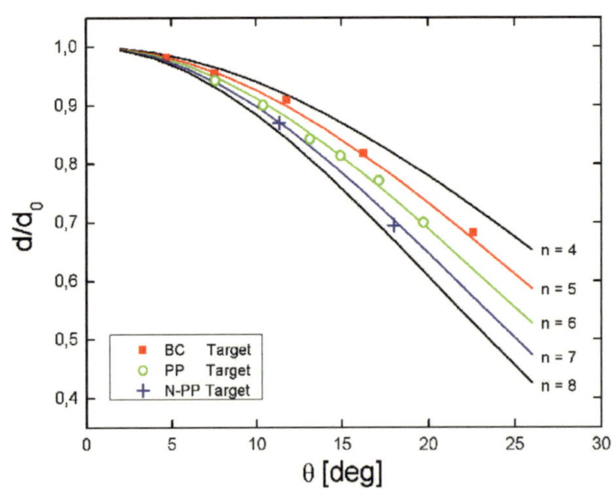

Fig. 14. Dependence of the relative film thickness d/d_0 on the polar angle θ for the three kinds of CdTe targets. Solid lines show the dependence predicted by Eq. (11) for various values of the exponent m. The values of d_0 for BC, PP and N-PP targets respectively are: 1.1 μm, 0.7 µm and 1.5 μm.

Comparing the experimental point with the theoretical curve we obtain the value of the exponent m. It is seen that the experimental points for CdTe BC target match with the curve obtained for $m = 5$, the experimental points for CdTe PP target are in agreement with the curve obtained for $m = 6$, and the experimental points for CdTe N-PP target are well described by the curve obtained for $m = 7$. It appears that the increase in the exponent m is correlated with the stream intensity, which is the less intense for BC target, and is the most intense for N-PP target.
Similar angular distribution of particles is determined for CdSe PP and ZnTe PP targets. In these experiments the energy of laser pulses is 160 mJ and the pulse frequency is 25 Hz. The result is that the exponent in Eq. (11) is m = 6. Thus, independent on material, the ablation of a PP target leads to the value of m = 6.
The investigations of the stream angular dependence show that the results are well described by Eq. (11) with $m > 4$. The value of 4 corresponds to the evaporation from a Knudsen cell and also from a flat target with no collisions in the gas phase. The obtained values $m = 5 - 7$ are considerably smaller than those expected for very forward-directed laser ablation with number of collisions in the gas phase (Saenger, 1994). A possible explanation of the small values of the exponent is the roughness of the target arising in the ablation process. High values of the exponent corresponding to very forward-directed evaporation are possible only for flat targets.

5. Final conclusions

Results of the investigations of the laser ablation of CdTe, CdSe and ZnTe with pulsed Nd:YAG laser are presented. They are performed with 100 μs pulses having energy from 130 mJ to 280 mJ. The laser pulses fall into the low power regime that allows diminishing effectively the splashing effect. The main goal of the studies is to elucidate the ablation process dependence on the target preparation method and the pulse energy.
The main investigation method is the determination of the vapour stream intensity and chemical composition, as well as their real time evolution with the help of a quadrupole mass spectrometer synchronized with the laser action. The studies are performed for three kinds of targets: target cut from crystalline CdTe, target made of powdered material formed into a pellet under high pressure and target made of loose powder.
The investigation reveals that the ablation has a thermal character and is very effective in this low power regime, due to the high volatility of the 2-6 compounds. The target erosion is large and is characterized by a considerable surface roughness, especially in the case of the powder targets. The surface roughness increases with the laser pulse power and with the degree of spatial overlapping of subsequent laser pulse spots on the target surface. With the development of the roughness, the ablation effectiveness, as measured by the quadrupole mass spectrometer strongly decreases. The decrease in the particle emission is apparently connected with a change in the angular distribution of the emitted particles introduced by the roughness. The less forward directed particle emission from the target is confirmed in the further investigations.
A large surface roughness is formed in the conditions of target overheating that results in intensive target vaporisation. In such a case the vapour stream consists of a component typical for the true laser vaporisation and a component typical for the thermal vaporisation associated with the target overheating.

The obtained experimental results can be a basis of a comparison of the targets from the point of view of their practical value. The best chemical composition of the vapour stream for obtaining high quality thin films can be obtained for a target made of a bulk crystal (BC) ablated with non-overlapping laser spots. A substantial drawback of that target, when ablated with the Nd:YAG laser, is an energetic threshold for the ablation and a delay in the ablation process

The main advantage of a pressed powder (PP) target over a BC target is the lack of any ablation threshold or delay. Also the costs of preparation of a PP target is substantially lower. However, the smaller effective heat conductance of the PP targets poses a problem of the target overheating. To avoid the generation of the thermal component of the vapour stream, the subsequent laser shots should be well spatially separated.

Mass spectrometer TOF measurements of the particle velocity distributions revealed their substantial narrowing, i. e. decrease in the FWHM of the distribution curves. The velocity distributions are well described by the shifted M-B distribution, of which parameters are: the stream velocity u in the laboratory system and the most probable velocity v_0 in the centre-of-mass (stream) system. According to the existing theory, with increase in the number of collisions in the gas phase, the half-range M-B distribution describing the collisionless gas phase evolves into the shifted M-B distribution. The available experimental data computer simulations, suggest that this should also be true for multicomponent gases consisting of several sorts of particles having different masses. This means that on reaching thermodynamic equilibrium, all the particles should form a common system having the same u, and v_0 should scale with particle mass as $m^{-0.5}$. However, our measurements show that at lower pulse energies the values of u decrease with particle mass, but with energy increase the values of u level. For the explanation of the velocity levelling, we adopt a model emerging from the matrix-assisted laser ablation of molecular solids computer simulation (Zhigilei et al.,1998; Zhigilei et al., 2003). In this model, the larger masses having smaller concentration are entrained into the stream of the smaller masses and hence level their velocities to a common stream velocity.

In our experiments the stream velocities are in the range 220÷650 m/s and the most probable in the centre-of-mass velocities are in the range 140÷240 m/s. The velocity in the laboratory system, which is 450÷700 m/s depends both on the laser pulse energy and the target preparation method. The dependence on the target preparation clearly results from the difference in their heat conductivities. Higher heat conductivity leads to higher effective temperature of evaporation and thereby to higher particle velocities. The highest heat conductivity has the bulk crystal target.

The obtained from the measurements angular dependence of the stream is smaller than expected. It seems to be associated with the roughness of the target arising in the ablation process. The confirmation of the relatively weak angular dependence is important for practical reasons, because it is the strong angular dependence that is characteristic for the laser ablation that causes difficulties in obtaining thin films with uniform thickness on large substrate surfaces.

Therefore, the present studies show that the use of the YAG:Nd laser in the low power regime should be an effective preparation method of stoichiometric 2-6 compound thin films.

6. References

Anisimov, S. I. (1968). Vaporization of Metal Absorbing Laser Radiation. *Soviet Phys. JETP* 27, 182–183

Bardi, G.; Jeronimakis, K.; Trionfetti, G. (1988). Vaporization enthalpy of cadmium selenide and telluride obtained by vapor pressure measurements. *Termochim. Acta* 129, 341 - 343

Cheung J. T. (1987). Role of atomic tellurium in the growth kinetics of CdTe (111) homoepitaxy. *Appl. Phys. Lett*, 51, 1940–1942

Cheung J. T. &.Sankur, H. (1988). Growth of thin film by laser-induced evaporation. *Critical Rev. Solid St. Mat. Sci.* 15,1, 63–109

Chrisey, D. B. & Hubler, G. K. (1994). *Pulsed Laser Deposition of Thin Films*, John Wiley & Sons, New York

Dubowski, J. J. (1991). Pulsed laser evaporation and epitaxy. *Acta Phys. Polon.*, A80, 221-244

Ignatowicz, S. &. Koblendza, A. (1990). *Semiconducting Thin Film of $A^{II}B^{VI}$ Compounds*, pp. 30-37, 73, 74, PWN, Warszawa

Kelly, R. & Dreyfus, R. W. (1988). On the effect of Knudsen-layer formation on studies of vaporization, puttering and desorption. *Surf. Sci.*. 198, 263–276

Kelly, R. (1992). Gas dynamics of the pulsed emission of a perfect gas with applications to laser sputtering and to nozzle expansion. *Phys. Rev.* A46, 860–874

Kelly, R. & Miotello, A. (1994).Mechanisms of Pulsed Laser Sputtering, *In: Pulsed Laser Deposition of Thin Films* , Chrisey, D. B. & Hubler, G. K., 55-85, John Wiley & Sons, New York

Kools, J. C. S.; Baller, T. S.; De Zwart, S. T .& Dieleman, J. (1992). Gas flow dynamics in laser ablation deposition. *J.Appl. Phys.*, 71, 4547–4556

Namiki, A.; Kawi, T.& Chige, I. (1986). Angle-resolved time-of-flight spectra of neutral particles desorbed from laser irradiated CdS. *Surf. Sci.*, 166, 129–140

Nasar, A. & Shamsuddin, M. (1990). Thermodynamic properties of ZnTe. *J. Less Common Met.*, 161, 93–99

Oszwałdowski, M.; Berus, T.; Sydorczuk, P. & Rzeszutek, J. (2003). Apparatus for pulsed laser deposition of semiconductor thin films. *Review of Scientific Instruments*, 74, 3160–3163

Rzeszutek, J.; Oszwaldowski, M. &. Savchuk, V. (2008). Ablation of CdTe with 100μs Nd:YAG laser pulses: dependence on target preparation method. *Cryst. Res. Technol.* 43, 1, 32-43

Rzeszutek, J.; Oszwaldowski, M. & Savchuk, V. (2008b). Ablation of CdTe with 100μs pulses from Nd: YAG laser: Velocity distribution of emitted particles. *Nucl. Instr. And Meth.* 266, 4766–4774

Saenger, K. L. (1994).Angular Distribution of Ablated Material, *In: Pulsed Laser Deposition of Thin Films* , Chrisey, D. B. & Hubler, G. K., 199–225, John Wiley & Sons, New York

Zhigilei, L. V.; Kodali, P. B. S. & Garrison, B. J. (1998). A microscopic view of laser ablation. *J. Phys. Chem.* B, 102, 2845–2845

Zhigilei, L. V.; Leveugle, E.; Garrison, B. J.; Yingling, Y. G. & Zeifman, M. I. (2003). Computer simulations of laser ablation of molecular substrates. *Chem. Rev.* 103, 321-348

17

Photoinduced Structural Changes of Doped SiO_2 Glasses using Ultraviolet Laser Pulses

Hiroaki Nishiyama and Junji Nishii
Hokkaido University
Japan

1. Introduction

There is a growing need for photonic band gap (PBG) devices in various applications including high-Q resonators, energy-efficient solar cells, microfluidic devices, microelectromechanical systems and so on. The PBG devices, which have artificial structures whose refractive index is periodically modulated, can reflect or diffract propagating light at a wavelength comparable to lattice spacing. The optical properties of the PBG devices can be controlled by changing lattice patterns, refractive index of lattices, duty ratio, interaction length and so on. Because of high potential for applications, the PBG effects have been intensively studied in the last decade. However, the fabrication of their fine structures, which can work at optical communication wavelengths and visible wavelengths, remains a challenging task.

Most PBG devices are manufactured by the semiconductor fabrication technology including lithography and various etching techniques. Although this technology enables us to create precise structures with high spatial resolution, we must use complicated experimental equipments such as vacuum systems and clean-room environments. In particular, highly expensive electron beam lithography systems are necessary to obtain sub-wavelength photonic structures, which are important for practical applications.

Direct laser writing technique of photonic structures is an attractive alternative. When optical materials are exposed to laser light below ablation threshold intensity, photochemical reactions such as chemical bond breakage and creation or annihilation of lattice defects occur in the laser irradiated regions, , thereby changing the refractive index and density. In 1978, Hill *et al.* firstly reported the direct laser writing of Bragg mirrors (one dimensional PBG devices) inside a Ge-doped SiO_2 core optical fiber using Ar laser (Hill *et al.*, 1978). This writing technique enables creation of photonic band gap devices with a simple setup on or in various glasses with sub-wavelength resolution (Svalgraad et al., 1994). Moreover, no debris or cracks are generated, although they often occur with laser ablation or mechanical machining. Especially, pulse laser, which has high peak intensity, is effective for such structural changes. From the point of view of practical applications, the structural changes of SiO_2-based glass materials, which are used as optical fiber or waveguide cores, are more important. Therefore, so far, many researchers have reported the structural changes of glass materials using ultraviolet nanosecond lasers, near-infrared femtosecond laser and so on (Åslund, et al., 2000, Hirao, *et al.*, 1997).

Recently, we have reported that crystalline Ge nanoparticles can be patterned in GeO_2-B_2O_3-SiO_2 glass, which is often used as optical fiber or waveguide cores, by a combined process of ultraviolet nanosecond pulse laser irradiation and thermal annealing (Nishiyama, *et al.*, 2004). In general, the laser-induced changes of refractive index are reported to be small and thermally less stable. For example, the refractive index changes induced by ultraviolet laser exposure in optical fiber cores are as small as 10^{-4}, even though photosensitizing processes are used. Furthermore, the index changes start to decay by annealing at 150°C (Baker, *et al.*, 1997). Our combined process is effective for creation of much larger refractive index change with high thermal stability.
In this chapter, we present the fundamentals of pulse laser-induced structural changes of SiO_2-based glasses and the PBG device fabrication using the ultraviolet laser-induced precipitation of Ge nanoparticles in GeO_2-B_2O_3-SiO_2 glass.

2. Fundamentals of laser-induced structural changes

Many reports have described the mechanism of the laser-induced refractive index changes since the first report of Hill's group. However, the details of photochemical reactions related to point defects remain unclear. In this section, we present popular three mechanisms on the refractive index changes. These mechanisms are generally accepted in the field of the direct laser writing of glasses. However, discussion continues about the contribution ratio of each mechanism to the refractive index changes.

2.1 Color center model

GeO_2-SiO_2 glasses often have intense absorption bands around 240 nm in wavelength, which originated from Ge related oxygen deficient defects with relative high stability at high temperature of glass preparation process. When the glass is exposed to laser pulse operating ultraviolet region such as KrF excimer laser of 248 nm wavelength, the intensity of the absorption band bleaches and some other absorption bands appear. This spectral change means that point defects are induced and annihilated in the glass via electronic transitions by the laser irradiation. Hand *et al.* and Dong *et al.* proposed these absorption spectral changes are responsible for the refractive index changes thorough Kramers-Kronig relation (Hand, 1990, Dong, 1995). In fact, highly Ge-doped SiO_2 glasses exhibit larger index changes. Figure 1 shows schematic structures of some defects related to laser-induced refractive index changes.

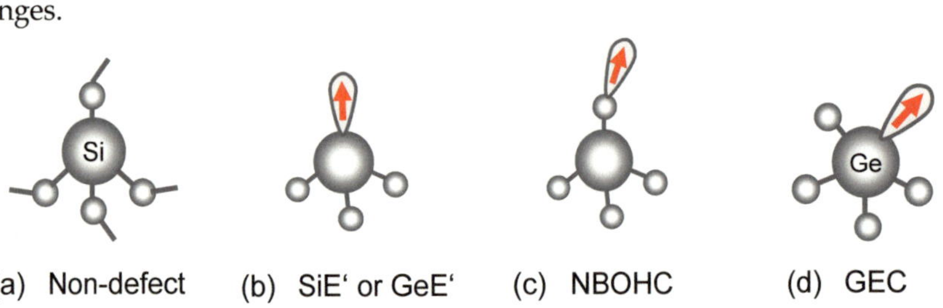

Fig. 1. Color centers in glasses.

2.2 Stress relaxation model

The stress relaxation model is based on the hypothesis that the refractive index changes originate from residual stress in optical fibers through the stress-optic effect (Dong, *et al.*,

2003). Such residual stress are generated mainly from a superposition of thermal stress caused by a difference in thermal expansion coefficients between core and cladding and mechanical stress induced by fiber drawing processes. The ultraviolet laser irradiation can promote the structural relaxation of point defects by ion rearrangements and thermal effect, resulting in refractive index changes of the glasses. In fact, there are several reports on formation of periodic structures using CO_2 laser irradiation and arc discharge as a heat source (Kim, *et al.*, 2001).

2.3 Densification model

Refractive index of glasses is affected by densification changes through Lorentz-Lorenz relation. One of the origins of the density changes is thermal relaxation by laser irradiation. Several groups reported the correlation between the refractive index changes and the densifications of fiber preforms and fibers by the irradiation (Poumellec, et al., 1996, Douay, et al., 1997, Kherbouche, et al., and Fonjallaz et al., 1995). They concluded that the densification mainly contributes to the refractive index changes by theoretical calculation and experimental results. Cordier *et al.* also showed direct evidence of the glass densifications from transmission electron microscope observations of laser irradiated glasses (Cordier *et al.*, 1994). On the other hand, other group reported results that the densification contribution to the refractive index changes is negligible by comparison of measurements and finite-element model calculation (Borrelli, *et al.*, 1999).

3. Direct laser writing methods

To obtain high Q resonance in PBG devices, precise microfabrication of the periodic structures is important. Therefore, to date, several interference techniques have been proposed. In this section, we describe two popular fabrication techniques of PBG devices with sub-wavelength periods using the direct laser writing.

3.1 Two-beam Interference technique

Meltz *et al.* firstly demonstrated the external laser writing of the periodic refractive index modulation on/in optical fiber cores, which can work as a PBG device, using interference techniques (Meltz *et al.*, 1989). The ultraviolet laser light is split to two beams and later recombined after propagation through different optical paths. This setup enables the formation of sub-wavelength fringe patterns without stitch errors. The grating period Λ can be determined by

$$\Lambda = \lambda/2sin\theta \tag{1}$$

where λ and θ are laser wavelength for direct laser writing and half-angle between the intersection beams (see Figure 2(a)). And PBG wavelength (Bragg wavelength) λ_B is given by $\lambda_B = 2N\Lambda$, where N is effective refractive index of fiber core. The advantage of this technique is that it is possible to various Bragg wavelengths by adjusting the interference angles and to obtain large periodic patterns without laser scanning. The long interaction region offers the fabrication of narrow band-pass filters. Furthermore, chirped gratings and curved patterns are also available by changing exposure geometry.

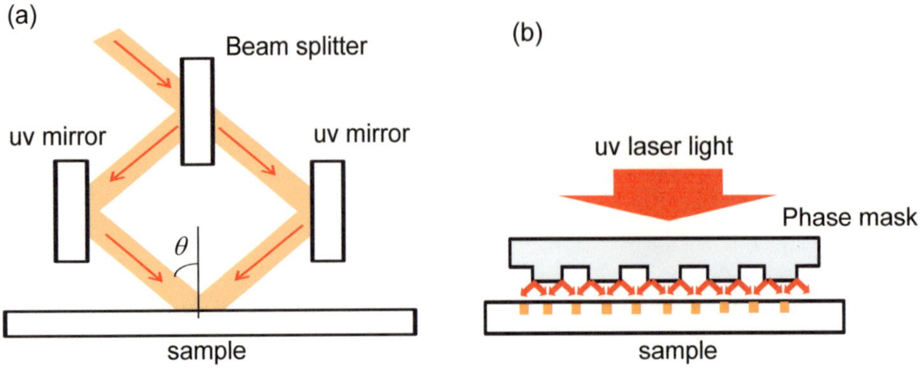

Fig. 2. Schematic depiction of techniques: (a) two-beam interference and (b) the phase mask.

3.2 Phase mask technique

The phase mask technique is one of the most popular writing processes for fine periodic structures. In this technique, diffractive optical elements, called phase masks, are used to modulate incident ultraviolet laser light (Hill, *et al.*, 1993). The phase mask has a one dimensional surface relief on a SiO_2 substrate. When the ultraviolet laser light is normally coupled to the phase mask, the first order diffracted beams are generated from the mask. The zero-order beam is designed to be suppressed to few percentages of the incident power. The samples such as optical fibers or waveguides are located just behind the phase mask. Typically, the distance between the samples and the phase mask is less than 100 μm. In this thin region, a near-field fringe pattern is generated by the interference of the plus and minus first order diffracted beams. The period of this fringe is identical to one-half that of the phase mask. To focus the fringe pattern into the fibers or waveguides, a cylindrical lens is often used. The optical setup of the phase mask technique is simple, compared to the two-beam interference technique. Therefore, it is effective for reproduction of the periodic patterns. KrF excimer laser is one of the popular light sources for the direct laser writing technique. The spatial and temporal coherence of this laser is low. However, low temporal coherence does not affect the writing capability of the phase mask technique.

4. Pulsed laser-induced precipitation of nanoparticles in glasses

4.1 Thermal stabilization of laser-induced refractive index changes

As mentioned in the section 1, the laser-induced structural changes are thermally unstable. Ultraviolet laser pulses can excite electrons of some defect species in glasses and release them. Such released electrons are then trapped in other defects. The structural changes are triggered by these electron excitations. However, thermal energy during annealing excites the trapped electrons again to return back to the original states (Erdogan, *et al.*, 1994). This thermal decay is unavoidable for the laser-induced structural changes in principle. In this section, we present a novel process for the formation of large refractive index changes with excellent thermal stabilities using ultraviolet laser pulses.

Figure 3 shows the changes of the first order diffraction efficiencies of laser-induced gratings (one-dimensional PGB devices) after isochronal annealing for 1 hour at each temperature in a nitrogen atmosphere. The laser-induced grating was written on GeO_2-SiO_2 and

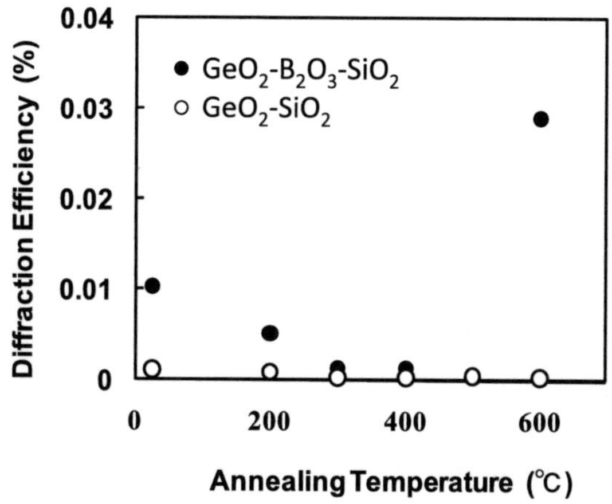

Fig. 3. Changes of diffraction efficiencies of laser-induced gratings after annealing up to 600°C.

GeO_2-B_2O_3-SiO_2 glass thin films using KrF excimer laser (λ = 248 nm) through phase mask of 1060 nm pitch. The glass films were prepared by the plasma-enhanced chemical vapor deposition method using $Si(OC_2H_5)_4$, $Ge(OCH_3)_4$ and $B(OC_2H_5)_3$ as raw materials. The diffraction efficiencies (ratio of diffracted power to incident power) were measured at a probe wavelength of 633 nm under phase matching condition. The diffraction efficiencies for GeO_2-SiO_2 and GeO_2-B_2O_3-SiO_2 decreased gradually after the annealing at a temperature up to 500°C, which was consistent with the previous reports. However, drastic increase of the efficiency for GeO_2-B_2O_3-SiO_2 was observed after annealing at 600°C. Such a drastic change is extremely exceptional phenomenon. This thermally-induced grating in GeO_2-B_2O_3-SiO_2 could not be removed out even by repeated heat treatment between room temperature and 600°C. The changes of laser-induced refractive index modulation Δn of the thermally-induced gratings were investigated as a function of the annealing times at 600°C. The Δn was estimated from the first order diffraction efficiencies. The Δn of the laser-induced grating before annealing was 6.9 x 10^{-4}, which was saturated after irradiation with 12 000 pulses. The Δn of the grating continuously increased after complete remove of the laser-induced grating by annealing up to 500°C. It saturated after annealing for 5 hours. The maximum Δn was 6.8 x 10^{-3}, which corresponds to a Δn value that is approximately ten times larger than that before annealing. The phase matching angle was not changed by this enhancement phenomenon, meaning that period of the grating was same. For conventional direct laser writing, a special photosensitizing process such as hydrogen loading is required to obtain such large Δn. Therefore, this enhancement phenomenon is useful for stabilization and for Δn enhancement.

4.2 Mechanism of thermally-induced refractive index changes

Figure 4(a) and 4(b) show scanning electron microscope images of the laser-induced gratings and the thermally-induced gratings after HF etching for 17 min. For this measurements, thermally-induced gratings were induced by annealing at 600°C for 80 min. Figure 4(a) presents that periodic surface relief can be observed. Reportedly, the etching rate of the SiO_2 surface increased after dense irradiation with vacuum ultraviolet light. For that reason, the formation of a relief structures in our samples is likely to be attributed to

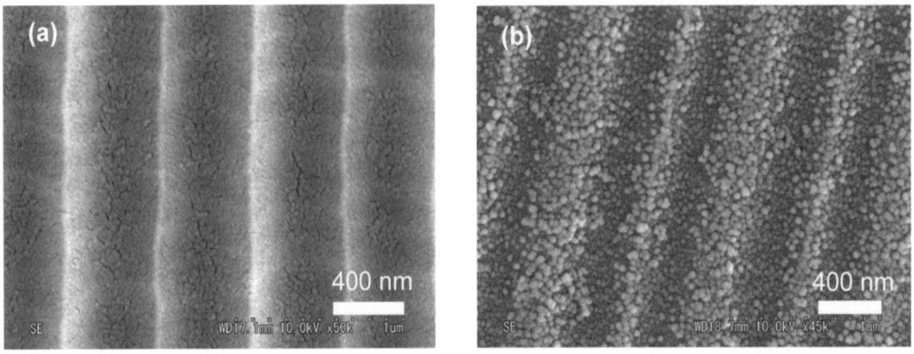

Fig. 4. Scanning electron microscope images of (a) a laser-induced grating and (b) a thermally-induced grating. Both samples were observed after HF etching. Reprinted with permission from Applied Physic Letters. Copyright 2004, American Institute of Physics.

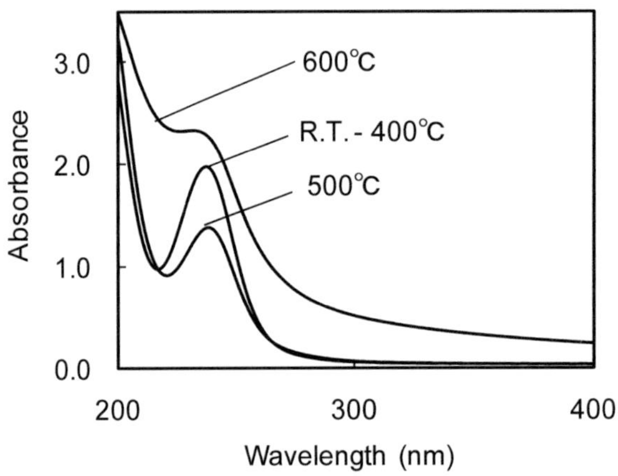

Fig. 5. Changes of absorption spectra of GeO_2-B_2O_3-SiO_2 thin films after annealing up to 600°C.

selective etching of the irradiated region. Figure 4(b) shows that nanoparticles of 20-40 nm diameters, which had higher resistance to HF solution than the glass matrix, were precipitated periodically on the thermally-induced grating surfaces. The pitch of the periodic structure was same as that of the laser-induced grating. From the XRD pattern of the glass film with the thermally-induced grating that was formed by annealing at 600°C for 80 min. Two signals at $2\theta = 45.3$ and 53.7 were detected. A slight diffraction peak at $2\theta = 27.3$ was also observable. These three signals in the pattern are assignable to the (220), (311) and (111) diffraction lines of cubic Ge crystal, respectively, which were never detected without nanoparticle precipitation.

Figure 5 shows the changes of absorption spectra of as-deposited GeO_2-B_2O_3-SiO_2 films before and after the annealing. The absorption band around 236 nm wavelength of the film before annealing is assignable to Ge related oxygen deficient defects such as neutral oxygen mono-vacancy or neutral oxygen di-vacancy related Ge ions. The absorption intensity due to the oxygen deficient defects did not change until 400°C, and then decreased at 500°C. However, marked increase of absorption was observed after the annealing at 600°C. Such changes were not observed for GeO_2-SiO_2 films. The spectral features after the annealing at 600°C were similar to those of glasses with Ge nanoparticles in previous report (Kawamura,

et al., 1996). As the diameter of the nanoparticels decreases, the absorption edge markedly shifts to shorter wavelength because of the size effect. The edge of the GeO_2-B_2O_3-SiO_2 films is located about 300 nm wavelength, and absorption shoulder spreads to visible wavelength region. Therefore, it is considered that the origin of the thermally-induced grating is the periodic precipitation of crystalline Ge nanoparticles. There have been many studies on space-selective precipitation of metal nanoparticles in glass using laser irradiation. In their reports, nanoparticles were always precipitated in the laser irradiated region because the thermal or photochemical reaction during laser irradiation promoted nucleation (Charles, *et al.*, 1968). However, the following experimental revealed that the Ge nanoparticles were precipitated predominantly in the unirradiated region of our samples.
Figure 6 shows a scanning electron microscope image of two-dimensional (2D) thermally-induced grating after the HF etching. The 2D gratings were formed in the same manner as that for the 1D gratings. At first, 2D laser-induced grating was written in the film. Then, the sample was rotated by 90 degree in its own plane midway through irradiation. The pulse number at each dimension was 12 000. Finally, the resulting 2D gratings were annealed at 600°C for 80 min to obtain the 2D thermally-induced gratings. Island structures, which consist of many nanoparticles, were can be observed. The pulse number at the

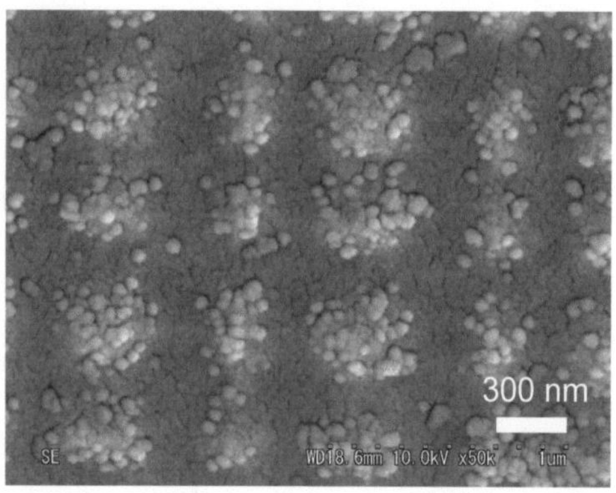

Fig. 6. Scanning electron microscope image of a 2D thermally-induced grating. The observation was carried out after HF wet etching. Reprinted with permission from Applied Physic Letters. Copyright 2004, American Institute of Physics.

crossover site of the 2D grating was twice as large as that at the noncrossover site. We confirmed that there was no marked difference in diffraction efficiencies of the laser-induced gratings and thermally-induced gratings by irradiation exceeding 12 000 pulses, meaning that the difference of the structural changes between the crossover and noncrossover sites are negligible. Therefore, the island-like precipitation in Figure 6 indicates that nanoparticles were precipitated predominantly in the unirradiated region.
We observed an etched surface of the unirradiated thin film after annealing at 600°C for 10 min. Although there were few nanoparticles, many dimples of approximately 300 nm diameters were observed on the surfaces. According to previous reports, the phase separation occurred in B_2O_3-SiO_2 glass below 600°C. Therefore, it is reasonable that the dimples represent the B_2O_3-rich phase, which is less stable against HF solution than other regions. It is notable that the phase separation occurred before the precipitation of Ge nanopartices. Assuming that the laser irradiation prior to annealing accelerated the phase separation that was induced by annealing at 600°C, the formation mechanism of thermally-

induced grating can be explained as follows. As a first step, the phase separation predominantly occurs in the laser irradiated region at the early stage of the annealing at 600°C. Formation of B_2O_3-rich phases increase Ge concentration in the surroundings. As a result, periodic distribution of chemical composition is induced in the grating region. In this case, Ge concentration in the irradiated region is lower than that in the unirradiated one. Therefore, the precipitation of Ge nanoparticles after sufficient annealing at 600°C might be suppressed in the irradiated region. Large refractive index modulation was likely to be induced by the precipitation of high refractive Ge nanoparticles. According to this model, the etch rate of the irradiated region will be higher than that in the unirradiated region because the irradiated region mainly consist of B_2O_3-rich phases. This prediction is consistent with the experimental result as shown in Figure 6.

5. Periodic structures consisting of nanoparticles in channel waveguides

Photonic band gap elements in channel waveguides have much higher potential for the integration with other functional components in single wafer. In this section, we present a periodic structures consisting of Ge nanoparticles, which have large PBG effect with high thermal stabilities, formed in the channel cores (Nishiyama, *et al.*, 2004).

5.1 Fabrication of periodic structures into channel waveguides

Figure 7 schematically shows the fabrication processes of the channel waveguides with thermally-induced gratings. At first, the grating structures with 530 nm pitch are written in GeO_2-B_2O_3-SiO_2 films with SiO_2 upper layers using KrF excimer laser with a phase mask. Annealing at 600°C for 20 min was carried out to induce the thermally-induced gratings. The Δn of the grating was 1.6 x 10^{-3} at 633 nm wavelength. The channel structures were subsequently fabricated using photolithography and plasma etching techniques. The width and height were of the core were 7.0 μm and 5.3 μm, respectively. Finally, the upper layers

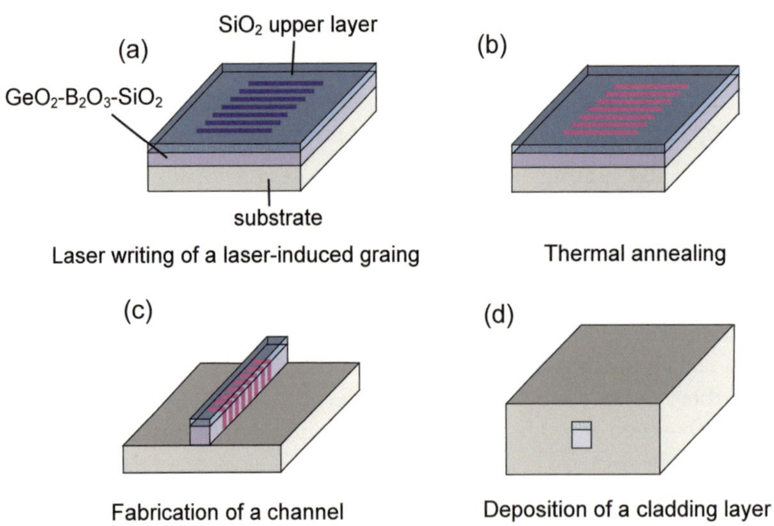

Fig. 7. Fabrication processes of channel waveguides with thermally-induced gratings.

of GeO_2-SiO_2 of 12 μm thickness were deposited to form single-mode waveguides. The interaction length of the grating was about 5 mm. Figure 8 shows the surfaces of as-fabricated channel cores, namely, before deposition of the cladding layer, after wet etching with 6% HF solution. For this observation, the SiO_2 upper layers were removed out by plasma etching process prior to the channel core fabrication. The channel core surfaces before HF etching in figure 8(a) were smooth in spite of the precipitation of nanoparticles. Especially, the sidewall roughness was also smooth. It is apparent that periodic relief patterns appeared on the channel surfaces after HF etching. The pitch of these periodic structures was 530 nm, which was identical to that of the laser-induced gratings written in the film before annealing. From Figure 8(d), the nanoparticles can be observed in the convex regions of the channels. In the section 5.2, we pointed out that the etch rate of the irradiated regions became higher than that of the unirradiated regions because of the predominant formation of B_2O_3-rich phases in the irradiated regions after annealing. In addition, Ge nanoparticles appeared mainly in the unirradiated regions. Therefore, these periodic relief structures are likely to be originated from the formation of thermally-induced gratings inside the channel. The periodic relief structures on the sidewall indicate that the Ge nanoparticles were precipitated not only near the film surfaces, but also across the films. The non-uniform relief patterns on the sidewall might result from the transmitted zero-order laser light, which was not completely suppressed by the phase mask during laser writing process.

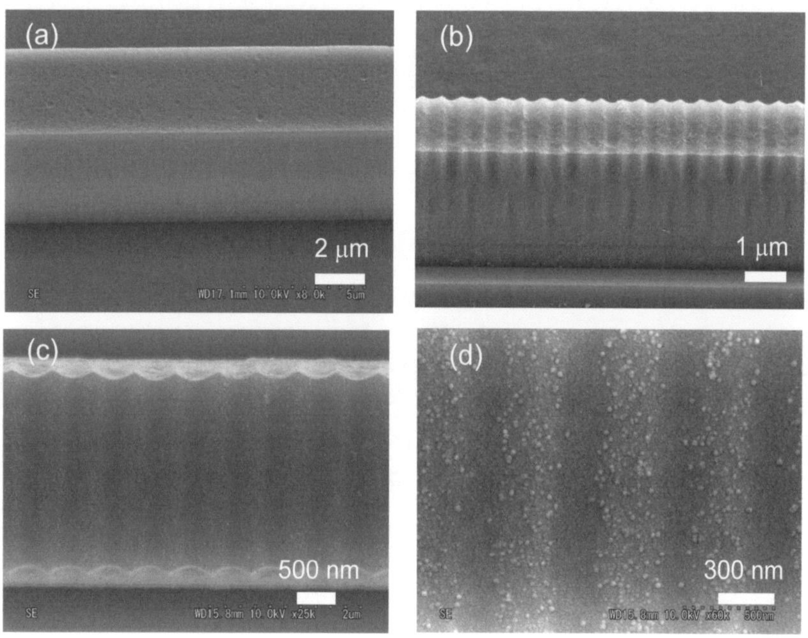

Fig. 8. Scanning electron microscope images of periodic structures consisting of nanoparticles (a) before and (b) after HF wet etching treatment, and images (c) from above and (d) of the enlarged one. Reprinted with permission from Optics Express. Copyright 2007, Optical Society of America.

Figure 9 shows transmission spectra for TE-like and TM-like modes when near-infrared wavelength light was coupled to the channel waveguides with the periodic structures consisting of Ge nanoparticles. These optical properties were measured with an optical spectrum analyzer using a wavelength tunable laser diode as a light source. The incident beam was butt-coupled into channel waveguides using commercially available single-mode optical fibers. The depths and positions of the PBG were 37.77 dB at 1536.2 nm and 38.72 dB at 1537.6 nm, respectively, for TE-like and TM-like modes. Only one dip appeared for each polarization mode, indicating that the channel structures worked as a single-mode waveguide. Such strong PBG effects indicate that the refractive index modulation by periodic structures is large not only in the visible wavelength region, but also in the optical communication wavelength. The peak depths of both modes were similar, indicating that the Ge nanoparticles were uniformly precipitated in the unirradiated regions across the channel core, which is consistent with the relief formation on the sidewalls in Figure 8. The polarization dependence of the PBG wavelength is likely to be originated from the asymmetrical core structures. In this PBG device, the periodic structures were induced by annealing only for 20 min. In the section 5.1, the amount of Δn continuously increased with

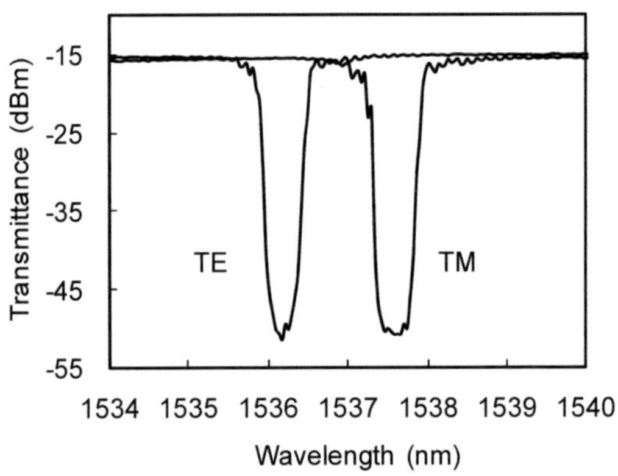

Fig. 9. Transmission spectra of PBG devices consisting of Ge nanoparticles in the channel.

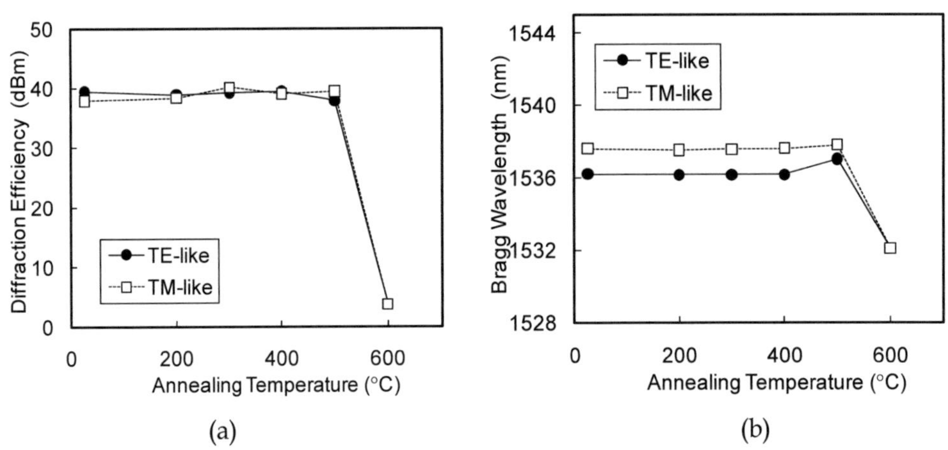

Fig. 10. (a) Diffraction efficiencies (PBG depths) and (b) Bragg wavelengths (PBG wavelengths) of thermally-induced gratings in channels.

annealing time at 600°C. Therefore, much stronger PBG effects are achievable after longer annealing time at the fabrication step (b) in Figure 7. Figure 10 (a) and 10(b) shows changes of the PBG depths and wavelength positions for both polarization modes after additional annealing from 200 to 600°C. The annealing time was 1 hour at each temperature. The conventional direct laser writing structures markedly decay even by annealing at 150°C. Compared to this, it is notable that the PBG depths for both modes exhibited no changes after annealing at 500°C. In particular, no changes of PBG depths and wavelengths were observed after annealing up to 400°C. The difference of the PBG wavelength between each polarization mode was changed from 1.4 nm to 0.8 nm without decay of PBG depths after annealing at 500°C. Such a decrease might be responsible for reduction of the residual thermal stress between the core and cladding layer of GeO_2-SiO_2 because the cladding layer was deposited at temperature lower than 500°C. These results indicate that the precipitation methods of Ge nanoparticles are effective to create much larger PBG effects with excellent thermal stabilities than those reported previously. The polarization dependence of the PBG wavelengths can be reduced between modes. Photonic band gap structures with higher efficiencies, thermal stability, and lower polarization dependence would be realized by optimization of fabrication conditions.

5.2 Direct laser writing of channel waveguides with periodic structures consisting of nanoparticle precipitation

In the last section, we presented the fabrication of PBG structures consisting of Ge nanoparticles in channel cores using photolithography and plasma etching processes. The devices exhibited strong PBG effects with high thermal stabilities. However, complicated vacuum processes and clean room environments are required. In this section, we present a fabrication process of PBG devices, which consist of Ge nanoparticles, inside channel waveguides only using direct laser writing and subsequent thermal annealing.

Figure 11 shows changes of refractive indices at 633 nm wavelength of the unirradiated GeO_2-B_2O_3-SiO_2 film and homogeneously irradiated film against an annealing time at 600°C. The laser fluence and pulse number for preparation of the irradiated film were 180 mJ/cm^2/pulse and pulse number of 27 000. By irradiation, refractive index increased by 3.1×10^{-3}. Annealing for 10 min markedly decreased this difference of refractive index between both films to 0.5×10^{-3}. Refractive indices of both films increased after annealing time longer than 10 min because of the precipitation of Ge nanoparticles with higher refractive indices than those of glass matrix. It is notable that the rate of the refractive index increase for the irradiated film was much lower than that for the unirradiated one. Due to this difference of increase rate, the refractive index of the unirradiated film became higher than that of the irradiated one after annealing after 10 min. Such thermally-induced refractive index difference exhibited high stabilities against repeated annealing up to 500°C.

Based on this unique behavior of refractive index of the GeO_2-B_2O_3-SiO_2 films of 4 μm thickness, we fabricate the PBG devices inside channel cores. Figure 12 shows the fabrication processes of the PBG devices and the expected refractive index distribution at each process. First, grating structures were written in a GeO_2-B_2O_3-SiO_2 film using KrF excimer laser through a phase mask. Then, the irradiation through a Cr mask with a straight line pattern of 7 μm width on a SiO_2 substrate was performed under the condition of the laser fluence of

180 mJ/cm²/pulse and pulse number of 27 000. Finally, the samples were annealed at 600°C for 20 min. The refractive index distribution was inverted after this annealing step, resulting PBG devices in channel cores. Propagation loss of the waveguides increased with the annealing time at 600°C. The loss before and after the annealing for 1 hour was 0.4 and 1.4 dB/cm at 1545 nm wavelength, respectively.

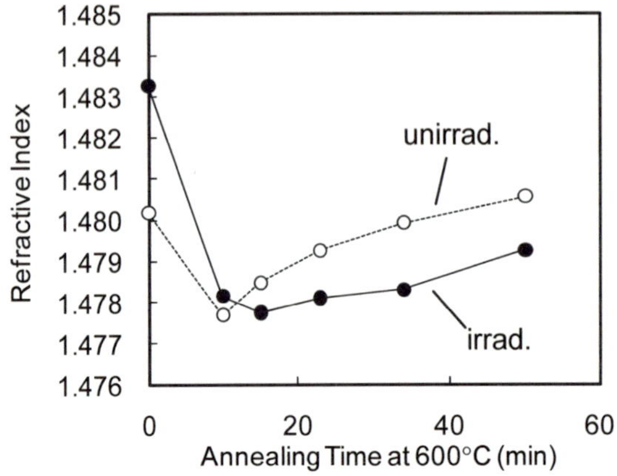

Fig. 11. Changes of refractive indices of unirradiated and irradiated films.

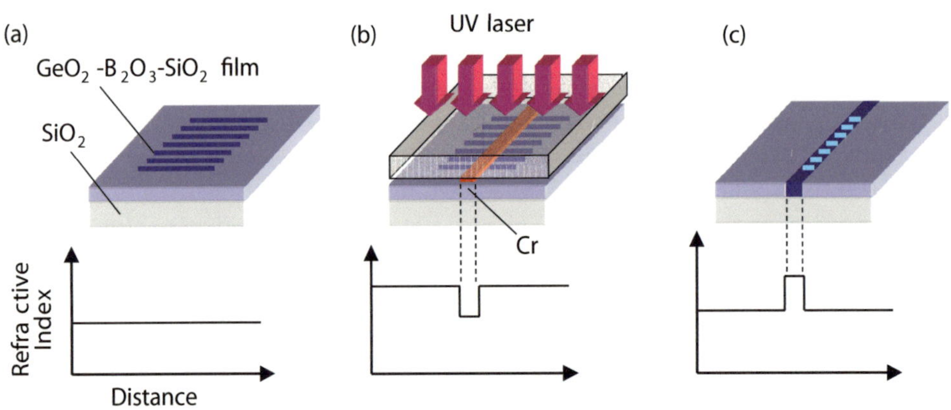

Fig. 12. Fabrication processes for the periodic structures consisting of Ge nanoparticles in channels. (a) Laser writing of the gratings, (b) laser irradiation through a Cr mask, and (c) thermal annealing at 600°C for 20 min

An optical microscope image of the waveguide structures is shown in Figure 13(a). We can see that a brown slight line was induced in the film. The line width was approximately 6 μm, which was consistent with the width of the Cr mask line. In the GeO_2-B_2O_3-SiO_2 film used in this study, intense absorption bands are induced in ultraviolet-visible wavelength regions by annealing at 600°C, which is responsible for the precipitation of Ge nanoparticles. The laser irradiation suppresses the nanoparticle precipitation. Therefore, we can consider that the nanoparticle concentration of the straight line area became higher than that of the

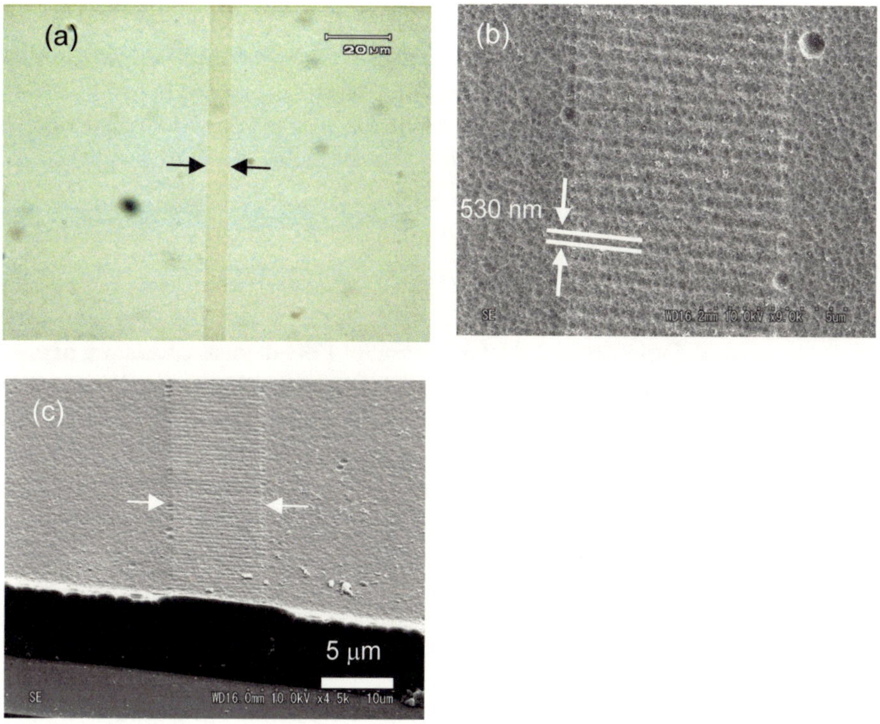

Fig. 13. (a) Optical microscope image of a channel core region and scanning electron microscope images of (b) top-view and (c) overall view of a PBG devices consisting of nanoparticles. Reprinted with permission from Journal of Physics. Copyright 2009, IOP publishing.

surroundings, and that brown-colored straight line is originated from the absorption shoulder of the nanoparticles. Figure 13(b) and (c) shows scanning electron microscope images of top view an overall view of the films after HF etching. Although it was difficult to recognize individual nanoparticle by this observation because the annealing time was as short as 20 min, the periodic structure and the channel core were observable in the film. No surface reliefs were detected before the HF etching because of their smooth surfaces. The channel width and period of the structures were 7 μm and 530 nm, respectively, which were identical with those of the photoimprinted patterns in Figure 12(b). Bright lines in the periodic structures are likely to be aligned aggregated nanoparticles. Note that the periodic structures at clad areas (outside areas of the cores in the film) were not visible. This erase is most likely due to irradiation with higher fluence in Figure 12(b) than that in Figure 12(a). By this irradiation, laser-induced refractive index contrast in Figure 12(b) decayed because larger refractive index changes can be induced by higher fluence. This confinement of the periodic structures in the core can avoid coupling of propagation light to clad and leaky modes. From Figure 11, the refractive index difference between core and cladding regions for the in-plane direction was approximately 1.2×10^{-3} at 633 nm wavelength. This periodic structures exhibited peaks of 18.0 dB at 1532.7 nm and 18.7 dB at 1533.1 nm for the TE-like mode and TM-like mode, respectively, in transmission spectra. These peaks indicate that the periodic structures worked as PBG devices inside the channels.These PBG devices exhibited high thermal stabilities compared to the conventional laser-induced PBG ones. The structural changes of the laser-induced gratings were almost completely erased after repeated annealing at 400°C. In contrast, it is notable that no remarkable changes of PBG depths and wavelengths were observed for the periodic structures consisting of Ge nanoparticles even after heat treatment at 500°C.

6. Summary

In this chapter, we presented the fundamentals of pulse laser-induced structural changes of SiO_2-based glasses and demonstrated the fabrication of PBG devices consisting of crystalline Ge nanoparticles using a combined process of direct laser writing and subsequent thermal annealing. Large refractive index modulations with high thermal stabilities were induced by this combined process. The periodic structures in channel cores exhibited intense PBG effects of nearly 40 dB for both polarization modes. And no changes of the PBG effects were observed even after repeated heat treatment up to 500°C. Optimization of pulse laser irradiation and phase change conditions can realize devices with much larger PBG effects using direct laser writing techniques.

7. References

Åslund, M. & Canning, J. (2000). Annealing properties of gratings written into UV-presensitized hydrogen-outdiffused optical fiber, *Optics Letters*, Vol. 25, 692-694.

Baker, S. R.; Rourke, H. N.; Baker, V. & Goodchild (1997). Thermal decay of fiber Bragg gratings written in boron and germanium codoped silica fiber, *Journal of Lightwave Technology*, Vol.15, 1470-1477.

Borrelli, N. F.; Allan, D. C. & Modavis, R. A. (1999). Direct measurement of 248- and 193-nm excimer-induced densification in silica-germania waveguide blanks, *Journal of Optical Society of America B*, Vol. 16, 1672-1679.

Charles, R. J. & Wagstaff, F. E. (1968). Metastable Immiscibility in the B_2O_3-SiO_2 System. *Journal of the American Ceramics Society*, Vol. 51, 16-20.

Cordier P.; Doukhan, J. C.; Fertein, E.; Bernage, P.; Niay, P.; Bayon, J. F. & Georges, T. (1994). TEM observation of structural changes in glass associated to Bragg grating inscription in a germanosilicate optical fiber preform, *Optics Communications*, Vol. 111, 269-275.

Cordier, P.; Dupont, S.; Douay, M.; Martinelli, G.; Bernage, P.; Niay, P.; Bayon, J. F. & Dong, L. (1997). Evidence by transmission electron microscopy of densification associated to Bragg grating photoimprinting in germanosilicate optical fibers, *Applied Physics Letters*, Vol. 70, 1204-1206.

Dong, L.; Archambault, J. L.; Reekie, L.; Russel, P. St. J. & Payne, D. N. (1995). Photoinduced absorption change in germanosilicate performs: evidence for the color-center model of photosensitivity, *Applied Optics*, Vol. 34, 3436-3440.

Douay, M.; Xie, W. X.; Taunay, T.; Bernage, P.; Niay, P.; Cordier, P.; Poumellec, B.; Dong, L.; Bayon, J. F.; Poignant, H. & Delevaque, R. (1997). Densification Involved in the UV-Based Photosensitivity of Silica Glasses and Optical Fibers, *Journal of Lightwave Technology*, Vol. 15, 1329-1342.

Erdogan, T.; Mizrahi, V.; Lemaire, P. J. & Monroe, D. (1994). Decay of ultraviolet-induced fiber Bragg gratings, *Jouranl of Applied Physics*, Vol. 76, 73-80.

Fonjallaz, P. Y.; Limberger, H. G. & Salathé, R. P. (1995). Tension increase correlated to refractive-index change in fibers containing UV-written Bragg gratings, *Optics Letters*, Vol. 20, 1346-1348.

Hand, D. P.; & Russell, P. St. J. (1990). Photoinduced refractive-index changes in germano-silicate fibers, *Optics Letters*, Vol.15, 102-104.

Hill, K. O. & Meltz, G. (1997). Fiber Bragg grating technology fundamentals and overview, *Journal of Lightwave Technology*, Vol. 15, 1263-1276.

Kawamura, K.; Hosono, H. & Kawazoe, H. (1996). Proton-implantation-induced nanosized Ge crystal formation in SiO_2:GeO_2 glasses. *Journal of Applied Physics*, Vol. 80, 1357-1363.

Kim, B. H.; Park, Y., Ahn, T. J.; Kim, D. Y.; Lee, B. H.; Chung, Y.; Peak, U. C. & Han, W. T. (2001). Residual stress relaxation in the core of optical fiber by CO_2 laser irradiation, *Optics Letters*, Vol. 26,1657-1659.

Kherbouche, F. & Poumellec, B. (2001). UV-induced stress field during Bragg grating inscription in optical fibers, *Journal of Optics: Pure and Applied optics*, Vol. 3, 429-439.Ky, N. H.; Limberger, H. G.; Salathé, R. P.; Cochet, F. & Dong, L. (2003). UV-irradiation induced stress and index changes during the growth of type-I and type-II fiber gratings, *Optics Communications*, Vol. 225, 313-318.

Miura, K.; Qiu, J.; Inouye, H.; Mitsuyu, T. & Hirao, K. (1997). Photowritten optical waveguides in various glasses with ultrashort pulse laser, *Applied Physics Letters*, Vol. 71, 3329-3331.

Nishii, J.; Kitaka, K.; Nishiyama, H. & Miyamoto, I. (2002). Thermally stabilized photoinduced Bragg gratings, *Applied Physics Letters*, Vol. 81, 2364-2366.

Nishiyama, H.; Kitaka, K.; Nishii, J. & Miyamoto, I. (2003). Thermo- and photosensitive GeO_2-B_2O_3-SiO_2 thin glass films, *Japanese Journal Applied Physics*, Vol. 42, 559-563.

Nishiyama, H.; Miyamoto, I.; Matsumoto, S.; Saito, M.; Kintaka, K. & Nishii, J. (2004). Direct laser writing of thermally stabilized channel waveguides with Bragg gratings, *Optics Express*, Vol. 12, 4589-4595.

Nishiyama, H.; Miyamoto, I.; Matsumoto, S.; Saito, M.; Fukumi, K.; Kintaka, K. & Nishii, J. (2004). Periodic precipitation of crystalline Ge nanoparticles in Ge-B-SiO_2 thin glass films, *Applied Physics Letters*, Vol 85, 3734-3736.

Nishiyama, H.; Miyamoto, I.; Hirata, Y. & Nishii, J. (2007). Periodic structures consisting of germanium nanoparticles in buried channel waveguides, *Optics Express*, Vol. 15, 2047-2054.

Nishiyama, H.; Nishii, J. & Hirata, Y. (2009). Periodic structures in GeO_2-B_2O_3-SiO_2 glass films fabricated using ultraviolet laser pulses, Journal of Physics: Conference Series, Vol. 165, 012049-1-4.

Poumellec, B.; Niay, P.; Douay, M. & Bayon, J. F. (1996). The UV-induced refractive index grating in Ge:SiO_2 preforms: additional CW experiments and the macroscopic origin of the change in index, *Journal of Physics D: Applied Physics*, Vol. 29, 1842-1853.

Poumellec, B.; Guènot, P.; Riant, I.; Sansonetti, P.; Niay, P.; Bernage, P. & Bayon, J. F. (1995). UV induced densification during Bragg grating inscription in Ge:SiO_2 preforms, *Optical Materials*, Vol. 4, 441-449.

Svalgraad, M.; Poulsen, C. V.; Bjarklev, A. & Poulsen, O. (1994). Direct UV writing of buried single mode channel waveguides in Ge-doped silica film, *Electronics Letters*, Vol. 30, 1401-1403.

Part 4

Chemical and Biological Applications

18

Hot Chemistry with Cold Molecules

Tapas Goswami and Debabrata Goswami
Department of chemistry, Indian Institute of technology Kanpur,
Kanpur-208016,
India

1. Introduction

Control of chemical reactions that focuses on the selective cleavage or on the formation of chemical bonds in a polyatomic molecule is a long-sought-after goal for chemists. Attempts have been made since the early days of lasers (1960) to make this dream come true (Rousseau, 1966). In most of the experiments, selectivity is lost because of the rapid intramolecular vibrational energy redistribution (IVR) which occurs on picosecond time scales. This simply leads to heated molecules. Supersonic molecular beam techniques have proven to be an excellent method for producing isolated cold molecules in the gas phase, where the molecules are in their lowest rotational and vibrational states and as a result several relaxation rates like the collisional and the IVR rates are much slower (Smalley et al., 1977; Levy, 1981). By combining ultrafast laser technology with supersonic molecular beam technique in a novel way, several control schemes known as 'coherent control', have been proposed that make use of the coherent nature of laser radiation (Zewail, 1980; Bloembergen & Zewail, 1984). Furthermore, study of control is typically pursued in molecular beams in order to isolate the elementary processes to be studied from surrounding solvent perturbations.

In a chemistry laboratory, however, the conventional control that we generally use in increasing the yield of the desired products in the chemical synthesis are macroscopic variables, such as, the temperature, pressure, concentration, etc. Sometimes catalysts are also used to control the chemical reactions. But the methods involved in this type of conventional cases are based on the incoherent collision between collision and we cannot get direct access to the quantum mechanical reaction pathway. On the other hand, in the quantum control of chemical reactions, molecular dynamics involved can be altered by specifically designed external light fields with different control parameters, namely, the intensity, phase, frequency, and polarization, which can vary with time. Using such methods, one can reach a user-defined chemical reaction channel more selectively and efficiently (Brixner & Gerber, 2003). The short temporal duration of ultrafast laser pulses results in a very broad spectrum (Fig.1). The output of our amplified laser system has pulse duration around 50 femtosecond and spectral bandwidth of around 18 nm. Possibilities of manipulating such an ultrafast coherent broadband laser pulses have brought forth the exciting field of ultrafast pulse shaping. Pulse shaping involves the control over amplitude, phase, frequency, and or inter-pulse separation (Goswami, 2003). Control of chemical reactions by laser can have various applications in diverse industrial and biological or medicinal settings e.g. microelectronic lithography, the fabrication of gene chips, and photodynamic therapy (Levis et al., 2001). Even in quantum information processing, controllability of the quantum system is the key issue (Ahn et al., 2000; Vivie-Riedle & Troppmann, 2007). All these probable applications of coherent control have made this a very 'hot' research area in chemistry and physics.

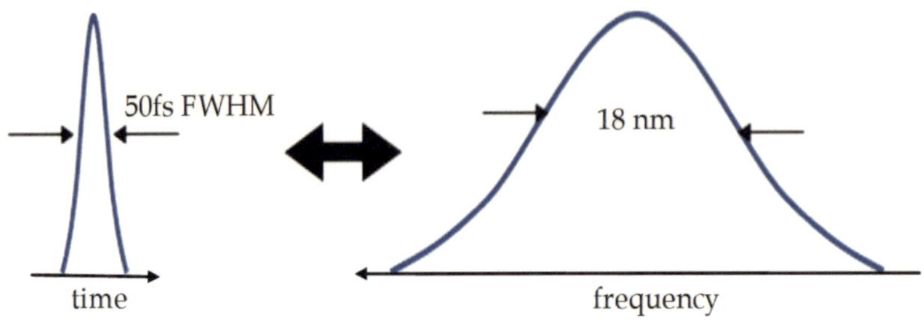

Fig. 1. An ultrafast laser pulse in time and frequency.

To begin with coherent control studies, it is essential to choose a simple atomic or molecular system, whose characteristics are fully known and can be understood very simply, such that we can easily analyze our results. We also need to have an easy detection system. Thus, the first and foremost objective is to design and build, for instance, a supersonic molecular beam chamber with linear time of flight mass spectrometer to produce an isolated cold molecular beam whose mass spectra can be detected.

Coincidentally, this year 2010 is considered as the hundred years of invention of molecular beam technique (Herschbach, 1987). Molecular beam is a large number of molecules in laminar flow, i.e., moving in straight and almost parallel trajectories in a practically collision free environment in a highly evacuated space. One of the essential elements in producing the molecular beam is the high vacuum pump. High vacuum pump was invented almost hundred years ago in the year 1910. Molecular beam study was started just after the invention of the high vacuum pump.

The control over the fragmentation and chemical reactions using shaped ultrafast laser pulses was demonstrated by different schemes of experiments by various groups of scientists (Assion et al., 1998; Levis et al., 2001; Daniel et al., 2001; Levis et al., 2002; Brixner et al., 2003; Daniel et al., 2003). All of their experiments are based on the closed loop approach using learning algorithms to control the laser field with feedback from the experimental signal. The resultant pulse shapes obtained by an adaptive approach were not always found to be a globally optimized solution (Assion et al., 1998; Levis et al., 2001; Weinacht et al., 1999; Itakura et al., 2003). Furthermore, the physical significance of such complex shaped pulses to the actual processes leading to discrimination among the fragmentation channels and their mechanisms are often too difficult to apprehend limiting the generalization of such feedback schemes. A continued effort, therefore, exists on the control of such fragmentation process from simpler pulse modulation concepts.

One of the easiest pulse modulation schemes is frequency chirping. Chirping essentially refers to the process of arranging the frequency components in a laser pulse with certain phase ordering. Linear ordering can be easily achieved by dispersing the ultrafast pulses through a pair of gratings. This ordering of frequency components results in the lengthening of an otherwise bandwidth limited ultrafast laser pulse. For positively chirped pulse leading edge of the pulse is red shifted and the trailing edge is blue shifted with respect to the

central frequency of the pulse. Negative chirp corresponds to the opposite effect. The first experimental demonstration of control using simple linear chirped laser pulses was by Warren and co-workers in the early 1990s (Melinger et al., 1992). Subsequently, several experimental and theoretical developments have made linear chirped pulse control a very attractive field of research (Krause et al., 1993; Cerullo et al., 1996; Pestrik et al., 1998; Yakovlev et al., 1998; Cao et al., 1998). Control of fragmentation with simple linear chirped pulses as a control scheme has also become an active field of research in recent years.
In this chapter on 'hot' chemistry with 'cold' molecules, we first present our experimental scheme of producing the 'cold' molecules in the supersonic beam chamber followed by the laser schemes to induce 'hot' chemistry. The resulting chemistry ensuing from such technological developments are discussed in the subsequent sections.

2. Experimental

Design of our molecular beam setup:
An engineering drawing of the system was prepared using AUTOCAD software (Fig.2). The design of our molecular beam setup (Fig.2) consists of three chambers with differentially pumping by turbo pumps. The three chambers were designed to attain a supersonic molecular beam region (source chamber), a laser interaction region (Fig.3) and a mass detection region. Each of the chambers is connected to turbo molecular pumps which are in

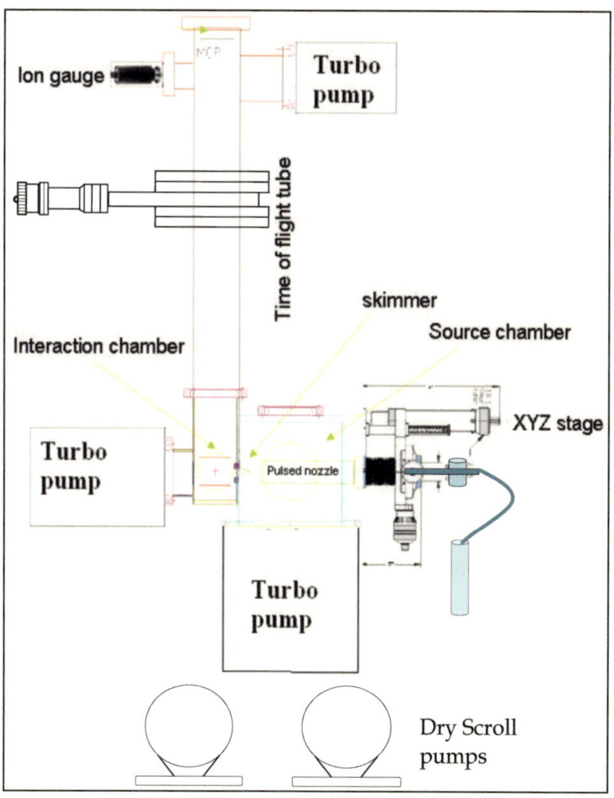

Fig. 2. Diagram of our Molecular beam chamber setup.

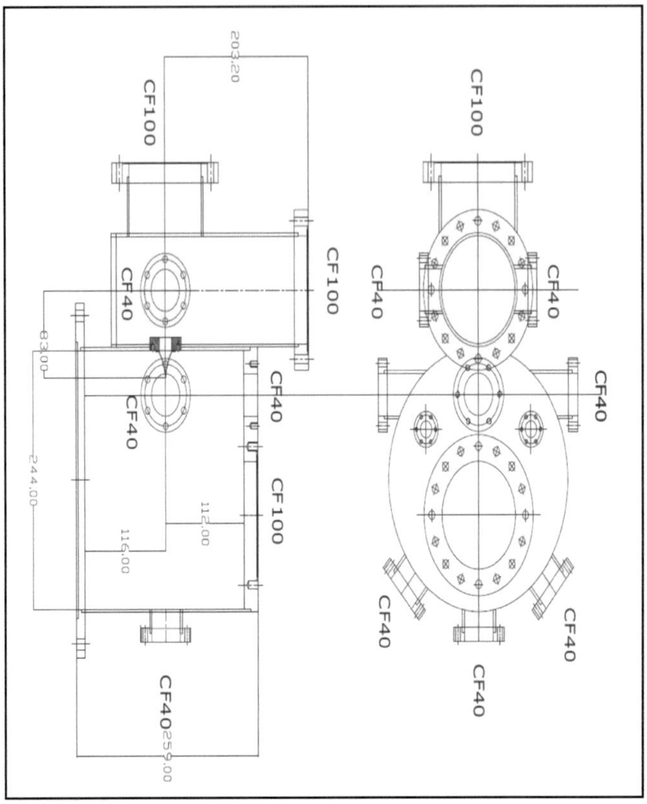

Fig. 3. Autocad diagram of the source and part of the interaction chamber. The flanges are also marked with their sizes. Diagram on top is the top view of the chamber what was designed and fabricated. Diagram in the bottom is the side view of the chamber with all viewports. CF- conflate flat copper gasket flange.

turn connected to dry scroll pump. Vacuum chambers were fabricated based on our design and requirements for complete oil free systems using SS-304 stainless steel and conflate connections. Design of our particular single beam experiments depends mainly on the considerations of signal-to-background noise ratio. The size and geometry was chosen to match the pulse duration of the source and the detector geometry. The chamber geometry suggests that a cylinder or cube with height and width approximately equal and with the beam source located so that the nozzle is about equal distance from the walls ahead and the sides of the beam.

It is sometimes advantageous in pumping speed requirement when the molecular beam source is kept directly into the throat of the vacuum pump. Because most high speed pumps are designed for mounting at the bottom of the chamber, this configuration implies mounting the molecular beam source directly above the pump. The source chamber or the supersonic molecular beam region is larger and it has the dimension of 10.19 inch (259 mm) height and 9.6 inch (244 mm) diameter. Source chamber is pumped by a turbo molecular pump (Varian Turbo V2K-G) which has a 2000 l/s pumping speed. The 2000 l/s turbo pump is connected to the source chamber with ISO250 F flange. This turbo pump is connected to the Dry scroll pump (Varian DS600) via stainless steel hose with a KF40 flange.

The source chamber contains a pulsed valve with 0.5 mm nozzle attached (Series 9 General valve) and a 1mm skimmer which is the only opening into the interaction chamber and it allows only the central portion of the expanded molecular beam. The pulsed valve is mounted on a compact X-Y-Z translation stage from MDC Inc. The Series 9 pulsed valve (General Valve, Inc.) provides a versatile way of introducing high backing pressures of gas into vacuum. The series 9 valve comes with its own Iota One pulser, which can be triggered either internally or externally (used in the experiments described in this chapter) and then be pulsed for a defined width. The valve consists of a magnetically actuated poppet and spring mechanism which requires sufficient rise time for opening and closing (250 μs each half cycle, which limits its maximum repetition rate to 35 Hz). We keep the repetition rate of the pulsed valve at 10 Hz and the pulse duration at 100 μs for the experiments described here. A poppet of tip size matched with the orifice (diameter 0.5 mm) of the nozzle is fitted inside the armature by a compressed stainless steel spring to close the orifice. The Iota One pulser generates a brief high voltage of 275-295 V and it pulls the armature with the tip to open the orifice and then the voltage drops to a 28 V holding voltage for the remainder of the pulse. To get the full benefit of a pulsed molecular beam, it is essential to have a leak tight pulsed valve. We used a PEEK (polyether-ketone resin) poppet material which can deform within few hours of use and needed to be replaced regularly. A leaking poppet will increase the foreline pressure on the roughing pump thermocouple gauge and a strong degradation of any type of collected ion signal and often large fluctuations in the ion gauge readings of source chamber pressure. For the electrical connection to the pulsed valve there is CF40 electrical feed through flange in the source chamber.

Our sample chamber is kept outside the vacuum chamber and is connected to the pulsed valve with a 0.5 inch diameter and 12 inch long stainless steel tube. The tube passes through the back flange of the source chamber with a conflate connection and is connected to the He gas supply cylinder.

The nickel skimmer of 1 mm diameter is used to separate the source chamber from the interaction chamber and it allows only the cold core of the supersonic molecular beam. The skimmer is a cone with a base diameter of 27.9 mm, an orifice diameter of 1mm and a length of 25.4 mm. It has a 2.5 mm clamping edge. Skimmers have ultra-thin walls and ultra-sharp orifice edges to minimize the skimmer interference with molecular beam.

The interaction chamber is pumped by a 500 l/s turbo molecular pump (TurboV-550). Turbo V-550 pump is connected to the chamber with CF100 conflat flange. It contains two sapphire view port (CF40) to pass the laser beam through the interaction region where the electric plates for the time of flight mass spectrometer is situated.

Fragment ion detection chamber with flight tube is pumped by a 300 l/s turbo pump (TurboV-300) and the pump is kept near the 18 mm dual MCP detector. The MCP detector is separated by a gate valve, when the experiment is not in operation, such that we can always maintain low pressure in the detection chamber. Other element in the detection chamber is the time of flight mass spectrometer, which will be described below.

Vacuum pressure is monitored in both source and detection chamber by Bayard-Alpert ionization gauge detector and convector gauge. After baking the vacuum chambers, we are able to obtain base pressure of around 10^{-8}torr and 10^{-9}torr in the source and interaction chamber respectively. The pumping speed of our turbo pumps are sufficient enough to maintain a chamber pressure of 10^{-6}torr and 10^{-7}torr, respectively, in the source and the interaction chamber, when the nozzle is in operation at 10 Hz repetition rate for a backing pressure of 2-3 atmosphere of He.

Our time of flight mass spectrometer:

Our time of flight tube (Fig. 4) was manufactured by R. M. Jordan Co. (C-677 lens stack assembly) according to our particular requirements. A skimmed (0.040 inch dia.) supersonic molecular beam is directed through the space (1 cm) between two plates and is ionized by a laser beam which intersects it at right angles. The ions thus created are repelled by the repeller plate (VA1 =+4100 Volts) and drawn through the extraction Grid (VA2 = +3500Volts). They are then accelerated through the Ground Grid (VA3) into the Flight tube. Total length of the drift tube is 56 inch from laser centerline to the detector.

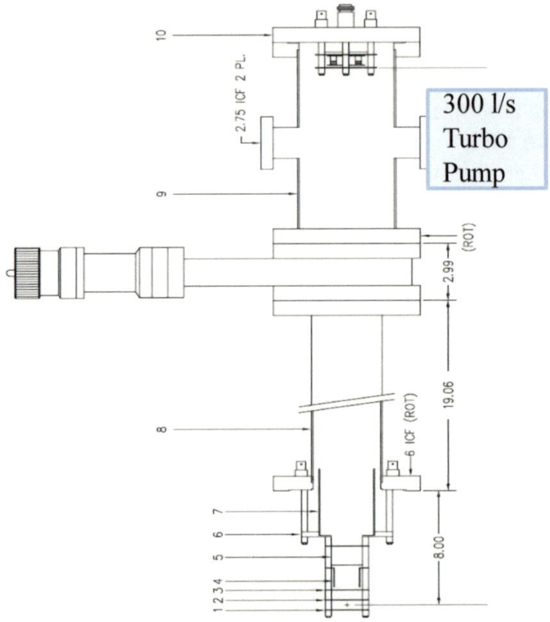

Fig. 4. Design of our Time of Flight tube with a manual Gate valve, 1. Repeller plate VA1. 2. Extraction grid VA2. 3. Acceleration grid GND. 4. X steering plates VX1, VX2. 5. Y steering plates VY1, VY2. 6. Mounting ring. 8. Flight tube 2.875" diam. 9. Side-port Tee. 10. C 16MM dual MCP detector. All measurements are shown in inches.

During this transit the ions pass through the Einsel lenses (no power supply is provided) and between the deflection plates VX1=+157 Volts, VX2=+176 Volts, VY1= -45 Volts, VY2= -21Volts. These plates steer the beam on to the repeller and Detector Grid and compensate for the transverse displacement due to any initial molecular beam velocity vector.

When each of these ions arrives at the detector it will impact the first microchannel plate (VD1= -2200 Volts) with about 6000 volts of energy. This impact will deliver approximately 104 electrons onto the face of the second microchannel plate (VD2=-1200 Volts). Each of these secondary electrons will generate another 104 electrons in the second plate. These electrons exit the bottom of the plate (VD3= -200Volts) and accelerate the final 200 Volts to the 50 Ohm anode which is at ground potential.

The MCP plates (Burle Electro-optics) are discs of lead glass ~0.75 mm thick and 5.2 cm^2 in diameter. (Fig. 5) It consists of around 106 cylindrical channels with a diameter of 10μm and an inter-channel separation of 12μm. Electrons emitted from the channel walls by ion impact are accelerated toward the other side of the plate by a differential electric field. Secondary ion emission from the channel walls initiates an electron avalanche which is collected by an anode and then output through a 50 Ohm termination circuit to an oscilloscope for observation. To increase MCP gain, additional MCPs can be placed in series. We used a 2 plate MCP stack with a total gain of 10^6 and a sub-nanosecond rise time. A grid on front of the MCP stack is held at ground potential to allow a field free drift region for ions in the flight tube. The cylindrical symmetric flight tube can be biased to a few hundred volts but it is usually held at ground. In order to protect the MCP against sparks causing short circuit, the pressure has always been kept below 10^{-5} torr. The minimal pressures in the TOF and source chamber were kept about 10^{-7} and 10^{-6} torr.

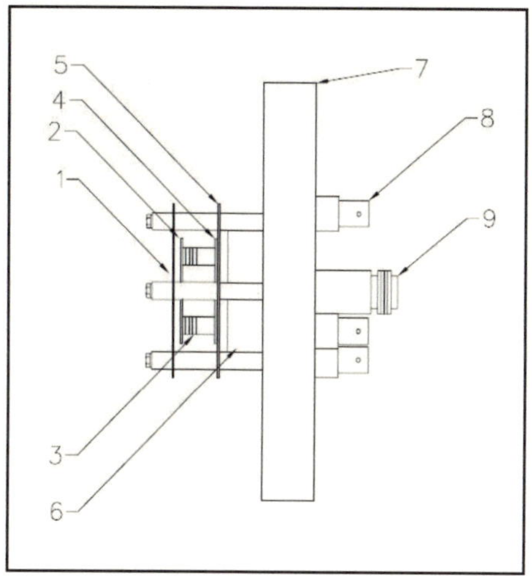

Fig. 5. Design of our microchannel plate detector on a conflate flange: 1- Input grid, 2-spring cap, 3-spring, 4-microchannel plate stack assembly, 5- base plate, 6- anode shield, 7- detector flange assembly, 9- 50Ω feedthrough.

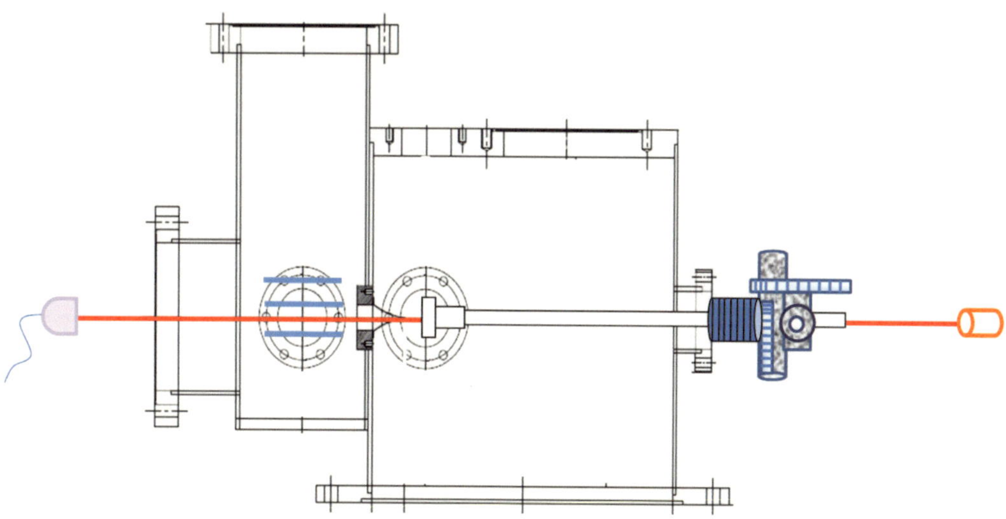

Fig. 6. Alignment of nozzle, skimmer and centre of the repeller and extraction plate.

Alignment of the nozzle, skimmer, and the centre of the repeller and extraction plates was achieved by aligning using a laser pointer with the pulsed valve's internal part removed and also with the turbo pump near the interaction region removed.(Fig. 6) The XYZ position of the pulsed valve mounting assembly was adjusted using the XYZ mount (MDC Vacuum Inc.) so that its centre falls in the line of the skimmer and the centre of the two plates (Repeller and Extraction). Slight adjustment of the position of the skimmer to pass the beam into the centre of the two electrical plates was done.

Data Acquisition:

Synchronization of 1 kHz, 50fs pulsed laser with 10 Hz, 100μs pulsed molecular beam was done by a homemade delay generator.(Fig. 7) Our delay generator can give a delay upto ~2 ms. We take the master clock of 1 kHz from our pockel cell and used as an external trigger for the delay generator and the delay generator is used to first count down the frequency from 1 kHz to 10 Hz and then delay it as per necessity to get the stable time of flight spectra. 10 Hz delay out is used to trigger the pulsed valve and the 10 Hz TTL out is used to trigger the MCP signal in Oscilloscope. Data acquisition procedure is shown pictorially in Fig. 8. Here we give the delay to the pulsed molecular beam by around 800 μs to make it synchronize with the next 1 kHz femtosecond laser pulse. TOF signal was collected using 1GHz digital oscilloscope. Each data was averaged with 500 laser shots. Fig. 9 is the picture of our setup.

Calibration of our linear time of flight mass spectrometer:

Calibration of our time of flight mass spectrometer was done by taking the mass spectra of some sample whose mass spectra is reported, then m/z vs t^2 was plotted and after fitting it linearly according to the equation $m/z=At^2+B$, we can find the calibration parameter A and B. For unknown sample we can easily convert flight time to m/z following the above equation where A and B are known. In our case, after fitting linearly according to the equation mentioned above we get calibration parameter A= -216.9 and B= 0.68. (Fig. 10) Mass spectra of n-propyl benzene was calibrated using the above calibration parameter and shown in Fig. 11.

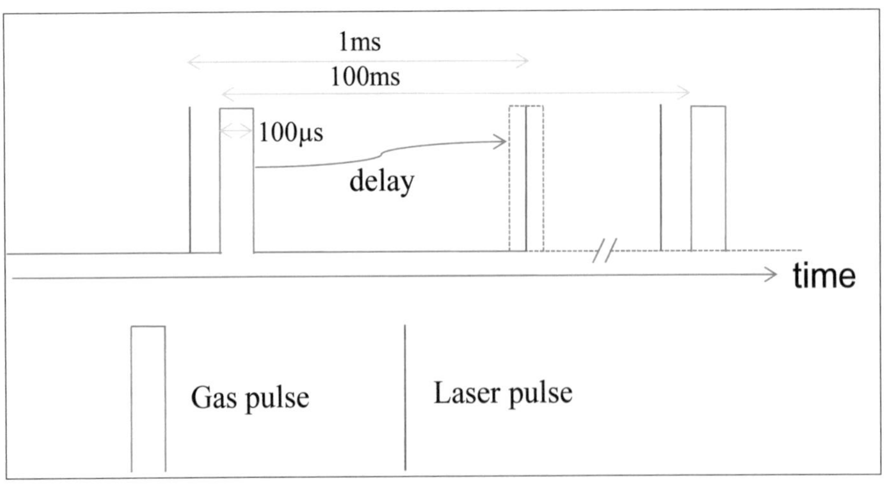

Fig. 7. Synchronization of pulsed molecular beam and laser pulses.

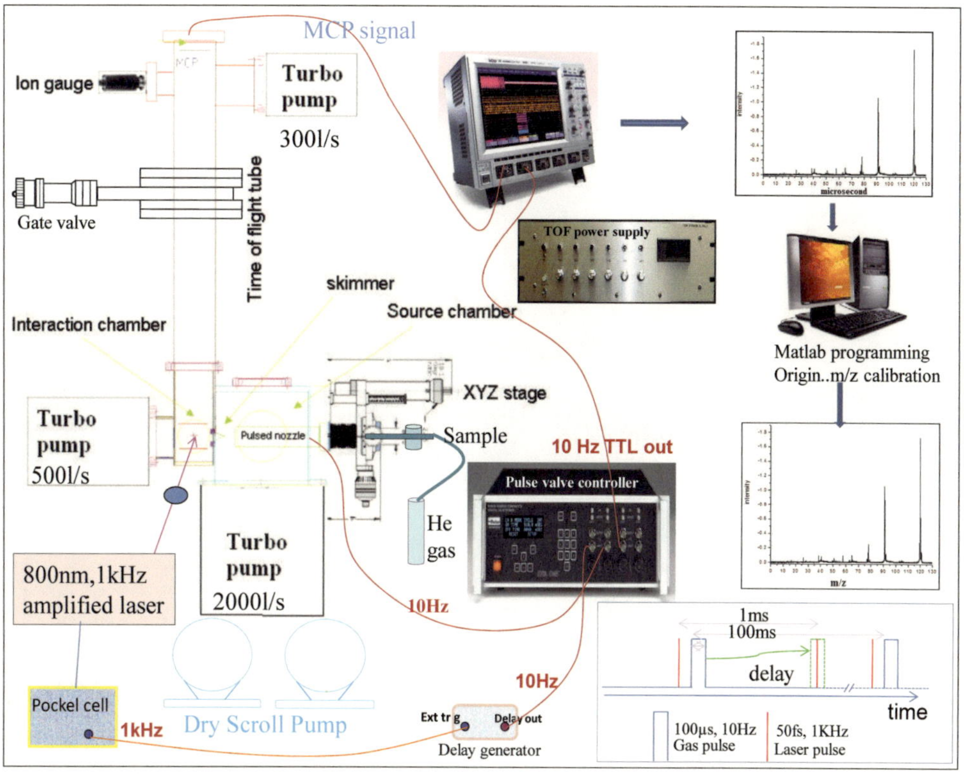

Fig. 8. Complete experimental setup for data acquisition.

Fig. 9. Picture of our supersonic molecular beam linear time of flight mass spectrometer.

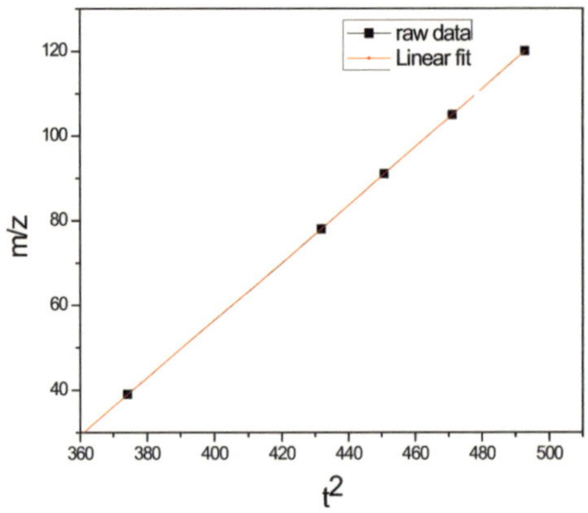

Fig. 10. Calibration line and its linear fit.

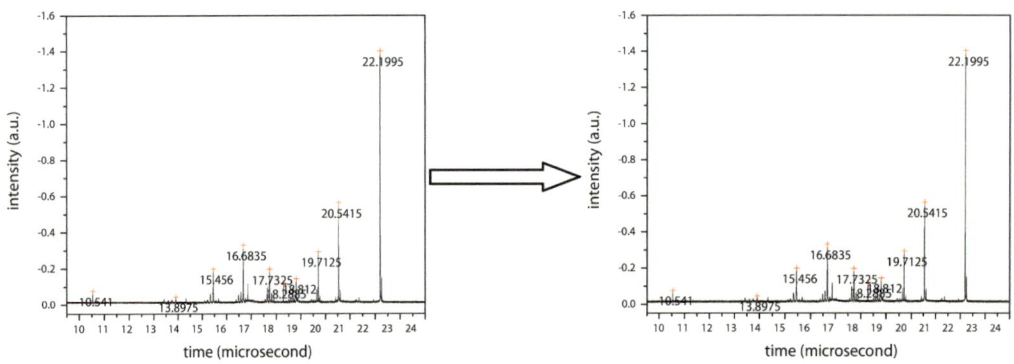

Fig. 11. Calibration of mass spectra of n-propyl benzene.

In mass spectrometry it is conventional to measure resolving power by the ratio of $m/\Delta m$ where Δm is a discernable mass difference. In TOFMS it is convenient to work in the time domain (Guilhaus, 1995; Opsal et al. 1985). Thus the resolving power $m/\Delta m$ can be measured in terms of $t/\Delta t$ as follows:

Since $m \propto t^2$, $m=At^2$, and $dm/dt=2At$ where A is constant;

$$dm/m=2dt/t$$

$$\text{Thus } m/\Delta m=t/2\Delta t$$

The finite time interval Δt, is usually the full width at half maximum (height) of the peak (FWHM).

Laser System:

The Laser system used in this experiment is a Ti:Sapphire multipass amplifier (Odin, Quantronix Inc.), which operates at 800 nm with 50 fs FWHM pulses at 1 kHz with energy of ~1 mJ. It is seeded with a homebuilt Ti:Sapphire Oscillator (starting point: K&M Labs Inc. oscillator kit). The oscillator is pumped by a Nd:YVO4 (Verdi 5, Coherent Inc.), resulting in femtosecond pulses with a centre wavelength of 800 nm and a spectral bandwidth of 50 nm FWHM at 94 MHz repetition rate and an average power of 400 mW. The oscillator output is stretched and then fed into the Ti:Sapphire multipass amplifier which is pumped by the second harmonic of Nd:YAG laser operating at a repetition rate of 1 kHz (Corona, Coherent Inc.). The chirped ultrafast laser pulses can be easily produced from our suitably modified compressor setup for the amplified laser system. As we increase the spacing between the compressor gratings relative to the optimum position for minimum pulse duration of 50 fs, we generate a negatively chirped pulse. Conversely, we obtain the positively chirped pulse by decreasing the inter-grating distance. Pulse durations were measured using a home-made intensity auto-correlation for the transform-limited pulse, as well as for the various negatively chirped and positively chirped pulses (Fig. 12). The pulses were further characterized by second harmonic frequency resolved optical gating (SHG-FROG) technique. Fig. 13a shows a typical SHG-FROG trace of our near transform-limited pulse that was collected using GRENOUILLE (Swamp Optics Inc.). In Fig. 13b retrieved spectra and phase of the transform limited pulse is shown. From this Fig. 13b is found that phase of the laser pulse is constant within the bandwidth of the laser pulse and hence it is made sure that at optimal grating distances we get the shorter pulse which is transform limited. In Fig. 14a, we show the SHG spectrum of the of the transform limited pulses after frequency doubling with 50 µm type-1 BBO crystals as well as the spectrum of the transform-limited pulse collected with a HR-2000 spectrometer (Ocean Optics Inc.). From the Fig. 14b, FWHM of the near transform limited pulse is found to be ~18 nm. The minimum time duration of a transform limited pulse giving a spectrum with $\Delta\lambda$ (18 nm) at FWHM, central wavelength (800 nm) and the speed of light (m/s) c: $\Delta t = K\frac{\lambda_0^2}{\Delta\lambda \cdot c}$ where K is the time-bandwidth product (K=0.441 for Gaussian pulse), Δt is found to be 50 fs for 18 nm bandwidth pulse which is our transform limited pulse.

The laser pulses are then focused with a lens (focal length = 50 cm) on a supersonically expanded molecular beam of n-propyl benzene at the centre of a time of flight chamber. The polarization of the laser was horizontal as it enters the mass spectrometer and is perpendicular to ion collection optics. The Mass spectra from our particular beam chamber constructed with dry-scroll pumps and turbo-molecular pump as described above has the advantage that it does not contain any extraneous water and hydrocarbon peaks and thus has better sensitivity for organic samples as reported here.

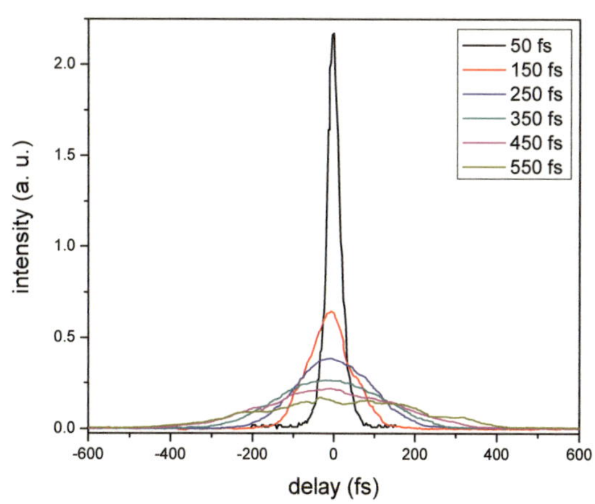

Fig. 12. The various autocorrelation traces of the pulse for different chirps.

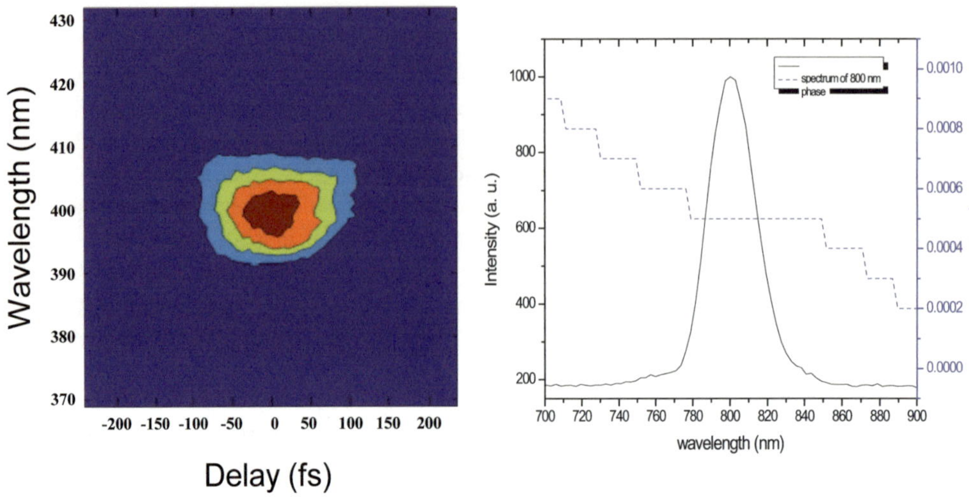

(a) (b)

Fig. 13. (a) SHG-FROG trace of our near transform limited 50 fs pulse, (b) retrived spectra and phase of the transform limited pulse.

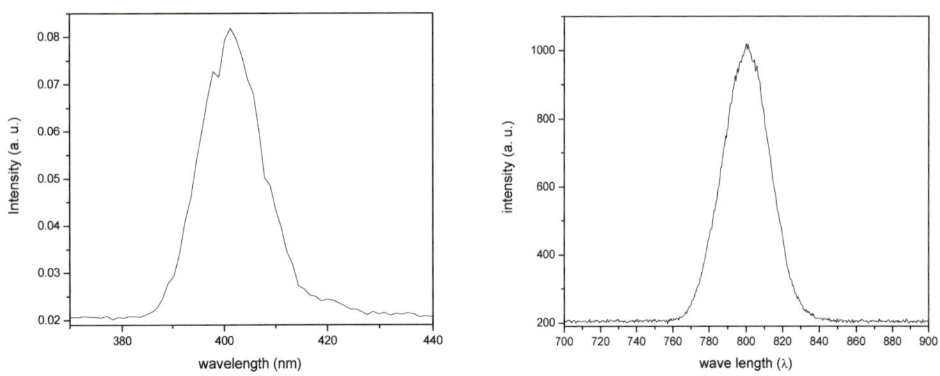

Fig. 14. (a) SHG Spectrum of transform limited pulse, (b) spectrum of the transform limited pulse.

Generation of Femtosecond chirped pulses:

Femtosecond chirped pulse can be easily generated by the compressor setup. This pulse increases its frequency linearly in time (from red to blue). In analogy to bird sound this pulse is called a "chirped" pulse. Our compressor setup (Fig. 15) consists of a pair of grating and a high reflective mirror. Our compressor gratings have 600 grooves/mm (Newport), and have throughput efficiency of 60%. One of the gratings is placed on a translation stage. By changing the distances between the two gratings and carefully aligning the optical paths at some optimal position a shorter near transform limited 50 fs laser pulse is generated. The transform limited condition is confirmed by characterizing the laser pulses at transform limited condition by measuring the SHG-Frog and autocorrelation trace. As we increase the spacing between the two gratings relative to the optimum position we can generate negatively chirped pulse. Conversely if we decrease the inter-grating distance we can generate positively chirped pulse.

An incident laser pulse of spectrum E_0 (ω) will be shaped by the compressor setup and spectrum of the output will be E_0 (ω)$e^{i\varphi\omega}$. For a light pulse which is centered around ω_0 having a reasonably small bandwidth, the total phase can be expanded around ω_0 to second order in ω:

$$\varphi(\omega) \approx \varphi(\omega_0) + \frac{1}{1!}\frac{\partial \varphi}{\partial \omega}\bigg|_{\omega=\omega_0}(\omega-\omega_0) + \frac{1}{2!}\frac{\partial^2 \varphi}{\partial \omega^2}\bigg|_{\omega=\omega_0}(\omega-\omega_0)^2$$

Here, the second order term is responsible for group velocity dispersion. In fact, $\beta = \frac{\partial^2 \varphi}{\partial \omega^2}\Big|_{\omega=\omega_0}$ is linear chirp coefficient (chirp parameter in the frequency domain) introduced by the compressor and is defined as second derivative of the spectral phase at the center frequency. The linear chirp coefficient (β) can be calculated using the equation:

$$\tau^2 = \tau_0^2 + \left[\frac{4\beta \ln 2}{\tau_0}\right]^2$$

where τ is the pulse duration of the chirped laser pulse and τ_0 is the chirp-free pulse duration of the transform limited pulse in FWHM. Pulse durations were measured by intensity autocorrelation technique. The experimental error in the chirp value calculated from the equation mentioned above is about ±9%.

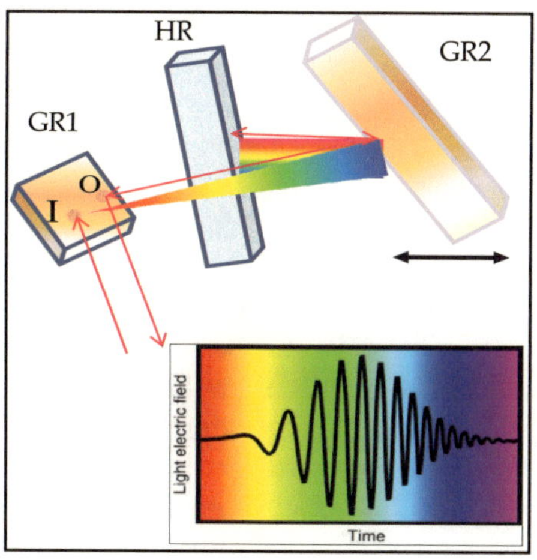

Fig. 15. Compressor setup, GR1, GR2- Garting, HR-high reflective mirror.

3. Results and discussions

Our TOF design provides high resolution when it is used in conjunction with our skimmed supersonic molecular beam chamber. Fig. 16 shows the time of flight mass peak of n-propyl beanzene cation. This was taken using femtosecond laser photo-ionization at 800 nm. The resolution is defined as R = t/(2Δt), where t is the total flight time of the ion packet and Δt is the FWHM of the peak. In the case n-propyl benzene cation resolution R is found to be around 1113. (Fig. 16) With careful adjustment of the voltages of the time of flight power supply and nozzle to skimmer distance to prevent turbulence on the skimmer, we have observed ion packet widths of ~10 ns. Resolution R = t/2Δt = 22.1995/(2×0.0097) = 1113, our mass spectra can resolve the adjacent mass peaks in the range of m/z from 0 to 1113 amu.

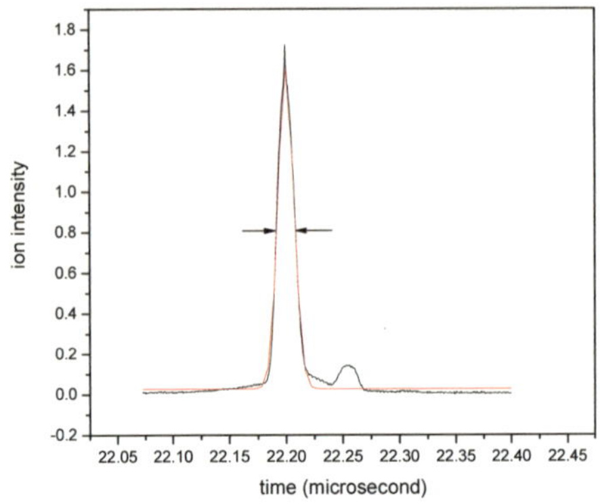

Fig. 16. Ion peak of n-propyl benzene cation

It has been known that the fragmentation processes in polyatomic molecules induced by an intense ultrafast laser field can sometimes exhibit sensitive dependence on the instantaneous phase characteristics of the laser field (Itakura et al., 2003; Pestrik et al., 1998; Mathur & Rajgara, 2003; Lozovoy et al., 2008, Goswami et al., 2009). Depending on the change in sign the chirped laser pulses, fragmentation could be either enhanced or suppressed. Controlling the outcome of such laser induced molecular fragmentation with chirped femtosecond laser pulses has brought forth a number of experimental and theoretical effects in the recent years. However, efforts are continuing for a specific fragment channel enhancement, which is difficult since it also is a function of the molecular system under study. Here we report the observation of a coherently enhanced fragmentation pathway of n-propyl benzene, which seems to have such specific fragmentation channel available. We found that for n-propyl benzene, the relative yield of $C_3H_3^+$ is extremely sensitive to the phase of the laser pulse as compared to any of the other possible channels. In fact, there is almost an order of magnitude enhancement in the yield of $C_3H_3^+$ when negatively chirped pulses are used, while there is no effect with the positive chirp. Moreover, the relative yield of all the other heavier fragment ions resulting from interaction of the strong field with the molecule is not sensitive to the sign of the chirp, within the noise level.

Study of aromatic hydrocarbons has indicated different fragmentation channels (DeWitt et al., 1997). A systematic study of aromatic compounds with increasing chain-length of substituent alkyl groups has indicated that the fragmentation process is enhanced as the chain-length of the alkyl substituent on the benzene ring increases. We have chosen to apply chirped pulse fragmentation control on certain members of these systematically studied aromatic compounds. In general, as reported previously for benzene and toluene, p-nitro toluene, we also find that chirping favors the formation of smaller fragments as compared to the heavier ones. However, n-propyl benzene has the unique property of enhancing a particular fragmentation channel under the effect of chirp.

We record the TOF mass spectra (Fig. 17) of n-propyl benzene using linearly chirped and unchirped ultrafast laser pulses with constant average energy of ~200 mW. Next, we compared the corresponding peaks in mass spectra by calculating their respective integrals under the peaks and normalizing them with respect to the molecular ion. These results also conform to the case when we just compare the simple heights of the individual peaks. When the linear chirp of the laser pulse is negative, the relative yields of the smaller fragment ion, such as, $C_3H_3^+$ (mass to charge ratio, i.e., m/z = 39) is largely different from those obtained using transform-limited pulses or from the positively chirped pulses, as reflected in the Fig. . The relative yield of $C_3H_3^+$ reaches maximum when the linear chirp coefficient (β, calculated by using the equation as mention earlier) is -8064 fs^2 and pulse duration is of 450 fs. We would like to point out that the fragment ion $C_6H_5^+$ (m/z = 77) yield is more when the chirp is positive (β=+6246fs^2), and this can be attributed to a different fragmentation pathway (Fig.18). However, the observation of enhancement for only one chirp sign implies that the observed enhancements are not due to the pulse width effects, they rather depend on the magnitude and sign of the chirp. Hence coherence of the laser field plays an important role in this photofragmentaion process. It is also seen that relative yields of the heavier fragments like $C_7H_7^+$ (m/z = 91) is not affected by the sign of the chirp.

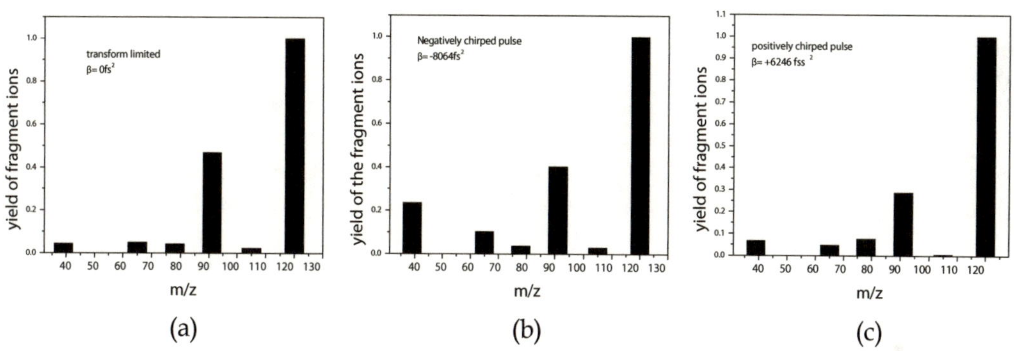

Fig. 17. Effect of chirping on mass spectra of n-propyl benzene, (a) β=0 fs^2, (b) β=-8064 fs^2, (c) β=+6246 fs^2.

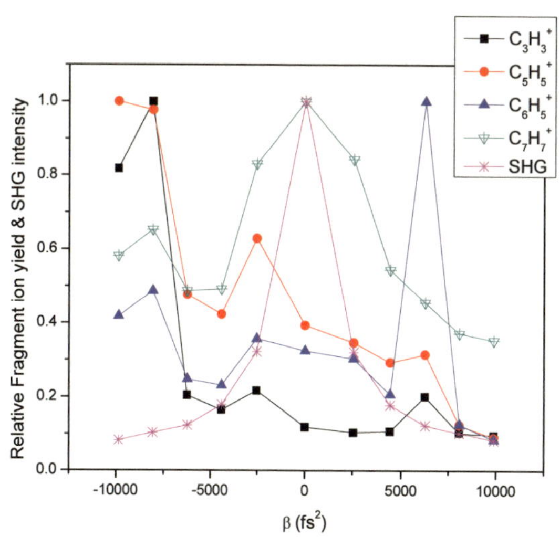

Fig. 18. Effect of chirping the laser pulse on the relative yield of different fragment ions shown in comparison to the integrated SHG intensity at the respective chirps.

The relative yield of $C_7H_7^+$ decreased in both the directions of the chirp and is at its maximum when the pulse is transform limited, indicating that the fragment yield only depends on the laser peak intensity as dictated by its pulse width. We have also seen that the integrated SHG intensity at different chirp is symmetrically decaying around 0 fs^2 (Fig. 18), which confirms that there is nothing systematic in the laser pulse causing the enhancements in the fragmentation process.

4. Conclusions

We have built a molecular beam chamber with linear time of flight mass spectrometer, which can be combined with femtosecond laser in a novel way to study the femtosecond coherent control of supersonically cooled molecules. Our system is constructed with dryscroll pumps and turbo-pumps and is completely free from oil, which is very necessary for any femtosecond laser laboratory. In fact, this oil-free capability ensures that the obtained mass spectra is free from extraneous hydrocarbon peaks, which, in turn, makes our system better sensitive in studying coherent control of organic samples as reported in this book chapter.

We have demonstrated that the phase characteristics of the femtosecond laser pulse play a very important role in the laser induced fragmentation of polyatomic molecules like n-propyl benzene. The use of chirped pulse leads to sufficient differences in the fragmentation pattern of n-propyl benzene, so that it is possible to control a particular fragmentation channel with chirped pulses. Overall, as compared to the transform-limited pulse, negatively chirped pulses enhance the relative yield of $C_3H_3^+$, $C_5H_5^+$, while the relative yield of $C_6H_5^+$ increases in case of positively chirped pulse. In fact, for the $C_3H_3^+$ fragment, the enhancement is almost 6 times for the negatively chirped pulse (β = -8064 fs^2) as compared to that of the transform-limited pulse.

5. References

Rousseau, D. L. J. (1966). Laser chemistry, *Chem. Educ.*, 43, 566-570.

Smalley, R. E.; Wharton, L.; Levy, D. H. (1977). Molecular optical spectroscopy with supersonic beams and jets, *Acc. Chem. Res.*, 10, 139-145.

Levy, D. H. (1981). The spectroscopy of very cold gases, *Science*, 214, 263-269.

Zewail, A. H. (1980). Laser Selective Chemistry: Is It Possible? *Phys. Today*, 33, 27.

Bloembergen, N.; Zewail, A. H. (1984). Energy redistribution in isolated molecules and the question of mode-selective laser chemistry revisited. New experiments on the dynamics of collisionless energy redistribution in molecules possibilities for laser-selective chemistry with subpicosecond pulses, *J. Phys. Chem.*, 88, 5459-5465.

Brixner, T.; Gerber, G. (2003). Quantum control of gas-phase and liquid-phase femtochemistry, *ChemPhysChem*, 4, 418-438.

Goswami, D. (2003). Optical pulse shaping approaches to coherent control, *Phys. Rep.*, 374, 385-481.

Levis, R. J.; Menkir, G. M.; Rabitz, H. (2001). Selective Bond Dissociation and Rearrangement with Optimally Tailored, Strong-Field Laser Pulses, *Science*, 292, 709-713.

Ahn, J.; Weinacht, T. C.; Bucksbaum, P. H. (2000). Information Storage and Retrieval Through Quantum Phase, *Science*, 287, 463-465.

Vivie-Riedle, R. de; Troppmann, U. (2007). Femtosecond Lasers for Quantum Information Technology, *Chem. Rev.*, 107, 5082-5100.

Herschbach, D. R. (1987). Molecular Dynamics of Elementary Chemical Reactions, *Angew Chem. Int. Ed. Engl.*, 26, 1221-1243.

Assion, A.; Baumart, T.; Bergt, M.; Brixner, T.; Kiefer, B.; Sayfried, V.; Strehle, M.; Gerber, G. (1998). Control of chemical reactions by feedback-optimized phase-shaped femtosecond laser pulses, *Science*, 282, 919-922.

Daniel, C.; Full, J.; Gonzalez, L.; Kaposta, C.; Krenz, M.; Lupulescu, C.; Manz, J.; Minemoto, S.; Oppel, M.; Rosendo-Francisco, P.; Vajda, Š.; Wöste, L. (2001). Analysis and control on η_5-CpMn$(CO)_3$ fragmentation processes, *Chem. Phys.*, 267, 247-260.

Levis, R. J.; Rabitz, H. A. (2002). Closing the Loop on Bond Selective Chemistry Using Tailored Strong Field Laser Pulses, *J. Phys. Chem. A*, 106, 6427-6444.

Daniel, C.; Full, J.; González, L.; Lupulescu, C.; Manz, J.; Merli, A.; Vajda, S.; and Wöste, L. (2003). Deciphering the reaction dynamics underlying optimal control laser fields, *Science*, 299, 536-539,

Weinacht, T. C.; White, J. L.; and Bucksbaum, P. H. (1999). Towards Strong Field Mode-Selective Chemistry, *J. Phys. Chem. A*, 103, 10166.

Itakura, R.; Yamanouchi, K.; Tanabe, T.; Okamoto, T.; and Kannari, F. (2003). Dissociative ionization of ethanol in chirped intense laser fields. J. Chem. Phys. 119, 4179.

Melinger, J.S.; Gandhi, S.R.; Hariharan, A.; Tull, J.X.; Warren, W.S. (1992). Generation of narrowband inversion with broadband laser pulses, *Phys. Rev. Lett.*, 68, 2000-2003.

Krause, J. L.; Whitnell, R. M.; Wilson, K. R.; Yan, Y.; Mukamel, S. (1993). Optical control of molecular dynamics: Molecular cannons, reflectrons, and wave-packet focusers, *J. Chem. Phys.*, 99, 6562.

Cerullo, G.; Bardeen, C. J.; Wang, Q.; Shank, C. V. (1996). High Power Femtosecond Chirped Pulse Excitation of Molecules in Solution, *Chem. Phys. Lett.* 262, 362-368.

Pastirk, I.; Brown, E. J.; Zhang, Q.; and Dantus, M. (1998). Quantum control of the yield of a chemical reaction, *J. Chem. Phys.*, 108, 4375-4378.

Yakovlev, V. V.; Bardeen, C. J.; Che, J.; Cao, J.; Wilson, K. R. (1998). Chirped pulse enhancement of multiphoton absorption in molecular Iodine, *J. Chem. Phys.*, 108, 2309.

Cao, J.; Che, J.; Wilson, K. R. (1998). Intra-pulse dynamical effects in multi-photon processes: Theoretical analysis, *J. Phys. Chem.*, 102, 4284-4290.

Guilhaus, M. (1995). Principles and instrumentation in time-of-flight mass spectrometry: Physical and instrumental concepts, *J. Mass Spectrometry*, 30, 1519-1532.

Opsal, R. B.; Owens, K. G.; Reilly, J. P. (1985). Resolution in the Linear Time-of-Flight Mass Spectrometer, *Anal. Chem.*, 57, 1884-1889.

Mathur, D.; Rajgara, F. A. (2004). Dissociative ionization of methane by chirped pulses of intense laser light, *J. Chem. Phys.*, 120, 5616.

Lozovoy, V. V.; Dantus, M. (2008). Femtosecond laser-induced molecular sequential fragmentation of para-nitrotoluene, *J. Phys. Chem. A*, 112, 3789-3812.

Goswami, T.; Karthick Kumar, S. K.; Dutta, A.; Goswami, D. (2009). Control of laser induced molecular fragmentation of n-propyl benzene using chirped femtosecond laser pulses, *Chem. Phys.* 360, 47-52.

DeWitt, M. J.; Peters, D. W.; Levis, R. J. (1997). The Photoionization/Dissociation of Alkyl Substituted Benzene Molecules Using Intense Near-Infrared Radiation, *Chem. Phys.* 218, 211-223.

UV-Laser and LED Fluorescence Detection of Trace Organic Compounds in Drinking Water and Distilled Spirits

Anna V. Sharikova and Dennis K. Killinger
University of South Florida
USA

1. Introduction

Current methods for the analysis of drinking water and many other liquids often call for the use of reagents and may require extensive sample preparation (American Public Health Association, 1989). For the case of water supplies and water treatment plants, this analysis is usually carried out once every few days or weeks (Killinger & Sivaprakasam, 2006). Most of the analysis is usually conducted using classical analytical chemical techniques, such as mass spectrometry, liquid chromatography, or fluorescence based or tagged reagents (Crompton, 2000). These analytical techniques are sensitive and provide accurate assessment of the chemistry related to the quality of the liquids. However, they often take considerable time and are usually not performed in real-time, especially for the case of a flowing process line. On the other hand, previous fluorescence spectroscopic measurements of ocean water by Coble showed that deep-UV excitation of naturally occurring organic compounds in water can yield significant and unique fluorescence signals in the near UV to visible wavelength range without the need to use additional reagents or sample preparation (Coble, 1996; Coble, 2007). As a result, we have been studying deep-UV laser-induced-fluorescence techniques for the detection of trace species in water and other liquids with the goal of using the natural fluorescence of trace species in the water or liquid samples and being able to provide readings within the time span of a few seconds.

Toward this goal, we have developed a reagentless deep-UV laser and UV-LED induced fluorescence (LIF) system to detect and continuously observe in real time trace levels of colored dissolved organic matter (CDOM) or Dissolved Organic Compounds (DOCs) in water and distilled spirits, such as drinking water, and related water/alcohol based liquids with a sensitivity exceeding that of commercial spectrofluorometers. Our system has been used to detect ppb trace levels of plasicizer Bisphenol-A (BPA) that have leached into drinking water, and has detected and monitored trace levels of DOCs within ocean currents (Killinger & Sivaprakasam, 2006; Sivaprakasam et al. 2003; Sivaprakasam & Killinger, 2003). Recently, our LIF system has been used to measure fluorescence of reverse osmosis processed water and different types of drinking water (Sharikova & Killinger, 2007; Sharikova & Killinger, 2010). These LED/LIF applications have now been extended to additional water related samples, including humic acid samples, tannic acid and chlorinated water samples, juices, coffee, and several wines and distilled spirits; these recent results are presented in this paper.

Our compact LIF system used either frequency tripled or forth harmonic diode pumped Nd:YAG lasers operating at 266 nm and 355 nm, or deep-UV LEDs (265 nm, 300 nm, 335 nm, and 355 nm) as UV excitations sources. The emitted fluorescence was measured over the range of 240–680 nm. Strong emission near 450 nm was observed for the DOCs in water, while emission bands near 340 nm were evident from distilled spirits and wine. It is important to note that one of the main advantages of using a deep-UV excitation wavelength, such as 266 nm, is that the emission fluorescence is separated in wavelength from the Raman emission of water (near 310 nm for 266 nm excitation), and thus yields greater sensitivity and wavelength selectivity than previous systems using lasers operating near 400 to 550 nm. In addition, as a point of reference, our laser based LIF system had a detection sensitivity for the fluorescence standard solution of quinine sulfate on the order of 0.1 ppb. The average laser power was approximately 30 times that of the LED, but differences in the signal intensity due to the difference in the laser and LED excitation intensity were consistent with theory.

Our studies show that deep-UV light emitting diodes (LEDs) are good alternative light sources for our LIF system, which would make the apparatus cheaper and more compact. It should be noted that the research presented in this paper is directed toward the development of new optical spectroscopic measurement techniques which have the potential to offer enhanced capabilities over conventional water monitoring and liquid analysis. However, while its sensitivity has been shown to be in the sub-parts per billion for standard fluorescing compounds used in fluorescence research, such as quinine sulfate, it needs to be further quantified and evaluated against conventional analytical chemistry instruments before it can be used as an on-line analytical instrument for water monitoring. Such comparisons are currently being conducted and will be reported later.

2. Experimental setup

Our Laser and LED induced fluorescence system is similar to that of a conventional spectrofluorometer, but has a sensitivity several orders of magnitude better. Commercial spectrofluorometers often use UV lamps and wavelength selecting spectrometers for the emission source, and single or double monochromators with Photo-multiplier Tubes or CCD detecting arrays for fluorescence detection (Albani, 2007). Often the signal processing is conducted using a chopped CW beam and lock-in amplifier signal detection. Our LIF system uses a high PRF (pulse-repetition-frequency) laser running at about 8,000 pulses/second as the excitation source (or a pulsed LED source running at about 330 pulses/second), and a high-speed boxcar integrator which detects and stores the fluorescence photon signal for each pulse. In addition, our system uses multiple excitation beams and double-pass collection optics to increase the fluorescence signal. Our past work has shown that this combination has enhanced the sensitivity of our laser-induced-fluorescence system by two to three orders of magnitude over conventional spectrofluorometers, depending upon the spectrometer and optical detector configuration used (Sivaprakasam & Killinger, 2003). Details of our current LIF system follow.

2.1 Description of the apparatus

Our fluorescence measurements were performed using a system shown in Fig. 1. The schematic diagram of the apparatus is shown in Fig. 2. The light source was one of the following: a microchip laser, 266 nm or 355 nm (JDS Uniphase, Models NU-10110-100 and NV-10110), or a LED operating at 265 nm, 300 nm, 335 nm or 355 nm (Sensor Electronic Technology, Inc., UVTOP® series). A silicon APD photodetector (New Focus, Model 1621) was used to trigger data acquisition with the laser source; the LEDs were software-triggered. The laser beam passed through the optical quartz cell (Spectrocell Inc., Model RF-3010-F) several times, for which plane mirrors on the sides of the cell were used. The fluorescence signal was collected at a right angle to the excitation beam with a concave mirror (Optosigma, Model 035-0130) and a fused silica lens (Optosigma, Model 014-0490). It passed through one of the 19 different narrow (10nm) bandpass optical filters ranging from 265 nm to 685 nm before being focused onto a PMT (Hamamatsu, Model H6780-03). One of the absorption cut-off filters (CVI Laser, Models CG-WG-280-2.00-2 and CG-WG-295-2.00-2) was used to block Rayleigh and Raman scattering. All filters were mounted on a stack of motorized filter wheels (CVI Laser, Models AB-302 and AB-304). The PMT signal was acquired by a gated integrator and boxcar averager unit (Stanford Research System, Model SR-250). Data collection and filter wheel control was handled by LabVIEW software through a computer interface unit (Stanford Research System, Model SR-245) and serial bus.

2.2 Laser and LED characteristics

The Q-switched microchip lasers produced 0.4 ns, 0.3-0.4 μJ pulses at a repetition rate of 8 kHz. The beam size was on the order of 1 mm.

The LEDs were operating in a 10 μs, 50 mA drive current regime with a 330 Hz repetition rate. The output light pulse energy was approximately 7 nJ (with the exception of 22 nJ for the 355 nm LED). It was noted that the LEDs also had an out-of-band emission in the visible region. As an example, this can be seen in Fig. 3 that shows the measured output power of the 265 nm LED (LED265) as a function of wavelength and of drive current. This is a log plot of the intensity and shows that the out-of-band LED emission had a peak value of about 1% compared to the peak emission at 265 nm.

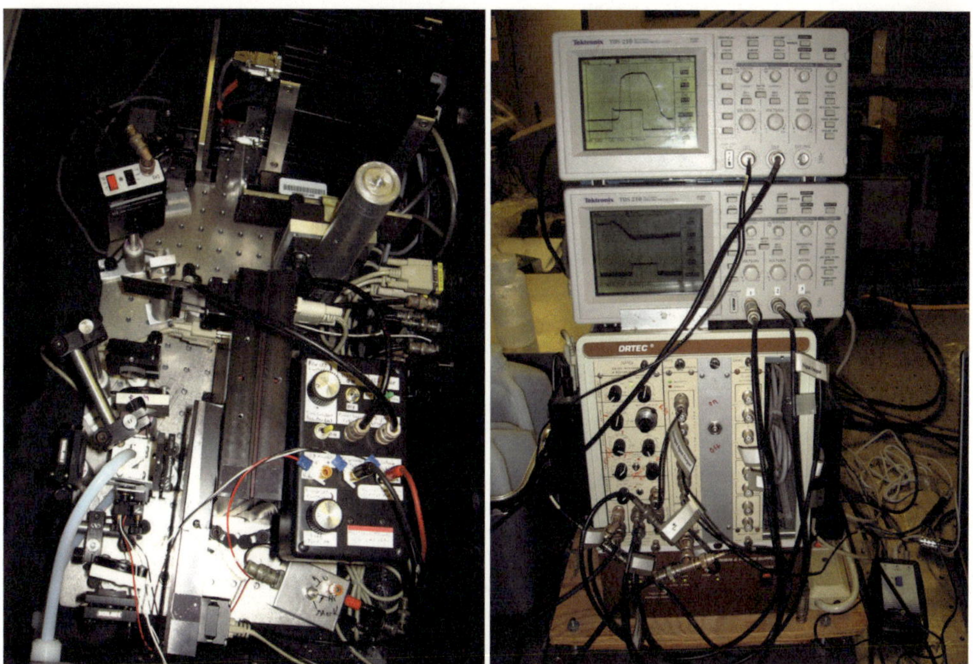

Fig. 1. Experimental system: optics box (left) and electronics box (right)

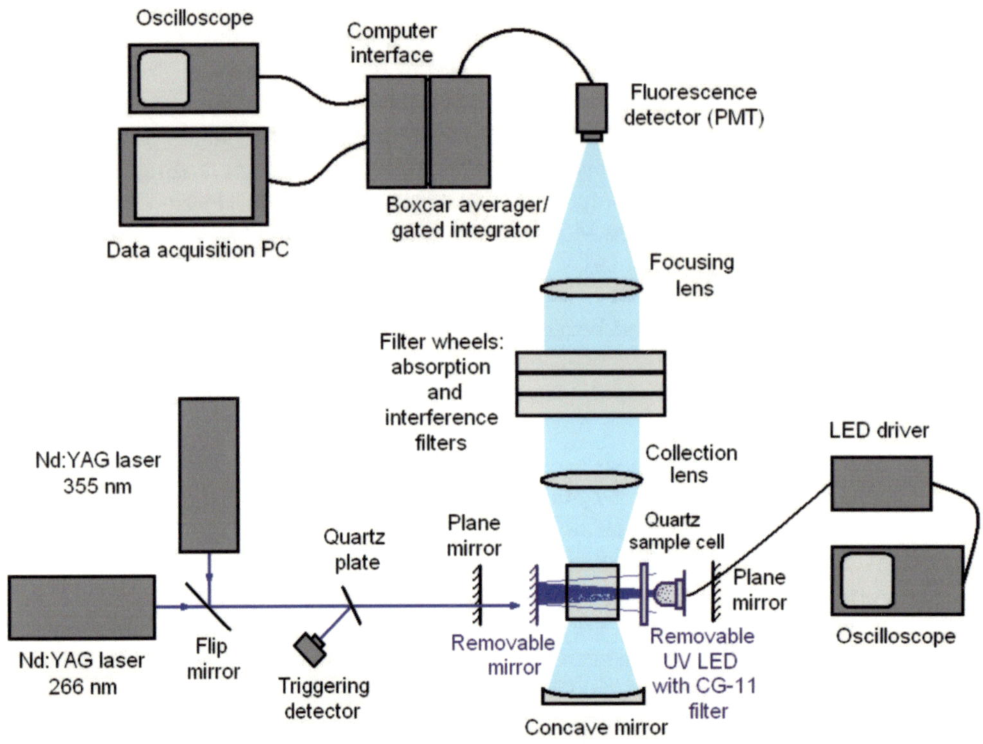

Fig. 2. Schematic diagram of the experimental apparatus

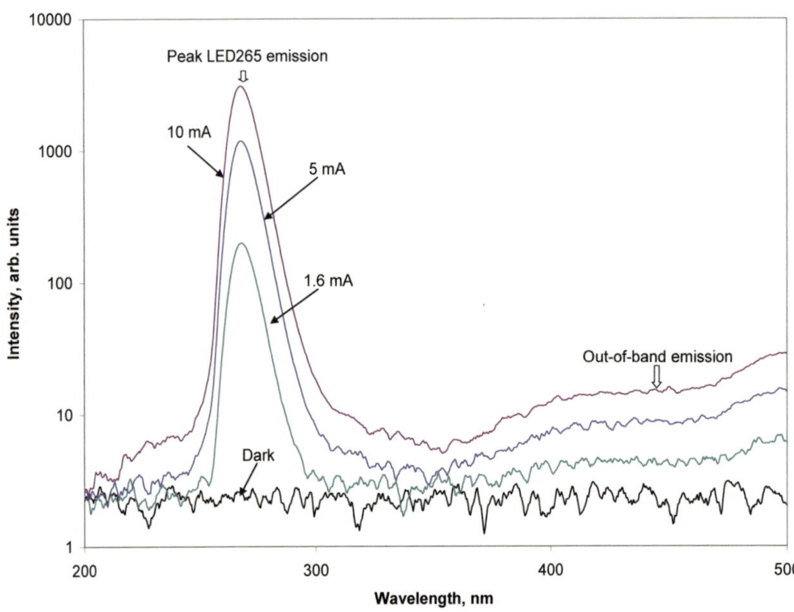

Fig. 3. Emission spectra of LED265 for different CW currents.

To block this out-of-band light, a VIS-blocking, UV-passing filter (CVI Model CG-UG-11) was used with our LED sources. Figure 4 shows the spectral output power of the LEDs with and without the filter using a linear scale for the intensity. The beam size within the sample cell was about 5 mm.

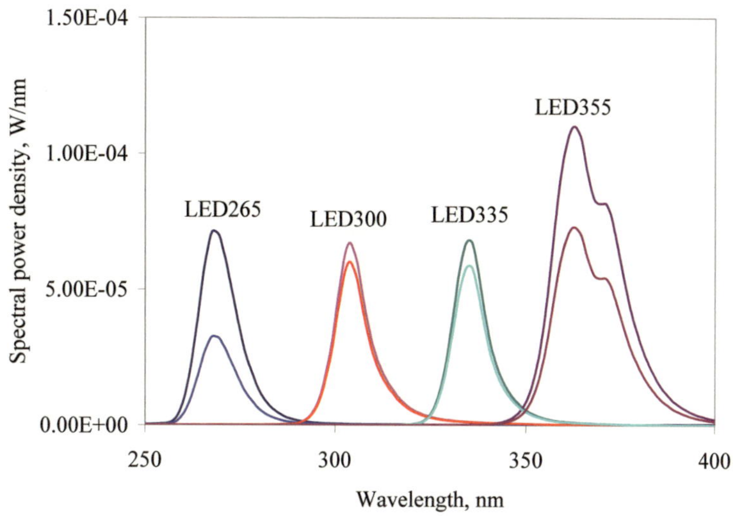

Fig. 4. LED spectral peak power density in 50 mA, 10 µs pulse regime (lower curves with CG-UG-11 filter)

3. Data and results

Our LIF system was used to measure a considerable variety of water related liquids including tap water, Reverse Osmosis (RO) treated ground water, and other water quality related substances. These results are shown in the following.

3.1 Experimental conditions and settings for liquid samples

The liquid samples were stored, when necessary, in the dark and cold. Water samples were not further processed. Wine samples were diluted to 10 mL per 1 L with distilled water. Humic substances (International Humic Substances Society, 1R101N, 1S103H, 1S104H) were prepared as 10 mg per 500 mL of distilled water.

A flow cell with a linear flow rate of 5 cm/s was used to minimize photobleaching. With each bandpass filter, 1000 measurements were taken, which lasted a few seconds per filter setting including filter switching time. Boxcar averaging setting was 300 samples. The sensitivity setting was adjusted for each sample to maximize the signal. The spectra were compensated for filter bandwidth and transmission, PMT quantum efficiency and gating integrator/boxcar averager sensitivity.

3.2 Reverse Osmosis processed ground water and drinking (tap) water

Ground water taken from a shallow well at USF was processed by a Reverse Osmosis unit at Prof. Carnahan's lab (Carnahan, 2006). The fluorescence spectra of ground water before and after RO treatment using 266 nm laser excitation is shown in Fig. 5. The broad peak observed around 470 nm in the untreated water is typical of the organic compounds usually present in such samples (Coble, 2007). After the RO treatment, the fluorescence signal decreased significantly, especially on the short-wavelength side. The signal from distilled water is shown for comparison.

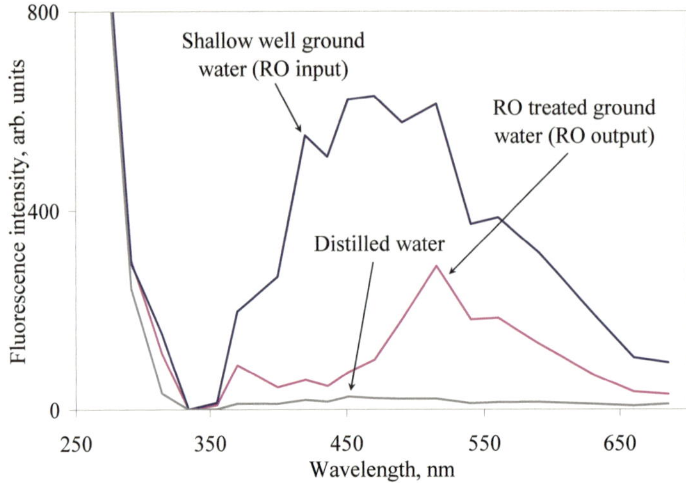

Fig. 5. Fluorescence of ground water before and after reverse osmosis treatment, 266 nm laser excitation

Tap water in our lab was continuously monitored for a period of a week (see Fig. 6). As can be seen, the flowing tap water, Fig. 6(a), had a greater range of variation than a sample recirculated through the system, Fig. 6(b). Certain repetitiveness of the running water signal might be indicative of the water usage patterns at our university. The initial growth in the recirculated signal was due to plastic leaching from the soft tubing used in the pump.

We have also tested tap water collected from different locations in the US (Fig. 7). All samples were taken directly from residential tap water except for the Tampa location, where an on-line water filter in a drinking fountain was present. For all samples, settings were the same during data acquisition. As can be seen, the fluorescence spectra obtained with 266 nm laser excitation, (a), and 265 nm LED, (b), were different only in overall intensity. Comparing the spectra from different locations, one can see that all possessed two large peaks centered around 420 and 460 nm, as well as smaller peaks on the sides. However, both the absolute and relative intensity of the peaks varied with the location, serving as an indication of the difference in both the total concentration and the species of organic compounds present in the sample.

The signal-to-noise ratio (SNR) was calculated for the set of tap water samples as a difference between peak fluorescence of the sample and the distilled water (reference) signals divided by double the standard deviation of 1000 measurements. The results were typically in the 300-900 range for the laser sources, and around 30-190 for the LEDs (Sharikova & Killinger, 2010). For example, at 266 nm, the laser pulse energy was about 100 times greater than the LED pulse energy, but the SNR values differed only by a factor of 10 being about 296 for the laser LIF and about 26 for LED excitation for the Ann Arbor water data. The reason for the stronger than expected signal with the LED excitation may be due to differences in the light excitation and fluorescence overlap volume in the LED configuration or sample photobleaching in the case of the lasers.

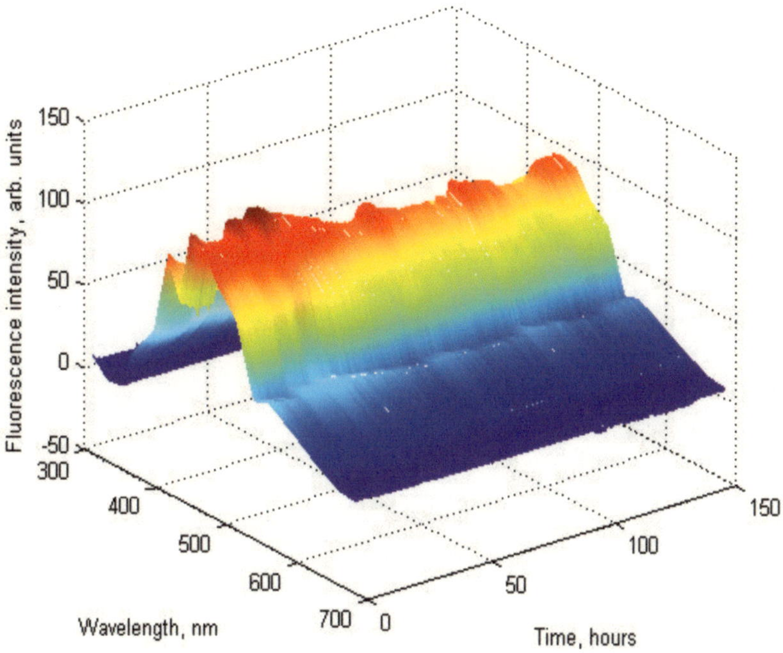

(a)

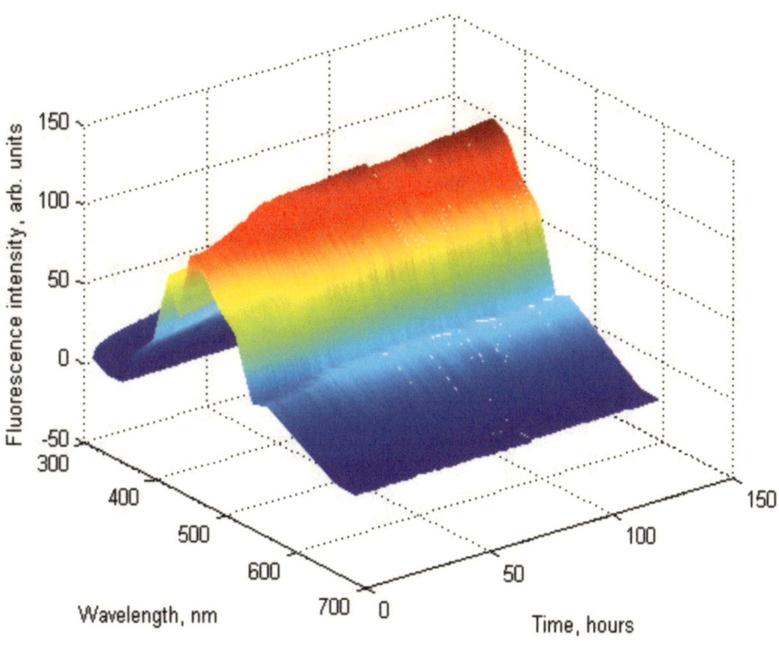

(b)

Fig. 6. Week long continuous tap water monitoring with 266 nm laser excitation: (a) running tap water; (b) recirculated tap water

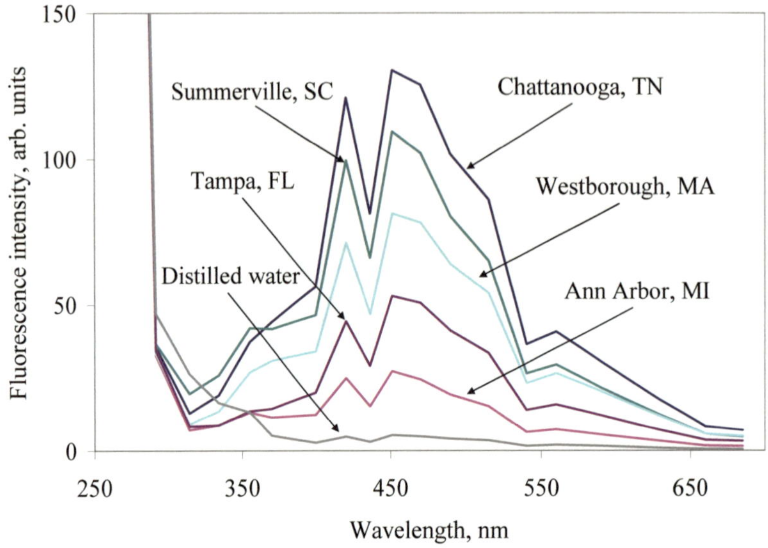

(a)

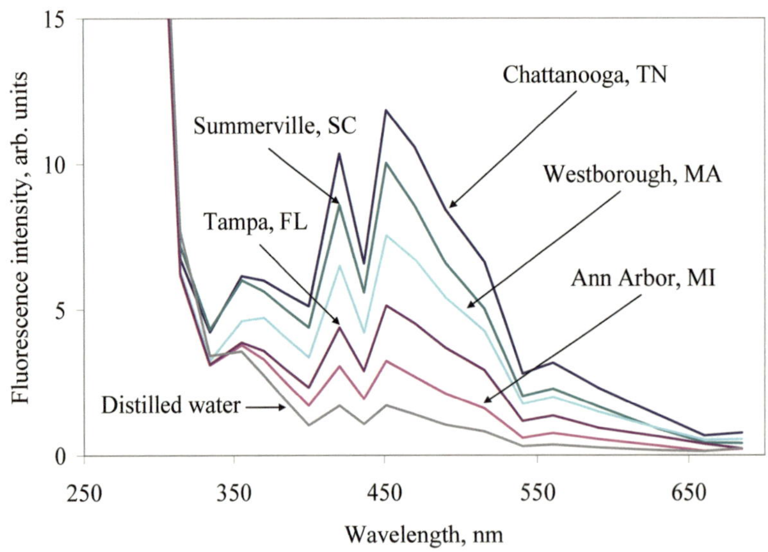

(b)

Fig. 7. Fluorescence of tap water with (a) 266 nm laser and (b) 265 nm LED excitation

3.3 Fluorescence spectrum of tannic acid and chlorine added to water sample

The LIF system was used to study long-term changes in the fluorescence of Total Organic Carbon (TOC) and the influence of chlorine on TOC fluorescence. Tannic acid is often used to represent TOC in water analysis measurements and is a specific type of tannin (plant polyphenol) (Hudson et al., 2007). Distilled water and trace solutions of organic compounds were recirculated in the portable LIF system for several hours. Tannic acid representing Total Organic Carbon (3 mg/L in deionized water) and chlorinated tannic acid (3 mg/L TOC and 12 mg/L Cl in deionized water) samples were prepared by Mr. Panagiotis Amitzoglou from the water processing laboratory of Prof. Audrey Levine (Levine, 2006). The solutions were produced by dilution of pure chemicals in nanopure deionized water. Concentrations of TOC and chlorine in the samples shown here were typical of those in drinking water.

The LIF signal for a sample of tannic acid is shown in Fig. 8 and was obtained using the 266 nm laser. As can be seen, the fluorescence in Fig. 8 had the strongest peak at 370 nm, the second-strongest at 420 nm, and a weaker peak at 451 nm. This made it quite distinct from the typical natural fluorescence of water, for example Fig. 7(a), which had a broad peak maximized at 451 nm.

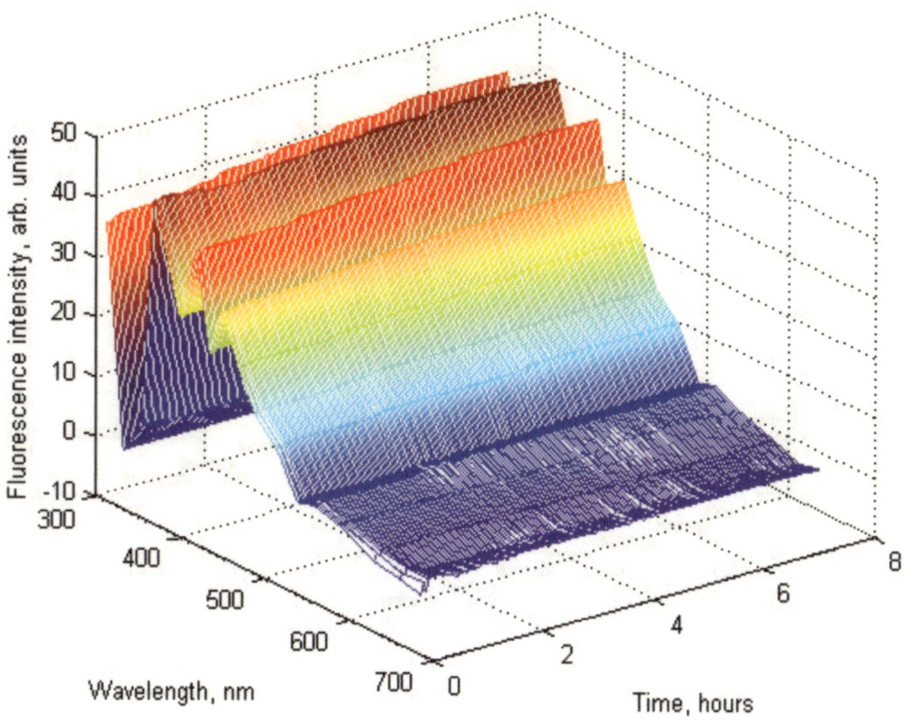

Fig. 8. LIF of tannic acid for 7 hours continuous monitoring; 266 nm excitation.

Fluorescence of a solution containing both tannic acid and chlorine is shown in Fig. 9 for 266 nm excitation; the solution was recirculated for about 5 hours. As can be seen, the spectrum for the 266 nm excitation was significantly different from that of tannic acid alone. The peak at 370 nm was completely suppressed, and the 420 nm peak was reduced drastically, so that the strongest fluorescence was observed at 451 nm.

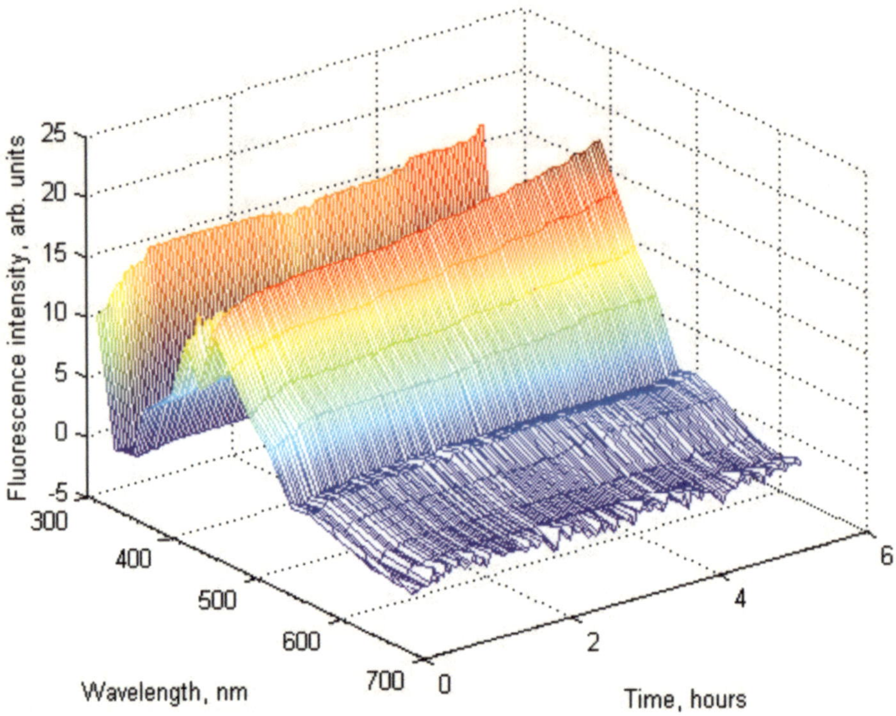

Fig. 9. LIF of tannic acid and chlorine for 5 hours continuous monitoring; 266 nm excitation.

3.3 Wine, juices, and distilled spirits

Several types of wine were diluted to ~1% with distilled water to minimize absorption and scattering in the sample cell. All spectra were taken under the same settings and conditions. Fluorescence spectra of these wine samples are shown in Fig. 10 (laser excitation) and Fig. 11 (LED excitation).

Comparing wine spectra (Fig. 10) with those of tap water (Fig. 7) one can see not only the difference in intensity, but also in the shape of the spectrum, particularly at 266 nm excitation. The emission peak at 350 nm dominates in the wine spectra of Fig. 10(a) using 266 nm excitation, but is possibly hidden by the scattering of the 355 nm excitation laser in Fig. 10 (b).

There is also a significant dependence on the excitation wavelength for the wine samples, as can be seen from Figs. 10 and 11. For example, at 266 nm excitation, the signal from sake is strongest, while fruit wines exhibit greater fluorescence at longer excitation wavelengths.

The laser- and LED-excited fluorescence spectra are generally consistent, with the exception of the Riesling wine that showed a much weaker secondary emission peak near 450 nm when the 265 nm LED was used. This might be due to the difference in the overlap between the sample absorption and the excitation line width, or be another indication of strong scattering at this excitation wavelength.

Finally, as part of applying our system toward common drinks, we also investigated the use of the LIF system for other drinkable liquids such as coffee and orange juice. Some of our results are shown in Fig. 12 using diluted samples. The data shows strong fluorescence emission near 350 nm and near 450 to 470 nm.

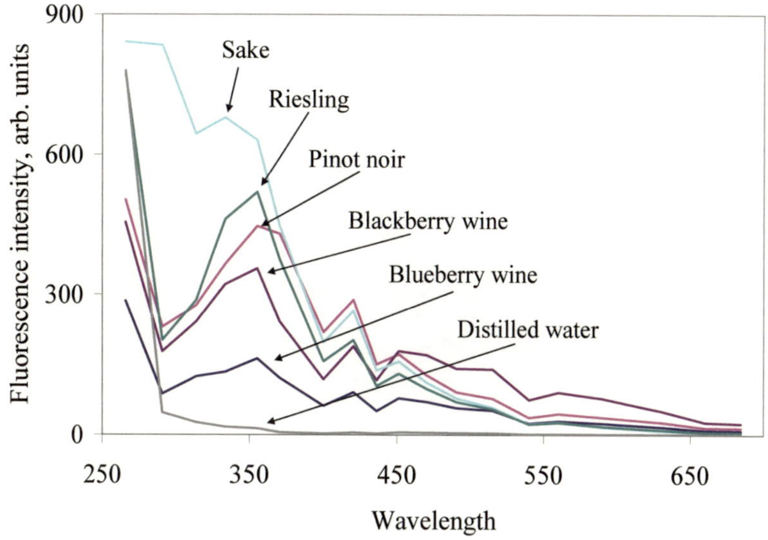

(a)

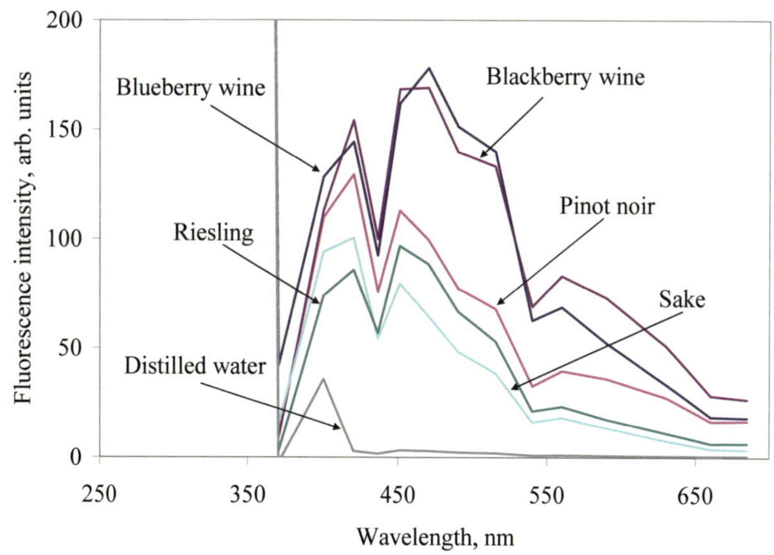

(b)

Fig. 10. Fluorescence of wine samples with (a) 266 nm and (b) 355 nm laser excitation

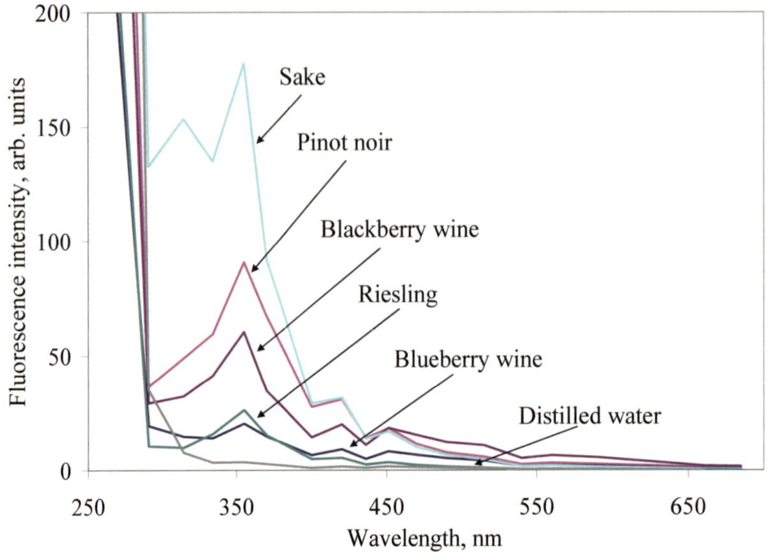

Fig. 11(a)

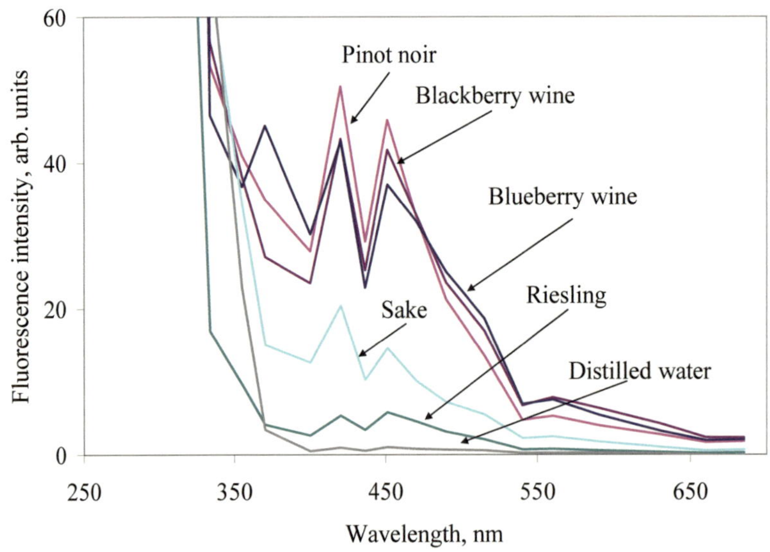

Fig. 11(b)

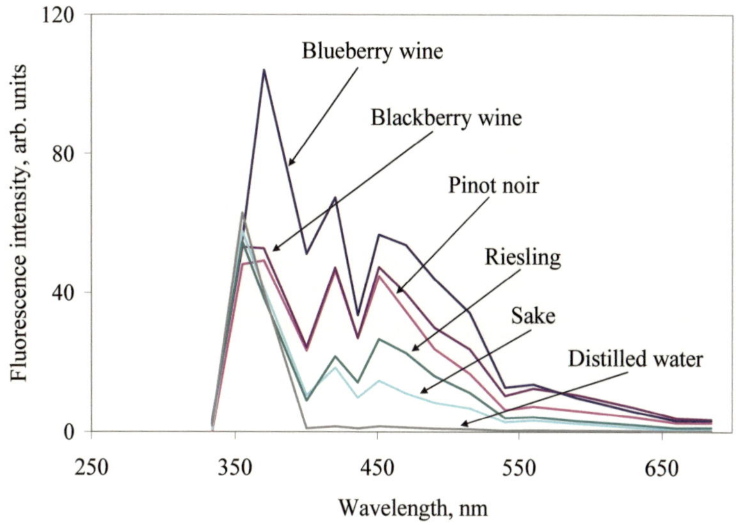

Fig. 11(c)

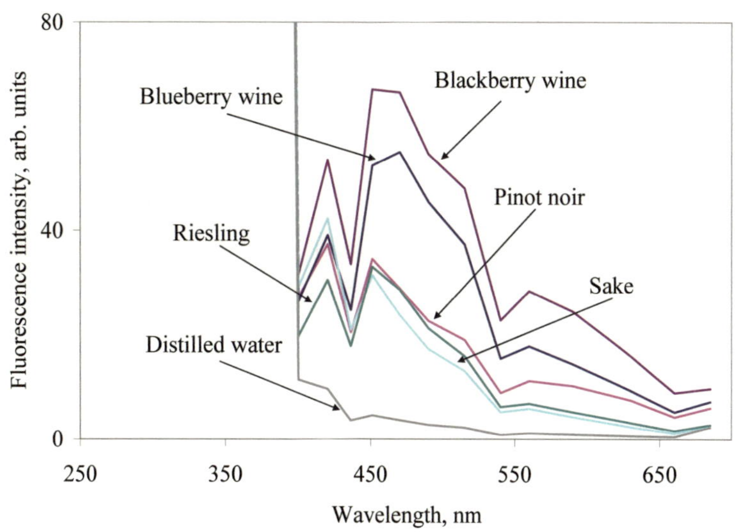

Fig. 11(d)

Fig. 11. Fluorescence of wine samples with (a) 266 nm, (b) 300 nm, (c) 335 nm, and (d) 355 nm LED excitation

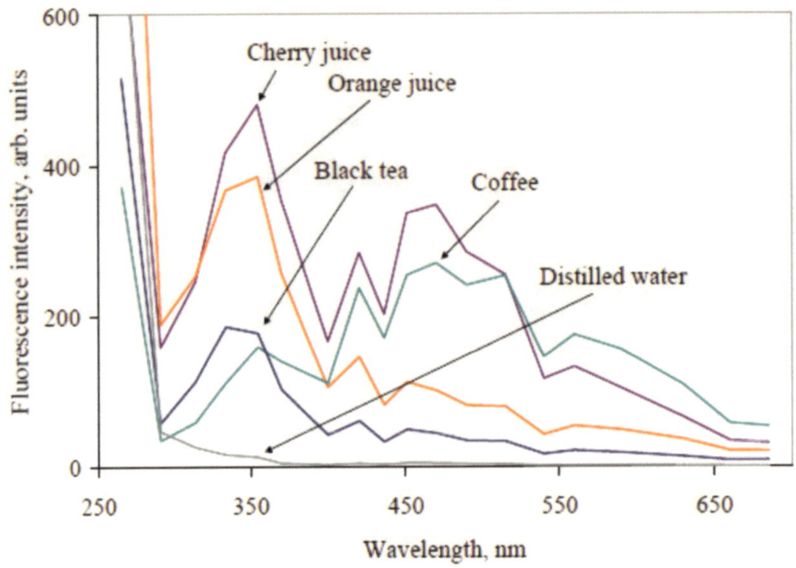

Fig. 12. Fluorescence of various drinks in distilled water; 266 nm laser excitation

3.4 Humic substance standards

Humic acid is a principal component of humic substances which are the major organic constituents of soil, humus, peat, coal, many upland streams, natural lakes, and ocean water. It is usually produced by biodegradation of dead organic matter. Samples of humic acid standards and natural organic matter were obtained from the International Humic Substances Society (www.ihss.gatech.edu). Figures 13 and 14 show fluorescence spectra of these substances with laser and LED excitation, correspondingly. Our data is in general agreement with excitation-emission spectroscopic properties of these materials as reported on the IHSS website (www.ihss.gatech.edu/spectra.html). It is interesting to note that the spectral peaks occur near 450 nm and 500 nm for these humic acid samples. Previous work by Coble has seen shifts in aged humic samples with fluorescence peaks gravitating toward the 500 nm wavelengths (Coble, 2007). Our results here are consistent with these earlier conclusions.

Further work on using LIF as a monitor of these humic substances is being conducted, and is also being explored as a way to monitor other hydrocarbon related substances.

4. Initial lab bench compact LIF measurement system

A compact version of our LIF system was also studied using a conventional compact spectrometer to replace the set of optical filters for fluorescence wavelength selection. Our earlier work had shown that such a system is several orders of magnitude lower in sensitivity than the system shown in Fig. 1, but the compact design and ease of use has many advantages especially for those cases where the concentration of the DOCs or CDOM is high. Toward this end, we developed a non-optimized compact system.

A schematic diagram of our laboratory bench-top LED-IF/LIF system is shown in Fig. 15. A microchip 266 nm laser or an UV LED equipped with a VIS-blocking CG-UG-11 filter were used interchangeably to illuminate a sample containing quartz cell. Fluorescence emission passed through a UV-blocking filter to eliminate the second-order peak of the scattered excitation wavelength, and was collected at 90° by an optical fiber connected to the compact spectrometer (Ocean Optics, Inc., USB2000). The spectrometer output was sent to the PC via an USB cable.

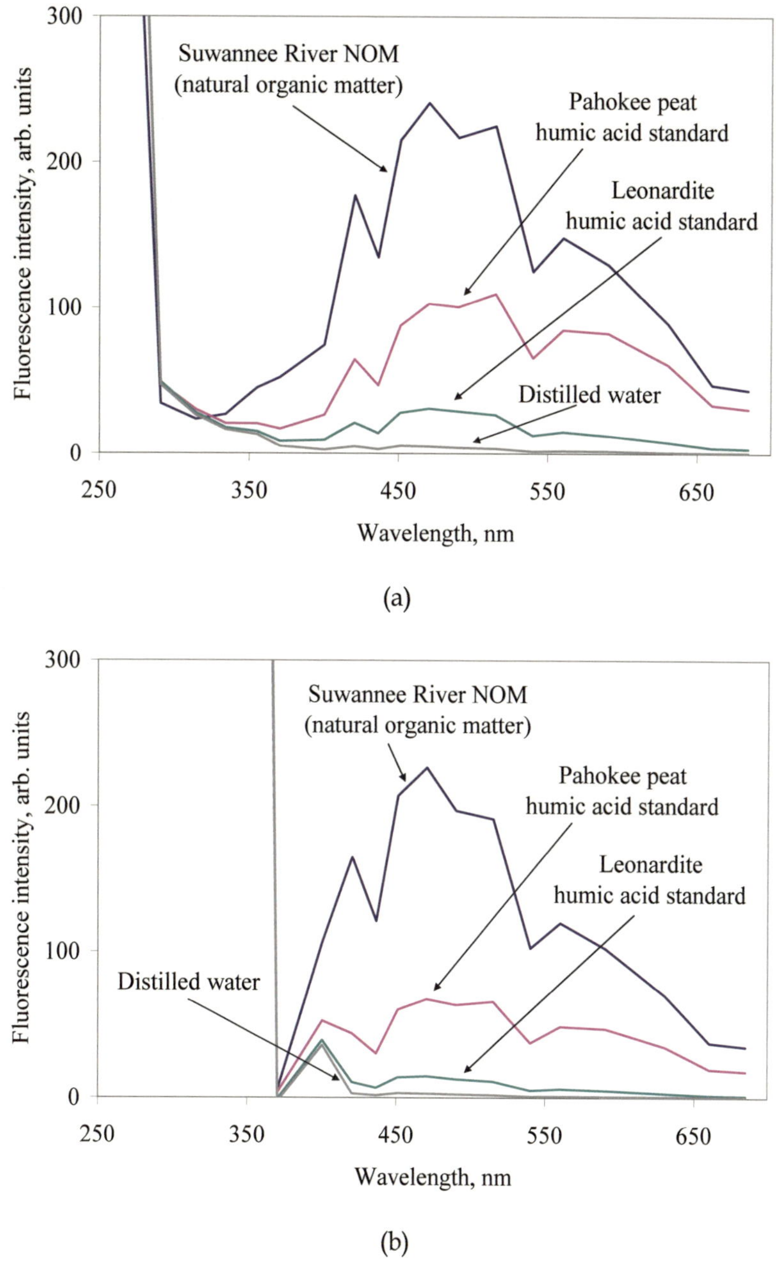

Fig. 13. Fluorescence of humic substances with (a) 266 nm and (b) 355 nm laser excitation

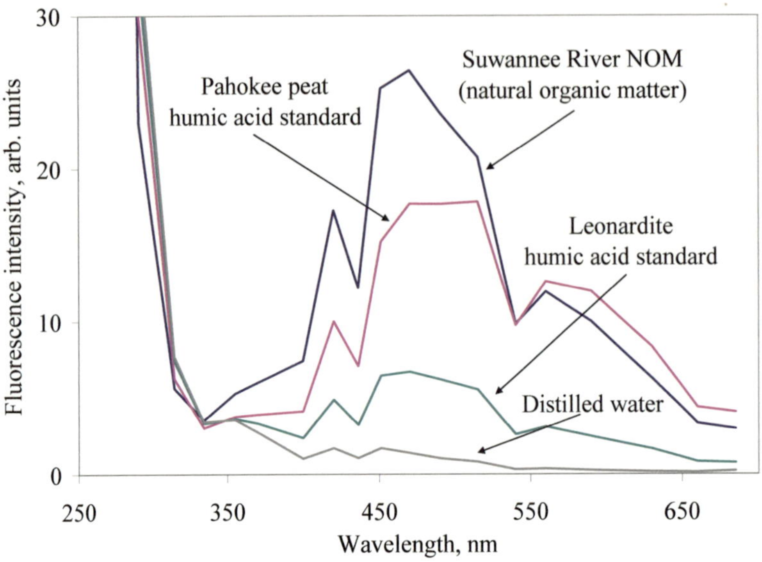

Fig. 14(a)

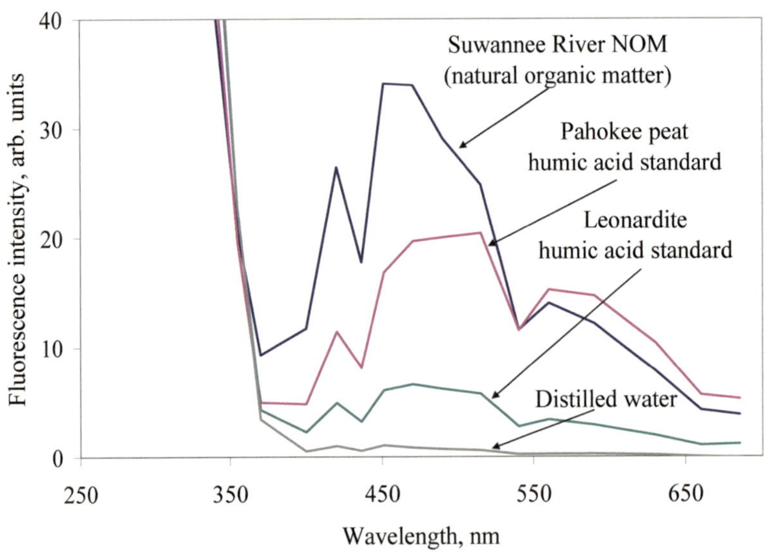

Fig. 14(b)

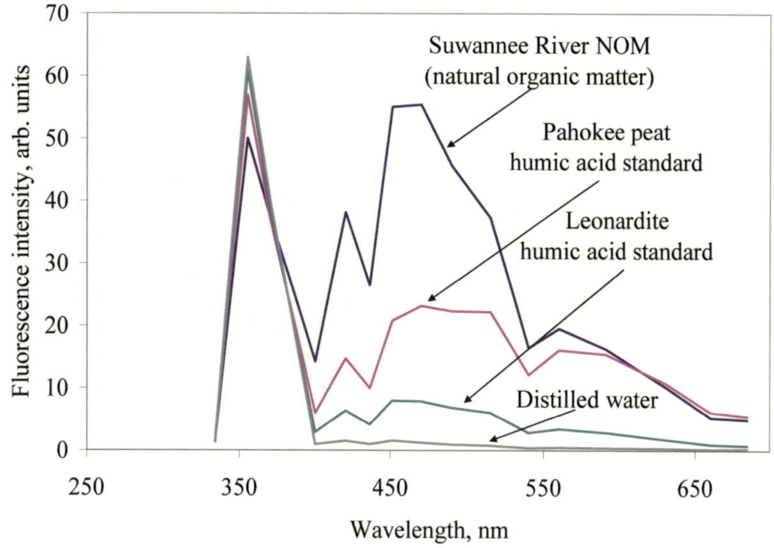

Fig. 14(c)

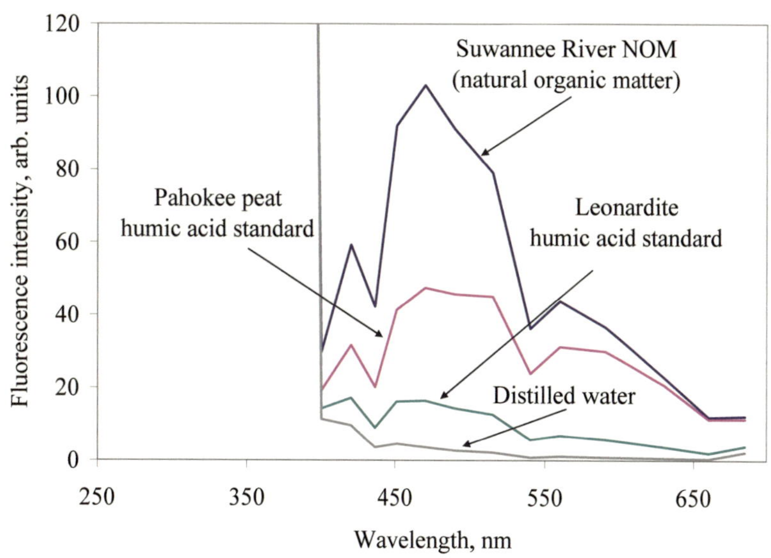

Fig. 14(d)

Fig. 14. Fluorescence of humic substances with (a) 266 nm, (b) 300 nm, (c) 335 nm, and (d) 355 nm LED excitation

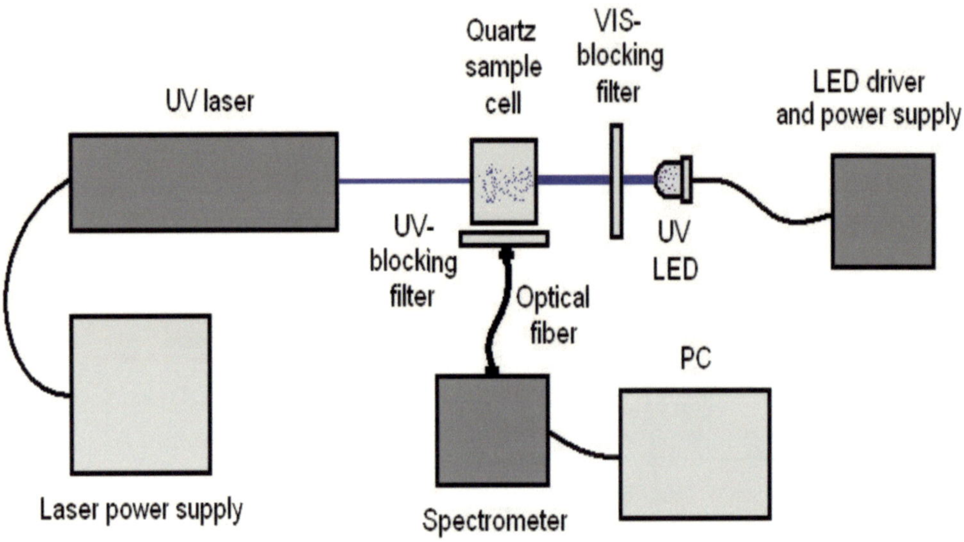

Fig. 15. Schematic diagram of the CW laboratory bench-top LED-IF / LIF system

Fluorescence of natural and drinking water samples were recorded using the system shown in Fig. 15 to compare the signal-to-noise ratio with the laser and LED excitation, and to determine the feasibility of CW LED sources in this compact setup. Lake water with its strong CDOM fluorescence was tested first, and then examples of drinking water were studied as well. As an example, Fig. 16 shows the fluorescence emission from lake water. As can be seen, the emission peak is near 500 nm, and is shifted somewhat from that observed for the tap water samples given earlier in Fig. 7. However, while the fluorescence signal is weak, the results are encouraging enough that more work is being conducted to better optimize the system and optical collection efficiency.

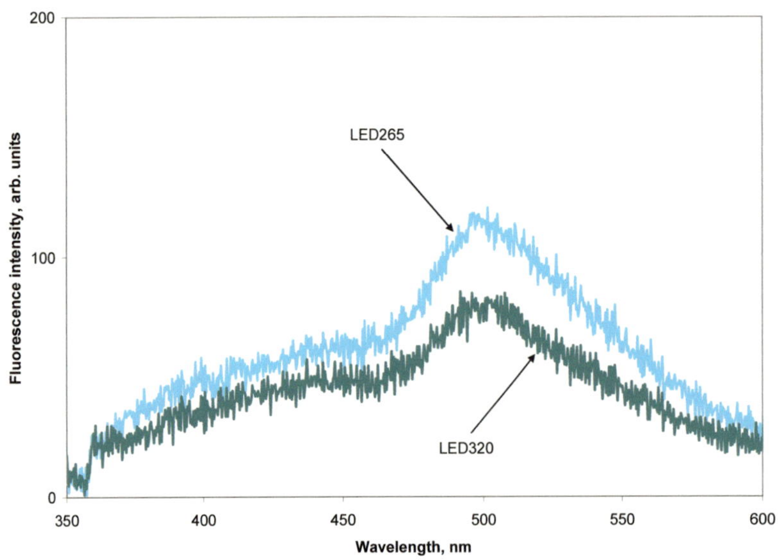

Fig. 16. Lake water fluorescence; 265 nm and 320 nm LED excitation; UV-blocking filter.

For comparison, a similar LIF spectrum for lake water, but using the 266 nm laser source, is shown in Fig. 17. As can be seen, the spectral features are similar, but the SNR is higher.

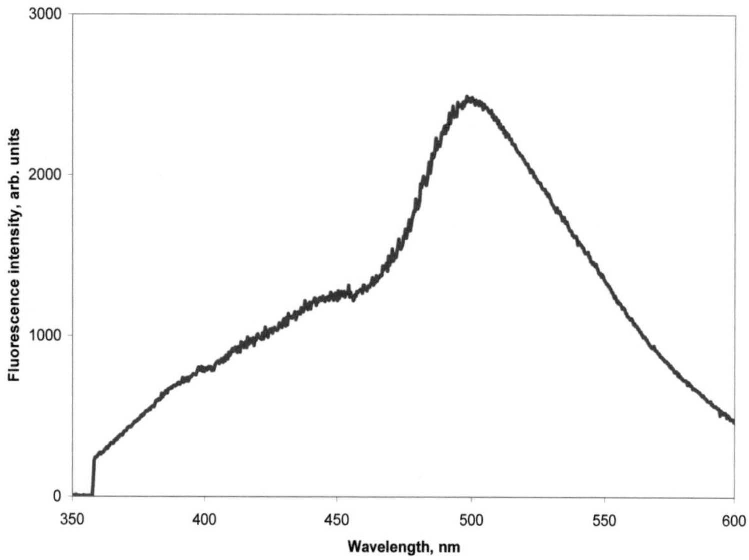

Fig. 17. Lake water fluorescence; 266 nm laser excitation; UV-blocking filter

5. Conclusion

The laser- and LED-induced fluorescence system developed in our lab has been used to study the spectra of organic contaminants in drinking water and other liquids. Spectra obtained using LED and laser excitation at the same wavelength exhibited great similarity, while differing in overall intensity. Greater than expected SNR observed with LED excitation is a good indication that a compact, less expensive LED based system might be used for detection and/or monitoring of trace organics and appropriate species in these liquids. Overall, our results are important in that they show that deep-UV LIF spectroscopy may offer some advantages for measurements of trace species in water in real time. In some cases the fluorescence spectrum is unique, but in many cases similar spectral peaks are observed. In the latter case, our LIF system may not be suitable for selective discrimination and identification of different trace species in water, but can still be used for real time monitoring of a known substance if correctly calibrated. It is anticipated that further work will be required to better quantify these techniques, and further cross-comparisons will have to be made using convention analytical water quality instrumentation. However, the deep UV laser and LED LIF spectroscopic techniques described in this paper offer potentially new instruments for such measurements.

6. References

American Public Health Association (1989), American Water Works Association, & Water Pollution Control Federation. *Standard methods for the examination of water and wastewater* (17th ed.)

Albani, J. R. (2007). *Principles and applications of fluorescence spectroscopy*. Oxford: Blackwell Publishing

Carnahan, Robert (2006); Private Communication (University of South Florida, Department of Chemical Engineering)

Coble, P. G. (2007), Marine optical biogeochemistry: the chemistry of ocean color. *Chem. Rev.*, Vol. 107, 402-418

Coble, P. G. (1996). Characterization of marine and terrestrial DOM in seawater using excitation-emission matrix spectroscopy. *Marine Chemistry*, 51, 325-346

Crompton, T. R.(2000). *Determination of organic compounds in natural and treated waters*. London: E & FN Spon

Hudson, N. J., Baker, A., & Reynolds, D. (2007). Fluorescence analysis of dissolved organic matter in natural, waste and polluted waters - a review. *Rivers Research*, 23, 631-649. DOI: 10.1002/rra.1005

Killinger, D. K. & Sivaprakasam, V. (2006). How water glows: Water monitoring with laser fluorescence. *Optics and Photonics News* Vol. 17, No. 1, 34-39

Levine, Audrey (2006); Private Communication (University of South Florida, College of Engineering)

Sharikova, A. V. & Killinger, D. K. (2007). LIF detection of trace species in water using different UV laser wavelengths. *International Journal of High Speed Electronics and Systems*, Vol. 17, No. 4, 689-695

Sharikova, A.V. & Killinger, D. K. (2010), Laser and UV LED induced fluorescence detection of dissolved organic compounds in water, SPIE, Conf. Defense & Security, Vol. 7666, May 2010

Sivaprakasam, V., Shannon, R. F., Luo C., Coble, P. G., Boehme, J. R. & Killinger, D. K. (2003). Development and initial calibration of a portable laser-induced fluorescence system used for in situ measurements of trace plastics and dissolved organic compounds in seawater and the Gulf of Mexico. *Appl. Opt.*, Vol. 42, No. 33, 6747-6756

Sivaprakasam, V., & Killinger, D. K. (2003). Tunable ultraviolet laser-induced fluorescence detection of trace plastics and dissolved organic compounds in water. *Applied Optics*, 42 (33), 6739-6746

Stender, H., Kurtzman, C.,Hyldig-Nielsen, J., Sorensen, D., Broomer, A., Oliverira, K., Perry-O'Keefe, H., Sage, A., Young, B., and Coull, J. (2001), Identification of Dekkera bruxellensis (Brettanomyces) from Wine by fluorescence in situ hybridization using peptide nucleic acid probes, *Applied and Environmental Microbiology*, Vol. 67, p. 938-941

20

Optical Coherence Tomography: Development and Applications

Anderson Zanardi de Freitas,
Marcello Magri Amaral and Marcus Paulo Raele
Nuclear and Energy Research Institute - IPEN-CNEN/SP
Brazil

1. Introduction

In early 90´s Huang *et al.*, applied the principles of low coherence interferometry to perform high resolution tomographic images of biological tissues. The technique became known as Optical Coherence Tomography (OCT) and opened new horizons in several areas of knowledge, e.g., ophthalmology (Hee, et al., 1995)(Meirelles, et al., 2005) and dermatology. In such areas other tomographic techniques does not reach resolutions good enough to study the structures of interest, see Fig. 1, in this way a whole new branch of imaging science was created. The OCT breakthrough was the capability of obtains high resolution histological images in vivo, in real time in non-invasive way and using harmless radiation (low power near infrared radiation, usually).

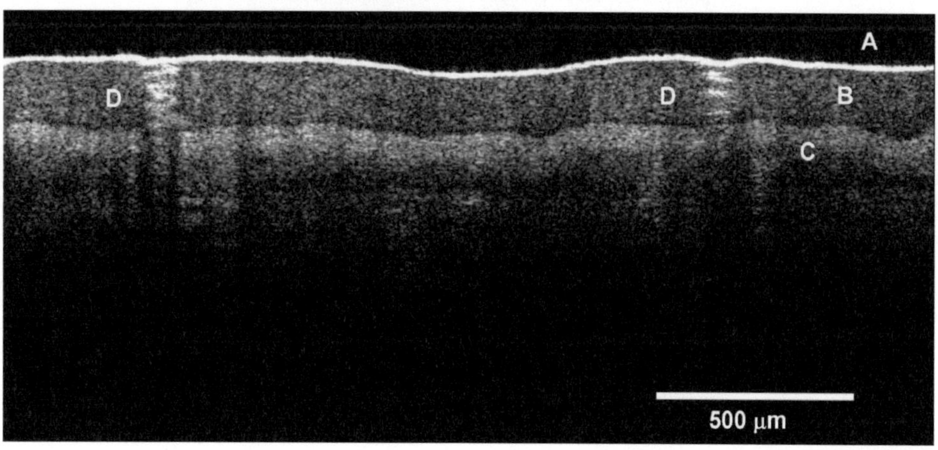

Fig. 1. Optical coherence tomography of a hand palm, the letters indentifies the structures: A) stratum corneum; B) epidermis; C) dermis and D) Sweat gland.

Nowadays OCT is routine in ophthalmologic clinic in many countries (approved by FDA in USA), and is spreading its applications in odontology (Freitas, et al., 2006)(Melo, et al., 2005) dermatology, industrial (Goode, 2008) applications among others. The image, presented in Fig. 1, shows an OCT image of the human hand palm.

The bibliographic data helps to convince the reader about OCT impact and its potential, a brief research was performed in the Web of Science database for the keyword "optical AND coherence AND tomography". Firstly, as shown Fig. 2, the record count, between years 1991 to 2009, has grown consistently since OCT has been proposed and has reached about 1800 documents only in 2009, and only for the keywords inserted in the search engine, for more dramatic comparison, this is 5 documents of OCT per day!

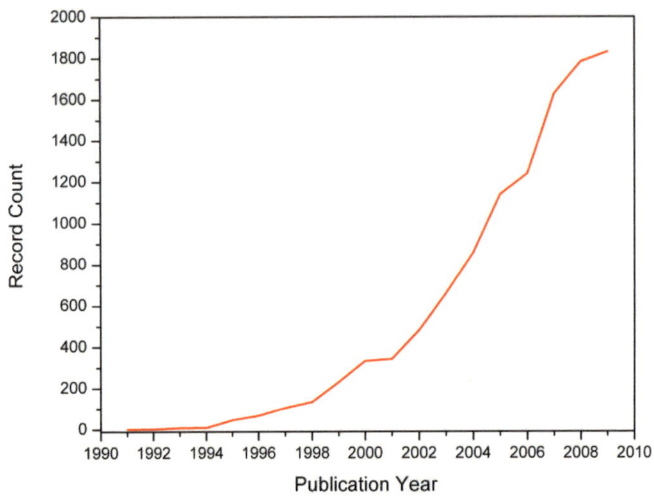

Fig. 2. Record count results for "optical AND coherence AND tomography" keyword in the Web of Science in July of 2010 organized by year.

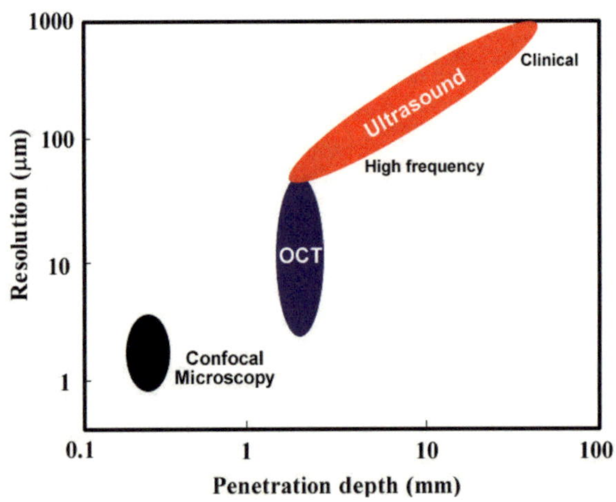

Fig. 3. Regions of actuation of tomographic systems.

These bibliographic productions are classified as shown in the Fig. 4, where worth to mention that patents are the third place in numbers of publications, showing the economic potential of OCT. Another highlight which can be done is about the reviews, this is a way to measure how consolidate a certain technique is, in other words, the expressive number of reviews shows that OCT can be faced as a established imaging tool.

Organizing this data by areas of knowledge and looking only for the most expressive ones, Fig. 5, it is possible to see the impact of OCT in ophthalmology, and also in cardiology, caused by OCT. That is because of the vacancy of technologies capable to perform images with resolution good enough to differ between structures, some of them with few microns sized.

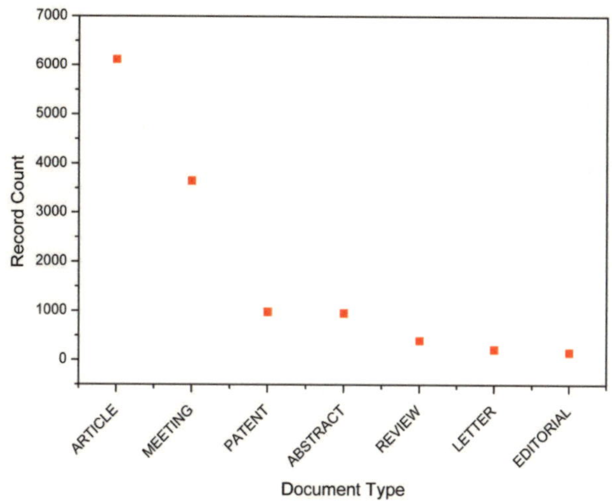

Fig. 4. Record count results for "optical AND coherence AND tomography" keyword in the Web of Science in July of 2010 organized by document type.

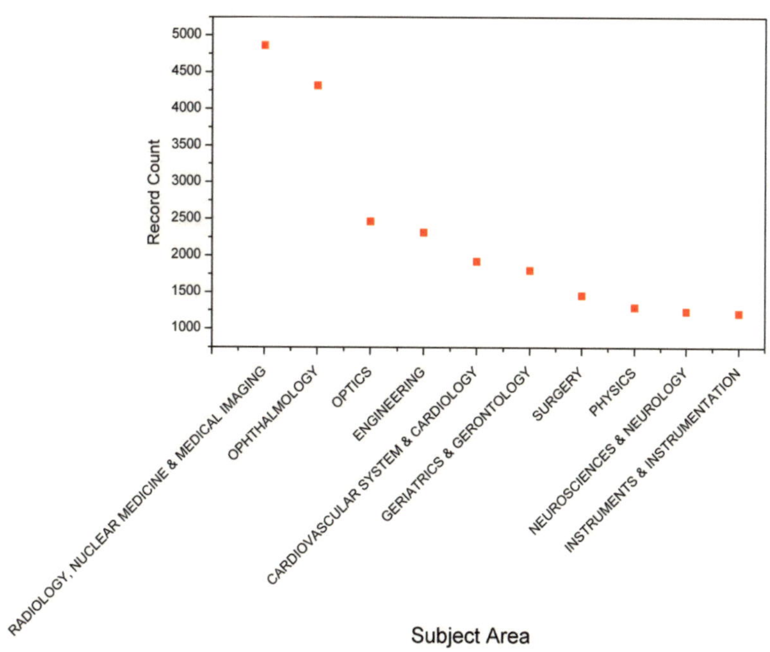

Fig. 5. Record count results for "optical AND coherence AND tomography" keyword in the Web of Science in July of 2010 organized by subject area.

1.1 OCT concept

The OCT setup it is generally mounted with a Michelson interferometer, and can be divided in the following main parts: light source, scanning system and light detector, see Fig. 6. These items define almost all crucial properties of the system. The light emitted from the light source is divided in two by a beam splitter, part of the beam is directed to the sample and the other part to the reference mirror, the light backscattered by the sample and the light reflected by the mirror are recombined at the beam splitter giving origin to an interference pattern collected by the detector. Because the broadband property of the light source, the interference pattern will occur only when the optical paths difference between this two arms are nearly the same. All this process will be discussed in more detail ahead.
The first OCT setup was implemented using femtosecond pulsed laser, due to its broadband spectral emission (Huang, et al., 1991), which implies in a low coherence length, this feature is the heart of the OCT system, and the image resolution is correlated to the light source coherence (as broader the spectral bandwidth, narrower the coherence length will be).

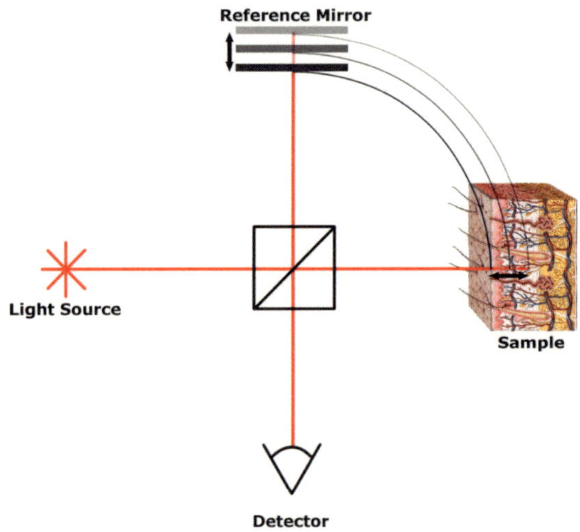

Fig. 6. Schematic representation of an OCT setup.

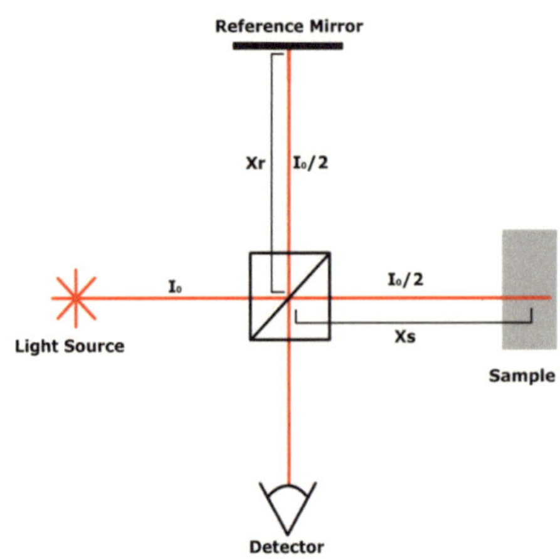

Fig. 7. Michelson Interferometer.

Nowadays many others light properties are explored too, like polarization sensitive OCT or Doppler shift OCT are already established, these techniques can extract information about fiber alignment or particles velocity within the sample, respectively. The efforts by the research groups for other approaches are being done continuously, since the OCT development, resulting in ways to extract more information of the sample by the analysis of the light: Mueller matrix OCT (MM-OCT), pumping-probe OCT (PP-OCT), autocorrelation OCT, are some examples of OCT approaches in current development.

2. Theory

2.1 Low coherence interferometry

The OCT technique is based on Michelson interferometer (Fig. 7) to produce tomographic images. A light source, expressed in terms of it electrical field amplitude (equation (1)) is introduced in the Michelson interferometer and is directed to the beamspliter. It splits the radiation in two components that are bounded to the reference arm (E_r) and to the sample arm (E_s). Using a beamspliter that divides the beam in two equal parts (50:50), e.g., E_r and E_s can be written as equation (2) and (3) respectively.

$$E_f(x) = E_{fo}e^{i(kx-\omega t)} \tag{1}$$

using $k = 2\pi / \lambda$, $\omega = 2\pi c / \lambda$ and considering just spatial wave propagation

$$E_S(x) = \left(1/\sqrt{2}\right)E_{fo}e^{i\left(\frac{\omega}{c}x_s\right)} \tag{2}$$

$$E_r(x) = \left(1/\sqrt{2}\right)E_{fo}e^{i\left(\frac{\omega}{c}x_r\right)} \tag{3}$$

The radiation is reflected by the reference mirror, and backscattered by the sample, the portion of radiation that returns is proportional to the mirror and sample capacity to reflect and backscatter this radiation. This coefficient $R(x)$, can vary in sample depth (depending on the sample features). It varies from 0 to 1, where 0 is total transmission and 1 is total reflection. So, the back reflected and backscattered field (light) suffers an amplitude modulation. Moreover, the resultant field will be equal to the sum of infinitesimal fields from different sample depth.

The field comes back to the beamspliter where they are recombined; the field from mirror and sample are described by equation (4) and (5) respectively. For the mirror $R_r(x)=R_r\delta(x-x_0)$, where $\delta(x-x_0)$ is the Dirac function and x_0 is the mirror position.

$$E_r(x_r) = \left(\frac{1}{2}\right)E_{fo}\int_0^{\infty} R_r\delta(x_r - x_0)e^{i\left(2\frac{\omega}{c}x_r\right)}dx_r = \left(\frac{1}{2}\right)E_{fo}R_re^{i\left(2\frac{\omega}{c}x_0\right)} \tag{4}$$

$$E_S(x_S) = \left(\frac{1}{2}\right)E_{fo}\int_0^{\infty} R_S(x_s^{'})e^{i\left(2\frac{\omega}{c}n\left(x_s^{'}\right)x_s\right)}dx_s^{'} \tag{5}$$

The factor two in the exponential is to take in account the optical path roundtrip, $n(x_S)$ is the refraction index as function of sample depth. The electrical field on the detector (E_D) will be a sum of the sample and reference arm electrical fields.

$$E_D = E_r + E_S = \left(\frac{1}{2}\right)E_{fo}R_r e^{i\left(2\frac{\omega}{c}x_0\right)} + \left(\frac{1}{2}\right)E_{fo}\int_0^\infty R_s(x_s)e^{i\left(2\frac{\omega}{c}n(x_s)x_s\right)}dx_s \tag{6}$$

The intensity on detector (I_D) is proportional to square modulus of electrical field E_D:

$$I_D \propto E_D E_D^* = \left|\left(\frac{1}{2}\right)E_{fo}R_r e^{i\left(2\frac{\omega}{c}x_0\right)} + \left(\frac{1}{2}\right)E_{fo}\int_0^\infty R_s(x_s)e^{i\left(2\frac{\omega}{c}n(x_s)x_s\right)}dx_s\right|^2 =$$

$$= \left(\frac{E_{fo}}{2}\right)^2\left[\begin{array}{l} R_r^2 + \int_0^\infty\int_0^\infty R_S(x'_S)R_S(x_S)e^{-i\left(2\frac{\omega}{c}\left(n(x_S)x_S - n(x'_S)x'_S\right)\right)}dx_S dx'_S \\ +\left(\int_{-\infty}^\infty R_r R_S(x_S)\cos\left(2\frac{\omega}{c}\left(n(x_S)x_S - x_0\right)\right)dx_S\right)\end{array}\right] \tag{7}$$

At the right side of this equation, the first two terms corresponds to the DC (constant) intensity from reference and sample arm, respectively, and they do not bring useful information. The third term is an oscillatory term, and is responsible of bringing information from the sample to generate OCT images.
Using a broadband spectral source, the equation (7) must be modified in such way that it comprehends an infinity number of frequencies. The interference only occurs between equal frequencies, so the total interference will be the sum of the infinitesimal interferences.

$$I_D \propto \int_0^\infty\left(\frac{E_{fo}(\omega)}{2}\right)^2\left[\begin{array}{l} R_r^2 + \int_0^\infty\int_0^\infty R_S(x'_S)R_S(x_S)e^{-i\left(2\frac{\omega}{c}\left(n(x_S)x_S - n(x'_S)x'_S\right)\right)}dx_S dx'_S + \\ +\left(\int_{-\infty}^\infty R_r R_S(x_S)\cos\left(2\frac{\omega}{c}\left(n(x_S)x_S - x_0\right)\right)dx_S\right)\end{array}\right]d\omega = \int_0^\infty I(\omega)d\omega \tag{8}$$

Where $I_D(\omega)$ is the intensity on detector for a given ω value, and the integral in frequency is equal to the total intensity on detector I_D.
The field that comes from reference and sample arm differs only by the optical path, as expressed by the interference term on equation (7)).
If t is the time that the radiation takes to travel from beamspliter to reference mirror, and $t+\tau$ is the time to travel from beamspliter to the scatter position in the sample, so τ is the temporal delay between the two arms. The interference term on equation (7) can be written as:

$$2\mathrm{Re}\Gamma_{rS}(\tau) \tag{9}$$

Where,

$$\Gamma_{rS}(\tau) = \left\langle E_r(t)E_S^*(t+\tau)\right\rangle_T \tag{10}$$

The $\Gamma_{rS}(\tau)$ is the coherence function or correlation function between E_r and E_S fields. The function:

$$\Gamma_{SS}(\tau) = \left\langle E_S(t) E_S^*(t+\tau) \right\rangle_T \tag{11}$$

It is known as autocorrelation function. From this definition it is possible to show that $\Gamma_{SS}(0)=I_S$ and $\Gamma_{rr}(0)=I_r$. For convenience the normalized form of coherence function will be used, it is called partial coherence degree:

$$\gamma_{rS}(\tau) = \frac{\Gamma_{rS}(\tau)}{\sqrt{\Gamma_{rr}(0)\Gamma_{SS}(0)}} = \frac{\Gamma_{rS}(\tau)}{\sqrt{I_r I_S}} \tag{12}$$

The $\gamma_{rS}(\tau)$ function is, in general, a periodic complex function of τ, so the interference pattern is obtained if the value of $|\gamma_{rS}(\tau)|$ is different from zero. The $|\gamma_{rS}(\tau)|$ can assume value between 0 and 1. If the value is equal 1 it says that complete coherence occur, if equal 0 it says that complete incoherence, for values between 0 and 1 partial coherence occurs.

$$\left|\gamma_{rS}(\tau)\right| = 1 \qquad \text{Complete coherence (13)}$$

$$0 < \left|\gamma_{rS}(\tau)\right| < 1 \qquad \text{Partial coherence (14)}$$

$$\left|\gamma_{rS}(\tau)\right| = 0 \qquad \text{Complete incoherence (15)}$$

2.2 Time domain

The Fig. 6 shows the basic components of an OCT system. The main part of the system comprehends an interferometer illuminated by a broadband light source.

The OCT system splits the broadband light source beam in reference field (E_R) and a sample field (E_S). They interfere at the detector by summing up the two electrical fields that are reflected by the optical scanning system (in general a mirror) and the sample. The intensity in the detector can be expressed by equation (8).

The oscillatory term on equation (8) can also be expressed as:

$$\text{Re}\left\{E_R E_S^{'*}\right\} = R_R R_S \cos(2\beta_R l_R - 2\beta_S l_S) \tag{16}$$

Where l is the optical path and β is the propagation constant (in this case the light source is highly coherent).

Defining $S(\omega) = R_S(\omega)R_R(\omega)^*$ and $\Delta\phi(\omega) = 2[\beta_S(\omega)l_S - \beta_R(\omega)l_R]$, and considering the case where the sample and the reference arms consists of a uniform, linear, no dispersive material and the light source spectral density is given by $S(\omega-\omega_0)$, which is considered to be bandwidth limited and centered at the frequency ω_0. The propagation constants $\beta_i(\omega)$ in each arm are assumed to be the same; the diffuse tissue material behaves locally as an ideal mirror leaving the sample beam unchanged. Propagating the $\beta_i(\omega)$ coefficient as a first-order Taylor expansion around the central frequency ω_0 gives

$$\beta_R(\omega) = \beta_S(\omega) = \beta(\omega_0) + \beta'(\omega_0)(\omega - \omega_0) \tag{17}$$

Then the phase mismatch $\Delta\phi(\omega)$ is determined solely by the optical length mismatch $\Delta l=l_S-l_R$ between the reference and the sample arms, and is given by

$$\Delta\phi(\omega)=\beta_0(\omega_0)(2\Delta l)+\beta'(\omega_0)(\omega-\omega_0)(2\Delta l) \tag{18}$$

Now, consider that the light source has a Gaussian power spectral density defined by

$$S(\omega-\omega_0)=\sqrt{\frac{2\pi}{\sigma_\omega^2}}e^{-\frac{(\omega-\omega_0)^2}{\sigma_\omega^2}} \tag{19}$$

which has been normalized to the unit power. Using this power spectrum and the phase mismatch is possible to find that detector signal is:

$$I\propto\frac{e}{h\beta\eta_0}\mathrm{Re}\left\{\underbrace{1}_{DC}+\underbrace{e^{-\frac{\Delta\tau_g^2}{2\sigma_\tau}}e^{-i\omega_0\Delta_{\tau p}}}_{AC}\right\} \tag{20}$$

In (20), the phase delay mismatch $\Delta_{\tau p}$ and the group delay mismatch $\Delta_{\tau g}$ are defined as:

$$\Delta\tau_p=\frac{\beta(\omega_0)}{\omega_0}(2\Delta l)=\frac{2\Delta l}{\mathrm{v_p}} \tag{21}$$

and

$$\Delta\tau_p=\beta'(\omega_0)(2\Delta l)=\frac{2\Delta l}{\mathrm{v_g}} \tag{22}$$

The detector signal given by equation (20) contains two terms, the first one is the mean (DC) intensities returning from the reference and sample arms of the interferometer, and the second one, which depends on the optical time delay (optical path mismatch) set by the position of the reference mirror, represents the amplitude of the interference fringes that carry information about the tissue structure, this is a Gaussian envelope with a characteristic standard deviation temporal width $2\sigma_\tau$, that is inversely proportional to the power spectral bandwidth: $2\sigma_\tau=1/2\sigma_\omega$, This envelope falls off quickly with increasing group delay mismatch $\Delta\tau_g$ and is modulated by interference fringes that oscillate with increasing phase delay mismatch $\Delta\tau_p$. Thus, the second term in equation (20) defines the axial resolving power of the OCT system. For a Gaussian shape function with standard deviation τ, the full width at half maximum (FWHM) is $2\sigma\sqrt{2\ln 2}$ then, the axial resolution of the system is:

$$\Delta l_{FWHM}=\frac{2\ln 2}{\pi}\frac{\lambda_0^2}{\Delta\lambda} \tag{23}$$

where λ_0 is the center wavelength.

2.3 Frequency domain

The Fourier domain optical coherence tomography (FD-OCT) uses a spectrometer, instead a single detector, to analyze the spectral interference pattern (Fig. 8).

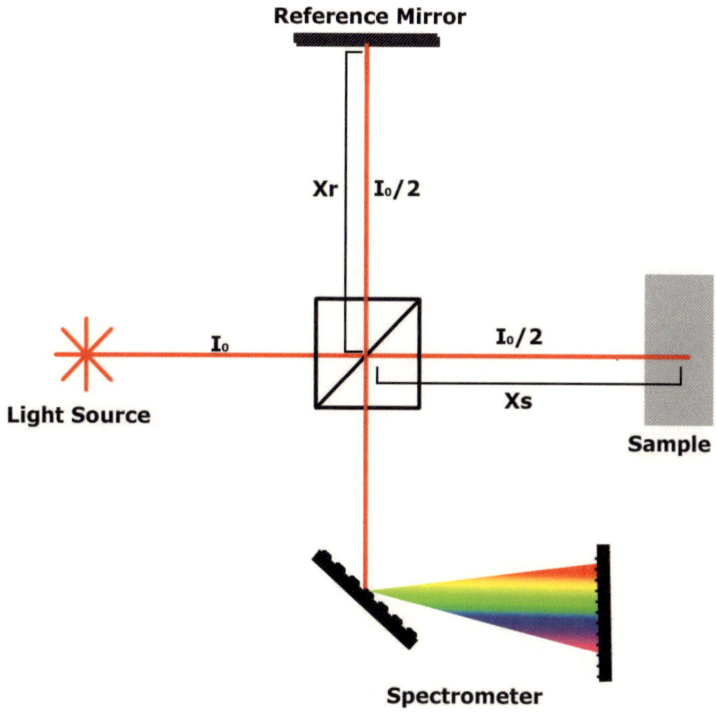

Fig. 8. Michelson Interferometer with a diffractive element an a CCD detector to spectral measurement.

The equation (7) can be written as a Fourier transform of $R_S(x_S)$. To write the equation as a function of wave number k instead ω (equation (24)). There is a correlation between reciprocal and direct space, given by Fourier transform. It correlates time (s) with frequency (1/s=Hz) and distance (m) with wave number k (1/m).

$$I(k)=\left(\frac{E_{fo}(k)}{2}\right)^2\left[R_r^{\,2}+\int_0^\infty\int_0^\infty R_S(x'_S)R_S(x_S)e^{-i\left(2k\left(n(x_S)x_S-n(x'_S)x'_S\right)\right)}dx_S dx'_S+\frac{1}{2}\Im_z\left[R_r R_S(z)\right]\right] \quad (24)$$

Where $z=n(x_S)x_S-x_0$ is the optical path difference between sample and reference arm. We can also rewrite the second term as a distance related to the reference mirror.

$$\begin{aligned}&\int_0^\infty\int_0^\infty R_S(x'_S)R_S(x_S)e^{-i\left(2\frac{\omega}{c}\left(n(x_S)x_S-n(x'_S)x'_S\right)\right)}dx_S dx'_S\\&=\int_0^\infty\int_0^\infty R_S(x'_S)R_S(x_S)e^{-i\left(2k\left[\left(n(x_S)x_S-x_0\right)-\left(n(x'_S)x'_S-x_0\right)\right]\right)}dx_S dx'_S\end{aligned} \quad (25)$$

Substituting z in equation (25), and as the auto correlation is a symmetric function, it has:

$$\int_0^\infty\int_0^\infty R_S(z')R_S(z)e^{-i(2k[z-z'])}dzdz'=\frac{1}{4}\int_{-\infty}^\infty\int_{-\infty}^\infty R_S(z')R_S(z)e^{-i(2k[z-z'])}dzdz'=\frac{1}{8}\Im_z[AutCorr(R_S(z))]' \quad (26)$$

Can be identified on this term a Fourier transform, using totally reflective mirror in the reference arm (R_r=1), the equation (24) can be rewritten as:

$$I(k)=\left(\frac{E_{fo}(k)}{2}\right)^2\left(1+\frac{1}{8}\Im_z[AutCorr(R_S(z))]+\frac{1}{2}\Im_z\left[R_S(z)\right]\right) \quad (27)$$

For the spectral signal ($I(k)$) analysis and $R_S(x_S)$ information attainment, an inverse Fourier transform is applied. Finally we obtain:

$$\Im_z^{-1}\left[I(K)\right]=\Im_z^{-1}\left[\left(\frac{E_{fo}(k)}{2}\right)^2\right]\otimes\left(\left[\delta(z)\right]+\frac{1}{8}AutCorr(R_S(z))+\frac{1}{2}R_S(z)\right) \quad (28)$$

$$\Im_z^{-1}\left[I(K)\right]=A\otimes\left(B+C+D\right) \quad (29)$$

Using a simplified notation (equations (29)) for equation (28), the $R_S(z)$ information is present in convolution of A and C (A⊗C). The convolution A⊗B brings information about radiation source properties. A⊗D brings information about interference between waves backscattered in different sample positions. This terms can be ignored for high reflective medium, since this signal is despicable related to A⊗C term. The signal A⊗B and A⊗D can be avoided by adequate reference mirror position, a mismatch of few tens of microns avoid the superposition of A⊗C and the last two terms.

2.3.1 Frequency domain and signal processing

As already discussed in the previous sections, the collected signal in the frequency domain needs to be processed to form images of interest, *i.e.*, processing the signal will make the signal direct related with the sample morphology.

Although the processing algorithm has in the core the Fourier Transform to retrieve the scattering profile (equation (28)), some mathematical manipulations are necessary on the interferometric pattern due to correction and refinements reasons, aiming images with good quality. Some of these corrections are necessary due to physical limitation of the equipment, for example the limited pixel number, or more basic corrections, like changes of unities, for instance.

Many algorithms can be implemented with different approaches, but this text will be focused in just three, they are: Direct Fourier Transform (DirFT), Interpolation (Int) and Zero-Filling (ZF), and they are more detailed explained ahead.

The direct Fourier transform (DirFT) method could be considered as the simpler one, consequently the more fast and robust. It perform just a change of unity, that is because spectrometers are calibrated in wavelength, and as OCTs gives information of depth (m), we need to change from wavelength to wavenumber ($k=2\pi/\lambda$). This process makes the spectrum, originally organized in crescent order in wavelength to a reversed order array, so the vector must to be inverted. After that the vector Fourier transform is done, resulting the scattering profile. The schematic diagram represents the process Fig. 9 (a). But this process, *i.e.*, 1/x, cause unequal sized bins, resulting in issues in the Fourier transform, leading to broadening of the structures and asymmetry of the peaks in respect to the it center. A method to avoid this problem is to perform an interpolation. After the changing of unities, the interpolation is done to retrieve equally sized bins, and then submitted to the Fourier transform, this process is schematic represented in the Fig. 9 (b). The last method (Fig. 9 (c)), Zero-Filling (ZF) is a technique more elaborated when compared with the two discussed

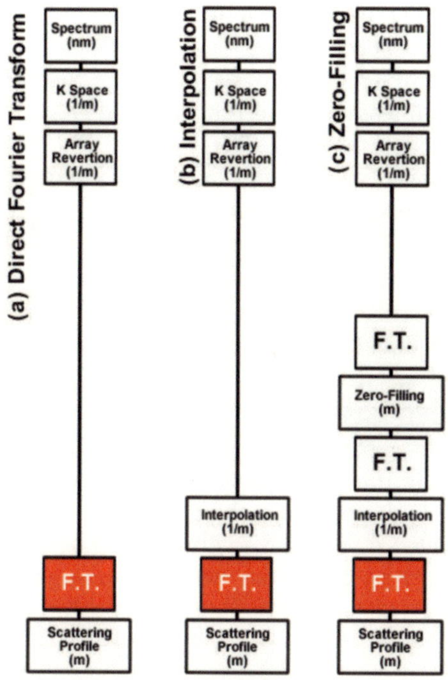

Fig. 9. Schematic representation of three types of spectral interferometry signal processing which results in the scattering profile. Between parenthesis dimensional unity.

previously, consequently more expensive computationally. The Zero-Filling technique is based in a mathematical gimmick, used to increase sampling without increase the data collection. In practice the (ZF) it is preformed applying the Fourier transform on the collected spectrum, then, in the reciprocal space, empty arrays (Zero-Filling) are added at the ends of the original array, the increased sampling of the original data will, according to the Nyquist theorem, allow to process higher frequencies resulting in less computational errors (Raele, et al., 2009).

3. Light source

3.1 Light source characteristics

The light source should attend, basically, four main desired characteristics: wavelength, spectral bandwidth, intensity and stability. Other features could be also listed, as portability, low cost and etc, but these first four are critical and the reasons for that follows.

First of all the wavelength must be compatible with the sample, mainly because scattering, absorption and dispersion are wavelength dependent, so if there is interest in measure inside a sample a wavelength that has low attenuation must be chose. To biological tissue studies, the region known as "diagnostic window" is often used. This spectral region is located between 800 nm and 1300 nm.

As shown by equation (23), the spectral band is related with the system resolution, naturally light sources with broad emission spectral will be preferred, but it is not usual to obtain broad spectral emission with high intensities. Also it is interesting to highlight that to maintain a resolution as the wavelength increases, the spectral band also needs to increases,

for instance an 800 nm with 28 nm of spectral band implies in 10 μm of resolution. To get the same resolution at 1600 nm the spectral band should be 113 nm.
The intensity of the light source must be intense enough to sensitize the detector giving a good signal to noise ratio, but as the OCT is often used in biological samples, the intensities should not overcome the maximum permissible radiation (MPR).
Finally the spectral profile and the intensity must be constant in time; any alteration can cause issues, like false structures, in the scattering image.

3.2 SLED, mode locked lasers, swept sources

Many kinds of light source can be used in OCT systems, as just they fill in the requirements described in the previous section, but let us highlight some features of each one of them.

3.2.1 Super iuminescent LED (SLED)

The SLED it is, perhaps, the most popular OCT light source nowadays due to its low cost and easiness of handling. It presents intensities high enough to perform tomographic images, and also presents high spectral stability. Another good thing about it is that is possible to acquire it pigtailed (connected to an optical fiber). The drawbacks are limited spectral band, about 30 nm and intensities not high enough to perform extremely fast scanning.

3.2.2 Lasers

Lasers, usually, are applied in OCT research, most of them using a Ti:Sapphire laser system operating in mode locked regime, because in this kind of operation a broadband radiation is promoted. Lasers are a most flexible, about spectrum and intensity, then system with SLED. Without doubt the major drawbacks of applying mode locked lasers is the cost.
Lasers systems allows intensities high enough to perform images at so high rates, in this way, the involuntary movements of the live system that are under study do not affect the image.
Mode locked lasers also can be used to generate a supercontinuum spectra by injecting it in a photonic fiber. In this type of fibers, nonlinear effects produces spectra large as 400 nm, allowing submicron of spatial resolutions.

3.2.3 Swept source

A Swept Source is a broadband laser with an intracavity optical narrowband filter. Only longitudinal modes with the exact frequency selected by the filter can oscillate, so the laser action occurs on a single frequency. This filter can be frequency tuned, sweeping the frequency laser action. The filtering tune is made so all the laser spectral frequencies be tuned on the photon cavity roundtrip. The output laser is not a sort pulse train, as a mode-locked laser, but a tuned frequency train with long pulses. The tuned frequencies have the same phase evolution and they are coherent between each other.

4. Scanning systems

Before entering in the subject itself, let us stress to the reader that is more than one type of scanning, usually we need a lateral scan and also a depth scan, be sure that are clear in mind before continue. The lateral scan can be easily done with a galvanometric system or even a linear translator that moves a sample perpendicularly to the incident light beam. Below are discussed the depth scan, also known as A-scan and have two different approaches to be performed: Time Domain and Frequency Domain, as detailed in the theory section, the issue, now, it is how to perform in practice this two types of A-scan.

4.1 Time Domain

In Time Domain OCT the optical path of the reference arm needs to vary in time so the scattering profile, of a single point of the sample, can be recorded. Change the optical path of an interferometer arm can be done with simple systems, simple as a mirror fixed in a speaker, of course that is not the most reliable and faster way, but it will do. When systems with more finesse are required more complex systems are needed. That are many type of scanning system, when using optical fibers, to stretch it with kHz repetition, usually, can be done using a piezoelectric device. This configuration assures high mechanical stability due to the use of the optical fiber. Another configuration reported is to place a rotating glass cube between the beam splitter and the mirror. As the cube rotates the optical path changes due to refraction, with this setup was achieved the A-scanning record (Bouma, et al., 2002), another used configuration is kwon as Fast Fourier Scanning (FFS) scanning, also achieves high repetition.

4.1.1 Frequency Domain

The main advantage of Frequency Domain OCT it is that, once that a CCD based spectrometer is used, there is no need of any mechanical variation in time. All the depth information, the scattering profile, is encoded in the spectral interference pattern, which can be recorded easily with the help of a personal computer. So in this case the reference arm stands still. In the other hands, as drawbacks about, about the FD-OCT, is the detector cost and complexity, still, is the configuration more used in commercial systems. Another issue to be mentioned is that FD signal needs more processing, i.e., more powerful computers.

4.1.2 Swept source

The swept source can be understood as a characteristic FD approaches, but due to the source features the spectrum is acquired as a function of time. There is a relationship between time and wavelength. Also the signal acquired needs to be processed in the same way that in the FD-OCT. So, what is the catch? Well, now the cost of having a spectrometer is avoided, in SS-OCT a single photo-detector is used (no gratings, no moving mirrors and CCDs). Mechanically the SS setup has no moving parts, as FD, which is a very desirable feature, however, in the SS system the swept source itself it has a high cost and complexity.
The Swept source applied to OCT (SS-OCT) allows images construction between 10 to 50 times quiker than traditional OCT and, due to SS be a laser, the SS intensity is greater than superluminescent LED, allows deeper tissue penetration.

5. PS-OCT

Light exhibits polarization states; due to the property that light vibrates orthogonally in respect to the propagation direction. Measure change in polarization in many cases can be considerate relatively simple process, but measure change in polarization as position function (inside a sample) it is not trivial. Using an OCT system it is possible to gather information from different polarizations states and perform not a scattering image, but it is possible to perform birefringence images (Hee, et al., 1992). PS-OCT needs some modification in the setup (Fig. 10), a polarized light source a polarization analyzer and a pair of quarter wave plate is needed.

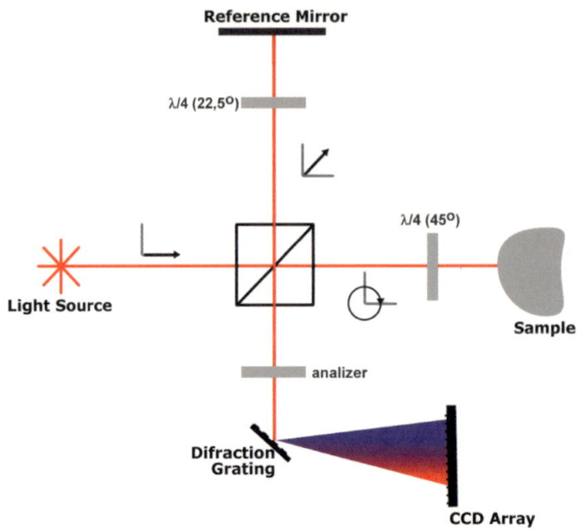

Fig. 10. Diagram to PS-OCT, a linear polarized light is spliced in two parts, in the sample arm the polarization is rotated to 45º, and in the sample arm is converted in circular polarized. The detector register the two orthogonal polarized light depending of the analyzer angle.

The polarized light hits the beam splitter, then a fraction is transmitted and other is reflected. Looking for the sample arm, one can think that the linear polarized light could be aligned with one optical axis of the sample (fast or slow axis) which would result a ordinary OCT scattering image, to avoid that a quarter wave plate, at 45°, is placed before the sample, causing a circular polarization state, in this way the light can be not aligned with any optical axis. The sample will cause some backscattering; the light will then, again, pass through the quarter wave plate. Remembering that the light has passed twice through the quarter wave plate at 45°, the light will return to the beam splitter with a linear polarization state rotated of 90° in respect of the original polarization state (light source).

So in the sample arm the light already contain all the information that is needed, the issue is now on the reference arm, that is because due to the reason that light interfere only when both beams have components of the same polarization state, i.e., a horizontally and a vertically polarized will not interfere, but a horizontally and a 45° polarized light will have a interference because the 45° state of polarization it is a superposition of vertical and horizontal polarization, and in this way will present interference pattern over a DC component. The polarization properties of the light can provide crucial information about the sample structure, and analyzing the polarization properties of a sample by the backscattered light as depth function allow to measure biological tissues and many other materials with strong scattering.

With a Polarization Sensitive OCT birefringence images can be performed (Fig. 11), in this way the differences between the refraction indices can be analyzed as an image, making diagnoses simple to be performed (Raele, et al., 2009).

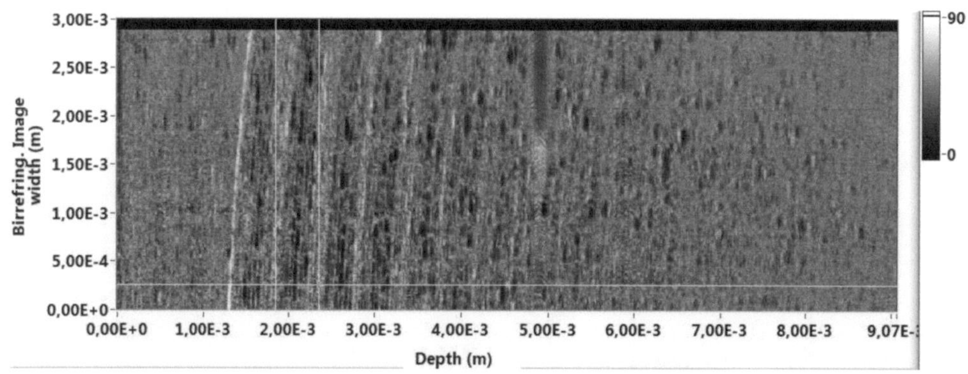

Fig. 11. Birrefringent image with PS-OCT of an adhesive tape, the birefringence of this tape was measured as $4.03(26)\times10^{-4}$

Besides PS-OCT images, a more complex, but also more complete way to study the polarization properties of light can be done using the Mueller Matrix theory (Bickel, et al., 1985).

5.1 Doppler OCT

A number of extensions of OCT capabilities for functional imaging of tissue physiology have been developed. Doppler OCT (Chen Z, et al., 1997), also named optical Doppler tomography (ODT), combines the Doppler principle with OCT to obtain high-resolution tomographic images of tissue structure and blood flow simultaneously (Fujimoto, et al., 2008). The Doppler OCT combine a technique developed in the 60´s, the Doppler velocimetry, with the traditional OCT high resolution images, mapping the fluids velocity and their localizations in the tissue open a new frontier as a diagnostic tool.

Considering the referential frame moving with a velocity v in this referential the frequency of light will be: $f_0 - \frac{1}{2}\pi K_i \cdot v$ (Fig. 12 (b)), and the scattered light field will be described by: The light frequency scattered by a moving object will be $2\pi f_0 t - (K_i - K_s)\cdot v$. In a Doppler OCT experiment the light and the scattered light share the same optical path in the sample arm like in (Fig. 12 (a)).

The Doppler shift can be determined measuring the phase shift between two consecutive spectra for A-scan, since in SD-OCT the A-scan are calculated with complex functions (Brezinski, 2006).

$$\tilde{i}_D(z'') = \mathrm{Re}[\tilde{I}_D(z'')] + iIm\left[\tilde{I}_D(z'')\right] = \left|\tilde{i}_D(z'')\right| e^{i\phi(z'')} \tag{30}$$

The phase of this complex function is the phase information of each A-scan described as:

$$\phi(z'') = \arctan\left\{\frac{Im[\tilde{i}_D(z'')]}{Re[\tilde{i}_D(z'')]}\right\} \tag{31}$$

the phase shift $\Delta\phi(z'')$ can be used to obtain the Doppler velocity.

$$V_D = \frac{\Delta\phi(z'')}{T4\pi k_0 n(k_0)\cos\theta} \tag{32}$$

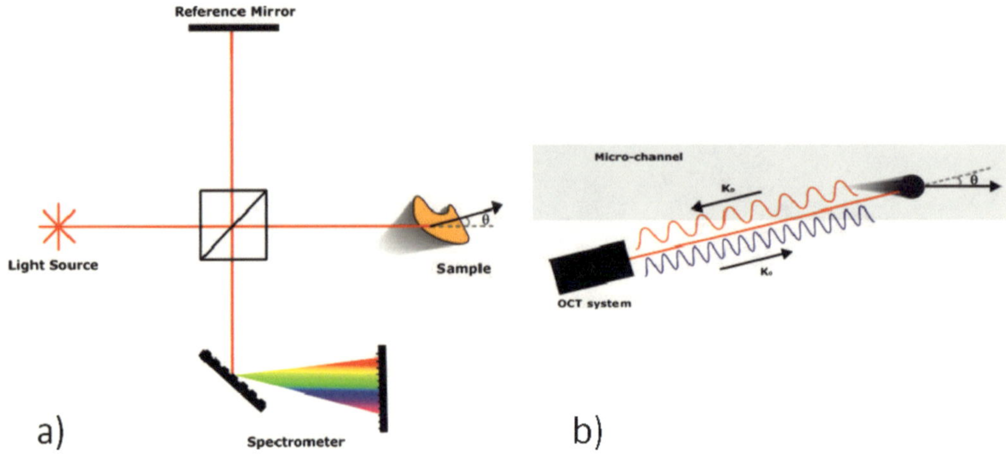

Fig. 12. (a) Diagram of Doppler-OCT and (b) the change in the wavelength of scattered light from a moving particle

6. Applications

6.1 Ophthalmology

OCT systems found it first application performing retina tomographies (Huang, et al., 1991), this was the beginning of what have become a revolution in the ophthalmology area, OCT allowed the specialists to exams the eye as the same way of histology does, but in a completely non invasive, non traumatic and painless way and also in real time. Latter OCT image resolution and depth characteristics matches the needs of ophthalmologists and the eye itself, especially when the OCT is operating at the NIR region, which has low attenuation and does not sensibilizes the vision cells. Besides retina the specialists also use OCT to examine the eye anterior segment and cornea, see Fig. 13. Nowadays OCT is routine in many ophthalmic clinics around the world, and researches are improving this tool continuously.

Diseases as glaucoma and retinal dystrophy among many other examples (Schuman, et al., 2004) can be diagnosed using OCT systems, some of them were a complicated issue to diagnoses, as macular degeneration, has now the OCT as a primary way to do it.

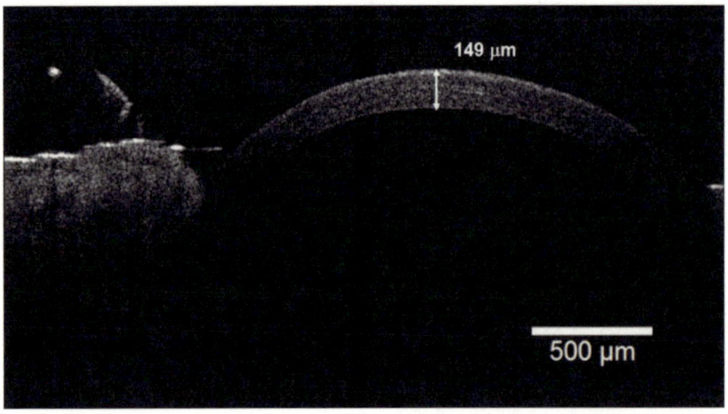

Fig. 13. Image of a mice cornea. The arrow indicates the place where a thickness of it was evaluated.

In terms OCT improvements, some outstanding studies has being done, for instance reported 3 μm of resolution (Wampler), this is refined enough to actually "see" the retina cells. The possibilities of applications are many, monitoring LASIK proceedings, study the blood flux, etc.

6.2 Dermatology

OCT also caused a significant impact dermatology for almost the same reasons that promoted OCT in ophthalmology. As shown in Fig. 1, OCT technique can perform images where is possible to indentify the different skin structures (A. stratum corneum; B. epidermis; C. dermis and D. Sweat gland), researches are studing many features of skin *in vivo,* impossible feat before OCT.

Diseases, even skin cancer, can be diagnosed by this tool (Mogensen, et al., 2009). Skin has been studied extensively also with PS-OCT (Hee, et al., 1992), that is because many compounds in skin presents birrefringence and the concentration of this compounds are related with skin health.

6.3 Odontology

In odontology, a series of reports first appeared (Colston, et al., 1998), (Feldchtein, et al., 1998) in 1998, with imaging of both hard and soft oral tissues. This led to several diagnostics of bucal diseases, including periodontal, early caries, among others. Another area in dentistry where OCT can have important findings is in dental restoration imaging.(Wang, et al., 1999), (Fried, et al., 2002), (Otis, et al., 2000), (Jones, et al., 2004), (Wang, et al., 1999) and (Fried, et al., 2002) exploited polarization-sensitive OCT to identify dental tissue/restoration interfaces. To date, there is no quantitative method capable to perform *in vitro* or *in vivo* analysis of dental restoration, particularly from the clinical point of view. Visual inspection and X-ray imaging are not precise enough to diagnose small gaps that result from bad restoration procedures. Dental tissues are high scattering media and infrared light can penetrate the full enamel extension (Fried, et al., 1995).

Although, in odontology, OCT is not yet broadly available, as in ophthalmology, the potential of the technique promises a fast technological development that requires more laboratory evaluation, prior to clinical trials.

One of important area of interest is the restorative procedures, the application of OCT to dental restoration, particularly analyzing failure gaps left after the restoration has been performed (Melo, et al., 2005).

Another import field of interest is the dental caries, and this disease is known as a multifactorial pathological process, characterized by hard tissue demineralization. Commonly, dentists evaluate the oral health of a patient through three main methods: visual, tactile examination and radiographic imaging (Bosh, 1993). The visual method cannot detect early caries lesion and depends of the dentist ability to identify these lesions. There are many caries detector dyes commercially available purported to assist the dentist in differentiation of infected tissue, but they are not specific and would result in unnecessary removal of healthy tooth structure (McComb, 2000). OCT in dentistry has been recently used to in vitro studies evaluating enamel interface restoration (Melo, et al., 2005), early caries diagnostics (Freitas, et al., 2006), and analysis of the performance of dental materials (Braz, et al., 2009). In 2006, the first OCT image of dental pulp was performed using rat's teeth (Kauffman, et al., 2006), and more recently, remaining dentin and pulp chamber from human's teeth were also imaged by OCT in vitro (Fonseca, et al., 2009).

The OCT can detect and quantify demineralization (FREITAS, et al., 2009). The Fig. 14 (a) shows an OCT image for a sample submitted to the demineralization process for 11 days, and in (b) the 3D image reconstruction.

The OCT system provides a powerful contact less and noninvasive diagnostic method that could be used to complement the traditional diagnostic methods such as X-ray radiography, avoiding potentially hazardous ionization radiation.

Polarization-sensitive optical coherence tomography (PS-OCT) is potentially useful for imaging the nonsurgical remineralization of dental enamel. The PS-OCT can image the effects of mineralization status and scattering properties of dental caries (Jones, et al., 2006), (Baumgartner A, et al., 2000).

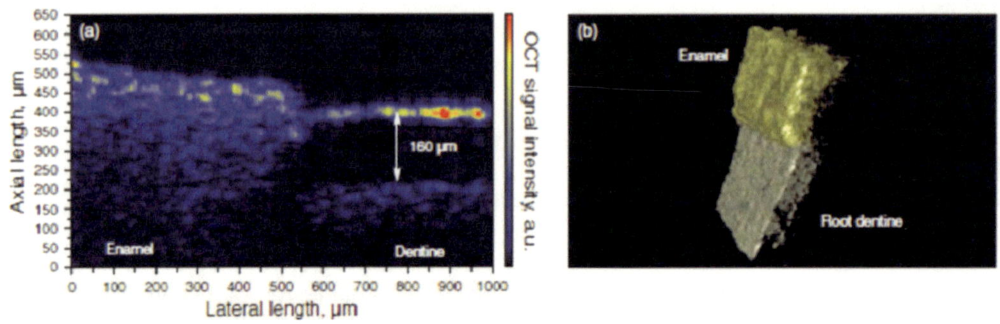

Fig. 14. (a) Transversal OCT image after the sample was submitted to the demineralization process for 11 days and (b) 3D image reconstruction allows another image sections analysis different from those obtained directly from the OCT system.

6.4 Non-biomedical applications

Despite the wide use of OCT for biological application, it noninvasive and nondestructive features are very attractive in other areas. The use of OCT technique has grown on art and cultural heritage artifacts studies (Targowski, et al., 2006) (Liang, et al.), e.g., for investigate various art objects such as oil paintings, see Fig. 15 and Fig. 16, on canvas to image pigments, glaze and varnish layers (Rouba, et al., 2008) or historical coins (Amaral, et al., 2009) for instance.

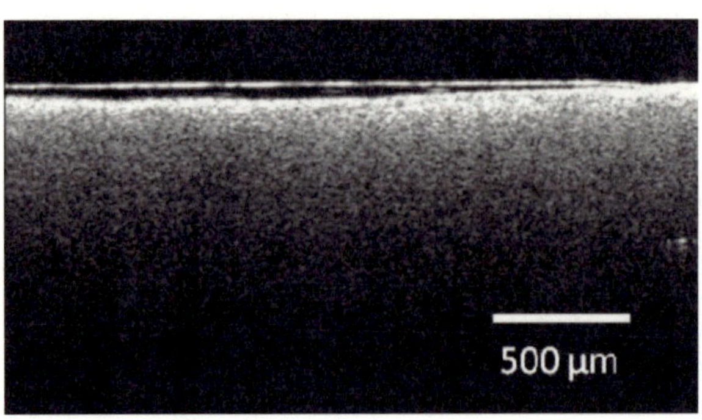

Fig. 15. Image of oil painting with varnish layer

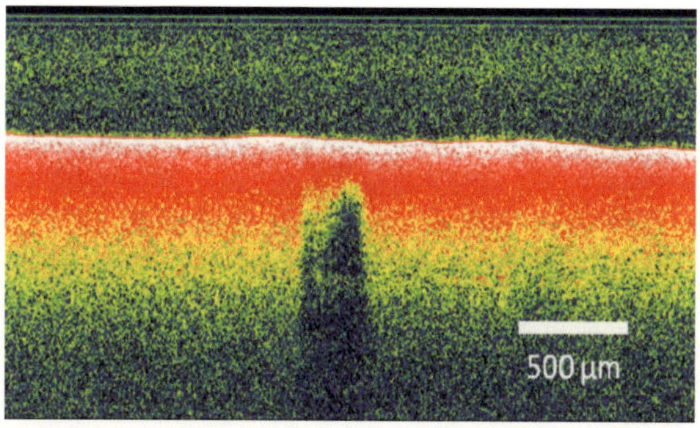

Fig. 16. Image of oil painting over sketch made with pencil.

It has been also used also for investigate wood from musical instruments structures, as fibers, rays, cell wall distribution. As well as varnished wood: morphological properties, such roughness, interfaces, and thickness; and compositional properties, such shape, size, distribution of pigments and fillers. And even more, understanding of the penetration of the varnish inside the wood or the wear processes of these materials, applying these kinds of study on a 18th century Italian violin (Latour, et al., 2009).
OCT finds on industry another field of actuation, due to the real time and noncontact feature. Moreover, some traditional technique, used in the industry, cannot be applied to some materials. For e.g., to measure textile roughness, see Fig. 17, traditional technique to roughness measurement, that uses contact cantilever, could cause erroneous values and even damage the textile. Is reported the development of a methodology and algorithm to give metrological parameter of roughness according DIN4768 (Amaral, et al., 2009), that can be applied to delicate samples. Applications also can be found in printed electronics products quality (Czajkowski, et al., 2010) and in paper industry (Prykär, et al., 2010).

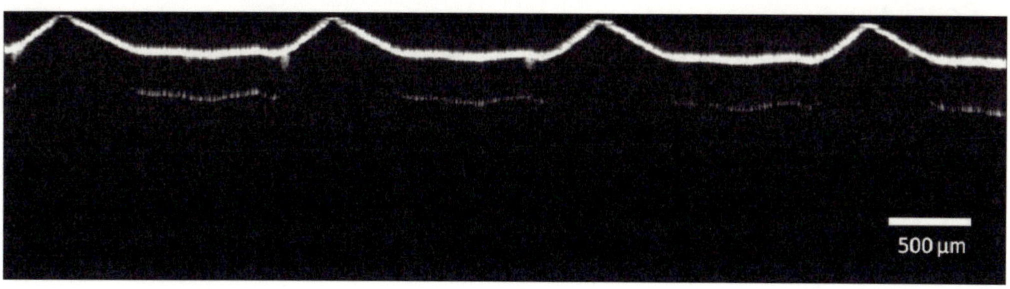

Fig. 17. OCT images can be used to perform roughness measurements without contact.

6.5 Dermatology

OCT also caused a significant impact dermatology for almost the same reasons that promoted OCT in ophthalmology. As shown in Fig. 1, OCT technique can perform images where is possible to indentify the different skin structures, researches are studing many features of skin *in vivo,* impossible feat before OCT.

Diseases, even skin cancer, can be diagnosed by this tool (Mogensen, et al., 2009). Skin has been studied extensively also with PS-OCT (Hee, et al., 1992), that is because many compounds in skin presents birrefringence and the concentration of this compounds are related with skin health.

Another important subject in dermatology is cosmetic science, which claim for a non invasive method to efficiency evaluation of anti-aging or anti-wrinkle products, for instance. Hair is a substructure of skin and it also can be studied with OCT. Fig. 18 (a) shows the cross-sectional image of Afro-Ethnic hair, where is possible to identify two hair structures: medulla and cortex (Lademann, et al., 2008), (Velasco, et al., 2009). The tridimensional image (Fig. 18 (b) was built starting from 601 cross-sectional images (slices) like Fig. 18 (a).

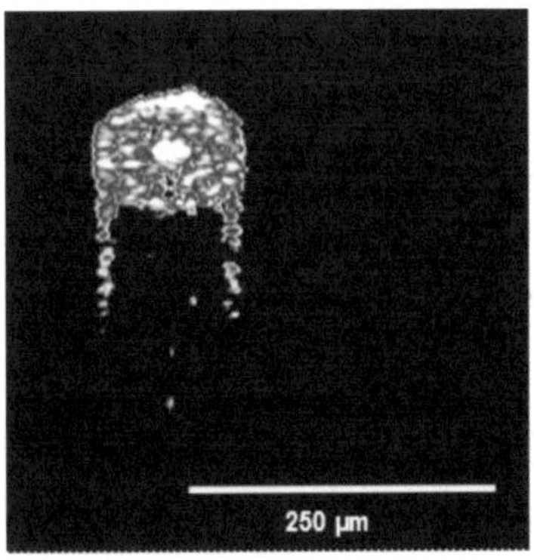

a)

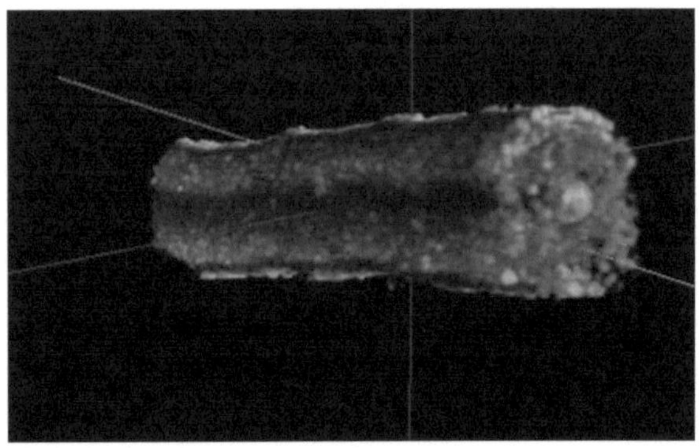

b)

Fig. 18. OCT image of an Afro-Ethnic hair sample. In (a) a sample of cross-section fiber and at (b) a tridimensional reconstruction of that fiber, showing the main hair structures: medulla, cortex and cuticle.

8. Conclusions

Using broadband light sources, as mode locked lasers, in an interferometric setup allows perform tomographic images with high resolution and non invasive way. This technique, known as optical coherence tomography (OCT), found, in many areas of knowledge, great potential. Worth mention that in ophthalmology OCT promoted a revolution in the way that diagnoses are performed.

Although OCT is a relatively new technique, created in 1991, the applications and more powerful and versatile systems are increasing rapidly in number.

9. References

A Baumgartner [et al.] Polarization–Sensitive Optical Coherence Tomography of Dental Structures [Artikel] // Caries. Res.. - 2000 . - Bd. 34.

Amaral Marcello M. [et al.] "Roughness measurement methodology according to DIN 4768 using optical coherence tomography (OCT) [Konferenz] // Proc. SPIE. - Munch : Spie, 2009. - Bd. 7390.

Amaral Marcello M. [et al.] Laser induced breakdown spectroscopy (LIBS) applied to stratigrafic elemental analysis and optical coherence tomography tomography (OCT) to damage determination of cultural heritage Brazilian coins [Konferenz] // Proc. SPIE / Hrsg. Spie. - Munch: Spie, 2009. - Bd. 7391.

Andrew M. Rollins Siavash Yazdanfar, Jennifer K. Barton, Joseph A. Izatt, " Real-time in vivo color Doppler optical coherence tomography [Artikel] // J. Biomed. Opt.. - 2002. - S. Vol. 7, 123.

Baumgartner A Dichtl S, Hitzenberger C K [et al.] Polarization–Sensitive Optical Coherence Tomography of Dental Structures [Artikel] // Caries. Res.. - 2000. - Bd. 34. - S. 59-69.

Bickel W.S. und Bailey W.M. Stokes vectors, Mueller matrices, and polarized scattered light [Artikel] // Am. J. of Phys.. - 1985. - S. v.53, n.5, p. 468-478.

Bosh B. Angmar-Mansson and J.J. Ten Optical Methods for the Detection and Quantification of Caries [Artikel] // Adv. Dent. Res. - 1993. - Bd. 7. - S. 70–79 .

Bouma Brett E. und Tearney Guillermo J. Handbook of Optical Coherence Tomography [Buch]. - NY : Marcel dekker Inc, 2002.

Braz A.K.S. [et al.] Evaluation of crack propagation in dental composites by optical coherence tomography [Artikel] // Dent. Mater.. - 2009. - Bd. 25. - S. 74-79.

Brezinski Mark E. Optical Coherence Tomography, Principles and Applications [Buch]. - Oxford : Elsevier Inc., 2006.

Chen Z Milner T, Dave D und Nelson [Artikel] // J. Opt. Lett.. - 1997. - S. 22 64–6.

Colston B. W. [et al.] Dental OCT [Artikel] // Opt. Express. - 1998. - Bd. 3. - S. 230-238.

Colston B. W. [et al.] Imaging of hard-and soft-tissue structure in the oral cavity by optical coherence tomography [Artikel] // Appl. Opt.. - 1998. - Bd. 37.

Czajkowski Jakub [et al.] Optical coherence tomography as a method of quality inspection for printed electronics products [Artikel] // Opt.l Review. - 2010. - 3 : Bd. 17. - S. 257-262.

Eden Alec The search for Christian Doppler [Buch]. - Wien : Springer-Verlag, 1992.

Feldchtein F. I. [et al.] In vivo OCT imaging of hard and soft tissue of the oral cavity [Artikel] // Opt. Express. - 1998. - Bd. 3. - S. 239-250.

Fercher A. F. [et al.] In vivo optical coherence tomography [Artikel] // Amer. J. Ophthalmol.. - 1993. - S. v.116, pp. 113-114,.

Fonseca D.D.D. [et al.] [Artikel] // J. Biomed. Opt.. - 2009. - Bd. 14. - S. 024009.

Freitas A. Z. [et al.] Imaging carious human dental tissue with optical coherence tomography [Artikel] // J. Appl. Physics. - 2006. - 024906 : Bd. 99.

FREITAS A.Z. [et al.] Determination of dental decay rates with optical coherence tomography [Artikel] // Laser Physics Letters. - 2009. - Bd. 6. - S. 896-900.

Freitas A.Z. [et al.] Imaging Carious Human Dental Tissue with Optical Coherence Tomography [Artikel] // Journal of Applied Physics. - 2006. - S. v. 99, n.2, p.24906.

Fried D. [et al.] Nature of light scattering in dental enamel and dentin at visible and nearinfrared wavelengths [Artikel] // Appl. Opt.. - 1995. - Bd. 34. - S. 1278–1285.

Fried D., Shafi S. und Featherstone J. D. B. Imaging caries lesions and lesion with polarization sensitive optical coherence tomography [Artikel] // J. Biomed. Opt.. - 2002. - Bd. 7. - S. 618–626.

Fujimoto J. G. und Drexler W. Optical Coherence Tomography [Buchabschnitt]. - Berlin : Springer, 2008.

Goode B.G. Optical Coherence Tomography: OCT aims for Industrial Applications [Online]. - 10. 9 2008. - 10. 9 2008. - http://www.laserfocusworld.com/display_article/368547/12/none/none/Feat/OPTICAL-COHERENCE-TOMOGRAPHY:-OCT-aims-for-industrial-application.

Hee M. R. [et al.] Polarization-sensitive low-coherence reflectometer for birefringence characterization and ranging [Artikel] // J. Opt. Soc. Am B. - 1992. - S. v. 9, p. 903-908.

Hee M.R. [et al.] Optical coherence tomography of the human retina [Artikel] // Arch. Ophthalmol.. - 1995. - S. Vol. 113, pp.326-332.

Huang D. [et al.] Optical Coherence Tomography [Artikel] // Science. - 1991. - 1178 : Bd. 254.

J. Wang T. E. Milner, and J. S. Nelson Characterization of fluid flow velocity by optical Doppler tomography [Artikel] // OPTICS LETTERS. - 1995. - S. Vol. 20, No. 11 / 1337-1339.

Jones R. S., Staninec M. und Fried D. Imaging artificial caries under composite sealants and restorations [Artikel] // J. Biomed. Opt.. - 2004. - Bd. 9. - S. 1297–1304 .

Jones Robert S. [et al.] Remineralization of in vitro dental caries assessed with polarization-sensitive optical coherence tomography [Artikel] // Journal of Biomedical Optics. - 2006. - 1 : Bd. 11. - S. 014016.

Jones Robert S. [et al.] Remineralization of in vitro dental caries assessed with polarization-sensitive optical coherence tomography [Artikel] // Journal of Biomedical Optics. - 2006. - 1 : Bd. 11.

Kato I. T. [et al.] Inhibition of enamel remineralization with blue LED: an in vitro study [Konferenz] // Proc. of Spie Lasers in Dentistry XV. - San José : Spie, 2009. - S. v. 7162.

Kauffman C.M.F. [et al.] [Konferenz]. - [s.l.] : Proc. SPIE, 2006. - 613707 .

Lademann J. [et al.] Application of optical non-invasive methods in skin physiology [Artikel] // Laser Phys. Lett.. - 2008. - 5 : Bd. 5. - S. 335-346.

Latour Gaël [et al.] Structural and optical properties of wood and wood finishes studied using optical coherence tomography: application to an 18th century Italian violin [Artikel] // Applied Optics. - 2009. - 33 : Bd. 48. - S. 6485-6491 .

Liang Haida [et al.] 2. En-face optical coherence tomography: a novel application of non-invasive imaging to art conservation [Artikel] // Optics Express. - 16 : Bd. 13. - S. 6133-6144 .

Martinez O. E. 3000 times grating compressor with positive group velocity dispersion: Application to fiber compensation in 1.3-1.6 pm regime [Artikel] // O. E. Martinez, IEEE J. Quantum Electron. QE-23, 59 (1987).. - 1987. - Bde. QE-23.

McComb D. Diagnoses of occlusal caries [Artikel] // J. Can. Dent. Assoc.. - 2000. - Bd. 66. - S. 195–198 .

Meirelles R.L., Aggio F.B. und Costa R.A. STRATUS optical coherence tomography in unilateral colobomatous excavation of the optic disc and secondary retinoschisis [Artikel] // Graefes Archive For Clinical And Experimental Ophthalmology. - 2005. - S. V. 243, n. 1, p. 76-81.

Melo L. S. A. [et al.] Evaluation of enamel dental restoration interface by optical coherence tomography [Artikel] // Journal of Biomedical Optics. - 2005. - S. v.10, n.6, p.1-5.

Mogensen Mette [et al.] OCT imaging of skin cancer and other dermatological diseases [Artikel] // Journal of Biophotonics. - 2009. - 6-7: Bd. 2. - S. 442 - 451.

Otis L. L. [et al.] Identification of occlusal sealants using optical coherence tomography [Artikel] // J. Clin. Dent.. - 2000. - Bd. 14. - S. 7–10.

Prykär Tuukka [et al.] Optical coherence tomography as an accurate inspection and quality evaluation technique in paper industry [Artikel] // Opt. Review. - 2010. - 3 : Bd. 17. - S. 218-222.

Raele Marcus Paulo and Freitas Anderson Zanardi Desenvolvimento de um sistema de tomografia por coerência óptica no domínio de Fourier sinsível á polarização e sua utilização na determinação das matrizes de Mueller [Report]. - São Paulo : Instituto de Pesquisas Energéticas e Nucleares IPEN-CNEN/SP, 2009.

Rouba Bogumiła J. [et al.] Optical coherence tomography for non-destructive investigations of structure of objects of art [Konferenz] // 9th International Conference on NDT of Art. - Jerusalem : [s.n.], 2008 .

Schuman Joel S., Puliafito Carmen A. and Fujimoto James G. Optical Coherence Tomography of Ocular Diseases [Book]. - NJ : SLACK, 2004.

Shuliang J., Gang Y. und Lihong Depth-resolved two-dimensional Stokes vectors of backscattered light and Mueller matrices of biological tissue measured with optical coherence tomography [Artikel] // Appl. Opt.. - 2000. - S. v. 39, n. 34, p. 6318-6324.

Targowski Piotr, Góra Michalina und Wojtkowski Maciej Optical Coherence Tomography for Artwork Diagnostics [Artikel] // Laser Chemistry. - 2006. - Bd. 2006.

Tearney G. J., Bouma B. E. und Fujimoto J. G. High-speed phase- and group-delay scanning with a grating-based phase control delay line [Artikel] // Opt. Lett.. - 1997. - Bd. 22.

Velasco Maria Valéria Robles [et al.] Prospective ultramorphological characterization of human hair by optical coherence tomography [Artikel] // Skin Research and Technology. - 2009. - Bd. 15. - S. 440-443.

Wampler Steve OCT News [Online] // OCT News. - 30. 07 2010. - https://publicaffairs.llnl.gov/news/news_releases/2010/NR-10-07-02.html.

Wang X. J. [et al.] Characterization of dentin and enamel by use of optical coherence tomography," , () [Artikel] // Appl. Opt.. - 1999. - Bd. 38. - S. 2092–2096.

Web of Science [Online] // Web of Science. - 7 2010, 7. - http://www.isiknowledge.com/WOS.

Zhongping Chen Thomas E. Milner, Digant Dave. und Nelson J. Stuart Optical Doppler tomographic imaging of fluid flow velocity in highly scattering media [Artikel] // Optics Letters. - 1997. - S. Vol. 22, Issue 1, pp. 64-66.

Zhongping Chen Thomas E. Milner, Shyam Srinivas, Xiaojun Wang, Arash Malekafzali, Martin J. C. van Gemert, and J. Stuart Nelson Noninvasive imaging of in vivo blood flow velocity using optical Doppler tomography [Artikel] // Optics Letters. - 1997. - S. Vol. 22, Issue 14, pp. 1.

21

High Resolution Biological Visualization Techniques

Pavlina Pike[1], Christian Parigger[2] and Robert Splinter[3]
[1]*University of Alabama at Birmingham*
[2]*University of Tennessee Space institute*
[3]*University of North Carolina at Charlotte*
USA

1. Introduction

Imaging with ionizing radiation is an integral part in today's medicine. For example, breast tissue is more radiosensitive during puberty and exposures to the chest during that time would impose a higher risk of radiation-induced breast cancer (Doody et al., 2000). The use of ionizing radiation for medical imaging has significantly increased in the past 20 years, which has caused the effective dose per individual in the US to increase from 0.53 mSv in the early 1980s to 3 mSv in 2006 as reported by the National Council on Radiation Protection and Measurement. [NCRP No. 160, 93] This has naturally raised concerns about the possible risks that are associated with multiple exposures from diagnostic sources of radiation. There is a huge effort towards creating imaging modalities that can visualize tissue with a reduced dose to the patient without loss of image quality. Therefore, it would be advantageous to develop visualization techniques that employ non-ionizing radiation and that provide comparable high-resolution image quality as the currently available modalities. This is one reason why Magnetic Resonance Imaging (MRI) technology is gaining so much popularity in spite of the cost and the length of the exams.

The non-invasive nature of light in the near-infrared region and its reduced attenuation in tissue are the reasons why optical imaging is gaining popularity. However, tissue scatter is a major limitation in optical imaging techniques. It is possible to combine optical imaging and ultrasound to produce images. This is accomplished by the use of ultra-short laser pulses that are absorbed in the medium, which cause the temperature of the material to rapidly rise and fall. Consequently, the irradiated volume becomes an internal source of ultrasonic waves, which can be used to obtain diagnostic information. Photo-acoustic spectroscopy has been shown to be a great tool to successfully visualize and characterize:

- blood vessels (Zhang et al. 2009; Hoelen et al. 1998; Sethuraman, S. et al. 2006);
- breast cancer (Ermilov et al. 2009);
- brain in vivo (Wang, X. D. et al. 2003; Yang, X. et al. 2007; Lingu et al. 2007);
- prostate lesions (Yaseen et al. 2010).

Photo-acoustic spectroscopy has also been considered for detection of caries in its early stage (Kim et al. 2006), analysis of dental materials (Coloiano et al. 2005) and potentially for visualizations under porcelain crowns. (Pike et al. 2007) New clinical demands for imaging are continuously emerging, placing different constraints on the imaging requirements and the imaging conditions, e.g., in-vivo versus in-vitro imaging. One such new development in biology is regenerative medicine, which addresses cell structure growth for transplant purposes. In regenerative medicine the organ is grown outside the body and requires quality control before it can be implanted in a living organism with a high degree of success for survival (Hendee W et al., 2008).

2. Examples of available imaging applications

The visualization of superficial vascular malformations, for instance has been shown to benefit most from thermography (Saxena AK, Willital GH, 2008). Another example is in dentistry: Dental X-ray images are problematical for clear display of the vitality of dental pulp, especially under ceramic crowns. Such an evaluation is important, because severe pain can originate from the pulp as a result of trauma or dental caries. In addition, for patients undergoing a hart valve replacement surgery or cancer patients undergoing treatment while on immunosuppressant medication such infections can be life threatening (Pike P et al., 2007).

Optical Coherence Tomography (OCT) can be applied in various formats, using the respective optical characteristics of laser- light: coherence, spectral compositions, polarization and time-of-flight. The specific OCT applications available are for instance: Time domain OCT; Frequency domain OCT (FD-OCT); Polarization-sensitive OCT; spatially encoded frequency domain OCT (spectral domain or Fourier domain OCT); and time encoded frequency domain OCT (also swept source OCT) in addition to combinations of their mechanisms. Each application reveals its own particular details about the tissue composition, pathological conditions and structural formation (Huang SA et al., 2007). Microwave images use Giga Hertz electromagnetic radiation to interact with the cellular structure on an electronic level (Fear EC et al., 2002).

Microwave imaging uses the fact that cells show dipole character with a membrane that act as independent electronic circuits. The electronic tissue details can provide histological and pathological information non-invasively and in-vivo. The mechanism of action relies on the cellular components' dielectric properties that will influence the wave propagation and hence act as a sensing mechanism (Semenov et al., 1995, Semenov et al., 1996).

Spectroscopic imaging for all wavelength ranges uses the alteration in the spectral structure of the input signal both in the time and frequency domain as a result of interaction with the biological medium. Time-resolved spectroscopy is gaining popularity as a result of the reduced price of sensors that are capable of the high speed, short-time response (femto-second) as well as the reduced size of the sensors.

In optical, microwave and acoustic imaging techniques the spectral performance will embed crucial tissue information. However, the most accurate spectral application of Raman spectroscopy can primarily be realized free-space since the delivery device usually introduces its own Raman spectral aberrations. In this case time-resolved spectroscopy may offer a solution for all non-ionizing imaging techniques.

The purpose of this chapter is to summarize both in-vivo and in-vitro techniques that involve image formation based on signals generated in tissue by the various visualization methods applying non-ionizing radiation for high-resolution signal generation and sensing/imaging. In this chapter three high resolution imaging techniques will be described each relying on the time-of-flight principle in various formats [both acoustic and optical] and the available options to enhance the level of detail that can be resolved. The importance of high resolution imaging for diagnostic clinical and laboratory applications is illustrated by the imaging requirements which is followed by the detailed description of the scientific basis of Photo-Acoustic imaging (PAI) and several potential applications, followed by the mechanism of action of ultra short laser pulses, generally crucial to the photo-acoustic technique that are particularly critical to establish non-destructive testing conditions. Additionally the coherence aspect of time-of-flight is described with Optical Coherence tomography and the available options within this technique. The chapter will conclude with a description of short-pulse measurements and in particular application of autocorrelation measurements from interference images.

3. Biological imaging requirements

One possibility is to generate acoustic waves with femto-second laser pulses and analyze the recorded signals. This process is called Photo-Acoustic (PA) spectroscopy and has been shown to be highly sensitive for variety of applications, including the determination of small concentrations of molecular species, and the analysis of photo-thermal expansion and contractions of molecules. One advantage of such techniques is that the signal generation is primarily a result of the absorption of electromagnetic radiation inside the tissue. Scattering is not a major factor in the signal degeneration. This method has been shown to be useful to detect breast cancer (Ermilov SA et al., 2009), to visualize vasculature in mouse brain by adding golden nano-spheres to improve optical contrast (Lu W et al., 2010), to evaluate tissue growth and regeneration (Yuan Z et al., 2010), and 3-D visualization blood vessels (Zhang EZ et al., 2009). In fact different variations of this technique have been developed for specific clinical applications.

The advantage of the PA technique is that the signal is generated primarily as a result of the absorption of light. Distortion of the signal due to scattering is not an issue as in other imaging methods. Recently, it has been shown to be viable for variety of biomedical applications, such as visualization of tumor angiogenesis and characterization of arterial walls.

4. Photo-acoustic spectroscopy

4.1 Photo-acoustic imaging mechanism of action

It was Alexander Bell who reportedly first described the phenomenon of sound generation in materials by light (Bell, 1880). As he shone chopped light on strongly absorbing substances he observed that audible sound was generated. Todays laser systems can deposit energy via ultra short optical pulses and cause the rapid thermal expansion and relaxation of the targeted volume in the tissue. That in turn will cause the generation of a pressure wave which will carry information about the thermal and mechanical characteristics of the tissue in which it originated. The ultrasonic wave is then detected by the use of a transducer or an array of transducers for parallel data acquisition.

The energy deposited by the laser puse will be absorbed and and then it will diffuse into the surroundings. The temperature will initially spike and then a gradual decrease will follow. The time it takes for it to reach 37% of its original value is called the **thermal relaxation time**, defined by equation 1:

$$\tau_r = \delta^2 / 4\alpha \tag{1}$$

where δ is the smaller of either the optical penetration depth [mm] or the dimesion of the absorbing lesion of interest, and α is the thermal diffusivity [cm²/s]. The ultrasonic wave that is generated will travel through the thickness of the target volume. The time it takes it to do so is called the **stress relaxation time**, shown by equation 2:

$$\tau_s = \frac{\delta}{V_a} \tag{2}$$

where V_a is the speed of sound in tissue. The energy deposition in the mterial will be defined spatially by the laser light distribution if the thermal confinement condition is fulfilled. This will happen when the laser pulse duration (τ_p) is less than the thermal relaxation time(τ_r). When ultra-short laser pulses are used, the stress relaxation is significantly faster than the thermal relaxation and $\tau_p < \tau_s < \tau_r$.

Although the temperature rise can be quite small, measurable pressure can be generated. The pressure rise ($p(r)$) can be calculated using material properties, such as the isothermal compressibility, κ [Pa^{-1}], the thermal coefficient of volume expansion and β [K^{-1}] as expressed by equation 3 (Gusev & Karabutov, 1993):

$$p(r) = \frac{\beta T(r)}{\kappa} \tag{3}$$

For example, the isothermal compressibilty of dental enamel can be calculated to be about $8.5 \times 10^{-12}\ Pa^{-1}$ using density of 3000 kg/m³ and speed of sound of 6250 m/s. If we take the volume expansion coefficient of enamel to be about $20 \times 10^{-6}\ K^{-1}$ and temperature rise after the interaction with a 70 fs laser pulse to be about 1 μK (Pike et al., 2007). The pressure generated is therefore about 2.35 Pa.

The generation of a pressure wave and its consequent propagation through the medium is governed by the nonhomogeneous wave equation or also called the "general photoacoustic equation" given by equation 4:

$$\nabla^2 p - \frac{1}{v^2}\frac{\partial^2 p}{\partial t^2} = -\frac{\beta}{C_p}\frac{\partial H}{\partial t} \tag{4}$$

where $H(x,y,z,t)$ is the absorbed electromagnetic energy density that is converted into heat per unit volume [W/cm³], β is the coefficient of thermal expansion [K^{-1}], C_p is the specific heat at constant pressure [J/kg.K], T_0 is the ambient temperature [K], α is the thermal diffusivity [cm²/s] and v is the speed of sound [cm/s] (McKenzie, 1990). The amount of energy that is absorbed as a result of the deposition of the laser energy pulse depends on the optical properties of the material and more specifically, the optical absorption coefficient μ_a [cm^{-1}] and the laser energy fluence, I [J/cm²]. Therefore, we can rewrite equation 4 as:

$$\nabla^2 p - \frac{1}{v^2}\frac{\partial^2 p}{\partial t^2} = -\frac{\beta \mu_a}{C_p}\frac{\partial I}{\partial t} \tag{5}$$

The source term in this equation can take different forms depending on the application of interest. For example, it has been modeled as a Gaussian distribution in space and time (Heretier, 1983), Lai, 1982) as follows:

$$I_{her}(r,t) = \frac{2E}{\pi^{3/2} w_0^2 \tau_L} \exp\left[-\frac{r^2}{2w_0^2} - \frac{t^2}{2\tau_L^2} \right] \tag{6}$$

where w_0 is the laser beam waist, E is the energy per pulse, and the pulse full-width at half-maximum (FWHM) is τ_L. The solution for the pressure therefore can be expressed in terms of Bessel functions $I_{1/4}$ and $K_{1/4}$, as displayed in equations 7 through 8:

$$p(r,t) = \frac{E\tau_e^{-3/2}}{8\pi^{1/2}} \left(\frac{v}{r}\right)^{1/2} \left(\frac{\mu_a \beta}{C_p} \frac{d\Phi(t')}{dt'} - \frac{\gamma \tau_e^{-1}}{2ncv^2} \frac{d^2\Phi}{dt'^2} \right) \tag{7}$$

$$\Phi(t') = \sqrt{t'} \exp(-t'^2 / 4) \left[\frac{\sqrt{2}}{\pi} K_{1/4}\left(\frac{t'^2}{4}\right) + 2\Theta(t') I_{1/4}\left(\frac{t'^2}{4}\right) \right] \tag{8}$$

$$t' = (t - r / v) / \tau_e \tag{9}$$

$$\tau_e = (\tau_p^2 + \tau_a^2)^{1/2} \tag{10}$$

4.2 Temperature distribution following femtosecond laser irradiation

Ultra short laser radiation induces sudden temperature and pressure spikes in the local volume because the pulse duration is much shorter than the thermal relaxation time of the medium. In a previously reported experimental study (Pike at al., 2007) rather than using single femto-second pulses pulse trains of femtosecond laser pulses were employed for measurement of the temperature rise. Figure 1 shows the experimental arrangement, and it illustrates a typical experimental result of the temperature rise due to repetitive exposure of dental material to 40 ms pulse trains of 70 fs laser pulses.

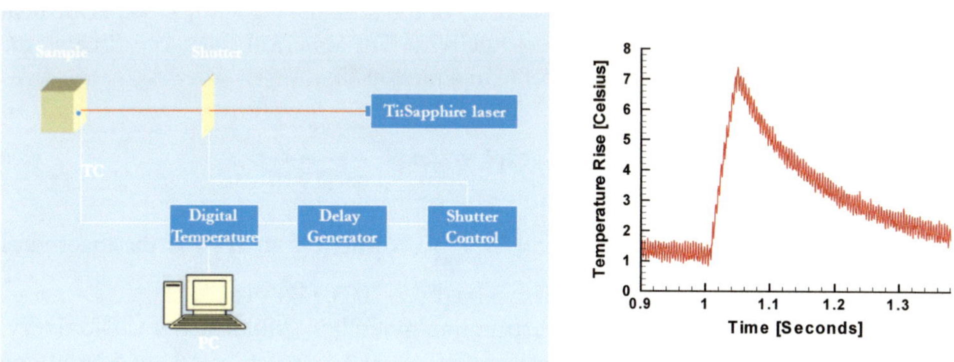

Fig. 1. (Left) Experimental arrangement for measurement of temperature rise due to exposure of dental material to 40 ms trains of femtosecond laser pulses that occur at a rate of 76 MHz (Spectra Physics model Tsuname Ti:Sapphire). (Right) Experimental results for the temperature rise during a 40 ms exposure - 3 Joules/cm^2 is delivered to the material causing a temperature rise of typically 6 degrees Celsius due to exposure to approximately 3 million femtosecond laser pulses.

From experimental data, see Fig. 1, a temperature rise of 2 μK can be inferred due to the 6 degree rise from 3 million laser pulses. Clearly the thermal relaxation time is much slower (order of 100 millisecond) than the duration of an individual laser pulse of 70 femtosecond. Modeling of the temperature rise due to ultra short laser radiation will be addressed in the following.
The heat transfer equation describes the temperature distribution following laser irradiation, indicated in equation 11

$$\kappa\nabla^2 T - \frac{\kappa}{\alpha}\frac{\partial T}{\partial t} = -T_0\beta\frac{\partial H}{\partial t} \tag{11}$$

Solution for the temperature can be found by modeling the source term on the right hand side of Equ. 11 in terms of the Dirac delta distribution (Qiu, 1993) in equation 12 below

$$T_0\beta\frac{\partial H}{\partial t} = \frac{(1-R)}{\kappa}\mu_a J\exp[-\mu_a z]\exp[-\frac{2(\vec{r}')^2}{\omega^2}]\delta(t), \tag{12}$$

$$\omega(z) = \omega_0\left(1+\left(\frac{\lambda(z-z_0)}{\pi\omega_0^2}\right)^2\right)^{1/2} \tag{13}$$

where R is the surface reflectivity, ω_0 is the laser beam minimum waist size [cm] located at the distance z_0. The analytic solution can be found utilizing Green's function, which can be written as follows in equation 14:

$$G(\vec{r},t;\vec{r}',t') = -\left(\frac{a^2}{4\pi(t-t')}\right)^{\frac{3}{2}}\frac{1}{a^2}\exp\left[-\frac{a^2}{4}\frac{|\vec{r}-\vec{r}'|^2}{t-t'}\right] \tag{14}$$

In this equation $a^2 = \rho c/\kappa$ with ρ - mass density of the material [kg/m³], c - specific heat capacity [J/kg.K], and χ is the thermal conductivity. For spherical polar coordinates the analytic solution yields the temperature profile in equation 15:

$$T = T_0 + \frac{2(1-R)\alpha J}{\rho c(1+\tau)^{3/2}}\exp[-\alpha z]\exp[-\frac{r^2}{w^2(1+\tau)}] \tag{15}$$

In Equ. 15, the scaled time τ mounts to $\tau = {}^{8t}\!/_{a^2w^2}$. Numerical analysis of the theoretical model shows agreement with experimental results (Pike, 2005; Pike et al. 2007).
Recent theoretical studies of temperature distribution modelling (Mukherjee & Chakrabotry, 2010) focus on developing a mathematical model of heat transfer based on identifying thermal resistances within the dental profile. Corresponding to the temperature rise, suitable laser powers are inferred for miniscule heat flux to occur.

4.3 Advantages of femtosecond laser radiation and femtosecond spectroscopy

The superiority of femto-second laser pulses for precise material processing has been demonstrated in the literature. (Chichkov et al., 1996; Serbin et al., 2002) The interaction of the ultra-short laser pulses with the target material is thermally confined, i.e., the thermal diffusion takes place within the irradiated area. Therefore, the femto-second laser pulses create minimal (when compared for example to nominal nanosecond laser pulses) thermal and mechanical damage to the surrounding area during laser imaging, drilling and/or ablation. This particular feature is favorable for use of femto-second laser pulses in dental practices. (Serbin et al., 2002; Frentzen & Hamrol, 2000) The temperature profile can be controlled temporally and spatially which is beneficiary for photo-acoustic imaging. The pressure amplitude during drilling of a dentine slice with femto-second pulses is 2 to 3 orders of magnitude lower than the acoustic transients created by nanosecond pulses. (Serbin et al. 2002) Diminished pressure amplitudes are important, because otherwise generation of micro-cracks of several tens of microns in the enamel would result in new carious attacks and thus defeat the treatment process.

Application of femto-second laser radiation has also seen a remarkable development in recent years (Hannaford, 2005). These developments include revolutionizing ultrahigh precision spectroscopy, development of soft X-ray femtosecond laser sources for biomedical imaging, generation of attosecond laser pulses for study of ultrafast phenomena such as electron dynamics, novel methods for measuring and characterizing ultrashort laser pulses and ultrashort pulses of light, coherent control of atomic- and molecular- and electron-dynamics, real-time spectroscopy of molecular vibrations, or multidimensional femtosecond coherent spectroscopies. However, in this chapter we focus on determination of the laser pulse-width using autocorrelation measurements, discussed in a later Section 6.

5. Optical coherent imaging

One mechanism of deep tissue sensing can be accomplished using the coherence properties of light. In order to obtain clinically relevant information for diagnostic purposes the suspected tissue may need to be surgically removed in the form of a biopsy for further examination. The successful treatment of pathological conditions will rely on expedient and effective diagnostic information. Whether this needs to be invasive (i.e. biopsy) or non-invasive (i.e. sensing/imaging) will depend on the level of intricacy that can be obtained by imaging, which will also be in-vivo and ideally will be instantaneous. The primary mechanism of diagnostic sensing using the coherence properties of light is referred to as Optical Coherence Tomography (OCT). The principle of OCT will be described next. Optical coherence tomography is a non-destructive and noncontact imaging modality that reconstructs depth-resolved images of biological tissue based on local optical property distribution in two- and three-dimensional layout (Gibson et al., 2006; Choma, 2004; Jiang et al., 1998; Jiang et al., 1997; Najarian, 2005; Splinter, 2006), whereas the physical reconstruction process can be found in signal processing literature (Najarian & Splinter, 2005; Splinter & Hooper, 2006).

5.1 Optical coherence tomography

Optical Coherence Tomography relies on interference between the probing source and a reference source to enhance the signal resolution. However, due to the scattering artefact in turbid media such as tissue, the signal strength quickly degenerates with depth, hence limiting the depth resolution.

Optical Coherence Tomography is the superposition of the following terms: optics, coherence and tomography. Each word has a root meaning that explains its place in the method name. Optics: using electromagnetic radiation, primarily in the wavelength range of

300 nm - 900 nm. Optics contains all the electromagnetic radiation properties such as amplitude, phase, frequency, polarization as well as time-of-flight. Coherence [cohaesus (Latin)]: fit together, synchronization in phase between waves of identical wavelength, allowing for interference. Tomography; contraction of tomo [tomos (Greek)]: slice/section and graphy [graphikos (Greek)]: depiction. In Optical Coherence Tomography an image is reconstructed from multiple adjacent one-dimensional signals derived from optical recordings executed in a grid pattern (Gibson et al., 2006; Choma, 2004; Jiang et al., 1998; Jiang et al., 1997; Najarian & Splinter, 2005; Splinter & Hooper, 2006). Since OCT uses optics as the signal carrier, it can be both free-space and through fiber-optics.

The fundamental of the mechanism of action of conventional OCT will be followed by more complex diagnostics including the polarization and frequency information as altered by the biological scatter.

5.2 Mechanism of action of optical coherence imaging

The mechanism of coherence relies on the use of two identical sources. Optical power from the light source, is divided approximately 50/50 between the two arms using a optical beam-splitter (Huang et al., 1991; Fercher et al., 1995; Splinter & Hooper, 2006). In the sample arm, light is backscattered/reflected is collected from different depths within the sample. At the splitter the light from the reference arm is recombined with the light from the sensing arm and interference will occur based on the phase of each wavelength of the returning light. The rudimentary description of OCT operation so far does not capture the physics involved. The basic principles involved in OCT are as follows:

- Coherence length and the generation of low coherence light.
- Low-coherence interferometer.

Since the rigorous physical analysis of partial coherence interferometry is extremely involved, a narrative description will be presented that preserves the important features while attempting to provide an intuitive picture. At first the low coherence source concept is outlined.

5.3 Light sources and coherence length

In order obtain the high resolution that is typically achieved in optical coherence tomography it is necessary to employ a light source that has a relatively long coherence length, compared to laser radiation. The axial resolution in OCT is limited by this coherence length (Huang et al., 1991).

The coherence length of a source can be defined as the physical length in space over which one part of the electromagnetic wave train bears a constant phase relationship to another part. The coherence length is thus the net delay that can be inserted between two identical waves that still allows observation of constructive and destructive interference. The coherence length: L_c of the source, is defined in equation 16:

$$L_c = {\lambda_{avg}}^2 / \Delta\lambda \tag{16}$$

Where: ${\lambda_{avg}}^2$ represents the square of the average wavelength of the spectral bandwidth of the source and $\Delta\lambda$ represents the bandwidth of the source. Long coherence length can be achieved by a super luminescent diode (SLD) or Q-switched laser as well as various other sources (Chen et al., 2002).

5.4 Theory of optical coherence tomography

Optical coherence tomography exploits the low temporal coherence of a light source to resolve, on the z-axis (i.e. in-line with the propagation direction), the position where backscattered light is being collected for identification (Choma, 2004). The diameter of the target point is related to the scanning beam diameter or the numeric aperture of the delivery/detection fiber, which also dictates the lateral resolution. A single mode fiber is generally used because the small numeric aperture reduces the solid angle from which light may be collected, providing high lateral resolution. Either the tissue or the fiber tip is then scanned in two dimensions to reconstruct point-by-point to form a two- or three-dimensional image of the specimen (Gibson et al., 2006; Choma, 2004; Jiang et al., 1998; Jiang et al., 1997; Najarian & Splinter, 2005; Splinter & Hooper, 2006). The single mode fiber satisfies the additional requirement of mandating coherence throughout the detection system. The interference signal can now be examined separate from the remainder of the returning light (Diels et al., 1985; Yan and Diels, 1991; Naganuma et al., 1989).

The principles of interferometry are discussed in two parts: the intensity transfer function of the interferometer, and the dependence of the detected signal on the optical path difference between the sample and reference arms. The following initial approximations are made for the theoretical description of the operating mechanism of OCT:

- Source is monochromatic.
- Dispersion effects are eliminated in analyzing the detected intensity as a function of reference mirror position.

The additional approximation that the source emits plane waves, which is not truly a good approximation to the Gaussian single mode beam profile, will not be revised here for the sake of simplicity (Huang et al., 1991; Splinter & Hooper, 2006).

The general source equation is the standard (scalar) wave expression shown in equation 17:

$$E_{source} = E_0 e^{i(kz-\omega t)} \tag{17}$$

where ω is the angular frequency of the electric field and E_0 is the real amplitude of the electric field of the wave. Equation 2 is specific to an electric wave propagating in the positive z-direction in a non-dispersive, dielectric material. The quantity k is the (pure real) wave number, which is related to wavelength of the source as defined in equation 18:

$$k = \frac{2\pi}{\lambda} n \tag{18}$$

where n is the material refractive index of the medium (n=1.36 for tissues at 623 nm) and λ the wavelength of the source and collected signal. The expression for the total electric field at the detector is given by the superposition of these two returning waves as shown in equation 19 (Splinter and Hooper, 2006):

$$E_{total} = E_{r0} e^{i(kz_r-\omega t)} + E_{s0} e^{i(kz_s-\omega t)} \tag{19}$$

With the expression for the returning signals: reference (E_r) and sample (E_s). Where the subscript: r indicates the returning reference signal and the subscript: s referrers to the returning sample signal. The terms z_r and z_s are the respective distances to and from the reference mirror and a specific reflection site within the sample. For the electric waves described above, the detected intensity at the interferometer can be calculated from the square of the electric field amplitude as given by equation 20:

$$I = \frac{n}{2\mu_0 c} E_{total} E_{total}^* = \frac{n}{2\mu_0 c}\left(E_{s0}^2 + E_{r0}^2 + \sqrt{E_{s0}E_{r0}}\cos(kz_r - kz_s)\right)$$
$$= I_{s0} + I_{r0} + 2\sqrt{I_{s0}I_{r0}\cos\left[\frac{2\pi}{\lambda_0}(z_r - z_s)\right]} \quad (20)$$

where μ_0 is the magnetic permeability, c is the speed of light in vacuum, and the star denotes complex conjugation of the electric field expression. The intensity collected by an optical sensor directly correlates to the output voltage used for identification (Splinter and Hooper, 2006; Jiang et al. 1991). The detected intensity depends on the relative distance travelled by each of the electric waves resulting from the interference (cosine) term. In coherence sensing from a biological sample the interference term is relatively small compared to the background due to the relative amplitudes of the reflected reference and sample signals.

Signal collection is in general affected with noise from various origins. This is particularly true for a direct-current (DC) signal (Jiang, 1998; Dunn, 1999). One way to eradicate noise is to convert to an alternating signal and more specifically by Doppler-shifting the light source by external means. Collecting the secondary source (i.e. generated inside the tissue) with the same modulation will dramatically enhance the signal-to-noise ratio. This principle is described as heterodyning. Heterodyning is the act of source modulation using a single frequency and detecting "locked-in" to that same frequency, this accomplishes a common mode rejection of any signal with random fluctuations or steady state signal (Splinter, 2006). The modulation can be based on path-length (mechanical) or amplitude (electronic) of an optical system. Heterodyning produces signal to noise ratio of up to 120 dB. Signal to noise ratios of at least 100 dB are necessary to obtain any useful depth resolution in highly scattering media (Najarian, 2006).

5.5 Coherence imaging for therapeutic feedback

One specific application of optical coherence imaging uses the changes in the optical path due to alteration made to the tissue while undergoing laser photocoagulation. Coagulation changes the cellular structure and hence the optical properties, thus changing the collected signal as a function of depth. Figure 2 illustrates the differing behavior. Additionally, different tissues have their own unique and specific optical characteristics. These respective optical properties can be used to curve-fit the optical path as a function of depth against a reference library to non-invasively determine the tissue type (i.e. fat, muscle, collagen, etc.) in an attempt to allow selection of the target for treatment.

All of the aforementioned imaging modalities can benefit from some form of spectral decoding. The frequency spectrum may act as a tuning fork to match a pattern to a look-up table with the listed "color-matrix" of pre-determined shapes, materials and biological conditions.

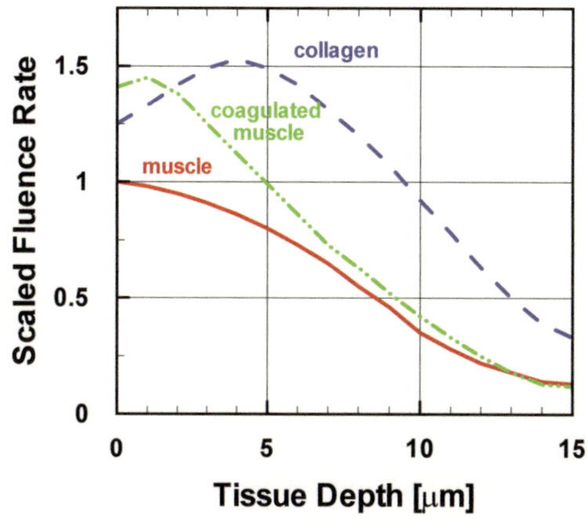

Fig. 2. Optical biopsy obtained by coherent imaging. Target recognition of collagen vs muscle as well as the influence of therapeutic coagulation on muscle. The depth profile of the fluence rate can be matched against a library of tissue characteristics for identification.

6. Measurement of ultra-short light pulses

Recent developments of ultra-short light pulses extend to use of Frequency-Resolved - Optical-Gating (FROG) that have evolved to measurements of intensity and phase so simple that essentially no alignment is required (Gu et al., 2005). In addition, with certain FROG variations it is now possible to measure more general light pulses, i.e., light pulses much more complex than common laser pulses. The new variation of FROG, called GRENOIULLE (O'Shea et al., 2001), has no sensitive alignment knobs and is composed of only a few elements. In this section measurements of nominally 70 fs laser pulses from interference images are discussed (Guan, 1999; Guan & Parigger, 2000) in the overlap region of two beams. The analysis makes use of Fourier transform techniques to extract the interference cross term in the spatial frequency domain. The autocorrelation function is obtained by systematically varying the time delay of the two beams. The laser pulse width can subsequently be determined for an assumed pulse shape.

6.1 Background for short-pulse measurements

The characterization of ultra-short laser pulses is generally required in experimental investigations with nominally 70 femtosecond laser pulses. The physical quantities of interest include wavelength, band width, coherence length and pulse width. A relatively simple characterization can be obtained by measuring the spectrum of the short pulse. The recorded spectrum yields the center wavelength and the bandwidth of the pulse. The pulse-width can be inferred for a transform-limited pulse.

The temporal characteristics of the laser pulses can be alternatively obtained from the measurement of the autocorrelation function, as indicated in early works on the subject. (Diels et al., 1985; Yan and Diels, 1991; Naganuma et al., 1989) The frequency domain method was developed to unambiguously determine the pulse shape and phase. (Chilla & Martinez, 1991a; Chilla & Martinez, 1991b; Chilla & Martinez, 1992) Almost simultaneously the technique of frequency-resolved optical gating was introduced. (Kane & Trebino, 1993; Trebino and Kane, 1993; Paye et al., 1993) Subsequent works are elaborated in the literature.

(Miyamoto et al., 1993; Chu et al., 1995; Meshulach et al., 1997; Li et al., 1997) A nonlinear medium is typically required to generate a second or third harmonic signal that is associated with the autocorrelation trace. The technique of interferometry or holography (Takeda et al., 1982; Macy, 1983; Kreis, 1986; Coobles, 1987; Zhu et al., 1989) may be used to measure the correlation function. The coherence length of a laser pulse can be determined from the correlation function.

In this section we present measurements of the autocorrelation function by scanning the relative delay of the two laser beams and by recording interference images. A nonlinear medium is not required in the overlap region. The information in an image is decomposed according to the spatial frequencies by the use of Fourier transform techniques. (Takeda et al., 1982; Macy, 1983; Kreis, 1986) The interference information is usually restricted to a specific range of spatial frequencies in Fourier transform space. This range may be separated from the low frequency background and high frequency noise. The interference information can be extracted through an appropriate filter to yield the autocorrelation coefficients for a particular time delay. The pulse-width is determined by fitting the autocorrelation function for an assumed hyperbolic secant pulse-shape.

6.2 Experimental arrangement for pulse-width measurements

The interference patterns of overlapping ultra-short pulses are directly recorded by the use of a CCD camera. Figure 3 shows the experimental arrangement. The Spectra Physics model Tsunami Ti:Sapphire laser pulses, produced at a repetition rate of 76 MHz, are specified to be as short as 60 femtosecond when dispersion due to transit through the output coupler is fully compensated.

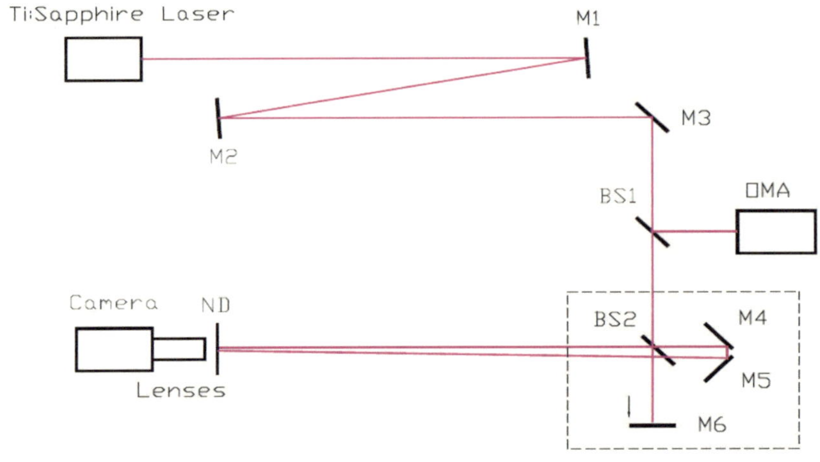

Fig. 3. Schematic experimental arrangement for short pulse interference measurements.

The laser beam is reflected by the mirrors M1, M2 and M3. A portion of the beam (see BS1) enters a Jarrell-Ash MonoSpec 27 spectrometer and the approximately 10 nm band-width spectra are monitored by the use of a Princeton Applied Research optical multichannel analyzer (OMA). For the measurement of interference images the laser beam was further split by the use of a wedge (BS2) as a beam splitter. The tilt of the wedge and the mirrors M4 and M5 were adjusted to obtain spatial overlap of the beams in the field of view of the camera.

Figure 4 shows details of the box indicated in Fig. 3. The reflected beam from the front-face of the wedge is re-directed and passed again through the wedge prior to entering the interference field. The mirror M6 is mounted on an AEROTECH translation stage to allow us to systematically vary the time delay between the beams. Two other relatively strong reflections are also illustrated in the figure. The two beams each pass the wedge once and are subject to similar dispersion. Therefore, it can be assumed that the pulses have the same temporal profile. The intensity ratio of the two beams is close to 1. However, the exact value is not important in our analysis. The translation stage is moved by a distance, $d/2$, in 0.5 μm steps which corresponds to a time-delay, d/c, of approximately 3.3 fs.

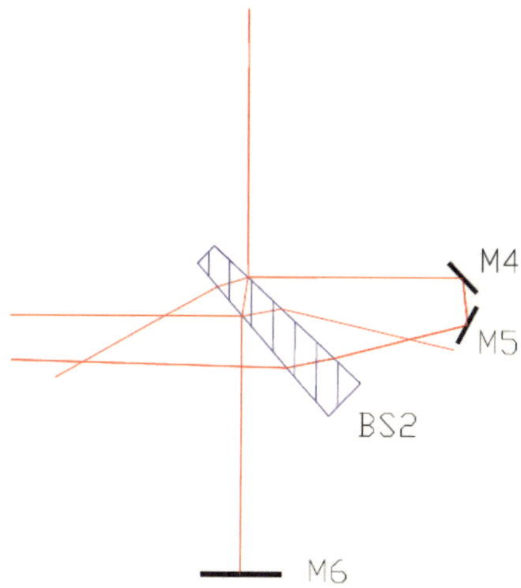

Fig. 4. Detailed arrangement of the beam path at the wedge.

A digital camera, model EDC-1000HR, is used to record the interference patterns. Neutral density filters with ND 2 to 3 are selected to adjust the light intensity below camera saturation levels. The magnification lens LEICA MONOZOOM 7 allowed us to record magnified interference patterns. The exposure time of the camera and the translator's motion is controlled by the use of a personal computer.

6.3 Experimental results of pulse-width measurements

In an individual experimental run, the two beam's temporal overlap is scanned, i.e., the step translator is moved to different positions and the images are recorded at these positions. The images are stored and are analyzed subsequently, although real-time recording/analyzing is possible in principle. The angle between the two beams is adjusted to $\alpha = 8.7\,\mathrm{m}rad$ (0.5°). During a 5 ms camera exposure time, interference patterns of approximately 380,000 laser pulses are generated. The individual images represent an average obtained from 380,000 spatio-temporal pulse superpositions.

Figure 5 displays two typical images. These images were recorded in two separate experimental runs. Each image consists of an array of 753×244 8-bit data. In the experimental runs with larger magnification the zoom lens was adjusted to a $\times 6$ larger than for the small magnification experimental runs. Only a small portion of the interference pattern is recorded and it shows details of the fringes. Figure 5 (b) shows the majority of the interference pattern. Both images were recorded for approximately zero time delay between the two beams.

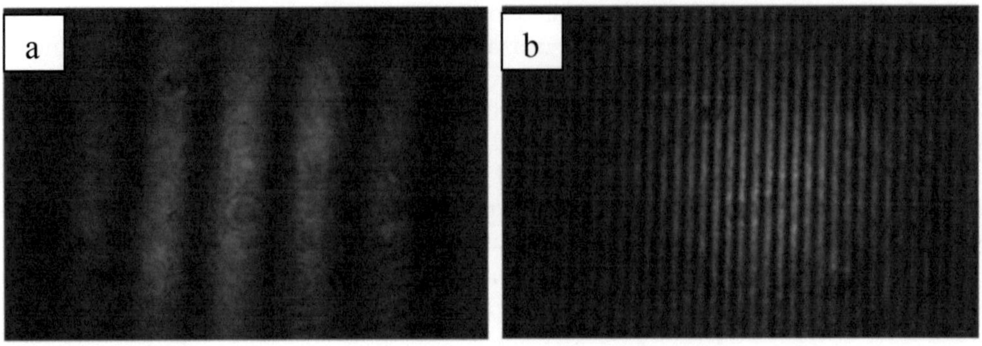

Fig. 5. Images of interference patters. The camera lens magnification used is ×6 larger for (a) the image on the left than (b) for the image on the right. The distance perpendicular to the fringes amounts to 0.6 mm and 3.6 mm, respectively.

6.4 Short-pulse interference details

The classical electromagnetic theory is used in the analysis of the ultra-short pulse interference measurements. The intensity distribution of the interference cross term is described by the first-order correlation function. For equal temporal pulse shapes the interference pattern is proportional to the autocorrelation function. This result is derived in the following.

First we formulate the electric field of a laser pulse that propagates in the $\hat{z}$ direction. In this formulation we use a wave packet: (Jackson, 1975)

$$E(t,x) = A(x)\int_{-\infty}^{\infty} F(\omega - \omega_0) e^{-i(\omega t - k\cdot x)} d\omega. \tag{21}$$

Here, k is the wave vector $k = \hat{k}\omega c$, c is the speed of light and $\hat{k}$ is the unit vector in the k direction ($k \uparrow\uparrow \hat{z}$). $F(\omega - \omega_0)$ describes the spectral distribution of the wave packet centered at ω_0, with the usual normalization

$$\int_{-\infty}^{\infty} | F(\omega - \omega_0)|^2 \, d\omega = 1. \tag{22}$$

The complex, spatial amplitude, $A(x)$, satisfies the spatial part of the wave equation, (Jackson, 1975; Milonni and Eberly, 1988)

$$\nabla^2 A(x) - 2ik\frac{\partial}{\partial z}A(x) = 0. \tag{23}$$

In the paraxial approximation, the laser beam distribution $A(x)$ can be described by a Gaussian beam that propagates along the $\hat{z}$ direction. (Kogelnik and Li, 1966; Siegmann, 1986; Möller, 1988; Milonni and Eberly, 1988)

Next we define the Fourier transform, f, of the spectral distribution F as:

$$f(t - \hat{k}\cdot x / c) = \int_{-\infty}^{\infty} F(\omega - \omega_0) e^{-i(\omega-\omega_0)(t-\hat{k}\cdot x/c)} d\omega, \tag{24}$$

with f also normalized to unity,

$$\int_{-\infty}^{\infty} | f(t-\hat{k}\cdot x / c)|^2 \, dt = 1, \tag{25}$$

because of the unitary property of Fourier transforms. By the use of the Fourier transform (Eq. (24)) the electric field (Eq. (21)) becomes

$$E(t,x) = A(x) f(t-\hat{k}\cdot x / c) e^{-i\omega_0 (t-\hat{k}\cdot x/c)}. \tag{26}$$

Now we proceed to describe the interference patterns. The two beams of identical, linear polarization propagate in directions $\hat{k}_1$ and $\hat{k}_2$, respectively, and are described by

$$E_1(t,x) = A_1(x) f_1(t-\hat{k}_1\cdot x / c) e^{-i\omega_0 (t-\hat{k}_1\cdot x/c)}, \tag{27}$$

$$E_2(t,x) = A_2(x) f_2(t-\hat{k}_2\cdot x / c - d / c) e^{-i\omega_0 (t-\hat{k}_2\cdot x/c-d/c)}. \tag{28}$$

The time delay between beam 1 and 2 equals d/c. In the overlap region of the two laser beams, the electric field is the sum of the individual fields:

$$\begin{aligned} E(t,x) &= E_1(t,x) + E_2(t,x) \\ &= A_1(x) f_1(t-\hat{k}_1\cdot x / c) e^{-i\omega_0 (t-\hat{k}_1\cdot x/c)}, \\ &\quad + A_2(x) f_2(t-\hat{k}_2\cdot x / c - d / c) e^{-i\omega_0 (t-\hat{k}_2\cdot x/c-d/c)}. \end{aligned} \tag{29}$$

The two pulses are assumed to have identical temporal pulse shape subsequently. This assumption is valid provided that the optical lengths of the two short pulses in a dispersive medium are equal. An ultrashort pulse can become chirped and hence stretched due to propagation through a dispersive medium or by reflection from a multilayer dielectric mirror. However, the dispersion due to propagation through the optical components results in negligible chirps in our experiment. The temporal pulse shape for both beams is described by $f(t)$, i.e., the subscripts for the pulse shapes $f_1(t)$ and $f_2(t)$ can be omitted. The intensity ($E(t,x)E^*(t,x)$) of the summed electric fields equals:

$$\begin{aligned} I(t,x) &= | A_1(x) f(t-\hat{k}_1\cdot x / c)|^2 + | A_2(x) f(t-\hat{k}_1\cdot x / c + \tau(x))|^2 \\ &\quad + 2\Re\left\{ A_1(x) A_2^*(x) f(t-\hat{k}_1\cdot x / c) f^*(t-\hat{k}_1\cdot x / c + \tau(x)) e^{i\omega_0 \tau(x)} \right\}, \end{aligned} \tag{30}$$

where the spatially dependent time-delay is

$$\tau(x) = (\hat{k}_1 - \hat{k}_2)\cdot x / c - d / c. \tag{31}$$

The interference intensity is integrated during each pulse in the ultra-short pulse interference measurements. The integration limits are extended to positive and negative infinity since the interval between subsequent pulses (generated at a repetition rate of 76 MHz) is more than five orders of magnitude larger than the pulse width. The integrated intensity, denoted by $\overline{I(x)}$, is calculated to be

$$\overline{I(x)} = \int_{-\infty}^{\infty} I(t,x)dt$$

$$= |A_1(x)|^2 + |A_2(x)|^2 + 2\Re e\left\{A_1(x)A_2^*(x)g(\tau(x))e^{i\omega_0\tau(x)}\right\}, \tag{32}$$

where we have used that the temporal pulse-shapes are normalized (see Eq.(5)). Also, the spatially dependent autocorrelation function $g(\tau(x))$ of f is defined by

$$g(\tau) = \int_{-\infty}^{\infty} f(t) f^*(t+\tau)dt. \tag{33}$$

Above equation (Eq.(32)) shows that the interference cross term (third term in Eq. (32)) is proportional to the autocorrelation function for identical temporal pulse shapes of the two beams. The autocorrelation function depends on the time delay d/c of the two beams and it also depends on the time delay introduced by the phase term that varies spatially according to $(\hat{k}_1 - \hat{k}_2)\cdot x/c$.

6.5 Determination of the autocorrelation function

Fourier transform techniques are utilized to separate the interference cross term from the other contributions in the spatial frequency domain, (Takeda et al., 1982; Macy 1983; Kreis, 1986) The terms $A_1(x)$ and $A_2(x)$ are transformed only into the domain of low spatial frequency. The interference cross term is transformed into the domain of high spatial frequency mainly due to the component $e^{i\omega_0\tau(x)}$. The Fourier transform in the x-y plane, denoted by $x_\perp$, of the interference cross term is:

$$\begin{aligned}\xi(\kappa) &= \int_{-\infty}^{\infty}\int_{-\infty}^{\infty} 2\Re e\left\{A_1(x)A_2^*(x)g(\tau(x))e^{i\omega_0\tau(x)}\right\}e^{i\kappa\cdot x_\perp}dx_\perp \\ &= \int_{-\infty}^{\infty}\int_{-\infty}^{\infty} A_1(x)A_2^*(x)g(\tau(x))e^{i\omega_0((\hat{k}_1-\hat{k}_2)\cdot x/c-d/c)}e^{i\kappa\cdot x_\perp}dx_\perp \\ &+ \int_{-\infty}^{\infty}\int_{-\infty}^{\infty} A_1^*(x)A_2(x)g^*(\tau(x))e^{-i\omega_0((\hat{k}_1-\hat{k}_2)\cdot x/c-d/c})e^{i\kappa\cdot x_\perp}dx_\perp.\end{aligned} \tag{34}$$

In the frequency domain, Equ.(14) represents two peaks at $\kappa^I = (\hat{k}_2 - \hat{k}_1)\omega_0/c$ and $\kappa^{II} = (\hat{k}_1 - \hat{k}_2)\omega_0/c$. The peak at κ^I can be isolated by using an appropriate filter, and one obtains

$$\xi^I(\kappa) = \int_{-\infty}^{\infty}\int_{-\infty}^{\infty} A_1(x)A_2^*(x)g(\tau(x))e^{i\omega_0((\hat{k}_1-\hat{k}_2)\cdot x/c-d/c)}e^{i\kappa\cdot x_\perp}dx_\perp. \tag{35}$$

The inverse Fourier transform of this peak is:

$$F^{-1}\left\{\xi^I\right\} = A_1(x)A_2^*(x)g(\tau(x))e^{i\omega_0((\hat{k}_1-\hat{k}_2)\cdot x/c-d/c)} \tag{36}$$

The autocorrelation function $g(\tau)$ can be determined from one image (for a relative time delay of approximately zero) for known $A_1(x)A_2^*(x)$. For an accurate measurement of $g(\tau(x))$, the spatial frequency distribution of $g(\tau(x))$ must be significantly broader than the spatial frequency distribution of $A_1(x)A_2^*(x)$. Equivalently, for approximately constant $A_1(x)A_2^*(x)$ and for sufficient temporal variation due to the phase term, $(\hat{k}_1 - \hat{k}_2)\cdot x/c$, across the detector surface, the autocorrelation function is proportional to the envelope of the fringe maxima, i.e., the visibility of fringes. (Milonni and Eberly, 1988)

The determination of the autocorrelation function with the time-delay method is elaborated in the following. For each image recorded at a set time delay between the beams, one integrates the intensity contribution of one interference peak in the frequency domain (the κ space). The integration limits are extended to $\pm\infty$ since the integration of the intensity contribution of one peak in the region beyond 3 times the full width from the peak center is negligible. By direct substitution or by use of the Wiener-Kintchine theorem, one finds:

$$\begin{aligned}\int_{-\infty}^{\infty}\int_{-\infty}^{\infty}\xi^{I}(\kappa)\xi^{I}(\kappa)^{*}d\kappa &= 4\pi^{2}\int_{-\infty}^{\infty}\int_{-\infty}^{\infty}|A_{1}(x)A_{2}^{*}(x)g((\hat{k}_{1}-\hat{k}_{2})\cdot x/c-d/c)|^{2}\,dx_{\perp} \\ &= 4\pi^{2}\,\Xi(d/c).\end{aligned} \tag{37}$$

This result shows that the square of the autocorrelation function is convolved with the square of the amplitude distributions. A temporal profile $\Xi(d/c)$ is obtained as function of the time delay d/c due to the spatial integration.

The Fredholm integral equation (Eq.(37)) for $g(\tau(x))$ can be inverted by the application of standard mathematical methods, for example, by the use of the Fourier convolution theorem to find the Fourier Transform Solution (Arfken and Weber, 2005) for 2-dimensional transforms

$$4\pi^{2}\,|g(d/c)|^{2}=F^{-1}\left\{F\{\Xi\}/F\{|A_{1}A_{2}^{*}|^{2}\}\right\}. \tag{38}$$

The use of this additional Fourier transform for the purpose of deconvolution is not necessary in the analysis of our experiments.

The convolution causes hardly any broadening for spatial phase terms that are significantly smaller than the autocorrelation function full-width at half-maximum (FWHM) τ_{auto},

$$(\hat{k}_{1}-\hat{k}_{2})\cdot x/c \ll \tau_{auto}. \tag{39}$$

The absolute value of the difference of the beams' wave-vectors equals the angle between the two beams, $|\hat{k}_{1}-\hat{k}_{2}|=\alpha$, for small angles ($\sin\alpha\cong\alpha$). In the experiment the phase terms are small, i.e., $(\hat{k}_{1}-\hat{k}_{2})\cdot x/c<\tau_{auto}$ for the images that were recorded with small and large magnifications of the zoom lens. Note the amplitude distributions $|A_{1}(x)A_{2}(x)^{*}|^{2}$ reduce the contributions to the convolution integral for larger distances from the center of the images.

For the images that were recorded with small magnification (see Fig. 5 (b)) the spatial variation across the image of the $|A_{1}(x)A_{2}(x)^{*}|^{2}$ term is significantly larger than the variation due to the square of the autocorrelation function. In this case, the amplitude terms can be modeled by a 2-dimensional delta-distribution, Area $\times$ $\delta^{(2)}(x_{\perp})$. Spatial integration with this delta-distribution immediately yields that the temporal profile $\Xi(d/c)$ is directly proportional to the square root of $g(d/c)$. The physical realization of the delta-distribution is indicated in Fig. 5(b) by the spatial variation of the beams' intensity profiles. The ratio of the $|A_{1}(x)A_{2}(x)^{*}|^{2}$ width and of the $|g(\tau(x))|^{2}$ width amounts to approximately 1:6. This difference in widths would yield a 1.5% broadening that can be estimated from the sum of the squares for the widths.

The experimental arrangement was designed to minimize effects from the convolution. For an angle of $\alpha = 0.5^\circ$, spatial dimensions of 0.6 mm × 0.6 mm and 3.6 mm × 3.6 mm, and a pulse-width of 70 fs, the numerically investigated broadening due to the convolution process would amount to 0.5% and 1.7 % of the FWHM of the autocorrelation function, respectively. For 70 fs pulse-width measurements this broadening would amount to typically 1 fs which is less than one half of an optical cycle (3 fs for a center wavelength of 0.9 μ m).

6.6 Image analysis and results

The analysis of the recorded images is accomplished by the use of discrete Fourier transforms for the finite image which is represented by a two-dimensional array. An array element h_{pq} that contains the integrated intensity of the corresponding pixel formally introduces the model:

$$h_{pq} = \Pi(x_{pq})\Delta_1\Delta_2 \qquad 1 \le p \le M, \qquad 1 \le q \le N, \tag{40}$$

where Δ_1 and Δ_2 are the pixel sizes in the two perpendicular directions. The total area of image is $a \times b$. The intensity of the electric field at the point x_{pq} may be of the same order as that in Eq. (40) for the interference pattern outside the recorded image region. In the numerical analysis, we set the array value h_{pq} to zero outside the active area of the camera. Corrections in κ-space due to the filter are negligible. Setting h_{pq} to zero outside the recorded area is equivalent with setting $A_1(x)A_2^*(x)$ to zero beyond the exposed area. In the numerical analysis we use Nyquist critical frequencies $\kappa_{1c} = \pi\Delta_1$ and $\kappa_{2c} = \pi\Delta_2$, and evaluate at the positions $\kappa_m = mM\kappa_{1c}$ and $\kappa_n = nN\kappa_{2c}$ the discrete Fourier transform set $\{H_{mn}\}$ of the set $\{h_{pq}\}$ given by

$$H_{mn} = \sum_{p=0}^{M-1}\sum_{q=0}^{N-1} h_{pq} e^{2\pi ipm/M} e^{2\pi iqn/N}. \tag{41}$$

The fast Fourier transform algorithm (Press et al., 2007) was used in the computation. The upper limits in the sums in Eq. (41) are equal to numbers of power of 2 in the fast Fourier transform algorithm.

Figure 6(a) shows the low-frequency or dc-component of the $\xi(\kappa)\xi(\kappa)^*$ distribution which corresponds to the small magnification experiment shown in Fig.5(b). The first two terms of Eq. (32) are included in the peak near the origin. Figure 6(b) shows details of the high-frequency peak near the pair ($m = 0$, $n = 50$); this distribution is associated with one of the interference cross terms. In the numerical evaluation of the temporal distribution $\Xi(d/c)$, the integration over κ reduces to a summation over m and n of the discrete values H_{mn} near the pair $(0,50)$, i.e., in the restricted domain (m,n) that corresponds to one of the specific interference cross terms for the applied filter.

Figure 7 shows the correlation coefficients (or the results of the above mentioned summation) versus time delay. The autocorrelation function can be obtained by the use of smoothing algorithms.

The temporal pulse shape can not be uniquely determined. This can be seen by considering the Fourier transform $F\{|f|\}$ since $|f|$ and $F\{|f|\}$ transform uniquely. The absolute value of $F\{|f|\}$ is determined by the Fourier transform of the autocorrelation function according to

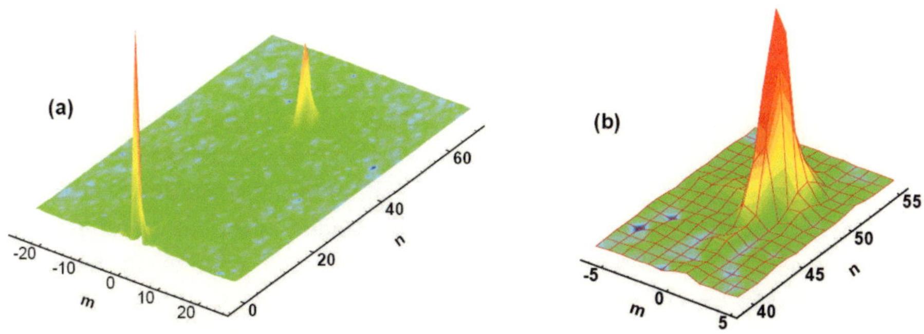

Fig. 6. Spatial frequency distribution of the recorded images . (a) Fourier transform of the intensity distribution. (b) Detailed viewof the interference cross term near the pair (m=0, n=50).

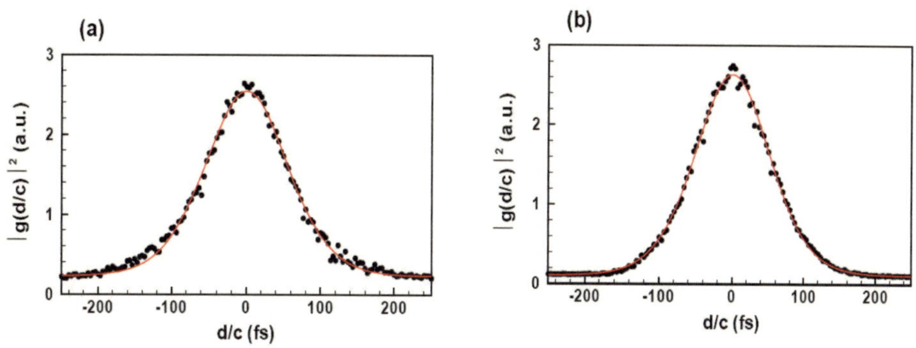

Fig. 7. Measured square of the scaled autocorrelation function (circles) and fitted curve for a hyperbolic-secant temporal pulse-shape (line) versus time delay d/c. (a) larger magnification (see Fig. 5 (a)) and (b) smaller magnification (see Fig. 5(b)).

$$F\{|f|\}F^{*}\{|f|\} = F\{Corr(|f|,|f|)\}, \tag{42}$$

while the phase of $F\{|f|\}$ is undetermined. We assume a hyperbolic secant temporal pulse-shape,

$$|f(t,\tau_p)| = \frac{2}{e^{1.76t/\tau_p} + e^{-1.76t/\tau_p}}, \tag{43}$$

which leads to an autocorrelation function of the hyperbolic secant that can be analytically obtained,

$$g(t,\tau_p) = \frac{1.76\,t/\tau_p}{e^{1.76t/\tau_p} - e^{-1.76t/\tau_p}}. \tag{44}$$

The pulse width τ_p (FWHM of the beams' intensity profile $|f(t,\tau_p)|^2$) is determined by fitting the experimental data with $C_1 + C_2 |g(t,\tau_p)|^2$ (C_1, C_2 and τ_p are the fitting parameters) by the use of the nonlinear least-square method. Figure 6 also shows these results.
The pulse widths amount to $\tau_p = 77$ fs and $\tau_p = 72$ fs for the larger and smaller magnification experiments, respectively. Note that these results are obtained from different experimental runs. The error bars are estimated from the distribution of the correlation coefficients to be at most 5% for the pulse-width τ_p. Smaller error bars result for the smaller magnification experiment. This can be seen, for example, by comparing the wings of the profiles illustrated in Fig. 7. The discrete Fourier transform can be more precisely determined from approximately 36 recorded fringes for each time delay. Lens aberration effect will also be smaller when using a smaller magnification of the zoom lens. The 5 fs disparity is attributed to the difference in the available pulse-widths for the separate experimental scans.

6.7 Discussion of the short-pulse measurements

The interferometric measurement of short pulses with the time-delay method involves in principle the convolution of the spatial amplitude distribution with the autocorrelation function. It is demonstrated that the autocorrelation function can be directly obtained from the fringes' spatial distribution when the small angle condition perpendicular to the fringe pattern, $\alpha |x|/c < \tau_{auto}$, is satisfied. Broadening from the convolution process is insignificant for an angle of 0.5° between the beams, the recorded image dimensions $|x|$ of 0.6 mm and 3.6 mm, and for pulse-widths τ_p of typically 70 fs. The broadening amounts to approximately 0.5% and 1.5% for the large and small magnification experiments, respectively. In the experiment, about 200 images are recorded in a scan of 663 fs. However, we infer from a detailed numerical analysis (not elaborated here) that the significantly smaller number of 16 images would have been sufficient for the determination of the autocorrelation function.
Measurement of the autocorrelation function from a single image would require that the amplitude distribution is approximately constant compared to the variation of the autocorrelation function across the array detector's surface and that many fringes would occur. An estimate for the number of fringes is obtained by taking the ratio of 3 times the FWHM of the autocorrelation function and of the optical cycle. For 70 fs pulses with a center wavelength of 0.88 μm, the number of fringes would be approximately equal to 200. Increasing the angle from 0.5° (for which we measured 6 and 36 fringes, respectively) to 3° between the laser beams would result in some 200 fringes across the detector. For a 3.6 mm $\times$ 3.6 mm camera area, the individual fringes would be separated by 18 μm for the 0.88 μ m center-wavelength beams.
For typical CCD cameras the pixel size amounts to approximately 10 to 20 μ m, therefore, measurement of the autocorrelation function from a single image is only indicated for shorter pulse widths. In turn, the presented time-delay method is suitable for the direct measurement of the autocorrelation function for Ti-Sapphire pulses of nominally $\tau_p = 60$ fs width or longer up to some 100 picosecond. The upper limit is obtained from the requirement that no additional spatio-temporal superposition or beam overlap occurs, and it

is estimated from the geometrical arrangement and the thickness of the wedge that was used to split the femtosecond laser beams.

7. Summary

The use of photo-acoustic imaging as well as coherent imaging and ultra-short optical pulse spectroscopy are showing significant potential for evolution into commercial sensing devices with likelihood for the procurement of detailed tissue information. With the continuously growing number of imaging devices there is still a gap between the clinical demands for detail and the capabilities of delivering these in a timely fashion and at a cost that will make the diagnostic devices available for a large part of humanity. However, more and more pathological data is becoming second nature due to the increasing histological and functional details that can be retrieved non-invasively and with minimal risk to the patient.

8. Acknowledgment

The authors thank Dr G. Guan for his contribution, discussions and interest in this work. The experimental material is based upon work that is in part supported by the National Science Foundation under Grant No. CTS-9512489.

9. References

Arfken, G.B., and Weber, H.J., (2005) *Mathematical Methods for Physicists, 6th edition*, ISBN: 978-0-12-059876-2, Oxford, Ohio, Academic Press.

Bell, A. (1880). Upon the production of sound by radiant energy. *American Journal of Science* 20, 305–324, ISSN: 0002-9599.

Chen, Q. and Zhang, X.-C., Electro-Optic THz Imaging, In: *Ultrafast Lasers: Technology and Applications*, Fermann, Galvanauskas, Sucha, Eds. Chapter 11, Marcel Dekker, Inc. ISBN: 0-8247-0841-5, 2002.

Chichkov, B.N., Momma, C., Nolte, S., von Alvensleben, F., Tünnermann, A.(1996) Femtosecond, picosecond and nanosecond laser ablation of solids," *Appl. Phys. A* 63, 109-115, ISSN: 0947-8396.

Chilla, J.L.A., and Martinez, O.E., (1991a) Analysis of a Method of Phase Measurement of Ultrashort Pulses in the Frequency Domain, *IEEE J. Quantum Electron.*, 27(5): 1228-1235, ISSN: 0018-9197.

Chilla, J.L.A., and Martinez, O.E., (1991b) Direct Determination of the Amplitude and the Phase of Femtosecond Light Pulses, *Optics Letters*, 16(1): 39-41, ISSN: 0146-9592.

Chilla, J.L.A., and Martinez, O.E., (1992) Frequency Domain Phase Measurement of Ultrashort Light Pulses: Effect of Noise, *Optics Communications*, 89(5,6): 434-440, ISSN: 0030-4018.

Choma, M.A., (2004) *Optical Coherence Tomography: Applications and Next-Generation System Design*, Ph.D. Dissertation, Duke University.

Chu, K.C., Heritage, J.P., Grant, R.S., Liu, K.X., Dienes, A., White, W.E., Sullivan, A. (1995) Direct Measurement of the Spectral Phase of Femtosecond Pulses, *Optics Letters*, 20(8): 904-906, ISSN: 0146-9592.

Cobbles, J.A., (1987) Picosecond Coherene Time Measurement with a Double Slit. *Applied Optics*, 26(19): 4048-4051, ISSN: 0003-6935.

Coloiano, E.; Rocha, R.; Martin, A.; da Silva, M.; Acosta-Avalos, D., and Barja, D. (2005). Photyo-acoustic analysis of dental raisin polymerization. *J. Phys. IV*, Vol. 125, pp. 793-795, ISSN: 1155-4339.

Diels, J.C.M, Fontaine, J.J., McMichael, I.C., and Simoni, F. (1985) Control and Measurement of Ultrashort Pulse Shapes (in Amplitude and Phase) with Femtosecond Accuracy, *Applied Optics*, 24(9):1270-1282, ISSN: 0003-6935.

Doody MM, et al. (2000) Breast Cancer Mortality after Diagnostic Radiography: Findings From the U.S. Scoliosis Cohort Study, *Spine*, 25 (16), 2052-2063, ISSN: 0362-2436.

Dunn, R.C., (1999) Near-field Scanning Optical Microscopy, *Chem. Rev.* 99, 2391-2927, ISSN: 0009-2665.

Ermilov SA, Khamapirad T, and Conjusteau A, (2009) Laser optoacoustic imaging system for detection of breast cancer, *J. Biomed. Opt.*, 14, 024007, ISSN: 1083-3668.

Ermilov, S. A.; Khamapirad, T., and Conjusteau, A. (2009) Laser optoacoustic imaging system for detection of breast cancer. *J. Biomed. Opt.*, Vol. 14, 024007, ISSN: 1083-3668.

Fear EC, Li X, Hagness SC and Stuchly MA, (2002) Confocal Microwave Imaging for Breast Cancer Detection: Localization of Tumors in Three Dimensions, *IEEE Trans. Biomed. Eng.* 49(8) 812-823, ISSN: 0967-3334.

Fercher, A.F., (1995) Measurement of intraocular distances by backscattering spectral interferometry, *Optics Communications*, 117, 43-48, ISSN: 0030-4018.

Frentzen, M., and Hamrol, D. (2000) Kavitätenpräperation mit dem Er:YAG-Laser - eine histologische Studie, Carl Hanser Verlag, München, *Dtsch. Zahnärztl. Z.* 55, ISSN: 0012-1029.

Gibson, A.P., Austin, T., Everdell, N.L., Schweiger, M., Arridge, S.R., Meek, J.H., Wyatt, J.S., Delpy, D.T., Hebden, J.C., (2006) Three-dimensional whole-head optical tomography of passive motor evoked responses in the neonate, *Neuroimage*, 30(2):521-528, ISSN: 1053-8119.

Gu, X., Akturk, S., Shreenath, A., Cao, Q., Trebino, R., (2005) The measurement of ultrashort light - simple devices, complex pulses, In: *Femtosecond Laser Spectroscopy*, Ed. P. Hannaford. New York: Springer Science and Business Media, ISBN: 0-387-23293-1.

Guan, G., (1999) *On the analysis of emission spectra and interference images*, Ph.D. dissertation, The University of Tennessee, Knoxville, Tennessee.

Guan, G., and Parigger, C. G., (2000) private communication.

Gusev, V. E., and Karabutov, A. A., (1993) *Laser Optoacoustics*. New York: American Institute of Physics, ISBN: 1-5639-6036-2.

Hannaford, P., *Femtosecond Laser Spectroscopy*. New York: Springer Science and Business Media, ISBN: 0-3872-3293-1, 2005.

Hendee W, et al., (2008) Bioengineering and Imaging Research Opportunities Workshop V: A white paper on imaging and characterizing structure and function in native and engineered tissues, *Med. Phys.*, 35(8), 3428-3435, ISSN: 0094-2405.

Heretier, J. –M. (1983) Electrostrictive limit and focusing effects in pulsed Photoacoustic detection. *Opt. Comm.*, Vol. 44, No. 4, pp. 267-272, ISSN: 0030-4018.

Hoelen, C. G.; de Mul, F. F. M.; Pongers, R., and Dekker A. (1998). Three-dimensional photoacoustic imaging of blood in tissue, *Opt. Lett.*, Vol. 23, No. 8, pp. 648-650, ISSN: 0146-9592.

Huang SA, Aguirre AD, Huber RA, Adler DC, and Fujimoto JG, (2007) Swept source optical coherence microscopy using Fourier domain mode-locked laser, *Opt. Exp.* 15, 6210-6217, ISSN: 1094-4087.

Huang, D., Swanson, E.A., Lin, C.P., Schuman, J.S., Stinson, W.G., Chang, W., Hee, M.R., Flotte, T., Gregory, K., Puliafito, C.A., (1991) Optical coherence tomography, *Science*, 254, 1178–1181, ISSN: 0036-8075.

Jackson, J.D., (1975) *Classical Electrodynamics, 2nd* Edition, Berkeley: John Wiley and Sons, ISBN: 0-4714-3132-X.

Jiang, H., (1998) Frequency-domain fluorescent diffusion tomography: a finite element algorithm and simulations, *Appl Opt*, 37, 5337-5343, ISSN: 0003-6935.

Jiang, H., Paulsen, K.D., Osterberg, U.L., Patterson, M.S., (1997) Frequency-domain optical image reconstruction for breast imaging: initial evaluation in multi-target tissue-like phantoms, *Medical Physics*, 25, 183-193, ISSN: 0094-2405.

Jiang, H., Paulsen, K.D., Osterberg, U.L., Patterson, M.S., (1998) Enhanced optical image reconstruction using DC data: initial study of detectability and evaluation in multi-target tissue-like phantoms, *Phys. Med. Biol*, 43, 675-693, ISSN: 0031-9155.

Kane, D.J., and Trebino, K., (1993) Characterization of Arbitrary Femtosecond Pulses Using Frequency-Resolved Optical Gating. *IEEE J. Quantum Electron.*, 29(2): 571-577, ISSN: 0018-9197.

Kim, K.; Witte, R.; Koh, I.; Ashkenazi, S., and O'Donnell, M. (2006). Early Detection of Dental Caries using Photoacoustics, *Proc. SPIE* Vol. 6086 (Vol.7, No 9),60860G1-9, ISSN: 0277-786X.

Kogelnik, H., and Li, T., (1966) Laser Beams and Resonators, *Proceedings of the IEEE*, Vol 54, pp. 1312 – 1329, ISSN: 0018-9219.

Kreis, T., (1986) Digital Holographic Interference-phase Measurement using the Fourier-transform Method, *J. Opt. Soc. Am. A*, 3(6): 847-855, ISSN: 1084-7529.

Lai, H. M., and Young, K. (1982) Theory of the pulsed optoacoustic technique. *J. Acoust. Soc. Am.*, Vol. 72, No. 6, pp. 2000-2007, ISSN: 0001-4966.

Li, Y., Baba, M., and Matsuoka, M., (1997) Femtosecond Measurement of Fluorescence by Two-photon Interference, *Physical Review A*, 55(4): 3177-3183, ISSN: 1050-2947.

Lingu, G.; Li, M.-L.; Xie, X.; Wang, L., and Stoica, G. (2007). *In vivo* imaging and characterization of hypoxia-induced neovascularization and tumor invasion. *Int. J. Oncol.*, Vol. 30, pp. 45-54, ISSN: 1019-6439.

Lu W, et al., (2010) Photoacoustic imaging of living mouse brain vasculature using hollow gold nanospheres, *Biomaterials*, 31 (9), 2617-2626, ISSN: 0142-9612.

Luo, M.S.C. et al., (1994) Generation of Terahertz Electromagnetic Pulses from Quantum-Well Structures, *IEEE J Quantum Electr*, 30, 1478-1488, ISSN: .

Macy, J.W.W., Jr., (1983) Two-dimensional Fringe-pattern Analysis, *Applied Optics,* 22(23): 3898-3901, ISSN: 0003-6935.

McKenzie, A. L. (1990) Physics of thermal processes in laser-tissue interaction. *Phys. Med. Biol.*, Vol 35, No. 9. pp. 1175-1209, ISSN: 0031-9155.

Meshulach, D., Barad, Y., Silberberg, Y., (1997) Measurement of Ultrashort Optical Pulses by Third-Harmonic Generation, *J. Opt. Soc. Am. B*, 14(8): 2122-2125, ISSN: 0740-3224.

Milonni, P.W., and Eberly, J.H., (1986) *Lasers*, ISBN: , New York: John Wiley and Sons.

Mirabella, F., Internal Reflection Spectroscopy, In: *Practical Spectroscopy Series,* vol. 15, Marcel Dekker, ISBN: 9780824780593, 1992, New York.

Miyamoto, Y., Kuga, T., Baba, M., and Matsuoka, M., (1993) Measurement of Ultrafast Optical Pulses with Two-photon Interference. *Optics Letters*, 18(11):900-992, ISSN: 0146-9592.

Möller, K.D., (1988) *Optics,* ISBN: 978-0-9357-0245-3, Mill Valley, California: University Science Books.

Mukherjee, J., and Chakraboty, S., (2010) Modelling of temperature distribution within a dental profile on account of laser irradiation, *Int. J. Biomedical Engineering and Technology*, Vol. 3: 186-198, ISSN: 1752-6418.

Naganuma, K., Mogi, K., and Yamada, H. (1989) General Method for Ultrashort Light Pulse Chirp Measurement. *IEEE J. Quantum Electron.*, 25(6):1225-1233, ISSN: .

Najarian, K, and Splinter, R., (2005) *Biomedical Signal and Image Processing,* CRC-Press, ISBN: 978-0-8493-2099-6, Boca Raton.

O'Shea, P., Kimmel, M., Gu, X., and Trebino, R. (2001) Highly simplified device for ultrashort-pulse measurement, *Optics Letters* 26: 932 -934, ISSN: 0146-9592.

Paye, J., Ramaswamy, M., Fujimoto, J.G., and Ippen, E.P., (1993) Measurement of the Amplitude and Phase of Ultrashort Light Pulses from Spectrally Resolved Autocorrelation, *Optics Letters*, 18(22):1946-1948, ISSN: 0146-9592.

Pike, P., (2005) *Photo-acoustic analysis of dental materials and tissue,* Ph.D. dissertation, The University of Tennessee, Knoxville, Tennessee.

Pike P., Parigger C., Splinter R., Lockhart P., (2007) Temperature distribution in dental tissue after interaction with femtosecond laser pulses, *Appl. Opt.* 46(34):8374-8378, ISSN: .

Press, W., Teukolsky, S., Vetterling, W., and Flannery, B., (2007) *Numerical Recipes, The Art of Scientific Computing, 3rd edition,* ISBN: 978-0-5218-8068-8, New York: Cambridge University Press.

Qiu, T. Q. and Tien, C. L., (1993) Size effects on nonequilibrium laser heating of metal films, *J. Heat Trans.*, Vol. 115, pp. 842-847, ISSN: 1528-8943.

Rost, F.W.D., (1992) *Fluorescence microscopy.* Cambridge University Press, ISBN: 978-0-5214-2277-2, Cambridge.

Saxena AK, Willital GH, (2008) Infrared thermography: Experience from a decade of paediatric imaging, *Eur J Pediatr*, 167, 757-764, ISSN: 0340-6199.

Semenov, S.Y., Splinter, R., Svenson, R.H., Baranov, V.Y., Dezern, K.R., Tuntelder, J.R., Thompson, M., Tatsis, G.P., (1995) Dielectric properties of canine myocardium tissues in the microwave spectrum, *IEEE Trans Biomed Eng.* 41(9), 569-73, ISSN: 0967-3334.

Semenov, S.Y., Svenson, R.H., Boulyshev, A.E., Souvorov, A.E., Borisov, V.Y., Sizov, Y., Starostin, A.N., Dezern, K.R., Tatsis, G.P., Baranov, V.Y., Splinter, R., (1996) Microwave tomography: two-dimensional system for biological imaging, *IEEE Trans Biomed Eng.* 43(9), 869-77, ISSN: 0967-3334.

Serbin, J., Bauer, T., Fallnich, C., Kasenbacher A., Arnold, W.H., (2002) Femtosecond lasers as novel tool in dental surgery, *Appl. Surf. Sci.* 197-198, 737-740, ISSN: 0169-4332.

Sethuraman, S.; Aglyamov, S. R.; Amirian, H. J.; Smalling, R. W., and Emelianov, S. Y. (2006). Development of a combined intravascular ultrasound and photoacoustic imaging system. *Proceedings of SPIE*, Vol. 6086, pp 108-117, Photons Plus Ultrasound: Imaging and Sensing 2006: The 7th Conference on Biomedical Thermoacoustics, Optoacoustics, and Acousto-optics, edited by A. A. Oraevsky, et al., ISBN: 978-0-8194-6128-5.

Shu, X., Spears, K.G., Serafin, J., (1989) Ultrashort Pulsed Laser Coherence measurements by Single-pulse Holography and Four-wave Mixing, *J. Opt. Soc. Am. B*, 6(7):1356-1362., ISSN: 0740-3224.

Siegman, A.E. (1986) *Lasers*, ISBN: 0-935702-11-5, Stanford: University Science Books.

Simhi, R., Gotshal, Y., Bunimovich, D., Sela, B.-A., and Katzir, A., (1996) Fiber-optic evanescent-wave spectroscopy for fast multicomponent analysis of human blood, *Appl. Opt.* 39, 3421-3425, ISSN: 0003-6935.

Splinter, R., Hooper, B., (2006) *An Introduction to Biomedical Optics*, CRC-Press, ISBN: 978-0-7503-0938-7, Boca Raton.

Takeda, M., Ina, H., Kobayashi, S., (1982) Fourier-transform Method of Fringe-pattern Analysis for Computer-based Topography and Interferometry, *J. Opt. Soc. Am.*, 72(1):156-160, ISSN: 1084-7529.

Trebino, R., Kane, D.J., (1993) Using Phase Retrieval to Measure the Intensity and Phase of Ultrashort Pulses: Frequency-Resolved Optical Gating, *J. Opt. Soc. Am. A*, 10(5):1101-1111, ISSN: 1084-7529.

Wang, X. D., Pang, Y. J., Ku, G., Xie, X. Y., Stoica G., and Wang L. H. ,(2003) Noninvasive laser-induced photoacoustic tomography for structural and functional *in vivo* imaging of the brain. *Nat. Biotechnol.* Vol. 21, No. 7, pp. 803-806, ISSN: 1087-0156.

Yan, C., and Diels, J.C. (1991) Amplitude and Phase Recording of Ultrashort Pulses, *J. Opt. Soc. Amer. B*, 8:1259-1263, ISSN: 0740-3224.

Yang, X., Scrabalak, S., Li, Z., Xia, Y. and Wang, L., (2007) Photoacoustic Tomography of a Rat Cerebral Cortex *in vivo* with Au Nanocages as an Optical Contrast Agent. *Nano Lett.*, Vol. 7, No. 12, pp. 3798-3802, ISSN: 1530-6984.

Yaseen, M., Ermilov, S., Brecht, H.-P., Su, R., Conjusteau, A., Fronheiser, M., Bell, B., Motamedi, M., and Oraevsky, A. (2010). Optoacoustic imaging of the prostate: development toward image-guided biopsy, *J. Biomed. Opt.*, Vol. 15, 021310, ISSN: 1083-3668.

Yuan, Z., Zakhaleva, J., Ren, H., Liu, J., Chen, W., Pan, Y., (2010) Noninvasive and high-resolution optical monitoring of healing of diabetic dermal excisional wounds implanted with biodegradable in-situ gelable hydrogels, *Tissue Eng Part C Methods*, 16(2), 237-47, ISSN: 1937-3384.

Zhang, E.Z., Laufer, J.G., Pedley, R.B. and Beard, P.C., (2009) In vivo high-resolution 3D photoacoustic imaging of superficial vascular anatomy, *Phys. Med. Biol.*, 54, 1035-1046, ISSN: 0031-9155.

22

CO_2 Laser Pulse-Evoked Nocifensive Behavior Mediated by C-Fibers

Bai-Chung Shyu
Institute of Biomedical Sciences, Academia Sinica, Taipei, Taiwan ROC

1. Introduction

Most tests that assess pain in animals involve motor responses to noxious stimuli. This practice depends on the implicit premise of a strong relationship between nociception and motor activity [Le Bars et al., 2001, Smith, 1995]. Therefore, a behavior can and should be analyzed as thoroughly as any brain structure and function involving motor activity in neurological investigations. The importance of analyzing behavior has been demonstrated by several studies on the mechanisms of nociception. For example, Rousseaux et al. [Rousseaux et al., 1999] showed that patients with spinothalamic tract injury had a higher temperature threshold than normal in response to nociceptive stimuli and have pain but not heat sensations. Thermal and electrical stimulation of the affected side in these patients elicited a withdrawal reaction. Another example, demonstrated by Shyu et al. [2003], was conditioned CO_2 laser-evoked behavior in rats to distinguish emotional components from sensory conductive pathways [Kung et al., 2003].
The ideal model of pain stimulation should be nociceptive-specific, controllable, safe, and reproducible. Pain pathways are a part of the somatosensory system. Therefore, neurophysiological investigation of pain pathways is related to the recording of somatosensory-evoked responses. To test the integrity of nociceptive pathways with evoked nociceptive responses, the timing of the activation of nociceptive afferents must be precise for stimulus-locked averaging. Many experimental pain models that fulfill this requirement have been successfully used in pharmacological studies. Nevertheless, most of these models have limited clinical application for precise diagnosis. For example, pain has been elicited by cutaneous intense electrical stimulation, which activates peripheral fibers that respond to both noxious and innocuous stimulation. The drawback of electrical stimulation may be partially overcome by using various blocking techniques. Another approach is to induce nociception by activating tooth pulp, which is innervated mostly by Aδ- and C-fibers [Roos et al, 1982, Van Hassel, 1972]. Stowell [1974] used an electromechanically driven pin to produce cutaneous pain, and Mitchell and Hellon [1977] used gradual warming to induce cutaneous thermal stimulation. However, both methods can excite thermo- or mechanoreceptors in addition to nociceptors [Burgess & Perl, 1973].
Non-contact thermostimulation is required for the investigation of the neural mechanisms underlying pain sensation without concurrent activation of mechanoreceptors. To record thermal-evoked potentials, thermostimulation must be very brief to permit the time-locking of potential averaging to the sensory event. Many repetitions of the thermal stimulation

without habituation are necessary to improve the signal-to-noise ratio, and only nociceptors should be activated. Applying gas lasers, such as CO_2 and argon, to the skin satisfies most of the aforementioned criteria. A laser beam does not actually contact the skin; therefore, heat and thermosensitive nociceptors are selectively activated, while mechanoreceptors are unaffected [Kakigi et al., 1989]. Thus, for the experimental study of pain and clinical neurology diagnosis, only thermal stimulation of non-glabrous skin by brief laser pulses has gained widespread clinical use.

The use of short CO_2 laser pulses in the study of pain was introduced by Mor and Carmon [1975]. These pulses have several potential advantages as noxious stimuli. First, the laser is the source of very short and concentrated thermal energy that activates thermosensitive nociceptors. Second, in contrast to conventional radiant heat, short laser pulses induce a synchronous volley of impulses in afferent nerves, allowing repeated and simultaneous measurement of time-locked evoked neuronal responses at different levels of the nervous system. Their high power output produces steep heating ramps, which improve synchronization of afferent volleys and therefore allow the recording of time-locked events, such as laser-evoked potentials (LEPs) in the brain. The amplitude of LEPs on the human scalp is related to subjective reports of pain sensation [Bjering & Arendt-Nielsen, 1988, Bromm & Treede, 1984, Carmon et al., 1978]. These findings suggest that LEPs may be used as objective correlates of pain perception.

Laser-evoked cortical responses can be classified into two groups [Danneman et al., 1994, Isseroff et al., 1982, Kalliomaki et al., 1993]. The first group, mediated by Aδ-fibers, is more sensitive to pentobarbital-induced anesthesia [Shaw et al., 2001] and laser pulse energy level [Kalliomaki et al., 1993]. The second group, mediated by C-fibers, is widespread across the cortical surface and can be diminished by topical morphine application onto the lumbar spinal cord and reversed by naloxone [Kalliomaki et al., 1998]. Although the peripheral and central conductance mechanisms responsible for cortical LEPs have been investigated, the results have been inconsistent. For example, Devor et al. [1982] claimed that only C-fibers, and no Aδ-fibers, were activated by CO_2 laser pulses, according to single-unit recordings in peripheral nerves in rats. However, the Isseroff group [1982] studied cortical potentials in anesthetized rats and found that laser-evoked cortical potentials have a latency of approximately 40 to 55 ms, which is within the range of Aδ-fibers.

We previously administered short-pulse CO_2 laser stimulation to freely moving rats to measure nociceptive sensitivity [Shyu et al., 1995] and studied nocifensive behavior induced by a similar model [Fan et al., 1995]. Nocifensive behavior was subdivided into eight discrete response components, including flinching, foot elevating, licking, body movements, and foot movements. Following stimulation with higher laser energy, rats exhibited both an increase in the number of response types and frequency of specific responses. The results suggested that hierarchically organized responses in the nocifensive motor system are recruited to varying degrees by noxious stimuli with different intensities [Fan et al., 1995]. Additionally, the threshold for activation of the nocifensive behavior component "body movement" was lower than that of the behavior components "flinching" and "licking" [Fan et al., 1995]. Le Bar et al. [2001] proposed that stimulus intensity can determine which type of nerve fiber initiates the reaction. Treede et al. [1995] illustrated this concept by showing that the threshold of C-fibers was significantly higher, and conduction was slower, than that of Aδ-fibers. The differential activation of Aδ- and C-fibers and the difference in their physiology suggest that they play differential roles in nociceptive responses under specific stimulation conditions [Lawson 2002]. In humans, considerable evidence indicates that distinct secondary pain sensation is driven by C-fibers [Ploner et al., 2002, Torebjork & Ochoa, 1990] and is activated by C nociceptors distributed in the rostral hypothalamus [Lumb, 2002]. The supraspinal circuits activated by C-fiber inputs can be linked to nocifensive behavioral patterns related to the relative contribution of these types of nociceptive afferents [Le Bars et al., 1976]. Therefore, nocifensive behaviors can be differentially elicited and mediated by different classes of nociceptors.

The aim of the present study was to test the hypothesis that nocifensive behavioral elements elicited by brief laser pulse stimuli are mediated by C nociceptors. We applied the neurotoxin capsaicin, which acts specifically on primary sensory afferent C-fibers, to selectively block activity. Capsaicin has been locally applied to the sciatic nerve of the hind legs in rats to examine whether cortical LEPs are conducted by C-fibers [Sun et al., 2006]. In the present study, we used a previously established nocifensive behavioral model in rats to study the correlation between cortical LEPs and the response components evoked by a short-pulse CO_2 laser beam and examined whether those response elements are nocifensive responses mediated by C-fibers.

2. Materials and methods

2.1 Subjects

A total of 68 male Sprague Dawley rats, weighing 250-350 g, were used for these experiments. Food and water were available *ad libitum*. Rats were housed singly in standard wire mesh cages in a temperature- and humidity-controlled environment (21-23°C, humidity 50%) with a 12 h/12 h light/dark cycle (lights on at 0600 h). All experiments were conducted during the light period in a quiet room. Animal care adhered to the guidelines of the Academia Sinica Institutional Animal Care and Utilization Committee and the National Institutes of Health Guide for the Care and Use of Laboratory Animals (NIH publication no. 80-23).

2.2 Implantation of chronic cortical and electromyogram electrodes

Rats were surgically prepared before being subjected to neurophysiological and behavioral experiments. The rats underwent surgery while under pentobarbital anesthesia (50 mg/kg, i.p.; Abbott Laboratories, Chicago, IL, USA). For recording of cortical-evoked potentials, electrodes were chronically implanted into the rats' brains. Three holes (1.0 mm diameter) were drilled in the skull, one on each side (coordinates: 3 mm lateral to bregma, approximately in the region of the primary somatosensory cortex, S1) and the third 2 mm behind lambda as a reference lead. Stainless-steel screws were tightly inserted into the holes so that their tips made contact with the dura mater overlying the cortical surface. A thin Teflon-coated stainless steel wire was soldered on the edge of the screw head and connected with one of eight legs of an integrated circuit (IC) chip receptacle. The screws were used as electrodes and anchoring points for the IC chip receptacle, which was affixed to the skull with dental cement.

For electromyography (EMG), two multi-strand stainless steel microwires were inserted into the hamstring muscle of the left hindlimb. Dental cement was applied to fasten the connection receptacle to the surface of the skull. The measurements were collected 1 week after surgery. The body weights of the treated rats were monitored between surgery and the recording day, which occurred 1 week after surgery. Post-surgical body weight gain served as an indicator of recovery from surgical stress. Rats that did not gain weight 1 week following surgery were excluded from the study.

2.3 Laser stimulation

The glabrous skin of the foot pad was stimulated with a high-intensity CO_2 laser beam (Carbon Dioxide Surgical Laser System, model 20 CH, Direct Energy, Inc., CA, U.S.A.). The system generated a laser radiation beam in the infrared spectrum (10.6 μm wavelength) and had adjustable power capable of producing peak laser power greater than 20 watts (W). The duration of the laser pulse was controlled by the duration of an externally triggered TTL pulse from a pulse generator (A-M systems Inc., USA). The laser beam was focused approximately 3 cm in front of the laser head, and the size of the beam was measured by burning a spot on heat-sensitive coated paper that was set at a distance indicated by a rod. In the acute spinal cord recording experiment, the laser tube was held by an experimenter or a rigid stainless steel stand.

2.4 Tactile stimulation

A 1.5 mm diameter stainless steel rod was attached to the voice coil of an 8 Ω, 15 W loud speaker as a tactile stimulus. The tactile stimulus was triggered and controlled by a TTL pulse from a pulse generator (A-M Systems Inc., USA). The movement of the coil transmitted to the rod produced an outward excursion of 5 mm.

2.5 Recording devices

A cable from above the cage was connected to the receptacle on the head of the animal. Electrical activity was recorded either from the cortex of freely moving animals or from the spinal cord of anesthetized animals and was amplified by differential AC amplifiers (0.1 Hz to 5 KHz, Model DAM70, WPI, USA). Analog signals were either displayed on a digital storage oscilloscope or sent to a IBM-PC-based data acquisition system for online A/D conversion (Metrabyte DAS-16F AD/DA interface card, USA) and digital analysis using Quick Basic Language. The signals were also stored on a hard disk for offline data analysis using a PC-based data analysis system.

2.6 Application of capsaicin to the sciatic nerve

After obtaining the evoked cortical potentials and behavioral responses, the left legs of the rats were treated with either capsaicin or vehicle solution. The rats (n = 26) were first anesthetized by halothane inhalation (2-3%). The sciatic nerves of the left legs were exposed and separated from the muscles and vessels by placing paraffin paper under them. A piece of cotton soaked with either vehicle (3% Tween-80 dissolved in 100% alcohol) or 1% capsaicin (Sigma-Aldrich, St. Louis, MO, USA) dissolved in vehicle was applied directly to the nerve trunk for 45 min. After treatment, the cotton was removed, and the open muscle and skin were sutured. Rats were then given antibiotics intramuscularly and allowed to recover. Their cortical potentials and behavioral responses were tested 3 days after surgery.

2.7 Behavioral assessment and procedures

Control experiment. Stimuli had a fixed pulse duration of 30 ms and five graded intensities: 3, 5, 10, 15, and 20 W. Animals were randomly divided into five groups (n = 6/group), and each group received one of five stimuli. In each stimulation session, 20 laser stimuli of the same intensity were delivered. Stimulation protocols were chosen to establish a correlation between evoked cortical potentials and behavioral reaction patterns.

Capsaicin experiment. Twelve rats were randomly divided into two groups. One group received capsaicin treatment, and the other group received vehicle. The average laser- and tactile-evoked potentials were recorded before and after capsaicin or vehicle treatment. Both the treated and untreated hindpaws were stimulated with laser pulses (15 W intensity, 30 ms duration). The untreated side served as a control. The stimulus power was chosen based on our previous results [Shyu et al., 2003].

Testing procedures. Tests were performed on the day before capsaicin treatment (Day 0) and on Days 3, 5, and 14 after treatment. Animals were brought to the laboratory and placed into the test cage for 20 min before each test. This procedure familiarized the animal to the test environment to reduce exploratory activity during the test. The test cage measured 26.5 cm length × 21 cm width × 17 cm height and was constructed with transparent acrylic sheets with a 0.5 cm^2 stainless steel grid at the bottom. The cage was hung such that the bottom could be viewed by the experimenter to apply laser stimuli to the hindpaw. During a test session, the laser beam was projected from below the test cage to the plantar surface of the rat's hind footpad. The animals were unrestrained and free to react during the experiments. The experimental setup for laser stimulation, electrophysiological recording, and behavioral measurement in freely moving rats is shown in Fig. 1.

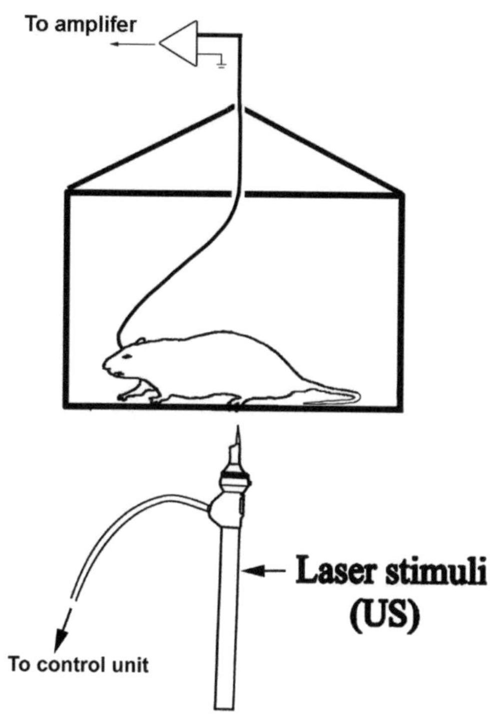

Fig. 1. Experimental setup for brief CO_2 laser pulse stimulation, electrophysiological recording, and behavioral assessment in unrestrained, freely moving rats.

Behavior analysis. Behavioral assessment was described in detail in our previous paper [Fan et al., 1995]. Briefly, behavior was observed for 2 min following each stimulus presentation because previous experience indicated that the animal returned to its resting status within 2 min. During this 2 min, whenever a response occurred, the frequency was counted as "1," no matter how many times it was repeated. The raw data following each stimulus consisted of whether a response occurred (1) or did not occur (0).

Five behavioral components were recorded and quantified: body movement, head turning, withdrawing, flinching, and licking. *Body movement* was defined as slow body motion, body turning, running, or exploration involving translocation of the body. *Head turning* included turning, shaking, or elevating the head. *Flinching* was recorded whenever a rat made a small abrupt jerking movement (< 1 cm) with its body. A *withdrawal* response was scored when an animal receded its body position by ≥ 1 cm away from the stimulus. *Licking* involved the rat retracting the paw by ≥ 1 cm and licking. Other complicated movements, such as tapping, that may have occurred together with a withdrawing or licking motion were recorded with the major movement that the complicated movement accompanied.

Twenty laser stimuli were applied during each session, and the maximum number of an observed behavioral component was 20 for each rat, even if the rat repeated the same behavior. The laser intensity used in the behavioral assessment was 10 W and 30 ms, based on our previous study [Fan et al., 1995].

2.8 Acute spinal cord preparation for electrophysiological recordings

Six rats that were not used for behavioral analysis were anesthetized by intraperitoneal injection of a mixture of ketamine (30 mg/kg) and rumpon (15 mg/kg). A PE-240 tube was inserted into the trachea for artificial respiration, and the left jugular vein was catheterized with a PE-50 tube for drug or fluid infusion. Body temperature was maintained at 37 ± 5ºC by a homeothermo blanket system. During the recording experiments, animals were paralyzed and received artificial respiration. The end tidal CO_2 concentration was monitored and maintained in a range of 3-4%. Because the surgery and experiments may be prolonged, supplementary doses of ketamine (10 mg/kg) were administered intravenously to maintain the depth of anesthesia, which was monitored by pinching the tail of the rat. If a withdrawal reflex occurred, then anesthesia was not sufficient. Laminectomy was performed to expose the lumbar spinal cord (L3-L6). The animal was then placed in a rigid spinal unit frame, and the dura mater covering the spinal cord was dissected. The skin flaps were used to form a pool that was filled with warm paraffin oil.

Surface potentials were recorded with bipolar Ag-AgCl ball electrodes with an active lead placed on the surface of the dorsal column of the spinal cord and a reference lead placed in the nearby connective tissue. The recording was made from the lumbar spinal cord (L4 and L5) because this is the area where the maximal evoked potential from stimulation of the hindlimb foot pad could occur. Capsaicin was applied acutely to investigate the contribution of C-fibers on spinal cord evoked potentials.

2.9 Acute preparation for cortical recording

The anesthesia procedure was the same as the spinal cord recording experiment. When the paralysis test was required, animals were immobilized with gallamine triethiodide (Flaxedil, 40 mg/kg, i.v.) and then artificially ventilated. End tidal CO_2 concentration was monitored and maintained at approximately 3-4%. Heart rate, respiratory rate, and pCO_2 were monitored throughout the recording session. Six rats were allowed to periodically recover from muscle paralysis to assess the arousal level of anesthesia by evaluating the corneal reflex or withdrawal reflex of the hindlimb. Supplementary doses of ketamine (10 mg/kg) were administered if reflexes were present.

2.10 Data analysis

The amplitudes of the laser-evoked cortical potentials and spinal cord potentials and the total frequency of each response over the 20 trials were obtained for each rat. Analysis of variance (ANOVA) was used to assess the effects of capsaicin treatment.

3. Results

3.1 Effects of capsaicin on spinal cord and cortical potentials evoked by mechanical and laser stimulation

Mechanical stimulation of the hindpaw evoked a short negative potential followed by a long-lasting positive potential in the dorsal lumbar column surface. Applying capsaicin to the sciatic nerve did not influence this potential (Fig. 2A, left panel). In contrast, a short laser pulse (10 W, 30 ms) applied to the hindpaw induced a prominent positive potential in the dorsal column of the lumbar spinal cord (L4-L5) (Fig. 2A, lower right panel). The LEP amplitude in the spinal cord was significantly lower in the capsaicin group (Fig. 2A, upper right panel) compared with the vehicle group, indicating that capsaicin specifically blocked the transmission of sensory information from the peripheral nerve to the spinal cord.

3.2 LEP conduction velocity

To estimate the conduction velocity of LEP in peripheral transmission, we recorded field potentials in the spinal cord evoked by laser pulses applied to two sites, 10 mm apart, on the hindpaw (Fig. 2B). To calculate the conduction velocity, the difference in the latency of both LEPs produced by stimulating the proximal and distal paw regions was divided by the distance (10 mm) between the sites. The LEP conduction velocity was 0.8 ± 0.3 m/s ($n = 5$), which was in the range of C-fiber conduction. The latency for spinal cord evoked potentials was in the range of 70 to 90 ms, which was close to the latency of the laser-evoked EMG response of the hindlimb. Thus, the EMG response recorded in the hindlimb was likely induced by C-fiber activation. The latency of cortical LEPs had a much longer latency than both the EMG responses and the spinal cord evoked potentials. Thus, the ascending spinal pathway appears to convey such nociceptive information to the supraspinal brain areas.

3.3 Laser-evoked cortical potentials

Short-pulse, low-intensity, CO_2 laser irradiation of the hindpaw in awake, freely moving rats produced distinct evoked cortical potentials in the S1 region. Prominent negative potentials were evoked and could be distinguished from background potentials. The peak latency of the LEP was 254.2 ± 11.9 ms, and the amplitude was -0.16 ± 0.05 mV. Laser-evoked potential components could be induced bilaterally, and their amplitudes increased as laser intensity increased (Fig. 3A, B). An earlier component was discernible at a latency of approximately 60 ms, but because of the variation in amplitudes and nonsignificant influence of capsaicin, this early component was not analyzed further.

3.4 Component profiles of nocifensive behavioral responses

Consistent with our previous findings [Fan et al., 1995], we found that as stimulus intensity increased, animals exhibited a greater variation in responses and a greater frequency of component responses (Fig. 3C). When the hindpaw was stimulated with 5 W laser power, the rats reacted with few withdrawal, flinch, and body movement responses. At 10 W, their

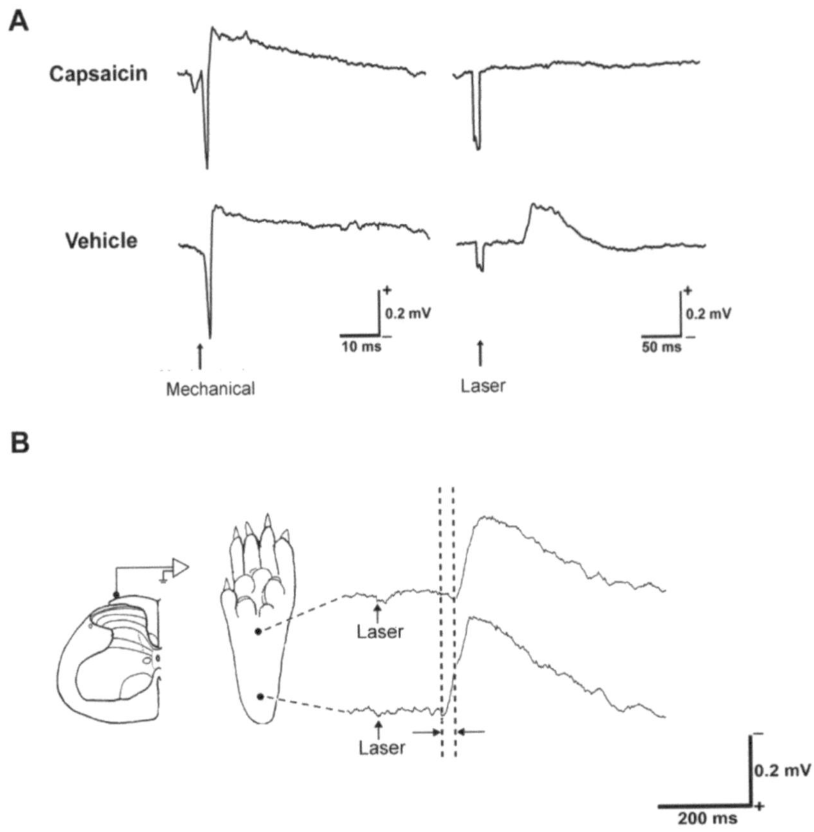

Fig. 2. Spinal cord evoked potentials and conduction velocity measurement. **A.** Spinal cord evoked potentials induced by either mechanical or CO_2 laser stimulation. Applying capsaicin to the left sciatic nerve in rats affected spinal cord LEPs but not mechanical-evoked potentials. No changes were found in vehicle group potentials evoked by either of the stimulation methods. **B.** Conduction velocity of spinal cord potentials evoked by CO_2 laser stimuli. The recording site was located in the dorsal column of the spinal cord. Laser stimulation was applied to the heel and middle regions of the hind paw. The representative evoked potential for each stimulated point is shown in the right panel. The difference in the response latency for both stimulation points was calculated and divided by the total distance from the hindpaw to the spinal cord.

withdrawal/flinch response frequency increased by more than 50%, with very few body movements and licking. Behavioral responses markedly changed when stimulated with 20 W, with withdrawal/flinch, body movement, and lick responses increasing by 40%, 30%, and 70%, respectively. At 25 W, a greater number of lick responses and 20% and 50% more withdrawal/flinch and body movement responses, respectively, were evident.

The of withdrawal/flinch, body movement, and lick response components corresponded to the cortical LEP amplitude (Fig. 3D). At 3 W laser power, LEPs were less than 0.1 mV. At 5 W, the rats reacted with withdrawal/flinch responses which produced ~0.2 mV of cortical LEPs. The amplitude increased to ~3-4 mV when body movements and licking were elicited by stimulation at the 10 and 20 W stimulus intensities.

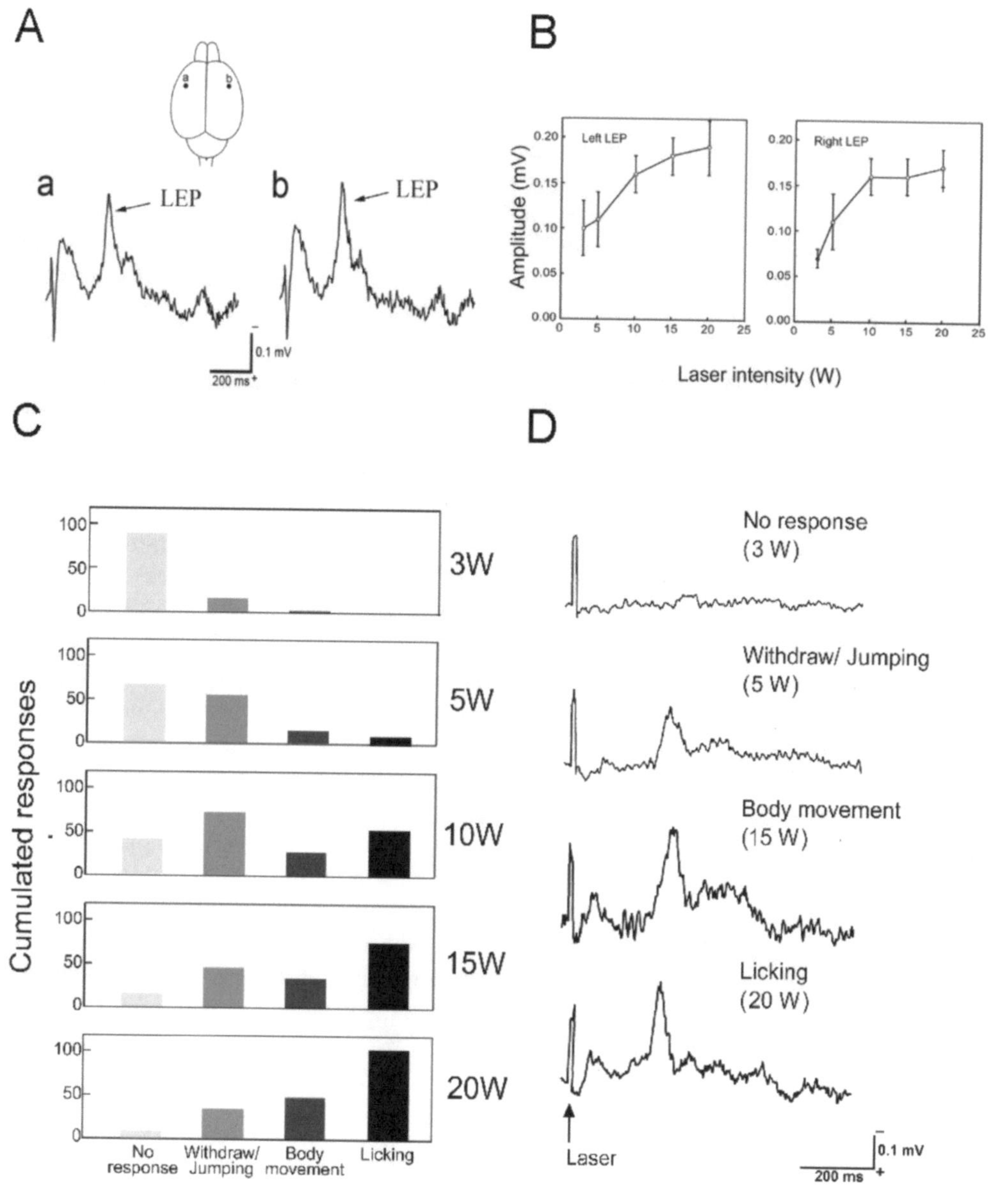

Fig. 3. Typical examples of cortical field potentials evoked by noxious laser pulses and nocifensive behavioral response component profile. **A.** Laser-evoked potentials were bilaterally recorded in the cortex. **B.** The graphs show the mean amplitude of LEPs recorded from the right or left side as a function of laser intensity. **C.** Average response frequency for each behavioral response component at different laser stimulation intensities ranging from 5 to 25 W. **D.** The representative evoked potentials corresponded to different behavioral elements: no response, withdrawal/flinch, body movement, and licking.

3.5 Effect of capsaicin on nocifensive behavioral components

Capsaicin treatment did not noticeably alter locomotion or the righting response during handling or open field exploration. Furthermore, capsaicin and vehicle treatment did not produce trophic or morphological changes in the lower extremities.

Laser stimulation of the footpad in freely moving rats induced several behavioral components, such as flinching, withdrawing, licking, moving, and head turning. The occurrence of some behavioral components were markedly altered after capsaicin treatment. The response frequency for flinching, withdrawing, and licking were significantly reduced after stimulation on the capsaicin-treated side, whereas the frequency for moving and head turning were not affected. Additionally, the reduced incidence of behavioral components after capsaicin treatment did not recover for at least 14 days.

Rats showed an average of three flinch responses induced by laser stimulation (10 W, 30 ms) before capsaicin treatment. After capsaicin treatment, this response decreased to an average of less than one time. This observation was confirmed statistically using a *post hoc* test, in which the flinch response significantly decreased compared with the vehicle treatment group ($F = 4.41$, $p < 0.05$) on days 3, 5, and 14 after treatment. Similarly, withdrawal responses decreased after capsaicin treatment. Before treatment, rats exhibited the withdrawal response 5-6 times, but following treatment, the capsaicin-treated group responded less compared with the untreated side and vehicle group ($F = 13.03$, $p < 0.001$) on days 3, 5, and 14 ($p < 0.01$) after treatment. After the laser pulse stimulation was applied to the hindpaws, the frequency of the licking responses was also reduced from 2-4 times before capsaicin treatment to approximately 1 time ($F = 6.56$, $p < 0.01$) on Days 3, 5, and 14 after treatment. In contrast to the above behaviors, rats only moved or turned their heads 1-2 times before treatment, and this frequency did not change after capsaicin treatment.

Typical effects of capsaicin and vehicle treatment on LEPs in behaving rats are shown in Fig. 4. Laser-evoked potentials were markedly diminished after the left sciatic nerve was treated with capsaicin but not with vehicle (Fig. 4A, B, and right panel). A prominent fast component could be elicited by mechanical stimulation, and this component was unaltered after capsaicin treatment. The amplitude of LEPs decreased by an average of 78% in the capsaicin group compared with the vehicle group and untreated paws of both groups. These observations were confirmed statistically. The amplitude of cortical LEPs produced by stimulating the left paw was significantly reduced in the capsaicin-treated group compared with the vehicle group ($F = 4.95$, $p < 0.05$; Fig. 4C). However, no significant differences were found for mechanical-evoked cortical potentials (Fig. 4D).

4. Discussion

This study showed that capsaicin, which selectively blocks activity of primary sensory C-fibers, inhibited specific components of nocifensive behavior. Applying capsaicin to the sciatic nerves of rats blocked flinch, withdrawal, and lick responses but did not affect movement and head turn responses. These results suggest that C-fibers mediate the nocifensive components of flinching, withdrawing, and licking. Interestingly, these components were temporally linked to activation of the nocifensive motor system and cortical-evoked potentials.

Nocifensive behaviors elicited by CO_2 laser stimulation were linked to LEPs recorded in freely moving rats. The average latency of late LEPs was 250 ms, which is in the range of C-

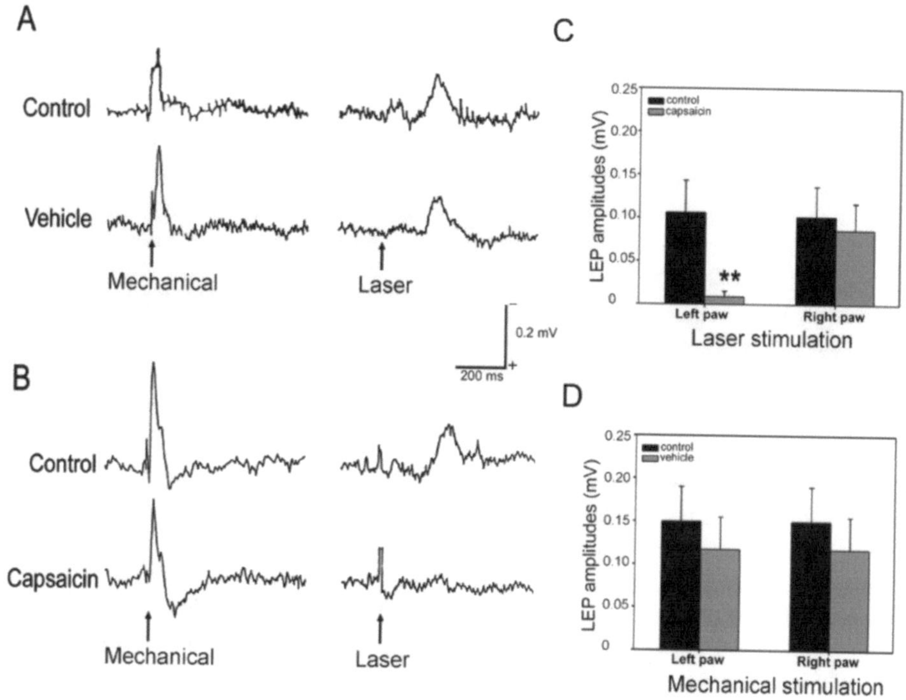

Fig. 4. Effect of capsaicin on cortical potentials evoked by mechanical and laser stimulation. The sciatic nerve of the left leg of rats was treated with capsaicin or vehicle ($n = 6$/group), whereas the right leg was untreated (Control). **A, B.** Traces of recorded potentials for the control, vehicle, and capsaicin groups. Mechanical and laser stimulation was applied to the right (A) and left (B) paws. **C, D.** Statistical analyses showed that the LEP (C) was significantly reduced in the capsaicin group compared with the untreated paw and vehicle group, but not with mechanical stimulation (D). $^{**}p < 0.01$.

fiber conduction [Bromm & Treede, 1987]. This result is closely related to C-fiber-mediated cortical responses evoked by laser pulses in anesthetized [Shaw et al., 2001] and awake, freely moving rats [Shaw et al., 1999]. Capsaicin, a neurotoxin known to block C-fiber activity [Wall & Fitzgerald, 1981], decreased the frequency of nocifensive responses, the amplitude of spinal evoked potentials, and LEPs when applied to the sciatic nerves of rats. Altogether, these results suggest that the nocifensive response components, including flinching, withdrawing, and licking, were mediated by C-fibers via the spinal cord and then linked to neurons in the cortex, which was demonstrated by LEP induction. We recently analyzed the current source density of LEPs in the S1 region of the cortex and found that intracortical current flows revealed late synaptic activation that initially occurred in layer VI and were relayed transsynaptically to both superficial and deep layers [Sun et al., 2006].

Evaluating the respective contributions of both Aδ- and C-fibers for a given test of nociception is challenging. Capsaicin administration primarily destroys C-fibers in rats and sometimes certain Aδ-fibers, albeit to a lesser extent [Lynn, 1990]. To exclude the possibility of Aδ-fiber destruction, we applied electrical stimulation on treated and untreated foot paws. Subsequently, we found that the animals had evoked spinal and cortical potentials. The latency of the evoked potentials was in the range of 60 ms, and their amplitudes were significantly higher than that of the capsaicin-treated group. These results demonstrate that faster conduction fibers, in the range of Aδ-fibers, were not destroyed. Therefore, we are confident that the behavioral components, including flinching, withdrawing, and licking, are nocifensive responses mediated by C-, but not Aδ-, fibers. In experiments conducted in humans, in which a brief and sufficiently intense single stimulus activates Aδ- and C-fibers, the emotional component of the secondary pain is much stronger than the primary pain [Handwerker & Kobal, 1993]. In the future, we suggest using an alternative experimental design that combines this pain-conditioning model with the current behavioral model to distinguish sensory components from affective components [Kung et al., 2003, Shyu et al., 2003].

Ideally, an animal model of behavior related to nociception should possess the following characteristics: (*i*) nociceptive stimulation ("input specificity"), (*ii*) a distinction between nociceptive and non-nociceptive stimuli, (*iii*) quantifiable responses that are correlated to stimulus intensity within a reasonable range (from thresholds for pain to pain tolerance), (*iv*) sensitivity to manipulations (notably pharmacological), which would specifically reduce nociceptive behavior, (*v*) differentiation between nonspecific behavioral changes (e.g., motility and attention) from those triggered by the nociceptive stimulus *per se*, (*vi*) consistent scores between identical or equivalent retests, and (*vii*) reproducible results. Most, if not all, of these characteristics are present in the model used in the present study. The behavioral model used here involved the use of a noxious stimulus, CO_2 laser, to elicit specific nocifensive response components. These nocifensive components (such as flinching, withdrawing, and licking) could be quantified, manipulated by capsaicin, and distinguished from non-nocifensive behaviors (e.g., head turning and moving). Moreover, these response components were consistently elicited when stimulated with CO_2 lasers and are reproducible because they have been described in reports from other laboratories [Shaw et al., 1999].

A laser pulse radiates intense and highly focused thermal energy and has been used as noxious stimulation in several human studies [Arendt-Nielsen & Bjerring, 1993, Kakigi et al., 2005]. Carmon et al. reported that an energy density of 1.64 to 1.94 J/cm ^{2} caused a slight burning pain sensation in subjects, and an energy density of 2.49 to 2.54 J/cm^2 caused a moderate pinprick pain sensation [Carmon et al., 1976]. In the present study, we used the same laser stimulus source and found that the energy densities that led to lick and foot movement responses were 1.93 and 3.11 J/cm^2, respectively. Thus, the range of energy density values obtained in the present animal study is comparable to that from human studies. In humans, when the stimulus intensity was increased to achieve progressive levels of pain, the amplitude of evoked cortical responses also increased, and the correlation between cortical responses and weighted subjective responses was significant (r = 0.707) [Carmon et al., 1976]. In the present study, we found that various intensities of CO_2 laser stimulation applied to the same location in anesthetized and freely moving rats induced dorsal horn activity and two cortical potentials, one with a latency of approximately 64 ms and a second with a latency of 253 ms. The spinal potential and the second cortical potential were sensitive to the stimulus intensity and susceptible to modification by capsaicin. These data further suggest that our laser stimuli were indeed noxious and specific to certain nocifensive behavioral elements and neuronal activity in rats.

The three response components of the nocifensive behavior model were most likely mediated by C-fibers, and we suggest that the present model is suitable for studying the neuronal mechanisms underlying the analgesic effects of morphine. Morphine reduces the responses of dorsal horn neurons produced by C-fibers more easily than it affects those produced by Aδ-fibers [Jurna & Heinz, 1979]. This observation may explain why experimental pain in humans, which is usually produced by Aδ-fibers, is little affected by morphine [Becher, 1957]. Some behavioral models using phasic stimulation methods predispose human subjects and animals to respond to pain as soon as it occurs (i.e., at the moment the first pain is produced by Aδ-fibers). The presence or absence of secondary pain will generally have no impact on the measurement. For example, animals withdrew their hindpaw after high-intensity electrical stimulation [Evans, 1961]. This test may involve the activation of both Aδ- and C-fibers, as well as some non-nociceptive fibers. Stimulation is stopped as soon as a response is observed. Yeomans and Proudfit [1996] suggested that most common nociceptive tests involving mechanical and thermal stimuli actually investigate only responses triggered by Aδ-fibers and thus are not sensitive to morphine, with the exception of very high doses. In a pain-induced audible and ultrasonic vocalization experiment in rats, a vocal response was clearly triggered by C-fibers and was very sensitive to morphine, with an ED_{50} five-fold less than when it is triggered by Aδ-fibers [Jourdan et al., 1998]. We therefore propose that our nocifensive behavioral model is suitable for studying the dynamic analgesic effects of morphine.

In conclusion, the present results suggest that nocifensive behavior has distinct components that can be analyzed, and the reaction pattern changes probabilistically, such that the greater the noxious stimulation, the more likely additional components will be evoked [Fan et al., 1995]. The nocifensive motor system may be viewed as a set of hierarchically organized responses, and a given subset of responses appear with a specific noxious stimulation, dependent on stimulus intensity. The study of the mechanism of pain must consider this pain response hierarchy to precisely define the neurological bases of sensory and motor aspects of the nociceptive system.

5. References

Arendt-Nielsen, L. & Bjerring, P. (1993). Sensory and pain threshold characteristics to laser stimuli. *J Neurol Neurosurg Psychiatry*,51, 35-42.

Becher, H.K. (1957). On misunderstanding the opportunities in anesthesia. *Anesthesiol*,18, 498-500.

Bjering, P. & Arendt-Nielsen, L. (1988). Argon laser induced single cortical responses: a new method to quantify pre-pain and pain perceptions. *J Neurol Neurosur Psychiatry*, 51, 43-49.

Bromm, B. & Treede, R.D. (1984). Nerve fibre discharges, cerebral potentials and sensations induced by CO_2 laser stimulation. *Human Neurobiol*, 3, 33-40.

Bromm, B. & Treede, R.D. (1987). Human cerebral potentials evoked by CO_2 laser stimuli causing pain. *Exp Brain Res*, 67, 153-162.

Burgess, P.R. & Perl, E.R. (1973). Cutaneous mechanoreceptors and nociceptors. In: *Handbook of Sensory Physiology: Somatosensory System.* Ainsley Iggo (Ed.), 29-78, Berlin, Springer-Verlag.

Carmon, A.; Mor, J. & Goldberg, J. (1976). Evoked cerebral responses to noxious thermal stimuli in humans. *Exp Brain Res,* 25, 103-107.

Carmon, A.; Dotan, Y. & Sarne, Y. (1978). Correlation of subjective pain experience with cerebral evoked responses to noxious thermal stimulations. *Exp Brain Research,* 33, 445-453.

Danneman, P.J.; Kiritsy-Roy, J.A.; Morrow, T.J. & Casey, K.L. (1994). Central delay of the laser-activated rat tail-flick reflex. *Pain,* 58, 39-44.

Devor, M.; Carmon, A. & Frostig, R. (1982). Primary Afferent and spinal sensory neurons that respond to brief pulses of intense infrared laser radiation: a preliminary survey in rats. *Exp Neurol,* 76, 483-494.

Evans, W.O. (1961).A new technique for the investigation of some drugs on a reflexive behavior in the rat. *Psychopharmacology,* 2, 318–325.

Fan, R.J.; Shyu, B.C. & Hsiao, S. (1995). Analysis of nocifensive behavior induced in rats by CO2 laser pulse stimulation. *Physiol Behav,* 57, 1131-1137.

Handwerker, H.O. & Kobal, G. (1993). Psychophysiology of experimentally induced pain. *Physiol Rev,*73, 639-671.

Isseroff, R.G.; Sarne, Y. Carmon, A. & Isseroff A. (1982). Cortical potentials evoked by innocuous tactile and noxious thermal stimulation in the rat: differences in localization and latency. *Behav Neural Biol,* 35, 294-307.

Jourdan, D.; Ardid, D.; Chapuy, E.; Le Bars, D. & Eschalier, A. (1998). Effect of analgesics on audible and ultrasonic pain-induced vocalization in the rat. *Life Sci,* 63, 1761–1768.

Jurna, I. & Heinz, G. (1979). Differential effects of morphine and opioid analgesics on A and C fibre-evoked activity in ascending axons of the rat spinal cord. *Brain Res,* 171, 573-576.

Kakigi, R.; Inui, K. & Tamura, Y. (2005). Electrophysiological studies on human pain perception.*Clin Neurophysiol,* 116, 743-63.

Kalliomaki, J.; Weng, H.R.; Nilsson, H.J. & Schouenborg, J. (1993). Nociceptive C fibre input to the primary somatosensory cortex (SI). A field potential study in the rat, *Brain Res,* 622, 262-270.

Kalliomaki, J.; Luo, X.L.; Yu, Y.B. & Schouenborg, J. (1998). Intrathecally applied morphine inhibits nociceptive C fiber input to the primary somatosensory cortex (SI) of the rat. *Pain,* 77, 323–329.

Kung, J.C.; Su, N.M.; Fan, R.J.; Chai, S.C. & Shyu, B.C. (2003). Contribution of the anterior cingulate cortex to laser-pain conditioning in rats. *Brain Res,* 970, 58-72.

Lawson, S.N. (2002). Phenotype and function of somatic primary afferent nociceptive neurones with C-, Adelta- or Aalpha/beta-fibres. *Exp Physiol,* 87, 239–244.

Le Bars, D.; Guilbaud, G.; Jurna, I. & Besson, J.M. (1976). Differential effects of morphine on responses of dorsal horn lamina V type cells elicited by A and C fibre stimulation in the spinal cat. *Brain Res,* 115, 518-524.

Le Bars, D.; Gozariu, M. & Cadden, S.W. (2001). Animal models of nociception. *Pharmacol Rev*,53, 597–652.

Lumb, B.M. (2002). Inescapable and escapable pain is represented in distinct hypothalamic-midbrain circuits: specific roles for Adelta- and C-nociceptors. *Exp Physiol*, 87, 281–286.

Lynn, B. (1990). Capsaicin: actions on nociceptive C-fibres and therapeutic potential. *Pain*, 41, 61-69.

Mitchell, D. & Hellon, R.F. (1997). Neuronal and behavioural responses in rats during noxious stimulation of the tail. *Proc. R. Soc. (London)*, 197, 194-196.

Mor, J. & Carmon, A. (1975). Laser emitted radiant heat for pain research. *Pain*, 1, 233-237.

Ploner, M.; Gross, J.; Timmermann, L. & Schnitzler, A. (2002). Cortical representation of first and second pain sensation in humans. *Proc Natl Acad Sci USA*, 99: 12444–12448.

Roos, A.; Rydenhag, B. & Andersson, S.A. (1982). Cortical responses evoked by tooth pulp stimulation in the cat. Surface and intracortical responses. *Pain*, 14, 247-265.

Rousseaux, M.; Cassim, F. ; Bayle, B. & Laureau, E. (1999). Analysis of the perception of and reactivity to pain and heat in patients with wallenberg syndrome and severe spinothalamic tract dysfunction. *Stroke*.30, 2223-2229.

Shaw, F.Z.; Chen, R.F.; Tsao, H.W. & Yen, C.T. (1999). A multichannel system for recording and analysis of cortical field potentials in freely moving rats. *J Neurosci Methods*, 88, 33-43.

Shaw, F.Z.; Chen, R.F. & Yen, C.T. (2001). Dynamic changes of touch- and laser heat-evoked field potentials of primary somatosensory cortex in awake and pentobarbital-anesthetized rats. *Brain Res*, 911, 105-115.

Shyu, B.C.; Han, Y.S. & Yen, C.T. (1995). Spinal pathways of nociceptive information evoked by short CO_2 laser pulse in rats. *Chin J Physiol*, 38, 27-33.

Shyu, B.C.; Chai, S.C.; Kung, J.C. & Fan, R.J. (2003).A quantitative method for assessing of the affective component of the pain: conditioned response associated with CO_2 laser-induced nocifensive reaction. *Brain Res Brain Res Protoc*, 12, 1-9.

Smith, G.P. (1995). Dopamine and food reward. *Prog Psychobiol Physiol Psychol*, 16, 83-144

Stowell, H. (1974). Human evoked responses to potentially noxious tactile stimulation. *Acta Nervosa Superior*, 17, 94-100.

Sun, J.-J.; Yang, J.-W. & Shyu, B.C. (2006). Current source density analysis of laser heat-evoked intra-cortical field potentials in the primary somatosensory cortex of rats. *Neuroscience*,140, 1321-1336.

Torebjork, H.E. & Ochoa, J.L. (1990). New method to identify nociceptor units innervating glabrous skin of the human hand. *Exp Brain Res*, 81, 509–514.

Treede, R.D.; Meyer, R.A.; Raja, S.N. & Campbell, J.N. (1995). Evidence for two different heat transduction mechanisms in nociceptive primary afferents innervating monkey skin. *J Physiol*, 483, 747-758.

Van Hassel, H.J.; Beidenbach, M.A. & Brown, A.C. (1972). Cortical potential evoked by tooth pulp stimulation in rhesus monkeys. *Archives of Oral Biology*, 17, 1059-1066.

Wall, P.D. & Fitzgerald, M. (1982). Effects of capsaicin applied locally to adult peripheral nerve. I. Physiology of peripheral nerve and spinal cord. *Pain*,11, 363-377.

Yeomans, D.C. & Proudfit, H.K. (1996). Nociceptive responses to high and low rates of noxious cutaneous heating are mediated by different nociceptors in the rat: electrophysiological evidence. *Pain*, 68, 141–150.